Insect Conservation Biology

Insect Conservation Biology

Proceedings of the Royal Entomological Society's 23rd Symposium

Edited by

A.J.A. Stewart

Department of Biology and Environmental Science
University of Sussex
Brighton, UK

T.R. New

Department of Zoology
La Trobe University
Melbourne, Australia

O.T. Lewis

Department of Zoology
University of Oxford
Oxford, UK

www.cabi.org

CABI is a trading name of CAB International

CABI Head Office
Nosworthy Way
Wallingford
Oxfordshire OX10 8DE
UK

Tel: + 44 (0)1491 832111
Fax: + 44 (0)1491 833508
E-mail: cabi@cabi.org
Website: www.cabi.org

CABI North American Office
875 Massachusetts Avenue
7th Floor
Cambridge, MA 02139
USA

Tel: + 1 617 395 4056
Fax: + 1 617 354 6875
E-mail: cabi-nao@cabi.org

A catalogue record for this book is available from the British Library, London, UK.

A catalogue record for this book is available from the Library of Congress, Washington, DC.

ISBN: 978 1 84593 254 1

Typeset by SPi, Pondicherry, India
Printed and bound in the UK by Biddles Ltd, King's Lynn.

Contents

Contributors

Paul Barrett, *Butterfly Creek, Tom Pearce Drive, PO Box 201 097, Auckland, New Zealand. tokoriro@xtra.co.nz*

Yves Basset, *Smithsonian Tropical Research Institute, Apartado 0843-03092, Balboa, Ancon, Panama City, Republic of Panama. bassety@si.edu*

Janne Bengtsson, *Department of Entomology (Landscape Ecology), Swedish University of Agricultural Sciences, PO Box 7044, SE-750-07 Uppsala, Sweden. jan.bengtsson@evp.slu.se*

Nigel Bourn, *Butterfly Conservation, Manor Yard, East Lulworth, Dorset BH20 5QP, UK. nbourn@butterfly-conservation.org*

Tom Brereton, *Butterfly Conservation, Manor Yard, East Lulworth, Dorset BH20 5QP, UK. tbrereton@butterfly-conservation.org*

Luisa Gigante Carvalheiro, *School of Biological Sciences, University of Bristol, Woodland Road, Bristol BS8 1UG, UK. Luisa.GiganteCarvalheiro@bristol.ac.uk*

Rebecca Chaplin-Kramer, *Department of Environmental Sciences, Policy and Management, University of California, Berkeley, CA 94720, USA. rchaplin@nature.berkeley.edu*

Oliver D. Cheesman, *108 Cholmeley Road, Reading, Berkshire RG1 3LY, UK. oliver@dipsacus.org*

Dave Clarke, *Zoological Society of London, Regent's Park, London NW1 4RY, UK. ppk@zsl.org*

Kelvin F. Conrad, *Plant and Invertebrate Ecology Division, Rothamsted Research, Harpenden, Hertfordshire AL5 2JQ, UK. kelvinconrad@trentu.ca; Current address: Department of Biology, Trent University, Peterborough, Ontario, K9J 7B8, Canada.*

Bexell Ayyachamy Daniel, *Zoo Outreach Organisation, PO Box 1683, Peelamedu, Coimbatore Tamil Nadu, 641004, India. icinsa@vsnl.net*

Zoe G. Davies, *Biodiversity and Macroecology Group (BIOME), Department of Animal and Plant Sciences, University of Sheffield, Sheffield S10 2TN, UK. z.davies@sheffield.ac.uk*

Roger L.H. Dennis, *NERC Centre for Ecology and Hydrology, Monks Wood, Abbots Ripton, Huntingdon, Cambridgeshire PE28 2LS, UK; and Institute for Environment, Sustainability and Regeneration, Mellor Building, Staffordshire University, College Road, Stoke on Trent ST4 2DE, UK. rlhdennis@aol.com*

Richard Fox, *Butterfly Conservation, Manor Yard, East Lulworth, Dorset BH20 5QP, UK. rfox@butterfly-conservation.org*

Rachel Gibson, *School of Biological Sciences, University of Bristol, Woodland Road, Bristol BS8 1UG, UK. Rachel.Gibson@bristol.ac.uk*

Ilkka Hanski, *Department of Biological and Environmental Sciences, University of Helsinki, PO Box 65, FIN-00014, Finland. ilkka.hanski@helsinki.fi*

Rúben Hüttel Heleno, *School of Biological Sciences, University of Bristol, Woodland Road, Bristol BS8 1UG, UK. r.heleno@bristol.ac.uk*

Kate Henson, *School of Biological Sciences, University of Bristol, Woodland Road, Bristol BS8 1UG, UK. Kate.Henson@bristol.ac.uk*

Patrick Honan, *Zoos Victoria, PO Box 74, Parkville, Victoria 3052, Australia. Invertebrates@zoo.org.au*

Roger S. Key, *Natural England, Northminster House, Peterborough PE1 1UA, UK. roger.key@naturalengland.org.uk*

David Kleijn, *Former address: Nature Conservation and Plant Ecology Group, Wageningen University, Bornsesteeg 69, 6708 PD Wageningen, The Netherlands. Current address: Alterra, Centre for Ecosystem Studies, PO Box 47, 6700 AA, Wageningen, The Netherlands. david.kleijn@wur.nl*

Claire Kremen, *Department of Environmental Sciences, Policy and Management, University of California, Berkeley, CA 94720, USA. ckremen@nature.berkeley.edu*

Owen T. Lewis, *Department of Zoology, University of Oxford, South Parks Road, Oxford OX1 3PS, UK. owen.lewis@zoo.ox.ac.uk*

Mitchell Magdich, *The Toledo Zoo, PO Box 140130, Toledo, OH 43614, USA. mitch.edu@toledozoo.org*

Melodie A. McGeoch, *Centre for Invasion Biology, Department of Conservation Ecology and Entomology, University of Stellenbosch, Private Bag X1, Matieland 7602, South Africa. mcgeoch@sun.ac.za*

Jane Memmott, *School of Biological Sciences, University of Bristol, Woodland Road, Bristol BS8 1UG, UK. Jane.Memmott@bristol.ac.uk*

Ian Middlebrook, *Butterfly Conservation, Manor Yard, East Lulworth, Dorset BH20 5QP, UK. imiddlebrook@butterfly-conservation.org*

Martha Lopezaraiza Mikel, *School of Biological Sciences, University of Bristol, Woodland Road, Bristol BS8 1UG, UK. ME.Lopezaraiza@bristol.ac.uk*

Randy Morgan, *Cincinnati Zoo and Botanical Garden, 3400 Vine St, Cincinnati, OH, 45220, USA. randy.morgan@cincinnatizoo.org*

Trevor Moxey, *Zoological Society of London, Regent's Park, London NW1 4RY, UK. ppk@zsl.org*

Timothy R. New, *Department of Zoology, La Trobe University, Melbourne, Victoria 3086, Australia. t.new@latrobe.edu.au*

Mark S. Parsons, *Butterfly Conservation, Manor Yard, East Lulworth, Dorset BH20 5QP, UK. mparsons@butterfly-conservation.org*

Sarina Pearce, *School of Biological Sciences, University of Bristol, Woodland Road, Bristol BS8 1UG, UK. S.Pearce@bristol.ac.uk*

Paul Pearce-Kelly, *Zoological Society of London, Regent's Park, London NW1 4RY, UK. ppk@zsl.org*

Lou Perrotti, *Roger Williams Park Zoo, Roger Williams Park, Elmwood Ave, Providence, RI 02905, USA. lperrotti@rwpzoo.org*

Juha Pöyry, *Finnish Environment Institute, PO Box 140, Helsinki, FIN-00251, Finland. juha.poyry@ymparisto.fi*

Ilik J. Saccheri, *Population and Evolutionary Biology Research Group, School of Biological Sciences, University of Liverpool, Crown Street, Liverpool L69 7ZB, UK. Saccheri@liv.ac.uk*

Michael J. Samways, *Department of Conservation Ecology and Entomology and Centre for Invasion Biology, University of Stellenbosch, Private Bag X1, Matieland 7602, South Africa. samways@sun.ac.za*

Tanja H. Schuler, *Plant and Invertebrate Ecology Division, Rothamsted Research, Harpenden, Hertfordshire AL5 2JQ, UK. tanja.schuler@gmx.com*

David A. Sheppard, *Natural England, Northminster House, Northminster Road, Peterborough PE1 1UA, UK. david.sheppard@naturalengland.org.uk*

Tim G. Shreeve, *School of Life Sciences, Oxford Brookes University, Headington, Oxford OX3 0BP, UK. tgshreeve@brookes.ac.uk*

Warren Spencer, *Clifton and West of England Zoological Society, Clifton, Bristol BS8 3HA, UK. wspencer@bristolzoo.org.uk*

Alan J.A. Stewart, *Department of Biology and Environmental Science, School of Life Sciences, University of Sussex, Falmer, Brighton BN1 9QG, UK. a.j.a.stewart@sussex.ac.uk*

Erin Sullivan, *Woodland Park Zoological Park Gardens, 5500 Phinney Ave, N, Seattle, WA 98103, USA. Erin.Sullivan@Zoo.org*

Chris D. Thomas, *Department of Biology (Area 18), University of York, PO Box 373, York YO10 5YW, UK. cdt2@york.ac.uk*

David J. Thompson, *Population and Evolutionary Biology Research Group, School of Biological Sciences, University of Liverpool, Crown Street, Liverpool L69 7ZB, UK. d.j.thompson@liv.ac.uk*

Teja Tscharntke, *Agroecology, University of Göttingen, Waldweg 26, D-37073 Göttingen, Germany. ttschar@gwdg.de*

Jason M. Tylianakis, *Former address: Agroecology, University of Göttingen, Waldweg 26, D-37073 Göttingen, Germany. Current address: School of Biological Sciences, University of Canterbury, Private Bag 4800, Christchurch 8020, New Zealand. jason.tylianakis@canterbury.ac.nz*

Ko Veltman, *Natura Artis Magistra, Plantage Kerklaan, 38–40, 1018 CZ Amsterdam C, The Netherlands. k.veltman@artis.nl*

Mark R. Wade, *National Centre for Advanced Bio-Protection Technologies, PO Box 84, Lincoln University, Canterbury, New Zealand.*

Martin S. Warren, *Butterfly Conservation, Manor Yard, East Lulworth, Dorset BH20 5QP, UK. mwarren@butterfly-conservation.org*

Phillip C. Watts, *Population and Evolutionary Biology Research Group, School of Biological Sciences, University of Liverpool, Crown Street, Liverpool L69 7ZB, UK. p.c.watts@liv.ac.uk*

Robert J. Wilson, *Área de Biodiversidad y Conservación, Escuela Superior de Ciencias Experimentales y Tecnología, Universidad Rey Juan Carlos, Tulipán s/n, Móstoles, E-28933 Madrid, Spain. robertjohn.wilson@urjc.es*

Ian P. Woiwod, *Plant and Invertebrate Ecology Division, Rothamsted Research, Harpenden, Hertfordshire AL5 2JQ, UK. ian.woiwod@bbsrc.ac.uk*

Steve D. Wratten, *National Centre for Advanced Bio-Protection Technologies, PO Box 84, Lincoln University, Canterbury, New Zealand. wratten@lincoln.ac.nz*

Introduction

Insects have played a key role in the development of the science of conservation biology. Their abundance and diversity in most terrestrial and freshwater ecosystems, and the rapidity of their responses to environmental changes make them attractive model organisms for conservation research and monitoring, and as indicators or surrogates for wider biodiversity. At a time of unprecedented human impacts on natural environments, insect conservation biology has an important role to play in assessing and ameliorating the impacts of anthropogenic habitat modification and climate change. Increasingly, insects are the targets of conservation action in their own right, guided by detailed autecological study.

The Royal Entomological Society's 23rd International Symposium was held at the University of Sussex, UK, from 12 to 14 September 2005 on the theme of 'Insect Conservation Biology'. In convening that symposium, we sought to build on the Society's previous symposium on this theme held in 1989 'The Conservation of Insects and Their Habitats' (Collins and Thomas, 1991) and, in particular, to explore how the discipline has matured and diversified in the intervening 16 years. Many of the world's leading workers in insect conservation accepted our invitation to participate, and we adopted three major themes to be treated in sequence, as reflected in this volume.

The first of three half-day sessions set up the broad themes in insect conservation. The session commenced with two contrasting 'scene-setting' papers to examine the state of insect conservation in major regions of the world and what the major avenues for progress, and hindrances, have been. The temperate regions (Stewart and New, Chapter 1) have benefited from the close attention paid to well-documented fauna by a relatively large number of resident entomologists, particularly in the northern hemisphere. This has allowed species-level conservation programmes to become a major focus of conservation need and advocacy, leading to well-defined protocols and approaches for insect conservation management. Many tropical insect faunas are much less tractable in that a large proportion of species remain as yet undescribed (Lewis and Basset, Chapter 2), with the consequence that approaches to conservation necessarily emphasize broader approaches, largely based on habitat. The next four chapters deal with these contrasting approaches to insect conservation. Pearce-Kelly and an international team of collaborators (Chapter 3) illustrate the increasing importance of *ex situ* conservation for insects – both in practical conservation and for advocacy – using examples from many different insect groups and from various parts of the world. Warren *et al.* (Chapter 4) examine the benefits gained from listing species for conservation priority, with particular reference to butterflies as the most thoroughly appraised insect group. Dennis *et al.* (Chapter 5) emphasize the central importance of habitats, assessed as both place and coincidence of critical resources, as a wider level of focus. Samways (Chapter 6) takes us to the landscape level and the features of landscape architecture and change so vital for wider-scale insect conservation in all parts of the world.

The theme of our second session was examination of insects as 'model organisms' in conservation biology, to show how they have been used not only to enhance their own well-being, but also to illustrate or facilitate progress on wider conservation agendas. McGeoch (Chapter 7) discusses the diverse and important roles of insects as 'indicators' of environmental condition and change, and the transition of theory into ever-diversifying practice. Hanski and Pöyry's (Chapter 8) pioneering work on understanding metapopulation structures and the effects of landscape fragmentation on insect populations emphasizes the importance of scale in considering the accessibility of isolated habitat patches, with important implications for wider conservation management. The central importance of monitoring insect population sizes and species distributions is discussed by Conrad *et al.* (Chapter 9) with long-term studies and monitoring sequences enabling sound assessments of recent and possible future changes. The central roles of insects in ecological interactions (Memmott *et al.*, Chapter 10) as 'ecosystem engineers' and providers of ecosystem services emphasize their importance in the maintenance of ecosystem dynamics and processes, as well as the wider importance of their conservation. While most of the threats to insects receiving attention in the past involved tangible factors such as habitat loss or the spread of alien species, future threats consequent upon global climate changes are universal, not readily predictable and will have wide impacts (Wilson *et al.*, Chapter 11). Although the details of different future climate scenarios are hotly debated, climate change is increasingly accepted as the most serious global threat to insects and indeed the whole of biodiversity. The final chapter in this session (Thompson *et al.*, Chapter 12) explores the emerging science of insect conservation genetics, and its roles and applications in effective conservation practice.

Our third session, entitled 'Future Directions in Insect Conservation Biology', looked to the future – how might the lessons learned so far be fostered and developed for the greater benefit of insect conservation, and what should our priorities be? New (Chapter 13) suggests ways in which insects might be elevated to being considered as core components in wider conservation programmes. Cheesman and Key (Chapter 14) then explore ways in which entomological expertise can be conserved, to assure continuity of the requisite knowledge, interest and commitment. The final three chapters focus more specifically on arenas of current interest and debate. Kremen and Chaplin-Kramer (Chapter 15) explore further the role of insects in ecosystem processes, using pollination as an example of one such process which people can see readily as being of major economic and functional importance in crop production. Tscharntke *et al.* (Chapter 16) affirm the central importance of managing agricultural systems and landscapes (accounting for ~36% of global land area) in ways that encourage insect conservation. Woiwod and Schuler (Chapter 17) summarize the complex issues arising from the increasing use of genetically modified crops, how patterns of usage may change in the future and the likely implications for beneficial and other non-target insects. Finally, we review just how far insect conservation has come in recent years and make some suggestions as to what the future might hold for this fast-moving field (Lewis *et al.*, Chapter 18).

As convenors of the Symposium and editors of this volume, we are well aware of the complexities of organizing such a meeting and bringing the proceedings to fruition. There are many people to thank for their contribution to a successful meeting. The participants – both speakers and attendees who contributed to the discussions – ensured that the Symposium was a scientific success. Each of the chapters was read by two reviewers, whose perceptive comments helped to ensure the integrity of the final volume. The president of the Society, Dr Hugh Loxdale, opened the Symposium and the vice-chancellor of the University of Sussex, Professor Alasdair Smith, co-hosted a wine reception on the first evening to welcome delegates. The Society's staff, Bill Blakemore (Registrar), June Beeson and Elena Lazarra, and a local team of postgraduates at the University of Sussex helped to ensure that the meeting ran smoothly. John Badmin and Dr Archie Murchie organized the concurrent Annual National Meeting of the Society, the afternoon sessions of which complemented the morning symposia. We are very grateful to them all.

Alan J.A. Stewart
Timothy R. New
Owen T. Lewis

Reference

Collins, N.M. and Thomas, J.A. (eds) (1991) *The Conservation of Insects and Their Habitats*. Academic Press, London.

1 Insect Conservation in Temperate Biomes: Issues, Progress and Prospects

ALAN J.A. STEWART[1] AND TIMOTHY R. NEW[2]

[1]*Department of Biology and Environmental Science, School of Life Sciences, University of Sussex, Falmer, Brighton BN1 9QG, UK;* [2]*Department of Zoology, La Trobe University, Melbourne, Victoria 3086, Australia*

1 Introduction

Insects present conservationists with a very different set of challenges in comparison with more popular groups such as vertebrate animals and vascular plants. These are a consequence of several aspects of their life histories that make them especially vulnerable to the types of environmental changes currently being experienced across many temperate regions (McLean, 1990; Kirby, 1992; UK Biodiversity Group, 1999). Many insects have highly specialized habitat (and often microhabitat) requirements that are further complicated by the fact that the discrete stages in the life cycle often require radically different resources. Most insects have comparatively short life cycles (often annual or more frequent) with no dormant stage in which they can escape adverse conditions, so that these habitat requirements have to be met without interruption. Finally, many species are incapable of dispersing more than trivial distances, or are behaviourally reluctant to do so, resulting in their complex habitat requirements having to be met within relatively small areas and an increased sensitivity to habitat fragmentation. Thus, maintenance of habitat quality, continuity, heterogeneity and connectedness are recurrent themes in insect conservation biology.

The field of insect conservation has undergone rapid development in the last 30 years or so, with particular acceleration of pace since the Royal Entomological Society last met to review this topic some 16 years ago (Collins and Thomas, 1991). Reasons are multifaceted but include a wider realization that: (i) for the reasons stated above, conservation of insect species and assemblages requires a different approach to that traditionally adopted by conservationists more concerned with plants and vertebrates, with the consequence that insects are often poorly served by the protective 'umbrella' of these more conspicuous and charismatic groups (McLean, 1990; Kirby, 1992; Hambler and Speight, 1995); (ii) insects are highly sensitive and useful indicators of habitat and environmental change (Woiwod, 1991; Harrington and Stork, 1995; Wright *et al.*, 2000; Thomas, 2005); (iii) many insects have already undergone serious declines that exceed those of other high-profile groups such as birds and plants (Thomas *et al.*, 2004; but see Hambler and Speight, 1996, 2004, and Shaw, 2005 for contrary views, and the convincing response to them by Thomas and Clarke, 2004); and (iv) insects arguably deserve to be conserved in their own right, for their intrinsic qualities, their utility to people, as providers of important ecosystem services and as part of overall biodiversity (Samways, 2005).

The major principles of insect conservation have been derived very largely from concerns for individual species and wider habitats in the northern temperate region, predominantly from northern and western Europe and parts of North America. Much of the effort elsewhere has drawn heavily on these experiences, sometimes uncritically, for both procedures and practices; progress has arisen from testing on other faunas the conservation lessons and paradigms learned in this part of the world. By contrast, the field of insect conservation has developed along a rather different path in tropical environments, where the sheer magnitude of species richness and a range of logistical constraints have forced a somewhat different approach (Lewis and Basset, Chapter 2, this volume). In this chapter, we examine the importance and relevance of these lessons, and their wider applications. We do this with the considerable benefit of hindsight, and largely through comparing and contrasting the interests and priorities for insect conservation in the better-documented and generally less species-rich northern temperate regions with the more poorly understood, but richer, biota of the southern temperate regions. Most examples are from the UK and Australia, the areas with which we are most familiar. Tracing the rapid recent development of the field of insect conservation, the ideas that motivate and underpin it, and its geographical distribution, allows us to place it in the wider context of the expanding modern science of conservation biology.

1.1 Temperate regions: the arena of concern

The northern and southern temperate regions (Fig. 1.1) show one immediate and important contrast: their extent. In the north, two large continental landmasses collectively occupy approximately 250° of longitude, whilst in the south three highly disjunct regions together span only 105° of longitude. The northern region is thus considerably the larger, and includes much of the Holarctic geographical zone, together with parts of northern Africa. The southern zones are southern Africa, southern South America, and Australia and New Zealand, with associated islands. Australia is the only designated megadiverse country spanning tropical to cool temperate regions under the same federal government and with a sufficient resident cohort of concerned biologists to address conservation across this variety of environments. The first two of these zones are linked trans-tropically with the northern regions by land, but no current land bridges occur between Australia and the Asian

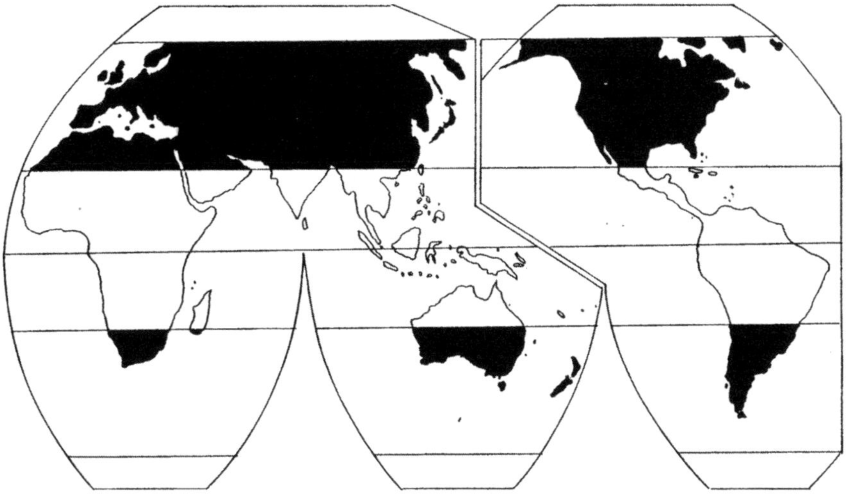

Fig. 1.1. The geographical extent of the temperate region (depicted in black), illustrating the contrast in total land area between the northern and southern hemisphere.

mainland. As Samways (1995) noted, the greater part of southern temperate land occurs north of about 40° S latitude, in marked contrast to the northern region, in which about half the land area occurs at latitudes higher than 40° N. For the most part, the northern and southern temperate regions are faunistically distinct. The least-documented southern area is that part of South America between the Tropic of Capricorn and about 40° S, mainly because the far south has attracted the interests of numerous visiting entomologists seeking to clarify Gondwanan relationships, particularly with New Zealand and southern Australia. Most biologists in South America have worked either in the tropics or the most southerly areas.

Patterns of local endemism are common, and many insect groups show southern concentrations of endemism or richness that are often coincident with the 'hotspots' of endemism and threat identified by Myers *et al.* (2000). The disproportionately elevated richness of southern Africa and Australia noted by Platnick (1991) reflects, in part, the extraordinarily rich floristic regions of the south-western Cape (the 'fynbos', for which the ecological importance of insects was evaluated by Wright, 1994) and south-west Western Australia, together with the wide variety of topography and habitats present. In contrast, the biota of far southern South America appear to be genuinely depauperate, but nevertheless important in supporting ancient and endemic lineages of insects, including significant Gondwanan taxa. The faunas of all southern areas need considerable further investigation, the recent discovery of the new insect order Mantophasmatodea in southern Africa (Klass *et al.*, 2002) attesting to the possibility of further novelty with considerable scientific interest.

Early developments in the field of insect conservation in some temperate regions were summarized by contributors to the earlier Royal Entomological

Society symposium (Collins and Thomas, 1991). Thus, Opler (1991) and Greenslade and New (1991) outlined the perspectives for North America and Australia, respectively; Mikkola (1991) and Balletto and Casale (1991) dealt with northern and Mediterranean Europe. With respect to the UK, McLean (1990) outlined broad themes, while Fry and Lonsdale (1991) and Kirby (1992) focused on habitat management principles. In a later symposium, Samways (1995) gave a broader perspective of southern hemisphere insect diversity, focusing mainly on southern Africa and Australia. Relevant topics for Australia are also discussed by Greenslade (1994, steppe-type landscapes), Rentz (1994, Orthoptera), New (1994, exotic species impacts), and for South Africa by Scholtz and Chown (1994, savannah) and Wright (1994, fynbos). These accounts refer to many of the early pioneering studies on British and other fauna, which remain highly pertinent in considering the emerging patterns of insect conservation. Some recent essays (such as those of McGeoch, 2002 on South Africa, and New and Sands, 2004 on Australia) demonstrate advances over the last decade or so. Symposia on invertebrate biodiversity and conservation both in South Africa (McGeoch and Samways, 2002) and the Australian region (Ingram *et al.*, 1994; Yen and New, 1997; Ponder and Lunney, 1999; Austin *et al.*, 2003) attest to the increasing interest and concerns in southern temperate regions. We are unaware of any parallel focus for southern South America, where there are few resident entomologists to appraise such problems and needs, but some recent surveys in Argentina (ants: Badano *et al.*, 2005; grasshoppers: Torrusio *et al.*, 2002) are important pointers to conservation focus.

1.2 Perspective: the tradition of conservation

Important regional differences in the levels of understanding of the insect fauna occur between the northern and southern zones. Perhaps the greatest geographical influence stems from a point discussed by Pyle (1995), namely that Britain, together with some parts of continental western Europe and North America, has long accepted natural history (including insect collecting and study) as a respectable activity. This tradition has led to the accumulation and documentation by professional and non-professional interests of vast amounts of information on insects based on well over a century of concerted endeavour. Thus, the diversity, specific biological and life history details, distribution patterns and their changes over a substantial period are reasonably well known for certain well-studied insect groups. Compendia such as the *Millennium Butterfly Atlas* (Asher *et al.*, 2001) and the analyses that continue to flow from it (e.g. Thomas *et al.*, 2004; Wilson *et al.*, 2004) demonstrate how detailed data on historical changes in species distribution patterns can inform conservation. Similarly detailed data-sets on the British insect fauna are steadily accumulating both for charismatic groups such as Odonata (Merritt *et al.*, 1996) and Orthoptera (Haes and Harding, 1997) and for groups, such as Carabidae (Luff, 1998) and Syrphidae (Ball and Morris, 2000), that have a more specialist following. The UK Biological Records

Centre has a long and venerable tradition of compiling and analysing distributional data for a wide range of insect groups, including those which have to rely on ad hoc accumulation of data rather than systematic surveys. These compilations represent the knowledge base for assessing the rarity status of individual species, even when based on only partial data coverage, and are critical in setting priorities for conservation on the most deserving targets. Such assessments are possible only for taxa for which information is reasonably adequate; Shaw and Hochberg (2001) make the point that around half the British parasitic Hymenoptera fauna cannot yet be identified reliably, if at all, other than by a handful of specialists, resulting in the almost complete neglect of this group in conservation assessments. Even in well-studied Britain, ecological knowledge of most insect species outside the popular groups is very fragmentary; precise habitat requirements are often unclear, so that appropriate management prescriptions are difficult or impossible to define for non-entomologist conservation practitioners who are charged with managing sites.

Major points of contrast between the northern and southern temperate zones relate to: (i) the much better documentation of many insect groups, particularly in parts of western Europe, than anywhere in the south; (ii) the longer history of conservation interests and concerns based on sound natural history; (iii) a larger population of resident concerned entomologists and other people, with wider support for conservation endeavours within (iv) a broader framework of ecological understanding and history of threats and their impacts on native species, communities and habitats. The less rich northern insect faunas have thus received far more attention, over a considerably longer period, than their southern counterparts. The fine-detail approach of species-focusing that has been possible for European butterflies, some beetles, dragonflies and others has led to these being 'global drivers' of insect conservation. The detailed and rigorous approach adopted by many of these studies has also been important in catalysing the wider development of insect conservation as a responsible and disciplined science (New *et al.*, 1995). Evidence for declines and losses of species (butterflies in particular) in the northern temperate zones has been provided because of the tradition of recording and monitoring species incidence and relative abundance. For example, both the Butterfly Monitoring Scheme (Pollard and Yates, 1993) and the Rothamsted Insect Survey of macro-moths (Woiwod, 1991) have drawn attention to dramatic recent declines in many species across Britain (see Conrad *et al.*, 2004). However, it is important to emphasize that this better information base for the north often relates to highly altered landscapes changed by many centuries of human impacts. By contrast, the major documented impacts in the southern zones are mostly more recent and can be compared more readily with conditions in relatively pristine environments in which human impacts have been minimal by comparison.

Levels of public sympathy and support for insect conservation, at least for the charismatic taxa, are much greater in Europe and North America than elsewhere; the Xerces Society in North America is a leading example. The recent establishment and growth of charities in Britain devoted to the conservation of specialist groups (e.g. Butterfly Conservation for the Lepidoptera; Buglife for invertebrates in general, British Dragonfly Society for the Odonata) is testament to this. Where there is a need to gather information on habitat needs or critical resources to guide management, interested people, support and expertise are often available or can be mustered relatively easily. Some species can command considerable resources over a long period to prevent their extinction. Campaigns to reintroduce the large copper *Lycaena dispar* to Britain, for example, extend over much of the 20th century, and continue (Pullin *et al.*, 1995), while the successful reintroduction to Britain of the large blue *Maculinea arion* after extinction in the 1980s (Thomas, 1999) has become a textbook example of how the fortunes of a single species can be turned around once its detailed ecological requirements are fully understood.

The level of this type of interest and commitment, and the information base which is necessary to inform conservation, can be considerably less elsewhere. Interest in conserving butterflies, or other insects, is still viewed in Australia as somewhat eccentric (New, 1984), although gaining impetus rapidly. Several state-based groups, mostly with few members, now focus on butterfly conservation in Australia, and some species have benefited from community involvement and the activities of local 'friends groups'. In much of the southern temperate region, insect conservation (together with many wider environmental issues) is viewed as low priority in relation to more pressing problems of human welfare, within social environments not intuitively sympathetic to such endeavours. This is not surprising in view of the pressures to establish, develop and sustain agriculture and other human-support systems and industries. Establishment and protection of agricultural or forestry crops and improved pastures (the latter often based on exotic pasture grasses, as in Australia) have traditionally taken priority over assuring sustainability of native biota, with insects ranked well below more charismatic and conspicuous wildlife in any conservation debates. Important exceptions include certain insects used as economic commodities such as human foods (e.g. caterpillars of *Imbrasia* [Saturniidae]; McGeoch, 2002) or for silk production (*Gonometa* spp. [Lasiocampidae]; Veldtman *et al.*, 2002), both in South Africa.

2 Limits to Species Focusing

The traditional single-species approach to insect conservation aims to set objective conservation priorities based on sound knowledge of the distribution and comparative status of all species in a group. Although elegantly demonstrated for certain well-studied insect groups in the northern temperate zone, this approach has not proved immediately transferable to all other temperate regions and taxonomic groups for a number of reasons. First, the number of formally described species is often only a fraction of the total number of species estimated to exist in a particular taxonomic group. Thus, a recent evaluation of the Australian insect fauna (Yeates *et al.*, 2003, building on the approach pioneered by Taylor, 1983) estimated the total insect fauna at 204,743 species, of which 58,491 (28.6%) are described, the authors noting that the fauna is likely to be far larger even than the highest figures cited. Austin *et al.* (2004) suggest that the conservative count for richness of Australian Hymenoptera (44,000 species) probably vastly underestimates the true size of the fauna which is 'difficult if not impossible to estimate with any accuracy given the current state of knowledge'. Comparative estimates are not always available for other temperate regions and often have high degrees of uncertainty attached to them. Scholtz and Chown (1994) suggest that 'between 5 and 50% of southern African insects are estimated to have been described'. Redak (2000) reports the North American insect fauna to comprise approximately 163,487 species of which about 72,500 (44%) remain undiscovered or inadequately described. Although not all species have been formally named even in the best-documented faunas, a stark contrast in this respect exists between the relatively well-documented northern faunas and the markedly less-studied southern temperate ones. It is sobering to contrast the relative excitement generated by the recent detection of a new butterfly *species* in Ireland (Nelson *et al.*, 2001), a comparatively unusual event in Europe for this well-studied insect group, with the equally recent discovery of a whole new insect *order* in southern Africa: the Mantophasmatodea (Klass *et al.*, 2002). Within southern temperate faunas, some insect groups are much better documented than others, with butterflies, some moths, some beetles, Odonata and some Orthoptera amongst the better-known. These, and some other groups differing between the continents, have high proportions of species described.

The inevitable consequence of these discrepancies in levels of knowledge between taxonomic groups is that they impact upon setting conservation priorities. Species richness increases the magnitude of the need, but also the difficulty of making such assessments reliable. For this reason, most insect species nominated or adopted for inclusion on protected species lists or national 'red lists' in southern temperate regions belong to the better-known groups, although other isolated species are sometimes present. In South Africa, by far the most advanced of the three southern zones in such compilations, databases have been compiled, and priority areas (such as centres of endemism) distinguished, for butterflies, termites, scarab and buprestid beetles, and Myrmeleontidae (references in McGeoch, 2002). A first *Red Data Book* exists for South African butterflies (Henning and Henning, 1989). Such works are important in helping to indicate some of the needs for species conservation, but for the southern zones can rarely be even reasonably representative of the real needs, because knowledge is generally insufficient to render such lists comprehensive for any taxonomic group other than butterflies. For this reason, butterflies are the best-represented group of protected insects in South Africa, with provincial lists of endangered insects for some areas consisting almost entirely of butterflies (Scholtz and Chown, 1994). Within the southern temperate zone, only in New Zealand has a reasonably comprehensive attempt been made to compile a preliminary listing of insects of conservation interest across a variety of orders (McGuinness, 2001), although less critical preliminary syntheses for Australia (Hill and Michaelis, 1988; Yen and Butcher, 1997) are also invaluable leads. McGuinness (2001) provided conservation profiles of 104 beetles and 13 moths, both groups assessed only by a small number of families, as well as other orders. Closer focus may be available for lower-level taxonomic categories: thus, again for New Zealand, Patrick and Dugdale (2000) profiled 114 species of Lepidoptera of conservation interest; a recovery plan for carabid beetles (McGuinness, 2002) dealt with 55 species; and a recovery plan for the most charismatic of all New Zealand insect groups, weta (Orthoptera), covered 15 species in some detail. Such formal action plans are rare for the southern temperate zones; an action plan for Australian butterflies (Sands and New, 2002) seems unlikely to be paralleled for other insect groups in the foreseeable future, although profiles for individual insects in isolation are appearing under various State Acts and more widely (Clarke and Spier, 2003).

Taxonomic bias in species listing is thus perhaps inevitable, even amongst the relatively well-studied European fauna, if we are to treat the process responsibly. For example, the British *Red Data Book* for insects (Shirt, 1987) lists representatives of only eight orders, and listings are dominated by Coleoptera (546 species, or 14% of the total fauna) and Lepidoptera. The latter are divided into three categories, which reflect relative popularity and knowledge: butterflies (12, 21%), 'macromoths' (99, 11%) and 'micromoths' (11, 0.7%), again emphasizing dependence on the more charismatic 'flagship groups' for conservation advocacy and advance, coupled with the relationship between good knowledge and improved ability to determine conservation status and management.

For most workers in temperate regions outside northern Europe, the challenge of dealing with the high proportion of undescribed species is exacerbated by the historical legacy that most taxonomic expertise and a high proportion of type material are housed in northern hemisphere museums. As Naskrecki (2004) noted 'access to those types is vital when studying these new faunas'. Fortunately, this discrepancy is now being countered by increasing deposition of type material in local (national or state) institutions. Furthermore, progressive development of the World Wide Web as a taxonomic tool is revolutionizing the ways in which information on such specimens can be communicated.

The main practical need is for consistent and replicable recognition of species or other taxonomic units, rather than necessarily for formal scientific binomials, so that the entities can be studied effectively to appraise conservation need and management. Although named species might appear more tangible, southern hemisphere workers have harnessed the concept of the 'morphospecies' (denoting a consistently recognizable entity without a formal binomial name) to address the challenge of incorporating numerous undescribed species into conservation assessment at both individual species and assemblage levels. This approach has given considerable power to analysing and appraising patterns of insect diversity and distribution in southern temperate regions. This approach to overcoming the 'taxonomic impediment' was pioneered in part through studies on Australian Orthoptera (see Taylor, 1983 for a discussion and the potential developments of the approach as then envisaged). Ecologists have been quick to adopt it as a short cut to studying assemblages or communities that are rich in species which are unnamed or difficult to identify. However, if this approach is to achieve its maximum value, vouchers of all designated entities need to be deposited in accessible reference collections so that future studies can be fully cross-referenced across different surveys and geographic regions. Whilst the progressive accumulation of specimens may eventually provide the basis for a more formal taxonomic appraisal of the group, it is invaluable for providing information on species distributions, diversity and ecology which is of massive importance for insect conservation.

3 Sailing on Flagships

The pioneering *Invertebrate Red Data Book* (Wells *et al.*, 1983) included representatives of 13 insect orders from the temperate regions. Northern zones were represented by 40 species (9 orders) but southern zones by only 13 species (7 orders), all of which were from Australia (9 species) or New Zealand (4 species, all weta). These initial suites raised awareness of insect conservation for many local scientists, largely as isolated cases deserving attention and advocacy. A number of these and other insects have achieved very high conservation profiles, sometimes elevated to the status of local or national emblems, and have thereby contributed enormously to wider understanding and awareness of insect conservation. Butterflies are amongst the most potent of these flagship taxa and have been instrumental in setting the paradigms of invertebrate species conservation. Thus, studies of the Bay checkerspot (*Euphydryas editha bayensis*) in North America (Ehrlich and Murphy, 1987; Opler, 1991; Ehrlich and Hanski, 2004) and the large blues (*Maculinea* spp.) in Europe (see summary in Wynhoff, 1998) have elucidated our understanding of butterfly ecology and conservation as well as more general ecological principles. *Maculinea* species, for example, captured public imagination not only because of their vulnerability, but also in drawing attention to the subtleties of interactions between butterflies, food plants and mutualistic ants and how these are affected by habitat change and management. This helped to emphasize the fact that conservation will be effective only if it is underpinned by sound science, and that successful rescue measures, such as the reintroduction of *M. arion* to Britain following its national extinction there, rely upon a detailed understanding of how to restore the right habitat conditions for a species. The North American Xerces blue (*Glaucopsyche xerces*), although extinct, has become an important flagship for wider invertebrate conservation interests, both as a reminder of what can happen if protective measures are not taken in time and in the name of a major invertebrate conservation pressure group (The Xerces Society) in North America.

Such prominent species are now some of the best understood of any non-pest insect species, and have become significant as models for ecological understanding and management procedures. Many of the papers in Boggs *et al.* (2003) are enviable examples for such wider emulation. These lessons from the north have been important drivers for conservation progress at the species level elsewhere. Although parallel levels of understanding are generally lacking for the southern hemisphere biota, a similar conservation focus on flagship species in South Africa and the Australian region has benefited from this prior knowledge, in spite of some major ecological differences. The Brenton blue butterfly *Orachrysops niobe* (Trimen) was rediscovered in 1977 for the first time after it was described 119 years earlier, and has become a national celebrity butterfly in South Africa (Steencamp and Stein, 1999), not only for its intrinsic worth but also as a political tool for emphasizing and countering the effects of building development on wildlife. Conservation recommendations were based on a detailed, although necessarily short-term, study of *O. niobe* at the single site where the butterfly is known to occur (Silberbauer and Britton, 1999), involving evaluation of site quality, population size, individual butterfly movements and an investigation of early stages.

Somewhat parallel roles have been promoted for two congeneric species of *Paralucia* (also Lycaenidae) in south-eastern Australia. *P. p. lucida* Crosby (the Eltham copper) and *P. spinifera* Edwards and Common (the Bathhurst copper) have become important flagships for insect conservation in Victoria and New South Wales, respectively. The former is important because it occurs on small isolated remnant urban sites within the greater Melbourne area (Yen *et al.*, 1990; New and Sands, 2003), whereas *P. spinifera* has been instrumental also in encouraging community involvement in practical butterfly conservation (Nally, 2003).

One further flagship Australian butterfly merits comment for helping to bridge conservation understanding in tropical and temperate regions, as possibly a unique example of this kind. The Richmond birdwing (*Ornithoptera richmondia*) has been the focus of a large community conservation effort in south-eastern Queensland and northern coastal New South Wales, where it is an outlier of a charismatic group of tropical butterflies in the Australian region. It has been used to introduce numerous young people and community groups to the subtleties of insect ecology, thereby helping to increase awareness that conservation is indeed possible through careful management of critical resources (Sands *et al.*, 1997; Sands and Scott, 2003). In addition, the lessons learned from *O. richmondia* have considerable relevance to other birdwings in northern Australia and New Guinea.

Flagship species have not been recruited solely from butterflies; species from other groups have helped to highlight particular issues. Thus, stag beetles (*Lucanus cervus*), which are quite numerous in suburban areas around London and south-east England, have drawn attention to the importance of the deadwood habitat and of retaining relatively unmanaged habitats in domestic gardens as well as forests for saproxylic invertebrates (Speight, 1989). The very rare, endemic, flightless *Colophon* stag beetles that are restricted to certain mountain peaks in South Africa have helped to raise awareness of the problem of illegal trade in specimens of endangered species (Geertsema, 2004). The hornet robberfly, *Asilus crabroniformis*, the largest Diptera species in the UK, has been used to focus attention on the rich insect community associated with dung (Holloway *et al.*, 2003). Finally, the giant New Zealand weta, some species of which are now reduced to single populations, have highlighted the problems that native species face when confronted with introduced predators, in this case rats, against which the local fauna has no innate defence.

All these examples demonstrate that certain charismatic species can be excellent instruments for raising public awareness of insect conservation issues in general, drawing attention to the fact that insects often have both complex and subtle requirements that can be met only through careful and scientifically based management. The intrinsic appeal of many of these species can also be used to engender public support and interest which can then be broadened to encompass other less charismatic species. It is likely that the spectrum of flagship insect species will continue to diversify.

4 Sheltering under Umbrellas, and Other Surrogate Measures

Insect conservation biologists have both the privilege and the challenge of investigating how to conserve a bewildering range of species. In order not to be overwhelmed completely by the task, entomologists have sought short cuts in the form of individual species or groups of species that can act as surrogates for a much wider set of species. In conservation biology, the principle of striving to conserve so-called umbrella species (often large conspicuous species with a requirement for large areas of habitat) on the assumption that a range of other taxa will also be automatically protected because they have similar habitat requirements has intrinsic appeal but has not been well supported by the evidence (Simberloff, 1998; Andelman and Fagan, 2000). Certainly, few convincing examples exist for insects and the evidence is contradictory. Ehrlich (2003) has suggested that 'not only do butterflies serve as a model system for research and function as individuals, but they can also serve as "umbrella groups" – ones whose preservation is likely, by protecting certain areas, to conserve many less charismatic organisms as well'. Thomas (2005) presents a carefully reasoned and convincing case for butterflies being imperfect but adequate indicators of change in many terrestrial insect groups, although this conclusion is not without its critics (see Hambler and Speight, 1995, 2004). Previously, Brown (1991) had suggested that, at least in the tropical context, the list of appropriate indicator groups could be extended from butterflies to include ants and certain Odonata and beetle groups. However, Ricketts *et al.* (2002) found that butterflies were poor predictors of diversity in a closely related but less well-studied group – moths – at least at the local scale, in Colorado, USA.

The principle of surrogacy covers a wide range of questions that have received much attention over the last 10 years or so. Conservation effort could be more efficiently focused geographically if species richness hotspots for different taxonomic groups: (i) coincided with each other; and (ii) encompassed foci of rare or endemic species. Perhaps not surprisingly, analysis of the UK fauna showed poor congruence between hotspots for butterflies and dragonflies (Prendergast *et al.*, 1993) while some studies have actually shown distributional complementarity rather than coincidence between groups. Furthermore, protection of butterfly richness hotspots in the UK and in Oregon, USA, did little to encompass sites with rare or threatened species (Prendergast *et al.*, 1993; Fagan and Kareiva, 1997). Even at the local scale, community-based rankings of sub-sites often do not run parallel for different insect groups. Painter (1999) found no correlation between species quality rankings of freshwater ditches based on beetles, snails and Odonata. This presents site managers with strategic dilemmas because it means that the habitat features and management options that are appropriate for one insect group may well be detrimental for another group.

Similar scepticism surrounds the issue of whether invertebrate conservation interest is coincident with, and predictable from, the composition of vegetation. The traditional conservationist's view that safeguarding the botanical interest of sites will ensure the protection of associated insect populations has long since been challenged and usually dismissed by entomologists (McLean, 1990; Kirby, 1992). Even exclusively phytophagous insects are reliant on more than the simple presence of their food plants, in many cases being equally dependent upon the physical structure of the habitat and how this is impacted by management. Thus, different grassland butterfly species have rather narrow preferences for particular vegetation heights (BUTT, 1986; Thomas, 1991) and they and other invertebrates respond rapidly to the seasonality, duration and intensity of grazing or cutting (Gibson *et al.*, 1992; Morris, 2000). Similarly, traditional woodland management practices in Britain such as coppicing, used by conservationists to promote a diverse ground flora, have profound effects on the associated fauna: whilst some butterflies associated with woodland clearings cue into the early stages of the coppice regeneration cycle, other invertebrates associated with shaded or deadwood habitats are adversely affected (Fuller and Warren, 1991; Hambler and Speight, 1995). Indeed, the creation and maintenance of bare patches within certain habitats such as heathland and grasslands, often regarded by botanists as unproductive ground or the result of mismanagement, is now recognized as crucial for certain thermophilous ground-nesting and predatory insect groups (Key, 2000). Thus, whilst vegetation composition may substitute for information on insects in certain narrowly defined habitats and taxonomic groups (Panzer and Schwartz, 1998), this 'coarse-filter' approach to site selection and monitoring is unlikely to be widely applicable except in very crude terms.

A related development has been to designate 'functional groups' of insects to aid ecological interpretation, sometimes accompanied by some form of taxonomic surrogacy, so that genera may be used in analysis instead of species and thus remove the need for the most labour-intensive level of taxonomic determination. This approach thus reduces the need for taxonomist input, other than for specialist advice, with the major advantage that interpretation may be achieved adequately for much reduced cost, and for insect groups which include numerous undescribed species. Ants in Australia are an important example of this approach. Following initial interpretation by

Greenslade (1978) and Andersen (1990, 1995) for Australia and subsequently developed for application in North America (Andersen, 1997) and South Africa (Andersen, 2003) (see background in Majer *et al.*, 2004), ant functional groups are designated at the genus or species level, and changes in the relative representation of those groups are used to indicate habitat condition, as a monitoring tool. Ants are used widely in this way in monitoring human impacts and subsequent habitat restoration in Australia.

Interestingly, one of the first of such approaches, and certainly now the most extensively developed, uses freshwater invertebrates for ecological evaluation of lotic systems and water quality assessment. Freshwater invertebrates have long been known to be sensitive to water quality. Originally developed in the UK to provide a simple monitoring system based on family-level identification of invertebrates that can be achieved without specialist knowledge (the Biological Monitoring Working Party score), the approach has since been extended to produce a standard method for assessing water quality for human consumption. The River Invertebrate Prediction and Classification System (RIVPACS) established a robust system for predicting freshwater invertebrate communities based on physical and chemical parameters of pristine UK watercourses; departures of communities in other rivers from these predictions are then used as an index of water quality (Wright *et al.*, 2000). Analogous systems have been implemented across several temperate countries (papers in Wright *et al.*, 2000). A substantial infrastructure has been developed in the UK to provide this annual monitoring service, but the disadvantage from a conservation standpoint is that identification rarely proceeds beyond family level. However, as Wright *et al.* (1993) point out, a species-level modification of the general approach could be developed to identify sites of potential conservation significance.

5 Rarity and Vulnerability

The various connotations of 'rarity' (Rabinowitz, 1981) have considerable importance in assessing conservation status, but can be interpreted only from sound and relatively comprehensive documentation. Thus, butterfly records from Britain and western Europe convey a reasonably, sometimes highly, accurate picture of distributions and patterns of local endemism. This is often supported by data on actual abundance and trends over time, together with detailed ecological information, all of which is helpful in assessing vulnerability of species or populations. This kind of detailed information is absent for most southern temperate insects, with the consequence that rarity is much more difficult to appraise. Many species are known from only single sites or localities and appear to be point or local endemics, but there is often considerable doubt over such interpretations, because substantial areas of apparently similar habitat have not been surveyed effectively. In such cases, 'rarity' may simply equate to 'under-recorded'.

Rarity and endemism are often incorporated uncritically as components of conservation status, but do not necessarily equate to vulnerability or threat of extinction, as Dennis (1997) noted for European butterflies. Simply because a species occurs (or appears to occur) over a very limited range does not render it threatened. Rare species attract attention, much of it emotional, not least because (paralleling Diamond's (1987) comment on birds) many people make special efforts to find rare or putatively extinct species: that effort is simply not available for surveying insects in southern temperate regions. Most insect groups have very few devotees in Australia or South Africa, particularly if Macrolepidoptera are excluded. Even for butterflies in Australia (approximating the land area of western Europe or the continental USA), only a few tens of people collect or study them with any view of contributing to scientific knowledge.

Caughley (1994) distinguished two different mechanisms by which species become vulnerable: the 'small population paradigm' that encapsulates the range of genetic and stochastic problems experienced by small populations by virtue of their restricted size, and the 'declining population paradigm' that includes all the factors that can drive population numbers down in the first place. There is still much uncertainty about the effective population numbers at which these processes become important. Soulé's (1987) 50:500 rule for minimum viable population sizes, proposed as population thresholds to avoid the effects of inbreeding depression and genetic drift respectively in vertebrates, probably has little application to insects although empirical tests are lacking. After an initial emphasis on rarity, the UK Biodiversity Action Planning and the conservation priority setting processes, prompted by the recently revised IUCN criteria, are now focusing more on species for which there is evidence of threat due to recent decline rather than rarity per se (e.g. Warren *et al.*, 1997; see also Warren *et al.*, Chapter 4, this volume). In Australia, there is increasing advocacy to focus on 'declining populations', not least because resources available for conservation are grossly insufficient to deal with all species that are regarded simply as 'rare' but without apparent threats to their well-being, and definition of threat provides a sound base for focused management. In contrast, the numerous 'rare' insects exhibiting small populations without apparent threat may not need active management other than to prevent them declining, such as by enhanced site buffering. It is difficult or impossible to formulate management to counter stochastic events, and the genetic consequences of existing in small populations (although potentially severe; Frankham *et al.*, 2002) are also difficult to predict confidently. In large and poorly documented faunas (such as Australia), so many insect species are regarded as rare (however the term is interpreted) that more tangible criteria are needed to help designate conservation priority, particularly as expertise and resources are grossly insufficient to treat all species in need of conservation attention individually.

6 Threats to Temperate Insects

Many action plans for insects throughout the temperate region necessarily include a substantial component of surveying to determine current status and distribution, and of research to define management needs more effectively. This reflects the paucity of information on many insects of conservation concern. A recent call in Australia for systematic inventory surveys of selected insect groups in national parks (Sands and New, 2003) to help address possibilities for species management in such areas is starting to be heeded, particularly in Queensland. In addition, threat evaluation is intrinsic to appraising vulnerability and chances of extinction. This process is central to the formulation of recovery or management plans, which must include clear objectives and periods for review and any necessary revision. However, statements about perceived threats, even in well-studied fauna such as in the UK, are often little more than very general pointers towards changes that would be detrimental. Although comparative details of threats in the northern and southern temperate regions are perhaps not constructive to investigate in detail, because of the enormous variety in both areas, some broad generalizations may be informative in helping to inform conservation strategy.

Vulnerable and threatened insect species are not evenly distributed across habitats. Thomas and Morris (1994) provided an illuminating analysis of 232 species listed in the British *Red Data Book* (Shirt, 1987). A striking pattern emerged in that the majority of endangered species are associated with either the very early or very late stages of succession. The early successional stages included bare ground, pioneer heathland, the early stages of the woodland coppice cycle and grassland that develops within 2 years of major disturbance, whilst the opposite end of the sequence was represented by deadwood habitats and their associated saproxylic fauna. As would be expected, the pattern is not universal across all taxonomic groups, being especially pronounced for Coleoptera and Diptera but less so for the Lepidoptera, Orthoptera and Hemiptera. Some of the emphasis on early succession habitats is undoubtedly because many of the associated species are at the northern edge of their range in Britain and are dependent upon the warm microclimates that these open habitats provide. Nevertheless, the general pattern highlights the fact that many entomologists attach high priority to habitats that are very different from those highly prized by conservationists who are concerned with other taxa. We know of no comparable analyses that have been carried out for other temperate countries, but similar studies elsewhere would be instructive.

Although not originally coined with invertebrates in mind, Diamond's (1989) 'evil quartet' – of habitat destruction, degradation and fragmentation, overexploitation, invasive alien species and chains of extinction – has plenty of relevance to insects. A fifth threat, climate change, has since gained equal potential significance, and has the potential to override more localized threats.

6.1 Habitat change

The topic of how habitat change impacts upon insect conservation encompasses change consequent upon natural processes such as succession, but also human-engendered degradation, fragmentation and wholesale destruction of habitats. Although the topic will not be dealt with in detail here because it is covered fully elsewhere (e.g. Thomas *et al.*, 2001; Warren *et al.*, 2001; Tscharntke *et al.*, 2002), it is worth drawing attention to two points. First, it is axiomatic that any change to a species' preferred or optimal habitat will have serious consequences. Since most insects are best envisioned as inhabiting microhabitats and their associated microclimates, even minor changes in the overall habitat, whether brought about by natural processes such as succession or by active management, may have far-reaching consequences for insects. Thus, even minor adjustments to the grazing pressure in grasslands can bring about substantial structural changes to the vegetation, which in turn have important effects on the microclimatic regime for temperature-sensitive insects. Second, it is worth highlighting the fact that many insects in the northern temperate region inhabit only remnant or restored habitats, or those altered substantially by people. Conservation attention is focused on minimizing loss of the remaining natural and semi-natural habitat, but may already be dealing with substantially impoverished biota, even though the extent of this impoverishment can only be speculative. Clearing of native vegetation in Australia and southern Africa has been imposed relatively recently on large areas of previously relatively undisturbed ecosystems, so that species losses can be more conspicuous and appear more dramatic because the near-natural remnant habitats that support higher proportions of the pre-disturbance taxa still remain for comparative study and evaluation.

6.2 Impact of introduced species

Although most introduced species fail to become established and spread, invasive species can have far-reaching consequences for communities and habitats. Inadvertent introductions, or cases of unexpectedly invasive spread by deliberately introduced insects, have occurred in most parts of the temperate region. Invasive ants are regarded as particularly severe threats to native species in Australia, South Africa and North America. The Argentine ant, *Lipepithema humile*, is native to South America but has been introduced to Mediterranean climates around the world. Sanders *et al.* (2003) showed how invasion by the Argentine ant caused a complete breakdown in the structure of the native ant community in California within 1 year, while Human and Gordon (1997) demonstrated strong effects on overall invertebrate diversity and the population sizes of many non-ant species and groups. Non-native *Vespula* wasps in New Zealand *Nothofagus* beech forests compete with native insects and birds that exploit the honeydew produced by endemic scale insects; additionally, predation by the wasps reduces and possibly eradicates populations of many native invertebrate species (Beggs, 2001).

Introduced plants, including weeds, exotic pasture grasses and crops, are important in displacing native vegetation and the specialized insects that depend upon it. Even non-herbivorous insects may be influenced by consequent changes in habitat. The impacts of exotic or invasive flora are of greatest concern when affecting restricted habitat types. McGeoch (2002) cited high-altitude montane grassland in South Africa as one such vulnerable environment supporting numerous endemic insect species. Invasive plants may significantly alter native insect diversity through changes in plant community composition. Himalayan Balsam, *Impatiens glandulifera*, is highly invasive along northern European watercourses where it outcompetes native riparian plant species that are hosts for an important and rich assemblage of insect herbivores, although the flowers are an important nectar source for pollinator species.

The deliberate planting of exotic forestry crops, often in very extensive stands, is widely regarded as detrimental to insect diversity. Certainly, *Pinus radiata* plantations in Victoria, Australia (Sinclair and New, 2004), and South Africa (Samways *et al.*, 1996) support very few native ant species in relation to the native forests they have replaced. The same is likely to be true where southern hemisphere trees have recently been introduced into northern temperate regions, for example, the widespread adoption of *Eucalyptus* spp. for plantation forestry in Iberia (Fernandez-Delgado, 1997). However, surveys in non-native conifer plantation forests in Britain have uncovered some unexpectedly diverse communities in which stand age, vertical structure and edge effects are important determinants of diversity (Humphrey *et al.*, 1999; Ozanne *et al.*, 2000; Jukes *et al.*, 2001).

Deliberate introduction of insects (e.g. as biological control agents or pollinators) to southern temperate regions has sometimes not been undertaken with due care, although increasing concerns in recent years are helping to overcome this through development of effective screening processes or other controls. For example, a current application has been presented to introduce bumblebees, *Bombus terrestris*, to the Australian mainland for pollination of greenhouse tomatoes. *B. terrestris* has been present in Tasmania since the early 1990s, and has spread over much of the state, including remote areas far from cropping systems and may be causing ecological harm through competing with native pollinators and damaging specialized native flora (Buttermore, 1997). Similar effects could possibly occur on the mainland, and such invasive species are regarded widely as important threats to native insects in the region, but capability to investigate these is limited. As an example of the contrasting attitude shown when an invading insect poses a direct threat to human interests, discovery of the red imported fire ant, *Solenopsis invicta*, in Queensland has led to 'perhaps the most ambitious and important effort ever undertaken to eradicate an invertebrate pest in Australia' (Vanderwoude *et al.*, 2003), with a funding commitment of AUS$120 million over 5 years.

6.3 Impacts of biological control agents on non-target species

Although the introduction of exotic predators and parasitoids in biocontrol programmes is often portrayed as an attempt to restore a balance between a pest and its natural enemies (e.g. Hoddle, 2004), impacts on other non-target species are often impossible to predict (Louda and Stiling, 2004) and are rarely adequately documented. Boettner *et al.* (2000) examined the effects of a generalist parasitoid fly that had been introduced into North America throughout most of the last century to control gypsy moth, *Lymantria dispar*. They reported 80% larval infestation rates by the parasitoid in a range of native saturniid moths, substantially explaining recent declines in these species, especially in the north-eastern USA. The Harlequin ladybird, *Harmonia axyridis*, is native to Asia but has been widely introduced into Europe and North America as a biological control agent of aphids and scale insects. As a very effective but also generalist predator, it is known to feed on the larvae of other Coccinellids as well; consequently, it has been implicated in the decline of certain native ladybird species in North America through both predation and competition for food resources (Koch, 2003). Its recent introduction into the UK has been taken sufficiently seriously to launch a government-funded national project to monitor both its spread and its impact on native ladybird species (Roy *et al.*, 2005). Alarmingly, screening for impacts of biocontrol agents on non-target species is not a requirement in many countries, including the USA. Thus, for example, no restrictions were placed on the recent importation and release of a dryinid parasitoid from North America into four separate provinces of Italy to control the flatid planthopper, *Metcalfa pruinosa*, even though no assessment had been made of whether it might impact on other native non-pest flatid species (Sala and Foschi, 2000).

6.4 Extinction cascades

Dunn (2005) has drawn attention to the threat of 'coextinction' of parasites (*sensu lato*) and mutualists as a consequence of the extinction of their hosts. Host-specific species are clearly more vulnerable in this respect than generalists. Such knock-on effects through ecological webs are likely to be common but may often go unnoticed. Perhaps the best example of an extinction cascade that led ultimately to the extinction of an insect (albeit only the local extinction of a subspecies) concerns the large blue butterfly, *M. arion*, in Britain. Ultimately, the loss of this species in Britain can be traced back to the successful biological control of rabbits, *Oryctolagus cuniculus*, using the *Myxoma* virus in the 1950s. The widespread collapse of the rabbit population caused open closely grazed grassland swards to be replaced by taller vegetation with consequent cooling of the soil surface layers. This, in turn, removed the hot microclimatic conditions required by the thermophilous host ant, *Myrmica sabuleti*, on which the butterfly larvae were dependent for food and protection (Elmes and Thomas, 1992). This is perhaps one of the best-documented cases of extinction of an insect, in which the links in the chain of extinction are well understood. However, it is unlikely to be an isolated case.

6.5 Insidious threats

Other more subtle, but possibly no less potent, threats also face insects in temperate zones. One that impacts particularly on temperate compared to tropical zones because it is directly associated with human population density is 'light pollution': artificial night lighting that interferes with the natural diurnal light cycle in ecosystems. The most obvious group in which effects might be expected is night-active moths, but many other insects respond to night illumination. Light pollution has been implicated in the decline of moth populations in the USA (Frank, 1988) and UK (Parsons *et al.*, 2005), but evidence is mostly anecdotal at present. A variety of reactions by insects (attraction/repulsion, orientation/disorientation) could be expected but very few have been investigated experimentally (Longcore and Rich, 2004). Long-term impacts on species distributions and population densities are unknown but could be profound and urgently need investigation.

The widespread prophylactic use of avermectins to treat intestinal parasitic infestations in grazing livestock means that the dung produced by such animals has a depauperate invertebrate fauna (Wall and Strong, 1987). This change, plus a general decline in low-intensity or 'extensive' livestock grazing as a traditional agricultural practice in many modern landscapes, has led to a general decline in the associated invertebrate specialist dung fauna, of which the hornet robberfly, *A. crabroniformis*, is a particularly vulnerable example.

7 Political Outliers

In the past, much conservation activity has been dictated by limited political jurisdictions, rather than by more global need. Insects common over much of Europe, or in some states of Australia, may receive considerable attention resulting from their rarity on the fringes of natural ranges, or in particular sites where they are deemed vulnerable, simply through the vagaries of their geography. The attention paid to such 'political outlier' insect taxa has been regarded by some as unduly parochial and misplaced in relation to more urgent needs, especially where taxa are relatively secure elsewhere (e.g. Hambler and Speight, 1995). The counterargument emphasizes that many such projects, in the process of unravelling the detailed ecology of individual species, have additionally been invaluable in developing general principles and methodology and in fostering local conservation interest, involvement and 'ownership'. In highly modified landscapes such as in Britain, there is also the consideration that often such species are not in decline solely as a result of natural edge-of-range processes but instead as a consequence of large-scale land use changes. As such, they may be indicative of declines across a wide range of unstudied taxa.

High-profile conservation reintroduction projects, often commanding considerable resources but not always delivering successful outcomes, have also sometimes been criticized for being too parochial. However, recent discussions and guidelines on the topic, both in general terms (Hodder and Bullock, 1997) and specifically in relation to insects (JCCBI, 1986; Oates and Warren, 1990), have encouraged greater scientific scrutiny of such projects, especially in relation to the global conservation status of the focal species. Thus, the reintroduction of the large blue butterfly to Britain was amply justified on the grounds that it is part of a group of globally threatened *Maculinea* spp. Additionally, evidence is accumulating that habitat restoration for the large blue is also benefiting other scarce butterflies, plants and even birds (Thomas, 1999). On the other hand, a long-established attempt to reintroduce the large copper *L. dispar* to Britain (Duffey, 1977) has been suspended following detailed autecological research (Pullin *et al.*, 1995) and the realization that its requirement for extensive fenland habitat is not currently met in the UK. Likewise, further investment in assessing the feasibility of reintroducing the chequered skipper *Carterocephalus palaemon* to England is being withdrawn (N. Bourn, 2006, *in litt.*) given its requirement for large areas of habitat (Ravenscroft, 1995) and the fact that it is both widespread and not threatened elsewhere.

Disproportionate attention to range edge butterfly species in Australia has caused concern over use of very restricted resources and has emphasized the need to differentiate between simple 'range edge' populations extending narrowly across political (State) boundaries, and so falling under different state legislations, and truly isolated populations separated from others by considerable distances. Sands and New (2002) attempted to distinguish these categories for butterflies, with the latter accorded higher conservation priority. Similarly, early tendencies in the UK to allocate resources to species which were on the northern edge of their range but widespread and unthreatened in nearby continental Europe have since given way to more global selection criteria that include consideration of the level of threat throughout the species' entire range. Of course, the latter approach is dependent upon good quality information on the distribution and status of species throughout their range, against which to assess the global significance of particular local populations, which has not always been available.

8 The Collecting Paradox

Collecting of butterflies and certain other insects is now prohibited or strongly discouraged in much of Europe, formally so in the case of protected species but also more widely. In large part this attitude reflects increasing conservation concern, but excessive zeal from the anti-collecting lobby can have undesirable consequences. The European protective legislations for insects, as reviewed by Collins (1987), included some extreme cases, extending far beyond the possible impacts of overcollecting on selected sensitive species or populations. All insect collecting is banned in Germany except with appropriate licences. Perhaps the most extreme case is for Laggintal, Switzerland, where (with the stated purpose of protecting the endemic satyrine butterfly *Erebia christi*) collection of all species of Lepidoptera and the carrying of butterfly nets are prohibited. Consequent acts of public ire over apparently innocuous and legal collecting activities elsewhere have perhaps deterred people from entering entomology as a hobby or lifelong interest. Fortunately, the attitude that collecting is incompatible with conservation, once particularly prevalent amongst less-well-informed site managers and nature reserve wardens, is now giving way to a realization that such activities are essential in order to build the biodiversity information base on which to make rational conservation decisions. Codes of conduct for collecting are now available in many temperate countries (e.g. in the UK; Invertebrate Link, 2002) and widely respected as pragmatic and responsible guidelines.

One argument commonly advanced is that for well-known insect groups (predominantly butterflies) in well-studied faunas further collecting is not needed for documentation, cannot be justified except in particular responsible scientific contexts and should be replaced by activities such as photography. This is not the case in the south, but regulatory approaches (and public opinion) in Australia and elsewhere have inherited the sentiment that collecting is a threatening process and should be curtailed. With relatively rare exceptions (including high-profile collectable species in demand by overseas dealers – such as *Colophon* stag beetles in South Africa; Geertsema, 2004), collecting is, at most, a subsidiary threat to habitat changes. Particularly for narrow-range endemic species, very small populations or populations with clear threats, any additional mortality may be undesirable and could provide an argument for prohibiting collecting. However, such cases are relatively unusual, and the common nexus of protecting a species by regulation or listing and banning collecting of butterflies in Australia has, in fact, retarded conservation progress:

1. Most knowledge of Australian butterfly biology and distribution has come from the activities of highly competent and enthusiastic hobbyists.
2. Collecting bans, or complex needs for permits, have deterred many such activities, eroding the badly needed goodwill of hobbyists to inform conservation, and driving much of the knowledge essentially 'underground' rather than being publicized freely, so that published information may be misleading and outdated.
3. Even when permits are granted, activities may be very restricted. For example, in Queensland until recently, permits applied only to particular places and dates, as well as to species. It was thus illegal to capture voucher specimens of possibly threatened species from other sites for verification of identity; many small lycaenids (such as *Hypochrysops piceatus* in southern Queensland; Sands and New, 2002) and hesperiids cannot be identified reliably from sight records alone.
4. More generally, such additional collecting is crucial in establishing the distribution and conservation status as well as the needs of insects, helping to overcome the under-recording so prevalent over the large areas involved. Any impediments to this endeavour are undesirable, particularly in the great majority of cases in which overcollecting cannot be considered credibly as a realistic threat.

In summary, the major need is to determine the cases in which collecting is indeed a threat and to ensure that appropriate safeguards are then implemented.

9 Species and Ecology

The best-studied insects in conservation, predominantly northern hemisphere butterflies as noted earlier, have highlighted the importance of understanding autecology when planning species-level conservation. This knowledge has indicated some valuable ways forward, and possible 'short cuts', as models for pursuing similar conservation measures in the southern zones. Parallel studies are indeed starting to occur; Kitching *et al.* (1999) summarized much earlier information on Australian butterfly biology, but relatively little information on population structure and dynamics of most species of conservation priority was then available. A full review of the importance of insect ecology in conservation is beyond the scope of this work, but one topic deserves particular mention in demonstrating the differing levels of information between north and south.

Perhaps the most significant of these ecological advances for conservation has been the development of the 'metapopulation concept' (Hanski and Gilpin, 1997). A number of rare insect species exist in small and substantially closed populations with minimal exchange of individuals with other local populations (Thomas and Harrison, 1992; Kindvall, 1996; Piper and Compton, 2003; see also Thompson *et al.*, Chapter 12, this volume). The metapopulation concept has revolutionized the ways in which extinctions of such local populations may need to be interpreted. Population or other extirpations were earlier interpreted largely as permanent loss of closed populations, but many such instances in Europe are now considered loss of metapopulation units, as part of a less unusual cycle of extinctions and colonizations that characterize the true spatial population structure of the species involved and so are less calamitous than 'true' extinction. Such considerations have had important influences on developing conservation management for butterflies, particularly in the northern hemisphere, and in helping to understand the aspects of landscape ecology that may be important to preserve or enhance in order to reduce the chance of more permanent losses (Ehrlich and Hanski, 2004). Unfortunately, the metapopulation concept has sometimes been applied too readily and uncritically to any species with spatial population structure; Harrison (1994) reviewed the evidence for metapopulation and related population spatial structures and their relevance to conservation. However, the metapopulation concept has been especially valuable in understanding and predicting the persistence of habitat specialists in modern fragmented landscapes (e.g. Thomas and Harrison, 1992) and how species can recover after range contraction (Davies *et al.*, 2005). Recent debates on the relative importance for overall persistence of metapopulation structure (the number and connectivity of suitable habitat patches) compared to habitat quality within sites (Thomas *et al.*, 2001; Bourn *et al.*, 2002) are of direct practical relevance to conservation managers. The same is true for the debate about the dimensions of habitat corridors and whether they function simply as dispersal conduits between local populations or represent usable habitat (Sutcliffe and Thomas, 1995; Pryke and Samways, 2001). These lessons have considerable potential for emulation as management models elsewhere, but the population structure of most butterflies in the south is not yet understood in comparable detail. Such studies would be significant in helping to confirm or contradict the general inferences from the north.

10 Extending Insect Conservation from Species

In regions with relatively small and well-known insect faunas and a relatively large number of concerned entomologists and conservationists, focus on individual species can play a leading role in insect conservation strategy. In the converse case of more insect species but fewer entomologists, this balance changes, and reliance on attention to individual species to drive conservation practice almost inevitably becomes less tenable. In this respect, the southern temperate regions are intermediate between the northern temperate regions and the tropics. Thus, in southern temperate regions, the predominating influence in conservation strategy has essentially switched from the species to the habitat or community level, with insects being conferred with roles as assessment tools as well as targets for individual attention, so that greater collective benefits accrue. Under Australia's federal legislation 'threatened communities' can be listed for protection in the same manner as for endangered species. Thus, 'Butterfly Community No. 1' is listed under state legislation in Victoria, although this entity has been defined solely in terms of a list of species (including several threatened Lycaenidae) occurring at one site (Jelinek *et al.*, 1994), and the extent to which this species list may need to differ from that at another site for that to be included in the same entity has not been defined. Many threatened vegetation types in Australia, some of them quite widespread, are important for insects, either notable species or wider diversity. For butterflies, Sands and New (2002) listed a number of vegetation-based communities that constitute important habitats to which notable species (some of them local endemics) are restricted. Sands and New also drew attention to the importance of 'topographical assemblages', to recognize the importance for butterfly conservation of features such as isolated hilltops in the landscape, utilized for hilltopping behaviour (see Britton *et al.*, 1995). Clearing of hilltops is now listed formally as a threatening process under New South Wales legislation.

In the UK, formal Species Action Plans have been prepared for some 219 insect species. Perhaps inevitably, these are unevenly distributed with respect to ordinal diversity: 4 Orthoptera, 64 Lepidoptera, 90 Coleoptera and 4 Hemiptera species, representing approximately 12.1%, 2.6%, 2.3% and 0.2% respectively of the total fauna in each order. Likewise, although the plans are somewhat formulaic (UK Biodiversity Group, 1999), varying amounts of resource have been devoted to the different species; some have not required or received much more than focused surveys to establish current status, whilst others have prompted major research projects (Piper and Compton, 2003; Purse *et al.*, 2003) and reintroduction programmes (Pearce-Kelly *et al.*, Chapter 3, this volume). In addition to addressing the conservation needs of individual species, the action plans collectively have served the useful purpose of drawing attention to gaps in knowledge (regarding status, threats, management, etc.) and the requirement for further research and monitoring.

Although the traditional emphasis on insect conservation at the species level remains strong in Britain, there is a growing realization that resources are grossly insufficient to deal adequately with all the deserving species. Greater emphasis is now turning to the identification and monitoring of ecologically based insect assemblages, including both common and rare species, that can be used for site assessment and for monitoring to assess habitat condition (Alexander *et al.*, 2004; Webb and Lott, 2006). This is a promising alternative to the traditional vegetation-based approach, since the UK National Vegetation Classification (NVC), now used almost universally as the template for much conservation assessment and monitoring, does not necessarily provide an appropriate classification for insect assemblages (Blake *et al.*, 2003; Maczey *et al.*, 2005).

Conservation strategies are often categorized as being either fine-filter or coarse-filter, reflecting respectively a focus on species or habitats (Samways, 2005). An extension to this dichotomy has recently been proposed which has some resonance with approaches now being adopted in Britain. Hunter (2005) adopted the term 'mesofilter' approach based upon identifying and prioritizing what he calls 'critical ecosystem elements': relatively small-scale habitat features that may be very important to individual species, including insects, but which are likely to be overlooked by more conventional habitat-based approaches to conservation focusing on higher-profile taxa such as plants and birds. This ties in with increasing focus in Britain on the conservation significance of specialized habitats and microhabitats harbouring important insect species. These include vegetated coastal shingle, soft-rock cliffs, quarries and 'brownfield' or post-industrial sites as habitat types that have conventionally received less attention for most taxa, although there is a growing realization of their importance for bryophytes, lichens, herpetofauna and invertebrates. Similarly, deadwood, bare ground, seepage, rot holes, temporary pools and river shingle banks are resources that have particular significance for insects in many other habitats. The challenge for conservation entomologists is to establish how best to create and maintain these habitat features sustainably and how to integrate them with the sometimes competing interests of other taxa.

11 Conclusions

Generally applicable patterns are elusive when faced with the very diverse canvas of insects and their habitats across temperate regions. However, some tentative conclusions are appropriate for developing future conservation strategies:

1. The past, present and future of insect conservation in temperate regions differ markedly between the northern compared to the southern hemisphere. In comparison to their southern hemisphere counterparts, northern temperate countries, especially in Europe, tend to have smaller and better-documented insect faunas, of which a higher proportion across many orders is formally described. Information is available for many groups in a non-specialist form, type material is largely accessible, a strong ecological and biological framework is available to support observations of species, and there are a relatively large number of entomologists sympathetic to a culture of conservation; the converse conditions pertain to much of the southern temperate region.

2. The single-species (fine-filter) approach that has been developed very successfully in northern regions, especially the UK, is normally impractical in southern temperate regions where the significantly larger number of species, a high proportion of which are undescribed, and the smaller number of workers have forced a more general, habitat- and community-based (coarse-filter) approach to sit alongside the species approach. This mirrors assemblage-based approaches that are now being actively developed in the UK. An intermediate (mesofilter) approach, which emphasizes the critical ecosystem elements that insects require, is helping to draw attention to habitat types and specialist habitat features that tend to be overlooked by conservationists focused on other taxa.

3. The single-species approach still has a role to play, especially where individual species can be presented as flagships for the general cause of insect conservation. Autecological studies have also done much to promote understanding of the unique requirements of insects and how these can be met in modern landscapes.

4. The sheer number of species of conservation concern precludes individual attention, so strategies will need to be developed for grouping species together in assemblages, communities or habitats that can be readily identified and conserved as higher groupings. Continuing emphasis will be needed on identifying, assessing and promoting indicator species and groups that can be used for routine monitoring of environmental change and human impacts.

Acknowledgements

We thank Dr Nigel Bourn and an anonymous reviewer for constructive comments on the manuscript.

References

Alexander, K.N.A., Drake, C.M., Lott, D.A. and Webb, J.R. (2004) Invertebrate assemblages on English SSSIs. English Nature Research Report 618. English Nature, Peterborough, UK.

Andelman, S.J. and Fagan, W.F. (2000) Umbrellas and flagships: efficient conservation surrogates or expensive mistakes? *Proceedings of the National Academy of Sciences of the United States of America* 97, 5954–5959.

Andersen, A.N. (1990) The use of ant communities to evaluate change in Australian terrestrial ecosystems: a review and a recipe. *Proceedings of the Ecological Society of Australia* 26, 347–357.

Andersen, A.N. (1995) A classification of Australian ant communities, based on functional groups which parallel plant life forms in relation to stress and disturbance. *Journal of Biogeography* 22, 15–29.

Andersen, A.N. (1997) Functional groups and patterns of organization in North American ant communities and comparison with Australia. *Journal of Biogeography* 24, 433–460.

Andersen, A.N. (2003) Ant biodiversity in arid Australia: productivity, species richness and community organization. *Records of the South Australian Museum*, Monograph series 7, 79–92.

Asher, J., Warren, M., Fox, R., Harding, P., Jeffcoate, G. and Jeffcoate, S. (2001) *The Millennium Atlas of Butterflies in Britain and Ireland*. Oxford University Press, Oxford.

Austin, A.D., Mackay, D.A. and Cooper, S.J.B. (eds) (2003) Invertebrate biodiversity and conservation. *Records of the South Australian Museum*, Monograph series 7.

Austin, A.D., Yeates, D.K., Cassis, G., Fletcher, M.J., La Salle, J., Lawrence, J.F., McQuillan, P.B., Mound, L.A., Bickel, D.J., Gullan, P.J., Hales, D.F. and Taylor, G.S. (2004) Insects 'down under' – diversity, endemism and evolution of the Australian insect fauna: examples from select orders. *Australian Journal of Entomology* 43, 216–234.

Badano, E.I., Regidor, H.A., Nunez, H.A., Acosta, R. and Gianoli, E. (2005) Species richness and structure of ant communities in a dynamic archipelago: effects of island area and age. *Journal of Biogeography* 32, 221–227.

Ball, S.G. and Morris, R.K.A. (2000) *Provisional Atlas of British Hoverflies (Diptera, Syrphidae)*. Biological Records Centre, Huntingdon, UK.

Balletto, E. and Casale, A. (1991) Mediterranean insect conservation. In: Collins, N.M. and Thomas, J.A. (eds) *The Conservation of Insects and Their Habitats*. Academic Press, London, pp. 121–142.

Beggs, J. (2001) The ecological consequences of social wasps (*Vespula* spp.) invading an ecosystem that has an abundant carbohydrate resource. *Biological Conservation* 99, 17–28.

Blake, S., McCracken, D.I., Eyre, M.D., Garside, A. and Foster, G.N. (2003) The relationship between the classification of Scottish ground beetle assemblages (Coleoptera, Carabidae) and the National Vegetation Classification of British plant communities. *Ecography* 26, 602–616.

Boettner, G.H., Elkington, J.S. and Boettner, C.J. (2000) Effects of a biological control introduction on three nontarget native species of Saturniid moths. *Conservation Biology* 14, 1798–1806.

Boggs, C.L., Watt, W.B. and Ehrlich, P.R. (eds) (2003) *Butterflies: Ecology and Evolution Taking Flight*. University of Chicago Press, Chicago, Illinois.

Bourn, N., Thomas, J., Stewart, K. and Clarke, R. (2002) Importance of habitat quality and isolation: implications for the management of butterflies in fragmented landscapes. *British Wildlife* 13, 398–403.

Britton, D.R., New, T.R. and Jelinek, A. (1995) Rare Lepidoptera at Mount Piper, Victoria: the role of a threatened butterfly community in advancing understanding of insect conservation. *Journal of the Lepidopterists' Society* 49, 97–113.

Brown, K.S. (1991) Conservation of neotropical environments: insects as indicators. In: Collins, N.M. and Thomas, J.A. (eds) *The Conservation of Insects and their Habitats*. Academic Press, London, pp. 350–404.

BUTT (Butterflies Under Threat Team) (1986) *The Management of Chalk Grassland for Butterflies*. Nature Conservancy Council, Peterborough, UK.

Buttermore, R.E. (1997) Observations of successful *Bombus terrestris* (L.) (Hymenoptera: Apidae) colonies in southern Tasmania. *Australian Journal of Entomology* 36, 251–254.

Caughley, G. (1994) Directions in conservation biology. *Journal of Animal Ecology* 63, 215–244.

Clarke, G.M. and Spier, F. (2003) *A Review of the Conservation Status of Selected Australian Non-marine Invertebrates*. Environment Australia, Canberra, Australia.

Collins, N.M. (1987) Legislation to conserve insects in Europe. *Amateur Entomological Society, Pamphlet* 13, London.

Collins, N.M. and Thomas, J.A. (eds) (1991) *The Conservation of Insects and Their Habitats*. Academic Press, London.

Conrad, K.F., Woiwod, I.P., Parsons, M., Fox, R. and Warren, M.S. (2004) Long-term population

trends in widespread British moths. *Journal of Insect Conservation* 8, 119–136.
Davies, Z.G., Wilson, R.J., Brereton, T.M. and Thomas, C.D. (2005) The re-expansion and improving status of the silver-spotted skipper butterfly (*Hesperia comma*) in Britain: a metapopulation success story. *Biological Conservation* 124, 189–198.
Dennis, R.L.H. (1997) An inflated conservation load for European butterflies: increases in rarity and endemism accompany increases in species numbers. *Journal of Insect Conservation* 1, 43–62.
Diamond, J.M. (1987) Extinct unless proven extant? Or, extant unless proven extinct? *Conservation Biology* 1, 77–81.
Diamond, J. (1989) Overview of recent extinctions. In: Western, D. and Pearl, M. (eds) *Conservation for the Twenty-first Century*. Oxford University Press, New York, pp. 37–41.
Duffey, E. (1977) The re-establishment of the large copper butterfly *Lycaena dispar batava* Obth. on Woodwalton Fen National Nature Reserve, Cambridgeshire, England, 1969–73. *Biological Conservation* 12, 143–158.
Dunn, R.R. (2005) Modern insect extinctions, the neglected minority. *Conservation Biology* 19, 1030–1036.
Ehrlich, P.R. (2003) Introduction: butterflies, test systems, and biodiversity. In: Boggs, C.L., Watt, W.B. and Ehrlich, P.R. (eds) *Butterflies: Ecology and Evolution Taking Flight*. University of Chicago Press, Chicago, Illinois, pp. 1–6.
Ehrlich, P.R. and Murphy, D.D. (1987) Conservation lessons from long-term studies of checkerspot butterflies. *Conservation Biology* 1, 122–131.
Ehrlich, P.R. and Hanski, I. (eds) (2004) *On the Wings of Checkerspots: A Model System for Population Biology*. Oxford University Press, Oxford.
Elmes, G.W. and Thomas, J.A. (1992) Complexity of species conservation in managed habitats: interaction between *Maculinea* butterflies and their ant hosts. *Biodiversity and Conservation* 1, 155–169.
Fagan, W.F. and Kareiva, P.M. (1997) Using compiled species lists to make biodiversity comparisons among regions: a test case using Oregon butterflies. *Biological Conservation* 80, 249–259.
Fernandez-Delgado, C. (1997) Conservation management of a European natural area: Donana National Park, Spain. *Quercus* 1(132), 50.
Frank, K.D. (1988) Impact of outdoor lighting on moths: an assessment. *Journal of the Lepidopterists' Society* 42, 63–93.
Frankham, R., Ballou, J.D. and Briscoe, D.A. (2002) *Introduction to Conservation Genetics*. Cambridge University Press, Cambridge.
Fry, R. and Lonsdale, D. (eds) (1991) *Habitat Conservation for Insects: A Neglected Green Issue*. The Amateur Entomologists' Society, Middlesex, UK.
Fuller, R.J. and Warren, M.S. (1991) Conservation management in ancient and modern woodlands: responses of fauna to edges and rotations. In: Spellerberg, I.F., Goldsmith, F.B. and Morris, M.G. (eds) *The Scientific Management of Temperate Communities for Conservation*. 31st Symposium of the British Ecological Society, pp. 445–471.
Geertsema, H. (2004) Some notes on *Colophon* (Coleoptera: Lucanidae) and peripherals. *Colophon* 4, 1–4.
Gibson, C.W.D., Brown, V.K., Losito, L. and McGavin, G.C. (1992) The response of invertebrate assemblies to grazing. *Ecography* 15, 166–176.
Greenslade, P. (1994) Australian native steppe-type landscapes: neglected areas for invertebrate conservation in Australia. In: Gaston, K.J., New, T.R. and Samways, M.J. (eds) *Perspectives on Insect Conservation*. Intercept, Andover, Massachusetts, pp. 51–73.
Greenslade, P. and New, T.R. (1991) Australia: conservation of a continental insect fauna. In: Collins, N.M. and Thomas, J.A. (eds) *The Conservation of Insects and Their Habitats*. Academic Press, London, pp. 33–79.
Greenslade, P.J.M. (1978) Ants. In: Lowe, W.A. (ed.) *The Physical and Biological Features of Kunoth Paddock in Central Australia*. CSIRO Division of Land Resources, Canberra, Australia, pp. 109–113.
Haes, E.C.M. and Harding, P.T. (1997) *Atlas of Grasshoppers, Crickets and Allied Insects in Britain and Ireland*. HMSO, London.

Hambler, C. and Speight, M.R. (1995) Biodiversity conservation in Britain: science replacing tradition. *British Wildlife* 6, 137–147.
Hambler, C. and Speight, M.R. (1996) Extinction rates in British non-marine invertebrates since 1900. *Conservation Biology* 10, 892–896.
Hambler, C. and Speight, M.R. (2004) Extinction rates and butterflies. *Science* 305, 1563.
Hanski, I. and Gilpin, M.E. (eds) (1997) *Metapopulation Biology: Ecology, Genetics, and Evolution*. Academic Press, San Diego, California.
Harrington, R. and Stork, N.E. (eds) (1995) *Insects in a Changing Environment*. Academic Press, London.
Harrison, S. (1994) Metapopulations and conservation. In: Edwards, P.J., May, R.M. and Webb, N.R. (eds) *Large-scale Ecology and Conservation Biology*. 35th Symposium British Ecological Society, pp. 111–128.
Henning, S.F. and Henning, G.A. (1989) *South African Red Data Book: Butterflies*. South Africa National Scientific Programmes Report no.158.
Hill, L. and Michaelis, F.B. (1988) *Conservation of Insects and Related Wildlife*. Australian Government Publishing Service, Canberra, Australia.
Hodder, K.H. and Bullock, J.M. (1997) Translocations of native species in the UK: implications for biodiversity. *Journal of Applied Ecology* 34, 547–565.
Hoddle, M.S. (2004) Restoring balance: using exotic species to control invasive exotic species. *Conservation Biology* 18, 38–49.
Holloway, G.J., Dickson, J.D., Harris, P.W. and Smith, J. (2003) Dynamics and foraging behaviour of adult hornet robberflies, *Asilus crabroniformis*: implications for conservation management. *Journal of Insect Conservation* 7, 127–135.
Human, K.G. and Gordon, D.M. (1997) Effects of Argentine ants on invertebrate biodiversity in northern California. *Conservation Biology* 11, 1242–1248.
Humphrey, J.W., Hawes, C., Peace, A.J., Ferris-Kaan, R. and Jukes, M.R. (1999) Relationships between insect diversity and habitat characteristics in plantation forests. *Forest Ecology and Management* 113, 11–21.
Hunter, M.L. (2005) A mesofilter conservation strategy to complement fine and coarse filters. *Conservation Biology* 19,1025–1029.
Ingram, G.J., Raven, R.J. and Davie, P.J.F. (eds) (1994) Invertebrate biodiversity and conservation. *Memoirs of the Queensland Museum* 36, 1–239.
Invertebrate Link (Joint Committee for the Conservation of British Invertebrates) (2002) A code of conduct for collecting insects and other invertebrates. *British Journal of Entomology and Natural History* 15, 1–6.
Jelinek, A., Britton, D.R. and New, T.R. (1994) Conservation of a 'threatened butterfly community' at Mount Piper, Victoria. *Memoirs of the Queensland Museum* 36, 115–120.
JCCBI (Joint Committee for the Conservation of British Insects) (1986) Insect re-establishment: a code of conservation practice. *Antenna* 10, 13–18.
Jukes, M.R., Peace, A.J. and Ferris, R. (2001) Carabid beetle communities associated with coniferous plantations in Britain: the influence of site, ground vegetation and stand structure. *Forest Ecology and Management* 148, 271–286.
Key, R.S. (2000) Bare ground and the conservation of invertebrates. *British Wildlife* 11, 183–191.
Kindvall, O. (1996) Habitat heterogeneity and survival in a bush cricket metapopulation. *Ecology* 77, 207–214.
Kirby, P. (1992) *Habitat Management for Invertebrates: A Practical Handbook*. RSPB, Sandy, Utah.
Kitching, R.L., Scheermeyer, E., Jones, R.E. and Pierce, N.E. (eds) (1999) *Biology of Australian Butterflies*. CSIRO, Melbourne.
Klass, K.D., Zompro, O., Kristensen, N.P. and Adis, J. (2002) Mantophasmatodea: a new insect order with extant members in the afrotropics. *Science* 296, 1456–1459.
Koch, R.L. (2003) The multicolored Asian lady beetle, *Harmonia axyridis*: a review of its biology, uses in biological control, and non-target impacts. *Journal of Insect Science* 3(32), 1–16.
Longcore, T. and Rich, C. (2004) Ecological light pollution. *Frontiers in Ecology and Environment* 2(4), 191–198.

Louda, S.M. and Stiling, P. (2004) The double-edged sword of biological control in conservation and restoration. *Conservation Biology* 18, 50–53.
Luff, M.L. (1998) *Provisional Atlas of the Ground Beetles (Coleoptera: Carabidae) of Britain*. Biological Records Centre, Huntingdon, UK.
Maczey, N., Masters, G.J., Hollier, J.A., Mortimer, S.R. and Brown, V.K. (2005) Community associations of chalk grassland leafhoppers (Hemiptera: Auchenorrhyncha): conclusions for habitat conservation. *Journal of Insect Conservation* 9, 281–297.
Majer, J.D., Shattuck, S.O., Andersen, A.N., and Beattie, A.J. (2004) Australian ant research: fabulous fauna, functional groups, pharmaceuticals and the Fatherhood. *Australian Journal of Entomology* 43, 235–247.
McGeoch, M.A. (2002) Insect conservation in South Africa: an overview. *African Entomology* 10, 1–10.
McGeoch, M.A. and Samways, M.J. (eds) (2002) Arthropod diversity and conservation in southern Africa. *African Entomology* 10, 1–159.
McGuinness, C. (2001) The conservation requirements of New Zealand's nationally threatened invertebrates. Threatened Species Occasional Publication no. 20. Department of Conservation, Wellington, New Zealand.
McGuinness, C. (2002) Threatened carabid beetles recovery plan (2002–2007). Department of Conservation, Wellington, New Zealand.
McLean, I.F.G. (1990) The 1998 Presidential Address – Part 2: what future for our entomological heritage? *British Journal of Entomology and Natural History* 3, 35–54.
Merritt, R., Eversham, B.C. and Moore, N.W. (1996) *Atlas of Dragonflies in Britain and Ireland*. HMSO, London.
Mikkola, K. (1991) The conservation of insects and their habitats in Northern and Eastern Europe. In: Collins, N.M. and Thomas, J.A. (eds) *The Conservation of Insects and Their Habitats*. Academic Press, London, pp. 109–119.
Morris, M.G. (2000) The effects of structure and its dynamics on the ecology and conservation of arthropods in British grasslands. *Biological Conservation* 95, 129–142.
Myers, N., Mittermeier, R.A., Mittermeier, C.G., da Fonseca, G.A.B. and Kent, J. (2000) Biodiversity hotspots for conservation priorities. *Nature* 403, 853–858.
Nally, S.C. (2003) Community involvement in the conservation of the endangered purple copper butterfly, *Paralucia spinifera* Edwards & Common (Lepidoptera: Lycaenidae). *Records of the South Australian Museum*, Monograph series 7, 217–224.
Naskrecki, P. (2004) The taxonomic impediment: a colonial legacy. *Wings* Fall, 20–24.
Nelson, B., Hughes, M., Nash, R. and Warren, M. (2001) *Leptidea reali* Reissinger 1989: a butterfly new to Britain and Ireland. *Entomologist's Record* 113, 97–101.
New, T.R. (1984) *Insect Conservation: An Australian Perspective*. W. Junk, Dordrecht, The Netherlands.
New, T.R. (1994) *Exotic Insects in Australia*. Gleneagles Press, Adelaide, Australia.
New, T.R. and Sands, D.P.A. (2003) Conservation concerns for butterflies in urban Australia. *Journal of Insect Conservation* 6 (2002), 207–215.
New, T.R. and Sands, D.P.A. (2004) Management of threatened insect species in Australia, with particular reference to butterflies. *Australian Journal of Entomology* 43, 258–270.
New, T.R., Pyle, R.M., Thomas, J.A., Thomas, C.D. and Hammond, P.C. (1995) Butterfly conservation management. *Annual Review of Entomology* 40, 57–83.
Oates, M.R. and Warren, M.S. (1990) *A Review of Butterfly Introductions in Britain and Ireland*. Joint Committee for the Conservation of British Insects and Worldwide Fund for Nature, Godalming, Surrey, UK.
Opler, P.A. (1991) North American problems and perspectives in insect conservation. In: Collins, N.M. and Thomas, J.A. (eds) *The Conservation of Insects and Their Habitats*. Academic Press, London, pp. 9–32.
Ozanne, C.M.P., Speight, M.R., Hambler, C. and Evans, H.F. (2000) Isolated trees and forest patches: patterns in canopy arthropod abundance and diversity in *Pinus sylvestris* (Scots Pine). *Forest Ecology and Management* 137, 53–63.

Painter, D. (1999) Macroinvertebrate distributions and the conservation value of aquatic Coleoptera, Mollusca and Odonata in the ditches of traditionally managed and grazing fen at Wicken Fen, UK. *Journal of Applied Ecology* 36, 33–48.

Panzer, R. and Schwartz, M.W. (1998) Effectiveness of a vegetation-based approach to insect conservation. *Conservation Biology* 12, 693–702.

Parsons, M.S., Fox, R., Conrad, K.F., Woiwod, I.P. and Warren, M. (2005) British moths: throwing light on a new conservation challenge. *British Wildlife* 16, 386–394.

Patrick, B.H. and Dugdale, J.S. (2000) Conservation status of the New Zealand Lepidoptera. Science for Conservation no. 136. Department of Conservation, Wellington, New Zealand.

Piper, R.W. and Compton, S.G. (2003) Subpopulations of *Cryptocephalus* beetles (Coleoptera: Chrysomelidae): geographically close but genetically far. *Diversity and Distributions* 9, 29–42.

Platnick, N.I. (1991) Patterns of biodiversity: tropical v. temperate. *Journal of Natural History* 25, 1083–1088.

Pollard, E. and Yates, T.J. (1993) *Monitoring Butterflies for Ecology and Conservation*. Chapman & Hall, London.

Ponder, W. and Lunney, D. (eds) (1999) The other 99%: the conservation and biodiversity of invertebrates. *Transactions of the Royal Zoological Society of New South Wales*. Mosman, New South Wales, Australia.

Prendergast, J.R., Quinn, R.M., Lawton, J.H., Eversham, B.C. and Gibbons, D.W. (1993) Rare species, the coincidence of diversity hotspots and conservation strategies. *Nature* 365, 335–337.

Pryke, S.R. and Samways, M.J. (2001) Width of grassland linkages for the conservation of butterflies in South African afforested areas. *Biological Conservation* 101, 85–96.

Pullin, A.S., McLean, I.F.G. and Webb, M.R. (1995) Ecology and conservation of *Lycaena dispar*: British and European perspectives. In: Pullin, A.S. (ed.) *Ecology and Conservation of Butterflies*. Chapman & Hall, London, pp. 150–164.

Purse, B.V., Hopkins, G.W., Day, K.J. and Thompson, D.J. (2003) Dispersal characteristics and management of a rare damselfly. *Journal of Applied Ecology* 40, 716–728.

Pyle, R.M. (1995) A history of Lepidoptera conservation, with special reference to its Remingtonian debt. *Journal of the Lepidopterists' Society* 49, 397–411.

Rabinowitz, D. (1981) Seven forms of rarity. In: Synge, H. (ed) *The Biological Aspects of Rare Plant Conservation*. Wiley, Chichester, UK, pp. 205–217.

Ravenscroft, N.O.M. (1995) The conservation of *Carterocephalus palaemon* in Scotland. In: Pullin, A.S. (ed.) *Ecology and Conservation of Butterflies*. Chapman & Hall, London, pp. 165–179.

Redak, R.A. (2000) Arthropods and multispecies habitat conservation plans: are we missing something? *Environmental Management* 26(Suppl. 1), 97–107.

Rentz, D.C.F. (1994) Orthopteroid insects in threatened habitats in Australia. In: Gaston, K.J., New, T.R. and Samways, M.J. (eds) *Perspectives on Insect Conservation*. Intercept, Andover, Massachusetts, pp. 125–138.

Ricketts, T.H., Daily, G.C. and Ehrlich, P.R. (2002) Does butterfly diversity predict moth diversity? Testing a popular indicator taxon at local scales. *Biological Conservation* 103, 361–370.

Roy, H., Rowland, F., Brown, P., Ware, R. and Majerus, M. (2005) Ecology of the Harlequin Ladybird: a new invasive species. *British Wildlife* 16, 403–407.

Sala, A. and Foschi, S. (2000) Allevamento e distribuzione di *Neodryinus typhlocybae* (Ashmead) in Italia. In: *La Metcalfa negli ecosistemi italiani*. ARSIA, Tuscany, Italy, pp. 119–130.

Samways, M.J. (1995) Southern hemisphere insects: their variety and the environmental pressures upon them. In: Harrington, R. and Stork, N.E. (eds) *Insects in a Changing Environment*. Academic Press, London, pp. 298–320.

Samways, M.J. (2005) *Insect Diversity Conservation*. Cambridge University Press, Cambridge.

Samways, M.J., Caldwell, P.M. and Osborn, R. (1996) Ground-living invertebrate assemblages in native, planted and invasive vegetation in South Africa. *Agriculture, Ecosystems, and Environment* 59, 19–32.

Sanders, N.J., Gotelli, N.J., Heller, N.E. and Gordon, D.M. (2003) Community disassembly by an invasive species. *Proceedings of the National Academy of Sciences of the United States of America* 100, 2474–2477.

Sands, D.P.A. and New, T.R. (2002) The Action Plan for Australian Butterflies. Environment Australia, Canberra, Australia.

Sands, D.P.A. and New, T.R. (2003) Coordinated invertebrate surveys in Australia's national parks: an important tool in refining invertebrate conservation management. *Records of the South Australian Museum*, Monograph series 7, 203–207.

Sands, D.P.A. and Scott, S. (eds) (2003) *Conservation of Birdwing Butterflies*. SciConEd Pty Ltd, Marsden, Queensland, Australia.

Sands, D.P.A., Scott, S.E. and Moffat, R. (1997) The threatened Richmond birdwing butterfly (*Ornithoptera richmondia* (Gray)): a community conservation project. *Memoirs of Museum Victoria* 56, 449–453.

Scholtz, C.H. and Chown, S.L. (1994) Insect conservation and extensive agriculture: the savanna of southern Africa. In: Gaston, K.J., New, T.R. and Samways, M.J. (eds) *Perspectives on Insect Conservation*. Intercept, Andover, Massachusetts, pp. 75–95.

Shaw, P. (2005) Estimating local extinction rates over successive time-frames. *Biological Conservation* 121, 281–287.

Shaw, M.R. and Hochberg, M.E. (2001) The neglect of parasitic Hymenoptera in insect conservation strategies: the British fauna as an example. *Journal of Insect Conservation* 5, 253–263.

Shirt, D.B. (ed.) (1987) *British Red Data Books: 2. Insects*. Nature Conservancy Council, Peterborough, UK.

Silberbauer, L.X. and Britton, D.R. (1999) Holiday houses or habitat: conservation of the Brenton blue butterfly *Orachrysops niobe* (Trimen) (Lepidoptera: Lycaenidae) in Kenya, South Africa. *Transactions of the Royal Zoological Society of New South Wales* 1999, 394–397.

Simberloff, D. (1998) Flagships, umbrellas, and keystones: is single-species management passé in the landscape era? *Biological Conservation* 83, 247–257.

Sinclair, J.E. and New, T.R. (2004) Pine plantations in south eastern Australia support highly impoverished ant assemblages (Hymenoptera: Formicidae). *Journal of Insect Conservation* 8, 277–286.

Soulé, M.E. (1987) *Viable Populations for Conservation*. Cambridge University Press, Cambridge.

Speight, M.C.D. (1989) *Saproxylic Invertebrates and Their Conservation*. Nature and Environment series 46, Council of Europe, Strasbourg, France.

Steencamp, C. and Stein, R. (1999) *The Brenton Blue Saga: A Case Study of South African Butterfly Conservation*. The Endangered Wildlife Trust, Parkview, Australia.

Sutcliffe, O.L. and Thomas, C.D. (1995) Open corridors appear to facilitate dispersal by ringlet butterflies (*Aphantopus hyperanthus*) between woodland clearings. *Conservation Biology* 10, 1359–1365.

Taylor, R.W. (1983) Descriptive taxonomy: past, present, and future. In: Highley, E. and Taylor, R.W. (eds) *Australian Systematic Entomology: A Bicentenary Perspective*. CSIRO, Melbourne, pp. 93–134.

Thomas, C.D. and Harrison, S. (1992) Spatial dynamics of a patchily distributed butterfly species. *Journal of Animal Ecology* 61, 437–436.

Thomas, J.A. (1991) Rare species conservation: case studies of European butterflies. In: Spellerberg, I.F., Goldsmith, F.B. and Morris, M.G. (eds) *The Scientific Management of Temperate Communities for Conservation*. 31st Symposium of the British Ecological Society, Blackwell Scientific Publications, Oxford, pp. 149–197.

Thomas, J.A. (1999) The Large Blue butterfly: a decade of progress. *British Wildlife* 11, 22–27.

Thomas, J.A. (2005) Monitoring change in the abundance and distribution of insects using butterflies and other indicator groups. *Philosophical Transactions of the Royal Society of London Series B – Biological Sciences* 360, 339–357.

Thomas, J.A. and Clarke, R.T. (2004) Extinction rates and butterflies: response. *Science* 305, 1563.

Thomas, J.A. and Morris, M.G. (1994) Patterns, mechanisms and rates of extinction among invertebrates in the United Kingdom. *Philosophical Transactions of the Royal Society of London Series B – Biological Sciences* 344, 47–54.

Thomas, J.A., Bourn, N.A.D., Clarke, R.T., Stewart, K.E., Simcox, D.J., Pearman, G.S., Curtis, R. and Goodger, B. (2001) The quality and isolation of habitat patches both determine where butterflies persist in fragmented landscapes. *Proceedings of the Royal Society of London B* 268, 1791–1796.

Thomas, J.A., Telfer, M.G., Roy, D.B., Preston, C.D., Greenwood, J.J.D., Asher, J., Fox, R., Clarke, R.T. and Lawton, J.H. (2004) Comparative losses of British butterflies, birds, and plants and the global biodiversity crisis. *Science* 303, 1879–1881.

Torrusio, S., Cigliano, M.M. and de Wysiecki, M.L. (2002) Grasshopper (Orthoptera: Acridoidea) and plant community relationships in the Argentine pampas. *Journal of Biogeography* 29, 221–229.

Tscharntke, T., Steffan-Dewenter, I., Kruess, A. and Thies, C. (2002) Characteristics of insect populations on habitat fragments: a mini review. *Ecological Research* 17, 229–239.

UK Biodiversity Group (1999) *Tranche 2 Action Plans*. Volume IV – *Invertebrates*. English Nature, Peterborough, UK.

Vanderwoude, C., Elson-Harris, M., Hargreaves, J.R., Haris, E.J. and Plowman, K.P. (2003) An overview of the red imported fire ant (*Solenopsis invicta* Buren) eradication plan for Australia. *Records of the South Australian Museum*, Monograph series 7, 11–16.

Veldtman, R., McGeoch, M.A. and Scholtz, C.H. (2002) Variability in cocoon size in southern African wild silk moths. *African Entomology* 10, 127–136.

Wall, R. and Strong, L. (1987) Environmental consequences of treating cattle with the antiparasitic drug ivermectin. *Nature* 327, 418–421.

Warren, M.S., Barnett, L.K., Gibbons, D.W. and Avery, M.I. (1997) Assessing national conservation priorities: an improved red list of British butterflies. *Biological Conservation* 82, 317–328.

Warren, M.S., Hill, J.K., Thomas, J.A., Asher, J., Fox, R., Huntley, B., Roy, D.B., Telfer, M.G., Jeffcoate, S., Harding, P., Jeffcoate, G., Willis, S.G., Greatorex-Davies, J.N., Moss, D. and Thomas, C.D. (2001) Rapid responses of British butterflies to opposing forces of climate and habitat change. *Nature* 414, 65–69.

Webb, J.R. and Lott, D.A. (2006) The development of ISIS: a habitat-based invertebrate assemblage classification system for assessing conservation interest in England. *Journal of Insect Conservation* 10, 179–188.

Wells, S.M., Pyle, R.M. and Collins, N.M. (1983) *The IUCN Invertebrate Red Data Book*. IUCN, Gland, Switzerland.

Wilson, R.J., Thomas, C.D., Fox, R., Roy, D.B. and Kunin, W.E. (2004) Spatial patterns in species distributions reveal biodiversity change. *Nature* 432, 393–396.

Woiwod, I.P. (1991) The ecological importance of long-term synoptic monitoring. In: Firbank, L.G., Carter, C., Darbyshire, J.F. and Potts, G.R. (eds) *The Ecology of Temperate Cereal Fields*. Proceedings of the 32nd Symposium of the British Ecological Society with the Association of Applied Biologists. Blackwell Scientific, Oxford, pp. 275–304.

Wright, J.F., Furse, M.T., Armitage, P.D. and Moss, D. (1993) New procedures for identifying running-water sites subject to envirnmental stress and for evaluating sites for conservation, based on the macroinvertebrate fauna. *Archives Hydrobiologiques* 127, 319–326.

Wright, J.F., Sutcliffe, D.W. and Furse, M.T. (eds) (2000) *Assessing the Biological Quality of Freshwaters: RIVPACS and Other Techniques*. Freshwater Biological Association, Ambleside, Cumbria, UK.

Wright, M.G. (1994) Insect conservation in the African Cape Fynbos, with special reference to endophagous insects. In: Gaston, K.J., New, T.R. and Samways, M.J. (eds) *Perspectives on Insect Conservation*. Intercept, Andover, Massachusetts, pp. 97–110.

Wynhoff, I. (1998) The recent distribution of European *Maculinea* species. *Journal of Insect Conservation* 2, 15–27.

Yeates, D.K., Harvey, M.S. and Austin, A.D. (2003) New estimates for terrestrial arthropod species-richness in Australia. *Records of the South Australian Museum*, Monograph series 7, 231–241.

Yen, A.L. and Butcher, R.J. (1997) *An Overview of the Conservation of Non-marine Invertebrates in Australia*. Environment Australia, Canberra, Australia.

Yen, A.L. and New, T.R. (eds) (1997) Proceedings of the conference 'Invertebrate Biodiversity and Conservation'. *Memoirs of Museum Victoria* 56, 261–675.

Yen, A.L., New, T.R., Van Praagh, B.D. and Vaughan, P.J. (1990) Invertebrate conservation: three case histories from south eastern Australia. In: Clark, T.W. and Seebeck, J.H. (eds) *Management and Conservation of Small Populations*. Chicago Zoological Society, Chicago, Illinois, pp. 207–224.

2 Insect Conservation in Tropical Forests

Owen T. Lewis[1] and Yves Basset[2]

[1]*Department of Zoology, University of Oxford, South Parks Road, Oxford OX1 3PS, UK;* [2]*Smithsonian Tropical Research Institute, Apartado 0843-03092, Balboa, Ancon, Panama City, Republic of Panama*

1 Introduction

In comparison with most temperate ecosystems, tropical forests are characterized by extraordinarily high but poorly inventoried insect diversity (perhaps 5–10 million species, with less than 1 million of them described), and by an absence of basic biological and ecological information for all but a handful of non-pest species (Godfray *et al.*, 1999; Novotny *et al.*, 2002). Rates of tropical forest habitat degradation and destruction are higher than in almost any other biome (Sala *et al.*, 2000; Pimm, 2001). In combination, these facts signal that the potential loss of insect diversity in tropical forests through human actions in the coming decades is enormous. In fact we are in danger of losing the vast majority of species before we have even documented them (Lawton and May, 1995).

Given the practical difficulties of gathering detailed ecological data in tropical environments where the species of interest may often occur at low levels of abundance (Folgarait *et al.*, 1995; Basset, 1999), and where the nature of the habitat often makes sampling or observation difficult, it is perhaps inevitable that efforts to conserve insects in temperate and tropical regions have typically involved rather different approaches. In temperate countries, at least in the northern hemisphere, conservationists have often focused on gathering detailed autecological information on threatened species, including their precise habitat requirements, local and global distributions, interactions with other species and dispersal ability (Stewart and New, Chapter 1, this volume). On the basis of such information, priority areas for the conservation of individual species have been designated, and management or recovery plans have been drawn up and implemented, often with great success (e.g. Collins and Thomas, 1991; Samways, 1994; New *et al.*, 1995). In contrast, there has been no consistent conservation approach for tropical insects. For a minority of rare, threatened or exploited tropical taxa we do have detailed ecological information that can help to guide conservation practice. These species tend to be members of what might be called the 'charismatic microfauna' – insects that are large, attractive or, ideally, both (e.g. *Ornithoptera alexandrae*; New, Chapter 13, this volume). These exceptions are representatives of a very large constituency: since at least 50% of terrestrial diversity occurs in the tropical zone, and at least 50% of the earth's species are insects, and since tropical habitats are often more threatened than temperate ones, it follows that the majority of threatened species are likely to be tropical insects.

Such exceptions aside, conservation studies of tropical insects are generally focused at the assemblage rather than the species level. An increasing number of studies are investigating how insect taxa respond to habitat disturbance and fragmentation, in terms of species richness, diversity or taxonomic or ecological distinctiveness. In this chapter, we elaborate on the potential and pitfalls of some of these approaches, focusing on three questions that we feel are key to tropical insect conservation: (i) How can we accurately make an inventory of insect diversity in tropical forests? (ii) What are the effects of human habitat exploitation or degradation on tropical insects? (iii) How critical are insects for ecosystem integrity in tropical forests? We conclude by considering some of the practical and methodological barriers to progress in answering these questions, and suggest some potential solutions; and we highlight additional areas of uncertainty, which may be fruitful areas for future investigation. Our focus is on humid tropical forests, the habitats with which we are most familiar, but many of our comments will be equally applicable to poorly studied, species-rich insect assemblages throughout the tropics and elsewhere at higher latitudes.

2 How Can We Accurately Make an Inventory of Insect Diversity in Tropical Forests?

A good understanding of the spatio-temporal distribution of insect biodiversity in tropical forests is fundamental information needed to guide conservation action. There are far too many tropical insect species to study them all, and so the goal of most conservation biology for tropical insects is to document patterns in diversity and community structure, and to assess the effects of anthropogenic disturbance on these patterns (Basset *et al.*, 1998). Such assessments can be undertaken at a hierarchy of spatial scales, from studies of vertical gradients from soil to canopy (Basset *et al.*, 2003b), through trends in richness along elevational gradients (e.g. Lewis *et al.*, 1998) to 'hotspots' analysis on a national or international scale (Bibby *et al.*, 1992). Depending on the spatial scale at which they are carried out, such studies may be used to identify the key habitat zones to conserve within a tropical forest, or to rank competing sites or regions in terms of conservation 'value'. Similar approaches can also be used to assess the effects of habitat fragmentation on tropical forest insect assemblages (Brown and Hutchings, 1997; Didham, 1997a,b), and the relative importance of 'undisturbed' or less-disturbed forests (Hamer *et al.*, 1997; Lawton *et al.*, 1998; Lewis, 2001), issues we discuss in more detail below. Whatever the precise goal of the investigation, the fundamental task for insect conservation biologists in tropical forests is to document the magnitude and spatial distribution of insect diversity; in essence, to produce comparable and representative inventories.

Conservationists interested in compiling species inventories for tropical forest sites face several major challenges. The very factor that makes tropical insect assemblages of such interest and concern – their extraordinary diversity – creates enormous practical and analytical difficulties. The long tail of species-abundance distributions typical of tropical forest habitats (Novotny and Basset, 2000) means that many species are encountered only infrequently, but the rare species least likely to be recorded in rapid assessments are often those of most conservation concern. Furthermore, if the pattern of species accumulation with sampling effort varies among habitats or sites, comparisons of diversity, species richness or other measures of conservation value based on restricted sampling may be unreliable. This makes ranking and comparing sites and treatments in terms of species richness or diversity problematic, unless intensive and long-term monitoring programmes are undertaken. Furthermore, the physical complexity of tropical forest habitats brings difficulties in sampling associated insects in a comprehensive or at least unbiased fashion (Kitching *et al.*, 2001). Finally, the challenge of identifying the material (Kitching, 1993) means that once the samples are collected the hard work is only just beginning.

Faced with these problems, there remains an urgent need to inform conservation decisions with data on species composition, species richness and diversity from tropical sites, without the need for expensive long-term and labour-intensive sampling. It is little wonder that (with some notable exceptions, e.g. Lawton *et al.*, 1998 (Fig. 2.1); project Investigating the Biodiversity of Soil and Canopy Arthropods (IBISCA): Didham and Fagan, 2003) the vast majority of such studies focus on a single taxon (e.g. Belshaw and Bolton, 1993; Eggleton *et al.*, 1996; Hill, 1999; Intachat *et al.*, 1999; Vasconcelos *et al.*, 2000; Davis *et al.*, 2001). Diurnal Lepidoptera are the most frequently studied group, by a substantial margin (e.g. DeVries *et al.*, 1997; Hamer *et al.*, 1997; Lewis, 2001; Ghazoul, 2002; Cleary, 2003; Cleary and Genner, 2004). Perhaps 80–90% of tropical taxa have never been the focus of tropical conservation studies, and it is an open question what the consequences of this taxonomic selectivity are likely to be. A full discussion of the choice of indicator taxa (and the question of what we might expect them to indicate) is beyond the scope of this chapter, but some key issues were covered in detail by Brown (1991) and are discussed by McGeoch (Chapter 7, this volume). More often than not the choice is more a function of the interests of the researchers involved, combined with selection of a group that has manageable levels of diversity, rather than 'megadiverse' taxa, such as weevils, leafhoppers and moths. An additional key reason for choosing a limited set of groups for study is the practical difficulties in identifying (even to morphospecies level) most taxa. In the tropics, insect surveys are continually hampered by the 'taxonomic impediment', something we return to later.

However, the single-taxon approach may be misleading: it is by no means certain that other insect taxa will show congruent patterns (Lawton *et al.*,

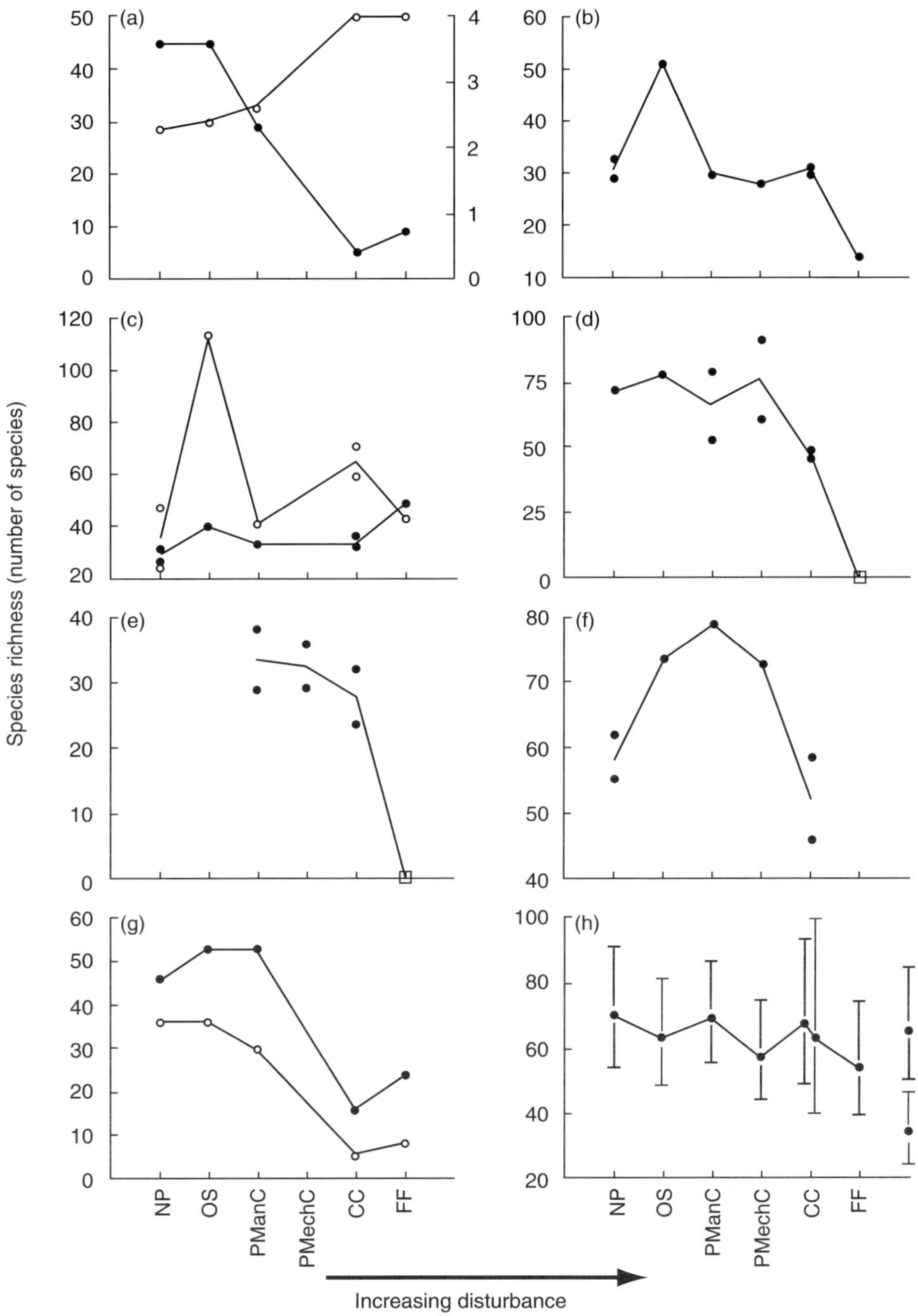

Fig. 2.1. Species richness of animal groups along a gradient of increasing habitat modification (left to right) in the Mbalmayo Forest Reserve, south-central Cameroon. (a) Birds (with mean habitat scores (open circles) on right ordinate); (b) butterflies; (c) flying beetles – malaise traps (filled circles), flight-interception traps (open circles); (d) canopy beetles; (e) canopy ants; (f) leaf-litter ants; (g) termites; (h) soil nematodes (with 95% confidence). (Reprinted from Lawton *et al.*, 1998, with permission from Macmillan Publishers.)

1998: Fig. 2.1). There is little consensus on the appropriate choice of 'indicator' species, especially in the tropics (Prendergast *et al.*, 1993; Hammond, 1994; Landres *et al.*, 1998; Lawton *et al.*, 1998; McGeoch, 1998; Kotze and Samways, 1999; Basset *et al.*, 2001b; Moritz *et al.*, 2001). A minority of tropical insect studies have a wider taxonomic focus, including whole orders or a few families from different orders, so that representatives of different guilds are included (e.g. Kremen, 1992; Didham *et al.*, 1998b; Kotze and Samways, 1999; Chung *et al.*, 2000; Kitching *et al.*, 2000). These studies may provide more representative results. Kitching (1993, 1996) and Didham *et al.* (1996) have advocated a more formal approach to widening the set of taxa included in such assessments through the use of 'predictor sets', including taxa from multiple functional groups or guilds (see also Kremen *et al.*, 1993). Such predictor sets are selected following statistical analysis of a larger data-set, including a wide range of taxa from multiple complementary sampling methods, and may give more reliable and general results.

Even for the best-studied taxa, little information is available to assess how much sampling is sufficient to provide a reliable indication of a site's conservation value. It would be extremely useful to generate 'rules-of-thumb' that may allow conservationists working on species-rich tropical assemblages to assess the completeness of their inventories, and whether a 'rapid' inventory approach can provide reliable information. Furthermore, guidelines on how best to employ the available effort would also be of value. For example, given a fixed period of time available to carry out surveys, is it more useful to concentrate sampling over a short period (perhaps during the season when abundance of the studied species is highest); or is it important to spread survey work throughout the year? Similarly, how useful is it to use multiple sampling methods, as opposed to a single method (Stork, 1994); and are comparisons among sites reliable if carried out at different times of year? We can start to answer some of these questions using the relatively restricted set of studies that have intensively surveyed particular taxa at individual sites. Structured inventories (Longino and Colwell, 1997) and the use of morphospecies or 'Recognizable Taxonomic Units' as surrogates for species level identifications (e.g. Netuzhilin *et al.*, 1999) provide a practical way forward, but additional work in this area is urgently needed. Although in many cases a morphospecies approach will be the only practicable way forward, we join the appeal for specimens to be assigned to morphospecies based on sound taxonomic methods (Wilson, 2000).

A related issue is the choice of metrics in such assessments. Diversity or species richness may seem a sensible metric to measure, but in practice in both tropical and temperate environments these measures often increase with disturbance, concurrent with a *decrease* in conservation value (Basset *et al.*, 1998). In many butterfly assemblages, for example, forest disturbance allows a suite of mobile, widespread and generalist taxa to colonize and coexist with much of the existing fauna (Thomas, 1991; Hamer *et al.*, 1997; Spitzer *et al.*, 1993, 1997; Lewis *et al.*, 1998), enhancing overall diversity. These newcomers are typically species of low conservation concern, and it does not make sense to give them equal weighting to restricted range habitat specialists in conservation assessments. One solution is to restrict analysis to endemics (e.g. Lewis *et al.*, 1998); or it may be possible to weight the conservation value of a species to reflect its geographic range or rarity, in a similar way to indices that take into account the taxonomic similarity of species for conservation assessments (Erwin, 1991; Vane-Wright *et al.*, 1991; Williams *et al.*, 1991). Alternatively, measuring the ratio of 'wider countryside' to forest specialist species might provide a rapid and approximate measure of human impacts on tropical forest ecosystems, although we are unaware of such studies. Of course, in order to use these approaches we do need some basic biological information in order to categorize taxa a priori as 'endemic' or widespread. Such information may be available for a surprisingly wide range of taxa, if the number of literature or museum localities for a taxon provides an approximate indication of its geographic range, although it is worth remembering that taxa can be both widespread and rare (Rabinowitz, 1981). It is quite uncertain how many tropical insect species are widespread yet rare: because of low levels of sampling for most taxa, if a species is locally rare then its recorded range is almost inevitably likely to be small. Many widespread and 'rare' species may prove to be much more common than has been assumed. A related issue concerns specialist versus generalist species: specialists will often (but not always) have relatively small geographic ranges (Gaston *et al.*, 1997; Gaston, 1999), but endemic generalists certainly exist, for example, many island taxa. If sufficient information is available to categorize species on both counts then specialists (in terms of food or habitat use) might perhaps be accorded more weight in conservation assessments than endemics, since they may be the species most endangered by habitat disturbance.

3 What Are the Effects of Human Habitat Exploitation or Degradation on Insects?

Approximately half of the earth's closed-canopy tropical forest has already been converted to other uses (Wright, 2005), and the population of tropical countries, having almost trebled since 1950, is projected to grow by a further 2 billion by 2030 (Wright, 2005). Inevitably, anthropogenic pressures mean that it will only ever be possible to maintain a small fraction of the world's tropical forests as reserves or parks, free from human disturbance. Most tropical forests are likely to remain subject to varying intensities of disturbance, which takes numerous interacting forms. Each year, approximately 5.8 million hectares of tropical forests are destroyed completely through conversion to pasture and plantation, habitats that are unlikely to support more than a fraction of the insect fauna present earlier. An equivalent area is degraded annually, to varying degrees and with less clear-cut effects on biodiversity (Mayaux *et al.*, 2005). Small-scale (often subsistence) agriculture is, in terms of the area affected, the most important single cause of tropical forest degradation, accounting for around 60% of deforestation. Commercial logging also typically results in degraded forest, rather than total forest loss since, with the exception of certain dipterocarp forests in South-east Asia, only a minority of tropical trees is economically viable for exploitation as timber. All of these human impacts, individually or in isolation, can result in a fragmented network of relatively intact patches, separated by a matrix that may vary from 'recovering' secondary forest, apparently rather similar to the pre-disturbance state of the system, through to pasture devoid of woody vegetation, or plantation monocultures. Few tasks can be more important for conservationists than assessing the impact of such human activities on tropical forest biodiversity. In order to minimize species extinctions globally, we need to know how we are altering the structure of these tropical communities, what degree of disturbance is consistent with the persistence of acceptable levels of tropical forest biodiversity and which groups of organisms are most seriously affected. Here, we consider disturbance and fragmentation separately, although one will rarely act entirely without the other.

3.1 Logging and other forms of disturbance

Can commercial timber extraction and other forms of tropical forest disturbance be reconciled with the maintenance of insect diversity? A growing set of studies throughout the tropics has investigated how human disturbance, in various forms and at varying intensities, is affecting the species richness or diversity of particular insect groups. The results of such studies have proved highly unpredictable, with disturbance shown to have a positive, negative or no effect on species richness in individual studies. Individual studies will be of local value, but generalizations are proving difficult to extract from the existing data. Are there general factors influencing whether species richness is observed to increase or decrease following disturbance? In particular, to what extent is the variability among studies real, and to what extent does it reflect variability in the sampling methods used, or idiosyncratic characteristics of individual study locations?

Replication is a troublesome issue for researchers trying to assess the effects of disturbance on tropical insect communities. Tropical rain forests have high spatial heterogeneity, which generates high beta diversity (Wolda, 1996; Vasconcelos *et al.*, 2000), so protocols should ideally partition the variance in insect response between forest disturbance and faunal turnover with increasing distance between study sites. Typically, researchers will compare insect diversity in a single area of 'disturbed' forest with diversity in a nearby 'less-disturbed' forest. If there are multiple sites within each habitat these are likely to be pseudoreplicates (Hurlbert, 1984) because they are clustered in space and effectively represent multiple samples from the same habitat unit: the true sample size for each habitat type is in fact one. When differences are detected between such areas, it is difficult to determine whether these are a consequence of disturbance, or if they simply reflect pre-existing differences in topography or geography. Such differences are likely to exist for practical reasons. For example, areas of forest are unlikely to be logged if they include steep slopes, major watercourses or low densities of timber trees, all factors that are likely to affect species composition in the absence of disturbance effects. There is no simple solution to this problem since the spatial scale necessary to sample truly replicated disturbed and undisturbed habit units is likely to be large and logistically challenging.

One opportunity for genuinely replicated sampling that has been taken advantage of rather rarely by tropical insect conservation biologists is the availability of silvicultural and logging plots in many tropical forests. These are typically set up by foresters to provide information on the effects of forest management on growth and yield of timber trees, and include before–after control impact (BACI) designs, which allow robust comparisons in the face of spatial and temporal variability (Stewart-Oaten and Murdoch, 1986). Such experiments provide excellent opportunities for insect conservation biologists to ask how the experimental treatments (which by definition are those under consideration for wider application in the area concerned) affect insect assemblages. A crucial advantage of such studies is that treatments have been allocated at random to experimental units, avoiding the risk of pseudoreplication. Basset *et al.* (2001a, b) provide an example of this approach for an unreplicated BACI protocol in Guyana. Experimental plots also provide an opportunity to assess the extent to which new logging protocols, such as 'Reduced Impact Logging', affect insect diversity, relative to conventional approaches. These protocols are typically designed with at least one of the following goals in mind: to reduce biodiversity loss from logging, to enhance sustainability of timber extraction, or to promote carbon sequestration by increasing the density of the residual stand (e.g. Bird, 1998; Davis, 2000). We have recently made use of such an experiment in Belize to assess the effects of an experimental selective logging regime on butterfly (Lewis, 2001) and dung beetle assemblages, and found that logging treatment effects were small relative to spatial block effects, highlighting the danger that spatial heterogeneity in species richness and species composition will generate misleading results in similar but non-experimental studies.

In reaching more general conclusions about the likely global effects of habitat modification on tropical insect assemblages it will be valuable to draw together information from many studies through meta-analysis. Individual studies in the literature should represent independent replicates, even if they are in themselves pseudoreplicated (Cottenie and De Meester, 2003). For the most widely studied taxon (Lepidoptera), sufficient studies are now potentially available to allow such analyses (Hamer and Hill, 2000; Hill and Hamer, 2004). Unfortunately and perhaps inevitably, because individual authors have had their own aims and methods specific to their particular studies, collating published investigations in a way that allows a meaningful meta-analysis is difficult. For example, the Lepidoptera studies vary considerably in the methods used to measure 'diversity'. Most present results for either species richness or for a diversity index, and rarely for both. Species richness is highly sensitive to sample size and many studies present 'raw' species richness values that have not been corrected for sample size.

Many of these problems could be avoided, and syntheses of published information could be made more rigorous and effective if individual authors included more information about their studies. The wider value of future individual studies can be increased through careful description of the methods employed and through consistent reporting of results. In Box 2.1 we present a 'wish list' for studies of the effects of disturbance on tropical insects. There are very many permutations in possible metrics for analysis, and some will be more appropriate than others for individual studies. Thus, rather than striving for standardization, we encourage authors to publish (perhaps as electronic appendices) summary tables of counts of species in each sampling unit. Analyses should take into account the numerous pitfalls inherent in comparisons of diversity measures (Gotelli and Colwell, 2001). In the tropics, insect species accumulation curves rarely saturate, but rarefaction or Coleman curves allow comparisons of species richness taking into account variations in sample size among sampling units. Where available, information on the nature, spatial extent and intensity of disturbance should be reported. In the context of logging, for example, Greiser Johns (1997) recommends a simple and consistent means of reporting the intensity of logging in terms of the percentage of the stand harvested and the time elapsed since harvesting. Disturbance from human activities other than logging may also vary markedly in its form and intensity, but this will be more difficult to quantify unambiguously and consistently. Additional complications are that the spatial scale of observation, sampling effort and sampling techniques may explain a large proportion of the

Box 2.1. A 'wish list' for studies of the effects of disturbance on tropical insects.

1. Take into account the geographical distribution/endemicity of taxa, rather than focusing solely on overall species richness or diversity values
2. Report both species richness and diversity measures, and control for the critical influence of sample size on species richness values through rarefaction
3. Be explicit about the nature of replication in the investigation
4. Document clearly the forms of habitat disturbance, and the time since disturbance events
5. Document the history of human and natural disturbance in the studied areas
6. To avoid publication bias, publish negative results (where no significant disturbance effect is found), as well as positive ones; this plea is addressed to editors, as well as authors
7. Consider employing or exploiting experimental protocols, such as before–after control impact (BACI)
8. Use sound concepts of taxonomy (where morphospecies correspond to unnamed species, rather than fuzzy groupings of unidentified specimens)
9. Use a multi-taxon approach to reach more general conclusions as to the impacts of disturbance on diversity; where this is not possible recognize clearly the limitations associated with individual study taxa
10. Include summary data on numbers or individuals of each species recorded from individual sampling locations (perhaps as electronic appendices), to facilitate subsequent meta-analyses

variation in outcomes observed across studies (Hamer and Hill, 2000; Hill and Hamer, 2004), and that sites that have a long history of 'natural' disturbance may be relatively insensitive to subsequent human disturbance (Balmford, 1996; Lewis, 2001).

Whether or not in a designed experiment, the scale of the study areas relative to the dispersal ability of the organisms studied is critical, and it may be important to take this into account when assessing the impacts of disturbance. Human-modified habitats are sometimes deemed to support a high proportion of the insect fauna associated with nearby, less-disturbed habitats. It is of course possible that these species have self-supporting breeding populations in disturbed habitats. However, if 'disturbed' sites are well within the dispersal range of 'less-disturbed' sites then for mobile insects like adult tropical butterflies, the nature of the surrounding habitat will almost inevitably influence the taxa recorded. Many may be 'tourists' from neighbouring, less-disturbed forest, which are not breeding in these habitats; others may breed there, but persist solely as 'sink' populations, dependent on repeated immigration for local persistence.

3.2 Habitat fragmentation

The creation of a patchwork landscape of forest fragments embedded in a matrix of habitats degraded to varying degrees is an inevitable consequence of deforestation (Wright, 2005). What effect does fragmentation have on tropical forest insect diversity? Habitat fragmentation has been a key focus of conservation research in temperate ecosystems over the last two decades, and insect studies have been key to the development, testing and application of metapopulation models in particular (Hanski and Poyry, Chapter 8, this volume). Fewer studies have investigated the effects of fragmentation on insect assemblages in tropical forests. A notable exception, on a large scale, is the experimental Biological Dynamics of Forest Fragments (BDFF) project in the Brazilian Amazon. The BDFF study is one of the most intensive habitat fragmentation assessments ever undertaken, and although much of the work there has focused on vertebrates, there has been intensive study of certain insect taxa, notably beetles (Didham *et al.*, 1998a,b) and butterflies (Brown and Hutchings, 1997). In fact there are compelling reasons to select insects as focal species in such studies. In particular, the relaxation period between fragmentation and species reaching equilibrium densities in the fragmented landscape is much lower for short-lived insects, allowing more rapid conclusions to be drawn about the true impacts of fragmentation. Existing data suggest that fragmentation has effects on insect communities that mirror in many ways the effects of logging and other forms of habitat degradation.

It remains to be seen how relevant single-species studies of fragmentation are to tropical situations. In particular, it is uncertain how many tropical insects have population structures that approximate to metapopulations, with local populations in patches of habitat subject to periodic extinction and colonization events, and a dependence on recolonization for long-term regional persistence. As for temperate insects, some tropical taxa may be forced into metapopulation-like situations by habitat fragmentation. Furthermore, tropical insects with high host specificity have breeding habitat that is defined by the spatial availability of host plants, which may represent patches of suitable habitat in a sea of unsuitable foliage. Resource fragmentation thus arises from two main factors: high host specificity (Janzen, 1973; Gilbert and Smiley, 1978; Basset, 1992; Marquis and Braker, 1993; Basset *et al.*, 1996; Barone, 1998) and high plant diversity (Novotny *et al.*, 2002, 2004). Individual plants of any one species are isolated in space, so host plant-specific tropical insects may occur as patchy populations or metapopulations on fragmented resource patches. Similarly, specialized predators or parasitoids will have a patchy spatial distribution determined by the distribution of their host herbivores. Will such fragmented populations act like 'true' metapopulations (Levins, 1969), with relatively independent demography in individual patches, and persistence dependent on dispersal among empty patches? Or will they operate more like 'patchy populations' (Harrison, 1991), where dispersal is high relative to the typical isolation between patches? If the former, then metapopulation models may be relevant to conservation planning; for example, selective logging, which removes individual trees may serve to increase patch isolation within the metapopulation. Since population densities for individual insect species are typically very low in tropical forests, establishing occupancy and local extinction of herbivores on whole trees is difficult, so it will be challenging to assess how widespread this type of population structure is in these habitats, and perhaps impossible to parameterize predictive, spatially realistic metapopulation models. However, many of the more general insights that have emerged from metapopulation biology may prove helpful in a tropical forest context, for example, the requirement for a landscape perspective, the importance of 'unoccupied' habitats and the fact that extinctions may be long delayed following fragmentation.

4 How Critical Are Tropical Insects for Ecosystem Integrity?

Whether humans alter insect assemblages through habitat modification or through fragmentation, then the consequences for ecological processes are of considerable interest, as are the likely direct and indirect effects of changes in the insect fauna on the wider community.

4.1 Ecosystem function

The relationship between biodiversity and ecosystem function has become a major preoccupation among ecologists and conservation biologists (e.g. Loreau *et al.*, 2002; Hooper *et al.*, 2005), and provides a widespread justification for conservation. The literature on this topic is dominated by studies of the relationship between diversity and productivity in temperate plants (e.g. Hector *et al.*, 1999), and studies of organisms at higher trophic levels (e.g. insects) are few. We join the call for an increasing emphasis on the impact of insect biodiversity loss on ecosystem processes (e.g. Didham *et al.*, 1996). Are insects really the 'little things that run the earth' (Wilson, 1987) by providing services that maintain the 'health' of ecosystems? A strong case can certainly be made for the key importance of several guilds, including dung beetles, termites and other arthropods involved in decomposition. More generally, insects play a key role in pollination (Kremen and Chaplin-Kramer, Chapter 15, this volume) and nutrient cycling via herbivory (Frost and Hunter, 2004). These and related topics are covered in more detail by Memmott *et al.* (Chapter 10, this volume) and Kremen and Chaplin-Kramer (Chapter 15, this volume), but here we briefly highlight tropical examples for a well-studied and ecologically important taxon: scarabaeid dung beetles.

The movement and burial of animal faeces by dung beetles for feeding and ovipositing results in soil fertilization and aeration, as well as nitrogen and nutrient cycling (Estrada *et al.*, 1999; Davis *et al.*, 2001; Andresen, 2002, 2003). The burial of dung also helps control important parasites of vertebrates, such as flies and hookworm. Furthermore, dung movement and burial is important for secondary seed dispersal: removing seeds from the surface of the soil protects seeds from predation and so is important for rainforest regeneration. The rate at which dung is buried can be measured in the field, and the correspondence between dung burial rates and diversity or species richness calculated. Klein (1989; see also Didham *et al.*, 1996), working in Amazonia, found a strong positive relationship between dung beetle diversity and rates of dung burial, and between fragmentation and diversity, such that forest fragments were characterized by low dung beetle diversity and reduced ecosystem function, compared to continuous forest (Fig. 2.2; see also Quintero and Roslin (2006) for recovery of these communities following

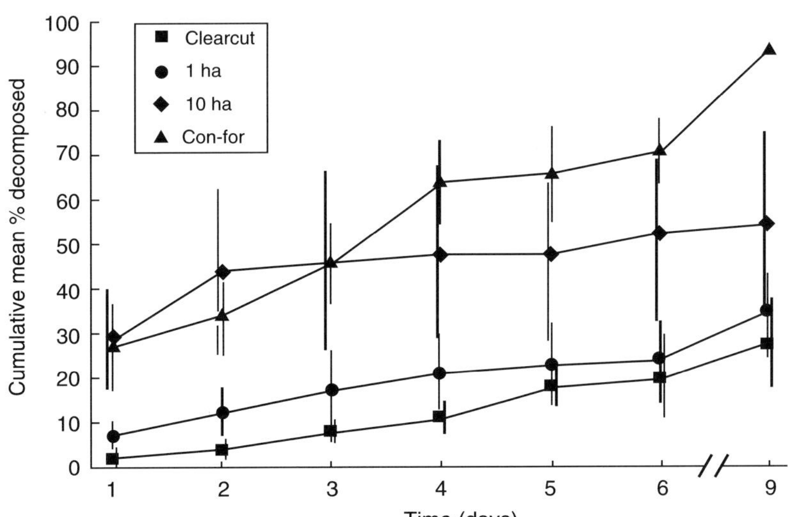

Fig. 2.2. Cumulative mean (*n* = 9, ± s.e.) percentage dung removal from experimental piles of cattle dung in Amazonian forest fragments of different areas, and in adjacent clear-cuts and continuous forest. (Reprinted from Klein, 1989, with permission from the Ecological Society of America.)

re-growth of secondary forests between fragments). However, Klein's (1989) methods appear not to rule out dung beetle abundance as the casual factor linking diversity and function (e.g. Andresen, 2003), and subsequent studies of dung beetle assemblages elsewhere in the tropics have found less clear-cut diversity–function relationships. In general, the field is ripe for further experimental and manipulative investigations of ecosystem processes (e.g. decomposition within litter bags: Fagan *et al.*, 2005) in relation to the diversity of the insect guilds involved in carrying out these functions.

4.2 Food webs and community interactions

Linked to ecosystem function is the study of trophic interactions among species. All species are embedded in complex webs of mutualistic and antagonistic interactions, and nowhere are these webs more complex and diverse than in tropical forest ecosystems (Janzen, 1983). Trophic interactions have been described as the glue that holds together ecological communities, and several authors have called for the conservation of trophic interactions as a goal for conservationists (Gilbert, 1980; Janzen, 1983; Memmott *et al.*, 2006). Through their high diversity and wide variety of feeding niches, insects are a key component of all tropical forest food webs and habitat modification can cause marked changes to food web structure (Tylianakis *et al.*, 2007). The effects of losing individual species from food webs can be unpredictable and may propagate some distance through interlinked chains of trophic linkages ('indirect effects'). One recent study of a tropical forest host-parasitoid community suggests that removal of a single species can have widespread cascading indirect effects through apparent competition (Morris *et al.*, 2004, 2005). Similarly, alterations in herbivore abundance can lead to trophic cascades (Letourneau and Dyer, 1998; Dyer and Letourneau, 1999). Given the major effects that insects can have on plant fitness (Marquis and Braker, 1993; Marquis, 2005) and potentially plant diversity (Janzen, 1970; Connell, 1971), alterations in insect assemblages may have major repercussions for the wider tropical ecosystem.

5 Unknowns, Practical Problems and Potential Solutions

5.1 The taxonomic impediment, and a role for parataxonomists

The 'taxonomic impediment' refers to the gaps of knowledge in our taxonomic system, the shortage of trained taxonomists and curators, and the impact these deficiencies have on our ability to manage and use biological diversity (Anon., 1998). The taxonomic impediment is perhaps at its greatest for tropical invertebrates, where the mismatch between taxonomic effort and biological diversity is at its greatest, and it greatly inhibits tropical insect conservation biology by making even the most taxonomically restrictive inventory a major undertaking.

Meeting the taxonomic challenge will require the use of new technologies (e.g. DNA barcoding and digital imaging) and the transfer of technologies and training to tropical countries, which harbour most biodiversity. Making taxonomic information available to entomologists around the world is increasingly possible with advances in information technology, but access to information in itself does not reduce the need for well-trained taxonomists and field workers.

Over the last decade or so, a new model has proved very successful in speeding the flow of biodiversity information from tropical ecosystems: working with parataxonomists (Janzen *et al.*, 1993; Basset *et al.*, 2000). Parataxonomists stand 'at the side' of conventional taxonomists: they collect specimens, prepare them, carry out preliminary sorting into morphospecies and enter the associated information onto databases. They are not an alternative to professional taxonomists in the field or laboratory, but enhance their activities and capacities. The advantages of working with local parataxonomists in the tropics were summarized by Basset *et al.* (2000, 2004) and include: (i) increased efficiency and replication of sampling with year-round activity in the field; (ii) rapid preparation of high quality specimens at low cost; (iii) enhanced integration of local ecological information associated with collected specimens; and (iv) enhanced public outreach and local interest in biodiversity. Parataxonomists may reduce greatly the time-lag between the initiation of the study and the publication of results, a particular advantage for conservation studies where there may be urgent need for action. With the help of parataxonomists, it may become feasible to include several taxa or guilds within the sampling protocol. As discussed in Section 2, we believe that this represents a promising alternative to the monitoring of species-poor taxa over relatively short periods. Training and employment of parataxonomists could profitably be put to use in conservation biology and in subsequent biodiversity management throughout the tropics.

5.2 The canopy

The tropical forest canopy – consisting of all the tree crowns in a forest stand – supports a diverse and poorly studied assemblage of insects, and has been described as the 'last biotic frontier' (Erwin, 1982a,b). At least 20% of tropical arthropods, most of them insect herbivores, are confined to the upper canopy (the canopy surface and the volume of vegetation within a few metres below it; Basset *et al.*, 2003b), where biotic and abiotic conditions contrast markedly with conditions in the understorey. Consequently, canopy insect assemblages are expected to show considerable differences in their composition, structure and function, compared with those in the understorey. The responses of canopy insects to anthropogenic habitat change are also likely to differ. Sound estimates of the effects of disturbance cannot be inferred from ground-based studies alone; data on the distribution and ecology of canopy arthropods are essential (e.g. Willott, 1999; Basset *et al.*, 2003b). Furthermore, most of the key ecosystem processes in which insects are involved (herbivory, parasitism, pollination) occur largely in the canopy.

A few conservation studies in tropical rainforests have specifically targeted canopy arthropods. The results of such studies have been mixed, a point that we illustrate with two recent examples, both from Malaysia, and both focusing on beetles sampled by insecticide knockdown ('fogging'). Speight *et al.* (2003) reported that loss of diversity in human-modified forests was small, compared with primary forests. They found that alteration in guild structure and loss of species was obvious only in plantations of exotic trees, and even these acted as partial refugia for the fauna, provided that the understorey was well developed (unlike in oil palm plantations). In contrast to this rather optimistic scenario, Floren and Linsenmair (2003) reported strong effects of anthropogenic disturbance. For example, 40 years after disturbance, the fauna of the disturbed forest they studied, including canopy inhabitants, still differed from that in the primary forest. They found a transition from deterministically structured communities to randomly assembled ones along a succession or disturbance gradient. In particular, assemblages of Coleoptera (and also Formicidae) showed patterns that were deterministic in disturbed forests, but random in primary forests, where non-equilibrium conditions may mediate species coexistence. Such conflicting results may result, in part, from the focus on beetles rather than a multi-taxa, multi-guild approach, and because of limited sampling of the fauna of the upper canopy, which may be rather specialized and therefore sensitive to disturbance (Basset, 2001). Future studies of the effects of disturbance on canopy arthropods should ideally address these two concerns.

What are the likely effects on arboreal arthropods of the opening of the canopy, after the creation of natural or anthropogenic gaps? Do the upper canopy and its fauna 'fall' to the ground? As far as insect herbivores are concerned, the short answer to this is most likely 'no', since forest gaps typically include sets of plant species (largely pioneers) different from those present in the mature canopy (largely shade-tolerant species), and many insect herbivores are relatively specialized. In addition, herbivores foraging on mature trees in Guyana tend not to attack conspecific seedlings in light gaps resulting after logging (Basset, 2001). Taxa less tied to resources occurring specifically in the upper canopy, such as dung beetles, may suffer less from canopy loss and survive well in the understorey of disturbed forests (Davis and Sutton, 1998). This and related issues warrant further investigation.

5.3 Climate change

Climate change remains a major unknown in the context of tropical insects, but the response of tropical forest insects to climate change is of some significance. Recent predictions that up to 15–37% of all biodiversity may be committed to extinction by climate change by 2050 (Thomas *et al.*, 2004) rely implicitly on tropical insects (which constitute the bulk of biodiversity) responding in a similar manner to better-studied temperate taxa (Lewis, 2006). It is debatable whether they will: most assessments suggest that tropical environments will be less affected by climate change than temperate biomes (Sala *et al.*, 2000), with habitat fragmentation and destruction rated as much greater threats. The steep environmental gradients from canopy to understorey in tropical forests may in part buffer populations against changes in climate. For example, specialized species of the upper canopy may move down to lower, cooler strata, although if their resources are less abundant in their new microhabitats, then extinctions are still likely (Basset *et al.*, 2003a). Certainly, we should not be complacent: the fact that existing examples of species responding to climate change are drawn entirely from temperate regions (e.g. Wilson *et al.*, Chapter 11, this volume) should not be surprising, given the limited monitoring data for tropical insects. Although predicting how tropical insects will respond to a warmer world is difficult, we may at least soon be in a position to detect the 'footprint' of climate change without the need for long time-series of survey data: recent work suggests that short-cuts may allow changes in species' status to be detected even from snapshot surveys (Wilson *et al.*, 2004).

6 Conclusions

The challenge to insect conservation biologists in the tropics is rather different from that facing many conservation biologists working on better-known taxa in better-studied parts of the world. In an influential paper, Caughley (1994) identified two paradigms in conservation biology: the small population paradigm (where conservationists seek to identify the measures needed to prevent small populations from going extinct) and the declining population paradigm (where conservationists seek to identify declining species and the causes of their decline). Conservation biology for the vast majority of tropical insects falls into neither category comfortably. We are not in a position to carry out – or act on – detailed population studies for the vast majority of rare tropical insects; and although we know that many species are likely to be declining, we rarely have information on rates of population or distribution decline. But the sheer magnitude of tropical insect diversity should not be allowed to stifle progress.

We have identified three main interlinked issues that we believe are fundamental to integrating insects fully into the conservation of tropical forests: undertaking reliable and comparable inventories, assessing the effects of disturbance and quantifying the wider role of insects within tropical forest ecosystems. We have also identified a series of challenges, which may impede progress towards these goals. These include the very diversity that we value, and the problems of identification, sampling and replication that it brings. Our suggested solutions are pragmatic ones: to design our studies more robustly to answer criticisms about replication; to improve reporting of results to allow more informative integration across studies and to speed the flow of biodiversity information from field to decision-maker through the work of parataxonomists.

As entomologists, we naturally rate the conservation of insects as an important goal; but we appreciate that, in practice, tropical insects will rarely, if ever, be the targets of conservation action in their own right. However, the danger is that they will be overlooked in setting conservation priorities and guiding habitat management practice. We feel that tackling the issues surrounding inventory, impacts and function should go a long way towards ensuring that the use of insects in conservation assessments in the tropics moves a step further towards reflecting their numerical and ecological importance in tropical forest ecosystems.

References

Andresen, E. (2002) Dung beetles in a Central Amazonian rainforest and their ecological role as secondary seed dispersers. *Ecological Entomology* 27, 257–270.

Andresen, E. (2003) Effect of forest fragmentation on dung beetle communities and functional consequences for plant regeneration. *Ecography* 26, 87–97.

Anon. (1998) *The Darwin Declaration*. Available at: http://www.deh.gov.au/biodiversity/abrs/publications/other/darwin/

Balmford, A. (1996) Extinction filters and current resilience: the significance of past selection pressures for conservation biology. *Trends in Ecology and Evolution* 11, 193–196.

Barone, J.A. (1998) Host-specificity of folivorous insects in a moist tropical forest. *Journal of Animal Ecology* 67, 400–409.

Basset, Y. (1992) Host specificity of arboreal and free-living insect herbivores in rainforests. *Biological Journal of the Linnean Society* 47, 115–133.

Basset, Y. (1999) Diversity and abundance of insect herbivores foraging on seedlings in a rainforest in Guyana. *Ecological Entomology* 24, 245–259.

Basset, Y. (2001) Communities of insect herbivores foraging on saplings versus mature trees of Pourouma bicolor (Cecropiaceae) in Panama. *Oecologia* 129, 253–260.

Basset, Y., Samuelson, G.A., Allison, A. and Miller, S.E. (1996) How many species of host-specific insect feed on a species of tropical tree? *Biological Journal of the Linnean Society* 59, 201–216.

Basset, Y., Novotny, V., Miller, S.E. and Springate, N.D. (1998) Assessing the impact of forest disturbance on tropical invertebrates: some comments. *Journal of Applied Ecology* 35, 461–466.

Basset, Y., Novotny, V., Miller, S.E. and Pyle, R. (2000) Quantifying biodiversity: experience with parataxonomists and digital photography in Papua New Guinea and Guyana. *Bioscience* 50, 899–908.

Basset, Y., Charles, E., Hammond, D.S. and Brown, V.K. (2001a) Short-term effects of canopy openness on insect herbivores in a rain forest in Guyana. *Journal of Applied Ecology* 38, 1045–1058.

Basset, Y., Charles, E., Hammond, D.S. and Brown, V.K. (2001b) Short-term effects of canopy openness on insect herbivores in a rain forest in Guyana. *Journal of Applied Ecology* 38, 1045–1058.

Basset, Y., Horlyck, V. and Wright, S.J. (eds) (2003a) *Studying Forest Canopies from Above: The International Canopy Crane Network*. Smithsonian Tropical Research Institute and UNEP, Panama City.

Basset, Y., Novotny, V., Miller, S. and Kitching, R. (2003b) *Arthropods of Tropical Forests – Spatio-temporal Dynamics and Resource Use in the Canopy*. Cambridge University Press, Cambridge.

Basset, Y., Novotny, V., Miller, S.E., Weiblen, G.D., Missa, O. and Stewart, A.J.A. (2004) Conservation and biological monitoring of tropical forests: the role of parataxonomists. *Journal of Applied Ecology* 41, 163–174.

Belshaw, R. and Bolton, B. (1993) The effect of forest disturbance on the leaf litter ant fauna in Ghana. *Biodiversity and Conservation* 2, 656–666.

Bibby, C.J., Collar, N.J., Crosby, M.J., Heath, M.F., Imboden, C., Johnson, T.H., Long, A.J.,

Stattersfield, A.J. and Thirgood, S.J. (1992) *Putting Biodiversity on the Map: Priority Areas for Global Conservation*. International Council for Bird Preservation, Cambridge.

Bird, N.M. (1998) *Sustaining the Yield. Improving Timber Harvesting Practices in Belize 1991–1998*. Natural Resources Institute, Chatham, UK.

Brown, K.S.J. (1991) Conservation of neotropical environments: insects as indicators. In: Collins, N.M. and Thomas, J.A. (eds) *The Conservation of Insects and Their Habitats*. Academic Press, London.

Brown, K.S.J. and Hutchings, R. (1997) Disturbance, fragmentation, and the dynamics of diversity in Amazonian forest butterflies. In: Laurence, W.F. and Bierregaard, R.O. Jr (eds) *Tropical Forest Remnants: Ecology, Management, and Conservation of Fragmented Communities*. University of Chicago Press, Chicago, Illinois, pp. 91–110.

Caughley, G. (1994) Directions in conservation biology. *Journal of Animal Ecology* 63, 215–244.

Chung, A.Y.C., Eggleton, P., Speight, M.R., Hammond, P.M. and Chey, V.K. (2000) The diversity of beetle assemblages in different habitat types in Sabah, Malaysia. *Bulletin of Entomological Research* 90, 475–496.

Cleary, D.F.R. (2003) An examination of scale of assessment, logging and ENSO-induced fires on butterfly diversity in Borneo. *Oecologia* 135, 313–321.

Cleary, D.F.R. and Genner, M.J. (2004) Changes in rain forest butterfly diversity following major ENSO-induced fires in Borneo. *Global Ecology and Biogeography* 13, 129–140.

Collins, N. and Thomas, J.A. (eds) (1991) *The Conservation of Insects and Their Habitats*. Academic Press, London.

Connell, J.H. (1971) On the role of natural enemies in preventing competitive exclusion in some marine animals and in rain forest trees. In: Boer, P.J.D. and Gradwell, G.R. (eds) *Dynamics of Populations*. PUDOC, Wageningen, The Netherlands, pp. 298–312.

Cottenie, K. and De Meester, L. (2003) Comment to Oksanen (2001): Reconciling Oksanen (2001) and Hurlbert (1984). *Oikos* 100, 394–396.

Davis, A.J. (2000) Does reduced-impact logging help preserve biodiversity in tropical rainforests? A case study from Borneo using dung beetles (Coleoptera: Scarabaeoidea) as indicators. *Environmental Entomology* 29, 467–475.

Davis, A.J. and Sutton, S.L. (1998) The effects of rainforest canopy loss on arboreal dung beetles in Borneo: implications for the measurement of biodiversity in derived tropical ecosystems. *Diversity and Distributions* 4, 167–173.

Davis, A.J., Holloway, J.D., Huijbregts, H., Krikken, J., Kirk-Spriggs, A.H. and Sutton, S.L. (2001) Dung beetles as indicators of change in the forests of northern Borneo. *Journal of Applied Ecology* 38, 593–616.

DeVries, P.J., Murray, D. and Lande, R. (1997) Species diversity in vertical, horizontal, and temporal dimensions of a fruit-feeding butterfly community in an Ecuadorian rainforest. *Biological Journal of the Linnean Society* 62, 343–364.

Didham, R.K. (1997a) The influence of edge effects and forest fragmentation on leaf litter invertebrates in central Amazonia. In: Laurece, W.F. and Bierregaard, R.O. Jr (eds) *Tropical Forest Remnants: Ecology, Management, and Conservation of Fragmented Communities*. University of Chicago Press, Chicago, Illinois, pp. 55–70.

Didham, R.K. (1997b) An overview of invertebrate responses to forest fragmentation. In: Watt, A.D., Stork, N.E. and Hunter, M.D. (eds) *Forest and Insects*. Chapman & Hall, London, pp. 303–320.

Didham, R.K. and Fagan, L.L. (2003) Editorial: IBISCA – Investigating the biodiversity of soil and canopy arthropods. *The Weta* 26, 1–6.

Didham, R.K., Ghazoul, J., Stork, N.E. and Davis, A.J. (1996) Insects in fragmented forests: a functional approach. *Trends in Ecology and Evolution* 11, 255–260.

Didham, R.K., Hammond, P.M., Lawton, J.H., Eggleton, P. and Stork, N.E. (1998a) Beetle species richness responses to tropical forest fragmentation. *Ecological Monographs* 68, 295–323.

Didham, R.K., Lawton, J.H., Hammond, P.M. and Eggleton, P. (1998b) Trophic structure stability and extinction dynamics of beetles (Coleoptera) in tropical forest fragments. *Philosophical Transactions of the Royal Society of London Series B – Biological Sciences* 353, 437–451.

Dyer, L.A. and Letourneau, D.K. (1999) Trophic cascades in a complex, terrestrial community. *Proceedings of the National Academy of Sciences of the United States of America* 96, 5072–5076.

Eggleton, P., Bignell, D.E., Sands, W.A., Mawdsley, N.A., Lawton, J.H., Wood, T.G. and Bignell, N.C. (1996) The diversity, abundance and biomass of termites under differing levels of disturbance in the Mbalmayo Forest Reserve, southern Cameroon. *Philosophical Transactions of the Royal Society of London Series B – Biological Sciences* 351, 51–68.

Erwin, T.L. (1982a) Tropical forests: their richness in Coleoptera and other arthropod species. *The Coleopterists Bulletin* 36, 74–75.

Erwin, T.L. (1982b) Tropical forests: their richness in Coleoptera and other arthropods. *Coleopterists' Bulletin* 36, 74–75.

Erwin, T.L. (1991) An evolutionary basis for conservation strategies. *Science* 253, 750–752.

Estrada, A., Anzuras, D.A. and Coastes-Estrada, R. (1999) Tropical forest fragmentation, howler monkeys (*Alouatta palliata*) and dung beetles at Los Tuxtlas, Mexico. *American Journal of Primatology* 48, 353–62.

Fagan, L.L., Didham, R.K., Winchester, N.N., Behan-Pelletier, V., Clayton, M., Lindquist, E. and Ring, R.A. (2005) An experimental assessment of biodiversity and species turnover in terrestrial vs canopy leaf litter. *Oecologia* October 15, 1–13 (Epub ahead of print).

Floren, A. and Linsenmair, K.E. (2003) How do beetle assemblages respond to anthropogenic disturbances? In: Basset, Y., Kitching, R., Miller, S. and Novotny, V. (eds) *Arthropods of Tropical Forests: Spatio-temporal Dynamics and Resource Use in the Canopy*. Cambridge University Press, Cambridge, pp. 190–197.

Folgarait, P.J., Marquis, R.J., Ingvarsson, P., Braker, H.E. and Arguedas, M. (1995) Patterns of attack by insect herbivores and a fungus on saplings in a tropical tree plantation. *Environmental Entomology* 24, 1487–1494.

Frost, C.J. and Hunter, M.D. (2004) Insect canopy herbivory and frass deposition affect soil nutrient dynamics and export in oak mesocosms. *Ecology* 85, 3335–3347.

Gaston, K.J. (1999) Implications of interspecific and intraspecific abundance-occupancy relationships. *Oikos* 86, 195–207.

Gaston, K.J., Blackburn, T.M. and Lawton, J.H. (1997) Interspecific abundance range size relationships: an appraisal of mechanisms. *Journal of Animal Ecology* 66, 579–601.

Ghazoul, J. (2002) Impact of logging on the richness and diversity of forest butterflies in a tropical dry forest in Thailand. *Biodiversity and Conservation* 11, 521–541.

Gilbert, L. (1980) Food web organization and the conservation of neotropical diversity. In: Soulé, M.E. and Wilcox, B.A. (eds) *Conservation Biology*. Sinauer, Sunderland, Massachusetts, pp. 11–33.

Gilbert, L.E. and Smiley, J.T. (1978) Determinants of local diversity in phytophagous insects. In: Mound, L.A. and Waloff, N. (eds) *Diversity of Insect Faunas*. Blackwell Scientific, London.

Godfray, H.C.J., Lewis, O.T. and Memmott, J. (1999) Studying insect diversity in the tropics. *Philosophical Transactions of the Royal Society of London Series B – Biological Sciences* 354, 1811–1824.

Gotelli, N.J. and Colwell, R.K. (2001) Quantifying biodiversity: procedures and pitfalls in the measurement and comparison of species richness. *Ecology Letters* 4, 379–391.

Greiser Johns, A. (1997) *Timber Production and Biodiversity Conservation in Tropical Forests*. Cambridge University Press, Cambridge.

Hamer, K.C. and Hill, J.K. (2000) Scale-dependent effects of habitat disturbance on species richness in tropical forests. *Conservation Biology* 14, 1435–1440.

Hamer, K.C., Hill, J.K., Lace, L.A. and Langan, A.M. (1997) Ecological and biogeographical effects of forest disturbance on tropical butterflies of Sumba, Indonesia. *Journal of Biogeography* 24, 67–75.

Hammond, P.M. (1994) Practical approaches to the estimation of the extent of biodiversity in speciose groups. *Philosophical Transactions of the Royal Society of London Series B – Biological Sciences* 345, 119–136.

Harrison, S. (1991) Local extinction in a metapopulation context: an empirical evaluation. *Biological Journal of the Linnean Society* 42, 73–88.

Hector, A., Schmid, B., Beierkuhnlein, C., Caldeira, M.C., Diemer, M., Dimitrakopoulos, P.G., Finn, J.A., Freitas, H., Giller, P.S., Good, J., Harris, R., Hogberg, P., Huss-Danell, K., Joshi, J., Jumpponen, A., Korner, C., Leadley, P.W., Loreau, M., Minns, A., Mulder, C.P.H., O'Donovan, G., Otway, S.J., Pereira, J.S., Prinz, A., Read, D.J., Scherer-Lorenzen, M., Schulze, E.D., Siamantziouras, A.S.D., Spehn, E.M., Terry, A.C., Troumbis, A.Y., Woodward, F.I., Yachi, S. and Lawton, J.H. (1999) Plant diversity and productivity experiments in European grasslands. *Science* 286, 1123–1127.

Hill, J.K. (1999) Butterfly spatial distribution and habitat requirements in a tropical forest: impacts of selective logging. *Journal of Applied Ecology* 36, 564–572.

Hill, J.K. and Hamer, K.C. (2004) Determining impacts of habitat modification on diversity of tropical forest fauna: the importance of spatial scale. *Journal of Applied Ecology* 41, 744–754.

Hooper, D.U., Chapin, F.S.I., Ewel, J.J., Hector, A., Inchausti, P., Lavorel, S., Lawton, J.H., Lodge, D.M., Loreau, M., Naeem, S., Schmid, B., Setaelae, H., Symstad, A.J., Vandermeer, J. and Wardle, D.A. (2005) Effects of biodiversity on ecosystem functioning: a consensus of current knowledge. *Ecological Monographs* 75, 3–35.

Hurlbert, S.H. (1984) Pseudoreplication and the design of ecological field experiments. *Ecological Monographs* 54, 187–211.

Intachat, J., Holloway, J.D. and Speight, M.R. (1999) The impact of logging on geometroid moth populations and their diversity in lowland forests of Peninsular Malaysia. *Journal of Tropical Forest Science* 11, 61–78.

Janzen, D.H. (1970) Herbivores and the number of tree species in tropical forests. *American Naturalist* 104, 501–28.

Janzen, D.H. (1973) Comments on host-specificity of tropical herbivores and its relevance to species richness. In: Heywood, V.H. (ed.) *Taxonomy and Ecology*. Academic Press, London, pp. 201–211.

Janzen, D.H. (1983) Food webs: who eats what, why, how, and with what effects in a tropical forest? In: Golley, F. (ed.) *Tropical Rain Forest Ecosystems*. Elsevier Scientific, Amsterdam, The Netherlands, pp. 167–182.

Janzen, D.H., Hallwachs, W., Jimenez, J. and Gamez, R. (1993) The role of the parataxonomists, inventory managers, and taxonomists in Costa Rica's national biodiversity inventory. In: Reid, W.V., Laird, S.A., Meyer, C.A., Gamez, R., Sittenfeld, A., Janzen, D.H., Gollin, M.A. and Juma, C. (eds) *Biodiversity Prospecting: Using Generic Resources for Sustainable Development*. World Resources Institute, Washington, DC, pp. 223–254.

Kitching, I.J. (1996) Identifying complementary areas for conservation in Thailand: an example using owls, hawkmoths and tiger beetles. *Biodiversity and Conservation* 5, 841–858.

Kitching, R.L. (1993) Biodiversiy and taxonomy: impediment or opportunity? In: Moritz, C. and Kikkawa, J. (eds) *Conservation Biology in Australia and Oceania*. Surrey Beatty, Chipping Norton, New South Wales, Australia, pp. 253–268.

Kitching, R.L., Vickerman, G., Laidlaw, M. and Hurley, K. (2000) *The Comparative Assessment of Arthropod and Tree Diversity on Old-World Rainforests: The Rainforest CRC/Earthwatch Manual*. Technical Report, Cooperative Research Centre for Tropical Rainforest Ecology and Management, Cairns, Queensland, Australia.

Kitching, R.L., Li, D.Q. and Stork, N.E. (2001) Assessing biodiversity 'sampling packages': how similar are arthropod assemblages in different tropical rainforests? *Biodiversity and Conservation* 10, 793–813.

Klein, B.C. (1989) Effects of forest fragmentation on dung and carrion beetle communities in central Amazonia. *Ecology* 70, 1715–1725.

Kotze, D.J. and Samways, M.J. (1999) Invertebrate conservation at the interface between the grassland matrix and natural Afromontane forest fragments. *Biodiversity and Conservation* 8, 1339–1363.

Kremen, C. (1992) Assessing the indicator properties of species assemblages for natural areas monitoring. *Ecological Applications* 2, 203–217.
Kremen, C., Colwell, R.K., Erwin, T.L., Murphy, D.D., Noss, R.F. and Sanjayan, M.A. (1993) Terrestrial arthropod assemblages: their use in conservation planning. *Conservation Biology* 7, 796–808.
Landres, P.B., Verner, J. and Thomas, J. (1998) Ecological uses of vertebrate indicator species: a critique. *Conservation Biology* 2, 316–328.
Lawton, J.H. and May, R.M. (eds) (1995) *Extinction Rates*. Oxford University Press, Oxford.
Lawton, J.H., Bignell, D.E., Bolton, B., Bloemers, G.F., Eggleton, P., Hammond, P.M., Hodda, M., Holt, R.D., Larsen, T.B., Mawdsley, N.A. and Stork, N.E. (1998) Biodiversity inventories, indicator taxa and effects of habitat modification in tropical forest. *Nature* 391, 72–76.
Letourneau, D.K. and Dyer, L.A. (1998) Experimental test in lowland tropical forest shows top-down effects through four trophic levels. *Ecology* 79, 1678–1687.
Levins, R. (1969) Some demographic and genetic consequences of environmental heterogeneity for biological control. *Bulletin of the Entomological Society of America* 15, 237–240.
Lewis, O.T. (2001) Effect of experimental selective logging on tropical butterflies. *Conservation Biology* 15, 389–400.
Lewis, O.T. (2006) Climate change, species–area curves and the extinction crisis. *Philosophical Transactions of the Royal Society, Series B* 273, 2909–2916.
Lewis, O.T., Wilson, R.J. and Harper, M.C. (1998) Endemic butterflies on Grande Comore: habitat preferences and conservation priorities. *Biological Conservation* 85, 113–121.
Longino, J.T. and Colwell, R.C. (1997) Biodiversity assessment using structured inventory: capturing the ant fauna of a tropical rain forest. *Ecological Applications* 7, 1263–1277.
Loreau, M., Naeem, S. and Inchausti, P. (eds) (2002) *Biodiversity and Ecosystem Functioning: Synthesis and Perspectives*. Oxford University Press, Oxford.
Marquis, R.J. (2005) Impacts of herbivores on tropical plant diversity. In: Burslem, D.F.R.P., Pinard, M.A. and Hartley, S.E. (eds) *Biotic Interactions in the Tropics: Their Role in the Maintenance of Species Diversity*. Cambridge University Press, Cambridge, pp. 328–346.
Marquis, R.J. and Braker, H.E. (1993) Plant-herbivore interactions: diversity, specificity, and impact. In: McDade, L.A., Bawa, K.S., Hespenheide, H.A. and Hartshorn, G.S. (eds) *La Selva: Ecology and Natural History of a Neotropical Rain Forest*. University of Chicago Press, Chicago, pp. 261–281.
Mayaux, P., Holmgren, P., Achard, F., Eva, H., Stibig, H. and Branthomme, A. (2005) Tropical forest cover change in the 1990s and options for future monitoring. *Philosophical Transactions of the Royal Society of London Series B – Biological Sciences* 360, 373–384.
McGeoch, M.A. (1998) The selection, testing and application of terrestrial insects as bioindicators. *Biological Reviews of the Cambridge Philosophical Society* 73, 181–201.
Moritz, C., Richardson, K.S., Ferrier, S., Monteith, G.B., Stanisic, J., Williams, S.E. and Whiffin, T. (2001) Biogeographical concordance and efficiency of taxon indicators for establishing conservation priority in a tropical rainforest biota. *Proceedings of the Royal Society of London Series B – Biological Sciences* 268, 1875–1881.
Morris, R.J., Lewis, O.T. and Godfray, H.C.J. (2004) Experimental evidence for apparent competition in a tropical forest food web. *Nature* 428, 310–313.
Morris, R.J., Lewis, O.T. and Godfray, H.C.J. (2005) Apparent competition and insect community structure: towards a spatial perspective. *Annales Zoologici Fennici* 42, 449–462.
Netuzhilin, I., Chacon, P., Cerda, H., López-Hernández, D., Torres, F., Paoletti, M.G. and Reddy, M.V. (1999) Assessing agricultural impact using ant morphospecies as bioindicators in the Amazonian savanna-forest ecotone (Puerto Ayacucho, Amazon

State), Venezuela. In: Reddy, M.V. (ed.) *Management of Tropical Agroecosystems and the Beneficial Soil Biota*. Science Publishers, Enfield, New Hampshire, pp. 291–352.

New, T.R., Pyle, R.M., Thomas, J.A., Thomas, C.D. and Hammond, P.C. (1995) Butterfly conservation management. *Annual Review of Entomology* 40, 57–83.

Novotny, V. and Basset, Y. (2000) Rare species in communities of tropical insect herbivores: pondering the mystery of singletons. *Oikos* 89, 564–572.

Novotny, V., Basset, Y., Miller, S.E., Weiblen, G.D., Bremer, B., Cizek, L. and Drozd, P. (2002) Low host specificity of herbivorous insects in a tropical forest. *Nature* 416, 841–844.

Novotny, V., Miller, S.E., Leps, J., Basset, Y., Bito, D., Janda, M., Hulcr, J., Damas, K. and Weiblen, G.D. (2004) No tree an island: the plant-caterpillar food web of a secondary rain forest in New Guinea. *Ecology Letters* 7, 1090–1100.

Pimm, S. (2001) *The World According to Pimm: A Scientist Audits the Earth*. McGraw-Hill, New York.

Prendergast, J.R., Quinn, R.M., Lawton, J.H., Eversham, B.C. and Gibbons, D.W. (1993) Rare species, the coincidence of diversity hotspots and conservation strategies. *Nature* 365, 335–337.

Quintero, I. and Roslin, T. (2005) Rapid recovery of dung beetle communities following habitat fragmentation in Central Amazonia. *Ecology* 86, 3303–3311.

Rabinowitz, D. (1981) Seven forms of rarity. In: Synge, H. (ed.) *The Biological Aspects of Rare Plant Conservation*. Wiley, New York.

Sala, O.E., Chapin, F.S., Armesto, J.J., Berlow, E., Bloomfield, J., Dirzo, R., Huber-Sanwald, E., Huenneke, L.F., Jackson, R.B., Kinzig, A., Leemans, R., Lodge, D.M., Mooney, H.A., Oesterheld, M., Poff, N.L., Sykes, M.T., Walker, B.H., Walker, M. and Wall, D.H. (2000) Global biodiversity scenarios for the year 2100. *Science* 287, 1770–1774.

Samways, M.J. (1994) *Insect Conservation Biology*. Chapman & Hall, London.

Speight, M.R., Intachat, J., Chey, V.K. and Chung, A.Y.C. (2003) Influences of forest management on insects. In: Basset, Y., Kitching, R., Miller, S. and Novotny, V. (eds) *Arthropods of Tropical Forests: Spatio-temporal Dynamics and Resource use in the Canopy*. Cambridge University Press, Cambridge.

Spitzer, K., Novotny, V., Tonner, M. and Leps, J. (1993) Habitat preferences, distribution and seasonality of the butterflies (Lepidoptera, Papilionoidea) in a montane tropical rain forest, Vietnam. *Journal of Biogeography* 20, 109–121.

Spitzer, K., Jaros, J., Havelka, J. and Leps, J. (1997) Effect of small-scale disturbance on butterfly communities of an Indochinese montane rainforest. *Biological Conservation* 80, 9–15.

Stewart-Oaten, A. and Murdoch, W.W. (1986) Environmental impact assessment: pseudoreplication in time? *Ecology* 67, 929–940.

Stork, N.E. (1994) Inventories of biodiversity: more than a question of numbers. In: Forey, P.I., Humphries, C.J. and Vane-Wright, R.I. (eds) *Systematics and Conservation Evaluation*. Clarendon Press, Oxford, pp. 81–100.

Thomas, C.D. (1991) Habitat use and geographic ranges of butterflies from the wet lowlands of Costa Rica. *Biological Conservation* 55, 269–281.

Thomas, C.D., Cameron, A., Green, R.E., Bakkenes, M., Beaumont, L.J., Collingham, Y.C., Erasmus, B.F.N., de Siqueira, M.F., Grainger, A., Hannah, L., Hughes, L., Huntley, B., van Jaarsveld, A.S., Midgley, G.F., Miles, L., Ortega-Huerta, M.A., Peterson, A.T., Phillips, O.L. and Williams, S.E. (2004) Extinction risk from climate change. *Nature* 427, 145–148.

Tylianakis, J.M., Tscharntke, T. and Lewis, O.T. (2007) Habitat modification alters the structure of tropical host-parasitoid food webs. *Nature* 445, 202–205.

Vane-Wright, R.I., Humphries, C.J. and Williams, P.H. (1991) What to protect? Systematics and the agony of choice. *Biological Conservation* 55, 235–254.

Vasconcelos, H.L., Vilhena, J.M.S. and Caliri, G.J.A. (2000) Responses of ants to selective logging of a central Amazonian forest. *Journal of Applied Ecology* 37, 508–514.

Williams, P.H., Humphries, C.J. and Vane-Wright, R.I. (1991) Measuring biodiversity:

taxonomic relatedness for conservation priorities. *Australian Journal of Systematics and Botany*, 665–679.

Willott, S.J. (1999) The effects of selective logging on the distribution of moths in a Bornean rainforest. *Philosophical Transactions of the Royal Society of London Series B – Biological Sciences* 354, 1783–1790.

Wilson, E.O. (1987) The little things that run the world: the importance of conservation of invertebrates. *Conservation Biology* 1, 344–346.

Wilson, E.O. (2000) On the future of conservation biology. *Conservation Biology* 14, 1–4.

Wilson, R.J., Thomas, C.D., Fox, R., Roy, D.B. and Kunin, W.E. (2004) Spatial patterns in species distributions reveal biodiversity change. *Nature* 432, 393–396.

Wolda, H. (1996) Between-site similarity in species composition of a number of Panamanian insect groups. *Miscellània Zoològica* 19, 39–50.

Wright, S.J. (2005) Tropical forests in a changing environment. *Trends in Ecology and Evolution* 20, 553–560.

3 The Conservation Value of Insect Breeding Programmes: Rationale, Evaluation Tools and Example Programme Case Studies

PAUL PEARCE-KELLY,[1] RANDY MORGAN,[2] PATRICK HONAN,[3] PAUL BARRETT,[4] LOU PERROTTI,[5] MITCHELL MAGDICH,[6] BEXELL AYYACHAMY DANIEL,[7] ERIN SULLIVAN,[8] KO VELTMAN,[9] DAVE CLARKE,[1] TREVOR MOXEY[1] AND WARREN SPENCER[10]

[1]Zoological Society of London, Regent's Park, London NW1 4RY, UK; [2]Cincinnati Zoo and Botanical Garden, 3400 Vine St, Cincinnati, OH, 45220, USA; [3]Zoos Victoria, PO Box 74, Parkville, Victoria 3052, Australia; [4]Butterfly Creek, Tom Pearce Drive, PO Box 201 097, Auckland, New Zealand; [5]Roger Williams Park Zoo, Roger Williams Park, Elmwood Ave, Providence, RI 02905, USA; [6]The Toledo Zoo, PO Box 140130, Toledo, OH 43614, USA; [7]Zoo Outreach Organisation, PO Box 1683, Peelamedu, Coimbatore, Tamil Nadu 641004, India; [8]Woodland Park Zoological Park Gardens, 5500 Phinney Ave, N, Seattle, WA 98103, USA; [9]Natura Artis Magistra, Plantage Kerklaan, 38–40, 1018 CZ Amsterdam C, The Netherlands; [10]Clifton and West of England Zoological Society, Clifton, Bristol BS8 3HA, UK

Keywords: insect conservation, species recovery programmes, conservation breeding, reintroduction, *Gryllus campestris, Decticus verrucivorus, Polposipus herculeanus, Motuweta isolata, Dryococelus australis, Nicrophorus americanus, Pareulype berberata, Lycaeides melissa samuelis, Motuweta isolata*

1 The Rationale for Species Conservation Breeding Programmes

For the majority of endangered species, across all taxa, landscape-scale habitat preservation represents the only realistic conservation measure. However, there are numerous instances where an *ex situ* breeding programme is essential for ensuring the continued survival of a species (IUCN, 1990; Rabb, 1994; WAZA, 2005). This is especially true when the immediate *in situ* threat includes such stress factors as invasive predators and competitors, disease, overharvesting and severe habitat alteration. Such threats are even more prevalent in the case of discrete genetic populations (Cheesman, 1999). The conservation potential of well-managed breeding programmes, as properly integrated components of wider species recovery programme effort, has been comprehensively detailed (Wilson and Stanley-Price, 1994; Mallinson, 1995; Pullin, 2004; Olney, 2005). In addition to providing secure populations for eventual field release, breeding programmes can inform the *in situ* management of a species by clarifying reproductive biology, life-history, behaviour, genetic and health data (WAZA, 2005). Breeding programmes can also raise public awareness and support for *in situ* species conservation. On a more fundamental level, as the general trend of habitat loss induced fragmentation of wild populations continues to increase, the metapopulation management strategies and methodologies being developed for *ex situ* populations are increasingly needed for effective *in situ* population management.

2 Species Threat Assessment and Breeding Programme Selection Tools

In addition to the IUCN Red List and its associated species threat evaluation criteria (Baillie *et al.*, 2004; Warren *et al.*, Chapter 4, this volume), a range of regional and national species threat assessment data are available to help prioritize species conservation focus. Examples include, the *Seychelles Red Data Book* (Gerlach, 1997), *British Insect Red Data Book* (Shirt, 1987), *Background Information on Invertebrates of the Habitats Directive and the Bern Convention* (van Helsdingen *et al.*, 1996) and the *Conservation Assessment Management Plan for Selected Soil Invertebrates of Southern India* (Daniel *et al.*, 1998). These assessment data are evaluated by the species specialist groups of IUCN's Species Survival Commission, including the Conservation Breeding Specialist Group. Conservation Assessment Management Plans provide a mechanism by which taxon-specific specialists can identify and prioritize species on a global level (Byers and Seal, 2003). Population and Habitat Viability Assessments evaluate factors affecting threatened species to develop *in situ* and, where appropriate, *ex situ* management strategies. Global Captive Action Plans (Seal *et al.*, 1994) formulate *ex situ* programme strategies for consideration by the Taxon Advisory Groups (TAGs) of regional zoo associations through their Regional Collection Planning (RCP) review process for adoption by zoological institutions as part of coordinated breeding programmes. The RCP formula and its associated Species Action Plan format have emerged as the principle mechanism by which TAGs and individual institutions review their species-level involvement and conservation focus. Although there are regional variances (Sullivan *et al.*, 2005), the essential elements of the RCP format are very similar and provide a relatively standardized evaluation tool applicable to all animal groups. The following RCP species assessment definitions are taken from the European Association of Zoos and Aquaria (EAZA) lower vertebrate and invertebrate TAG manual for collection planning (Visser *et al.*, 2005).

Category 1. Conservation Breeding Programmes

1. *Ark* – Species that are globally extinct in the wild and which would become completely extinct without *ex situ* management.
2. *Rescue* – Species that are in *imminent* danger of extinction (locally or globally) and are managed in captivity as part of the *recommended* conservation action.
3. *Supplementation* – Species for which *ex situ* breeding for release may benefit the wild population as part of the *recommended* conservation action.

Category 1 includes the potential for field release where appropriate and fully evaluated, and therefore must be managed accordingly (European Endangered Species Programme – EEP or equivalent) and have clearly defined field links or at least a plan to develop a field component.

Category 2. Research

1. *Conservation research* – A species undergoing specific applied research that directly contributes to the conservation of that species or a related species and/or their habitats in the wild.
2. *General research* – A species recommended for clearly defined pure or applied research that increases knowledge of natural history, population biology, taxonomy, husbandry, or disease and health management.

Category 3. Education

1. *Conservation education* – A species (or group of species) recommended for a clearly defined educational purpose of inspiring visitors, raising awareness or increasing knowledge of conservation issues or projects associated with that species or its habitat. Conservation education species can be used to promote positive behavioural changes in the general public and/or generate financial or other support for field conservation projects.
2. General education – A species (or group of species) recommended for clearly defined educational purposes based on novel or otherwise remarkable characteristics, such as appearance, natural history and behaviour.

A species may fit within one, two or all three of categories 1–3 provided its conservation needs are appropriately evaluated, and can be demonstrated to meet the necessary criteria.

Island faunas, which include relatively large numbers of endemic species, are particularly susceptible to the effects of introduced alien predators and competitors, and to anthropogenic induced habitat stress. It is therefore not surprising that island faunas register high on the list of priority target species (Howarth and Ramsay, 1991).

3 Rationale for Insect Conservation Breeding Programmes

Insects exemplify the assertion that landscape-scale habitat preservation is the only realistic option for the overwhelming majority of species, due to their sheer numbers (New, 1995; Hutchings and Ponder, 1999; Samways, 1999, 2005). This is especially so in the case of most tropical species (Sutton and Collins, 1989; Pullin, 1999; Lewis and Basset, Chapter 2, this volume). However, a growing number of insect breeding programmes support the contention that, with appropriate management regimes, many insect species can be technically feasible and cost-effective conservation breeding programme recipients (IUCN, 1991; Morton, 1991a,b; Pearce-Kelly, 1994; Balmford *et al.*, 1996). Although insects are among the first animals to suffer as a result of habitat disturbance and other pressures (Brown, 1991; Erhardt and Thomas, 1991; Samways, 2005), their often remarkable recovery powers mean that many insect species have a relatively high chance of being successfully established once the *in situ* stress has been effectively addressed (Morton, 1991b; Samways, 2005). The habitat requirements and associated management considerations for many insect species are relatively well understood, especially in temperate regions (Fry and Lonsdale, 1991; Stewart and New, Chapter 1, this volume) further enhancing the chance of realizing successful reintroduction outcomes.

The experience derived from culturing many terrestrial and aquatic insect species, largely through the development of zoo and aquarium invertebrate exhibits (Collins, 1986; Andrews, 1990; Hughes and Bennett, 1991; Pearce-Kelly *et al.*, 1991; Robinson, 1991), has provided both the skills-base and facility resources necessary for developing insect conservation breeding programmes. The typically modest accommodation requirements of insects, combined with often high reproductive rates and short generation times, enable large numbers of insects to be maintained in culture for relatively modest cost. Although insects and other invertebrates are not immune to inbreeding depression risk (see Thompson *et al.*, Chapter 12, this volume), the practical management considerations described above help minimize inbreeding risk through the ability to follow best genetic management practices (as described in Samways, 1994; New, 1995; Frankham *et al.*, 2004). There is also a considerable knowledge-base on insect disease (Rivers, 1990) and health management (Cooper and Cunningham, 1991; Rivers, 1991; Cunningham, 1996; Cunningham, 1997; Pizzi, 2004) available to help insect programmes comply with field introduction protocols and codes of practice (IUCN, 1987; Lees, 1989; English Nature, 1995; JCCBI, 1996).

In addition to providing secure genetic reservoirs and large numbers of animals for reintroductions, *ex situ* insect programmes can help clarify an array of life-history, reproductive and health-related information of great relevance to the conservation management of the *in situ* population. A further consideration in favour of insect breeding programmes is the speed with which they can be developed through to the field release stage (Pearce-Kelly *et al.*, 1998). Experience has also shown that insect conservation breeding programmes have as much potential for attracting media and public support as most vertebrate programmes. This awareness raising potential can have significant *in situ* conservation benefits (Yen, 1993).

4 The Wider Value of Invertebrate Conservation Programmes

Surviving wild populations of numerous species, vertebrate and invertebrate alike are becoming increasingly fragmented and confined to ever-smaller patches of suitable, secure habitat. Meta-population management strategies developed for *ex situ* population management are increasingly relevant to the conservation management of *in situ* populations. This is particularly so in the case of group-level demographic and genetic management tools that are being developed using invertebrate breeding programme model case studies (Amin *et al.*, 2005). Because invertebrate conservation programmes can progress through to the field release phase relatively quickly, their outcomes, successful or otherwise, can help inform the development of longer term conservation programmes typical of many vertebrate species.

5 Insect Conservation Breeding Programme Case Studies

The following case studies have been selected to help illustrate the current range, efficacy and wider value of insect conservation breeding programmes.

5.1 The field cricket, *Gryllus campestris*

5.1.1 Programme background

Due to alteration and fragmentation of its highly selective grassland habitat, by the late 1980s, the UK population of the field cricket, *Gryllus campestris*, was reduced to a single colony of fewer than 100 individuals in West Sussex (Edwards *et al.*, 1996). In 1991 the species was placed on English Nature's Species Recovery Programme (SRP). The SRP action plan called for the establishment of ten secure field populations in areas of the species' historic range (M. Edwards, 1995, unpublished data). Because the surviving population was too low to support direct translocations, the development of a conservation breeding programme was required, and in 1992 a breeding and rearing initiative was established at the Zoological Society of London. The strategy entailed collecting three pairs of subadult crickets from the surviving wild population each spring. These were to be bred at the London zoo to produce large numbers of late-instar F_1 generation nymphs for the establishment of new colonies in sites identified by the SRP ecological team.

5.1.2 Management summary

The management regime is detailed in Pearce-Kelly *et al.* (1998) and Jones *et al.* (1999). To help clarify natural health profiles, a faecal screening and post-mortem protocol was implemented for all field-collected founder crickets. Newly collected crickets were reared to the adult stage and paired up in standard aquarium tanks partly filled with a sandy soil mix topped with a sod of turf from the wild colony site. Hatching nymphs were transferred to nursery tanks furnished with egg cartons to optimize moulting conditions. Timer-controlled radiant basking bulbs helped synchronize nymphal development rates with those of the wild population. The crickets were housed in an isolated breeding room to reduce the risk of disease contamination from non-native insect species. Separate progeny lines were maintained to ensure maximum genetic diversity in the *ex situ* F_1 population prior to combining for field release.

5.1.3 Results

Overall breeding and rearing success has been high, with annual mortality rates ranging between 10% and 20% in the F_1 nymphs. To date, the breeding programme has provided in excess of 17,000 late-instar nymphs for the SRP field establishment programme. The importance of effective post-arrival and pre-release health-screening protocols was highlighted by the discovery in 1996 and 1997 of gregarine parasites in the captive population, preventing field releases in both those years (A. Cunningham *et al.*, 1996, unpublished data; Pearce-Kelly, 1997). This underlines the necessity of ensuring that adequate infection barriers are in place for all *ex situ* populations destined for reintroduction. Four of the seven field colonies established with zoo-bred crickets are still extant, the longest of which was shown to have persisted to the eighth generation without the need for reinforcement. In addition to providing large numbers of release stock, the breeding programme helped clarify fecundity ranges (D. Clarke, 2005, unpublished data). The knowledge derived from monitoring the fluctuation dynamics of the field-released *G. campestris* populations has informed optimal site management requirements for the species, and helped clarify the subtle environmental factors influencing colony survival. The breeding programme has also helped raise public awareness of the field cricket and its conservation issues and provides a model for developing similar recovery initiatives for the species in other range countries.

5.2 The wart-biter bush cricket, *Decticus verrucivorus*

5.2.1 Programme background

Although a relatively common species in areas of its mainland European range, the British population of *Decticus verrucivorus* is confined to a handful of isolated sites in the South of England, providing the necessary sheltered, lightly-grazed, chalk grassland habitat it requires (Cherrill, 1993; Cherrill and Brown, 1993). The cricket is omnivorous, feeding on a variety of plant and insect species. The embryo normally goes through two diapauses and the first may last several years (Ingrish, 1994). The cricket was placed on English Nature's SRP in 1991. The associated action plan required the establishment of additional colonies in areas of the species' historic range (Shaughnessy and Cheesman, 2005). To provide the large numbers of late-instar nymphs necessary for establishing new field populations, a breeding programme was established at the London Zoo using 500 eggs obtained from wild-caught females originally collected for a dietary research project at Imperial College.

5.2.2 Management summary

The management regime is described in Pearce-Kelly *et al.* (1998) and Jones *et al.* (1999). An environmental chamber was used to take the eggs through their summer and winter development cycle. To reduce the incidence of cannibalism and optimize moulting conditions, hatching nymphs were housed in low density groups of around 10 individuals. Timer-controlled radiant basking bulbs were used to synchronize nymphal development with the wild population. Adults were housed as breeding pairs in standard aquarium plastic tanks with a sandy soil substrate for oviposition. A predominately natural plant food diet was provided, supplemented with wax moth larvae. A pre-release health-screening protocol was implemented from the outset of the programme.

5.2.3 Results

The first year's breeding season produced in excess of 3000 eggs. Unlike the field cricket, the wart-biter's more demanding husbandry and diet requirements meant that relativity low numbers could be reared for field release (Jones *et al.*, 1999). The discovery, and successful eradication, of a fungal infection in the *ex situ* population (Cunningham *et al.*, 1997; Pearce-Kelly, 1997) highlights the importance of effective health monitoring. In excess of 500 late-instar crickets were provided to the SRP for several sets of field-releases into two sites, one of which also had translocations. Follow-up monitoring of these new populations has confirmed sustained colony persistence (Shaughnessy and Cheesman, 2005). The wart-biter cricket breeding programme provided additional information on the developmental biology of the species, in particular, the maximum egg developmental period was shown to be at least 2 years greater than the 7 years recorded by Ingrish (1994). Significant levels of media and public interest helped highlight the plight of the species and the importance of the wider SRP initiative.

5.3 Middle Island tusked weta, *Motuweta isolata*

5.3.1 Programme background

The New Zealand weta family Anostostomatidae, formerly Stenopelmatidae (Johns, 1998) demonstrates a high degree of endemicity to New Zealand (Gibbs, 1998). Many species are vulnerable to habitat loss or alteration and are extremely sensitive to introduced predatory fauna, especially mammals (Gibbs, 1998; McIntyre, 2001). One species in dire need of conservation management is the Middle Island tusked weta, *Motuweta isolata* (Johns, 1998). This species has only been found in certain areas of the 13 ha Middle Island situated off the Coromandel Peninsula on the east coast of the North Island. The Department of Conservation's (DoC) *M. isolata* recovery plan identified the need to establish the species on other offshore islands via a breeding and release programme (Sherley, 1998). Project Weta was initiated in 1986 and by 1991 had worked with a total of seven species (Barrett, 1991). Since this time, a further ten species had been worked on up until 2006 with a total of 12 species being bred to the first generation and some through to the fourth generation. This experience provided the confidence to develop a breeding and release programme for *M. isolata* between 1999 and 2001 in collaboration with Chris Winks of Land Care Research and Ian Stringer, then of Massey University. Three captive populations were subsequently established.

5.3.2 Management summary

The initial breeding group at Land Care Research, Mt Albert Auckland provided 60 first-instar nymphs to Auckland Zoological Park between August and November of 1999. These were raised through years 2000–2001. Two groups of nymphs were translocated to Double Island, in the Mercury Island group during the year 2000. The remainder were retained at the zoo and raised separately before being paired and subsequently translocated to Double Island between May and September 2001. The weta were kept in an air-conditioned room at a temperature of 17–20°C, with humidity levels at 60–90%. The animals were fed fish flakes, leaves and insects. A 2-l container of soil was provided in the breeding enclosures for oviposition. A succession of males were paired with each female.

5.3.3 Results

With only two mortalities, a total of 58 of the Project Weta stock were reared to suitable stages for field release. Initially 39 nymphs were released, followed by 19 adults after they had been mated and were laying eggs. These were added to animals from the other breeding programme groups to provide a total of 120 crickets for release on Double and Red Mercury Islands. They were established under special shelters prepared by Rob Shappell of DoC. Eggs were subsequently collected from all three captive populations and incubated at the Land Care Research facility with eclosion occurring in October 2001. The rearing of this second generation population resulted in a further 106 animals being translocated to the islands. Progeny that had completely developed *in situ* were confirmed on both islands in March 2003 (eight on Mercury and three on Double Island) and all were adult or large juveniles (I. Stringer, 2003, personal communication).

5.4 The Karner blue butterfly, *Lycaeides melissa samuelis*

5.4.1 Programme background

The Karner blue butterfly *Lycaeides melissa samuelis* is a resident of oak savannah, pine barren and sand barren habitats of the Midwest, mid-Atlantic and New England regions of the USA. Within these arid habitats resides its sole host plant, wild lupine *Lupinus perennis* (Dirig, 1994). In the last 25 years, the butterfly has suffered a dramatic population decline throughout its range primarily from habitat loss and fragmentation. Originally native to 12 states and one Canadian province, the species is now extant in Indiana, Michigan, Minnesota, New Hampshire, New York and Wisconsin. It was placed on the US Endangered Species Act in 1992. The species was reintroduced to Ohio in 1998 to a region of restored oak savannah and sand barren habitats near the western shore of Lake Erie. A recovery team was formed to spearhead the reintroduction effort. The team devised a seven-part strategy for recovery: (i) host plant propagation; (ii) reintroduction site selection, evaluation and management; (iii) post-management evaluation; (iv) breeding protocol development; (v) founder selection; (vi) captive breeding; and (vii) release and monitoring. The Nature Conservancy would manage habitat restoration and of the chosen release site. Staff from the Toledo Zoo would assess the habitat to determine its suitability for reintroduction. Zoo staff were also charged with host plant propagation, captive breeding and monitoring. The recovery plan specified that first generation adult female founders would be captured and placed on potted plants for egg deposition. Larvae would be reared on the plants through the life cycle to eclosion. Second generation adults would be transported to the introduction site and released. The species is bivoltine, producing two generations per season, the first May to June, the second July to August. The species over-winters in the egg stage, hatching the following April.

5.4.2 Management summary

Annually from 1998 to 2002, Toledo Zoo staff captured first generation adult females from sites in Michigan. Individual females were placed in a clear plastic container that was then positioned in a cooler for transport to the zoo. Each female was sequestered on a potted host plant covered with a cylindrical net. Adults were hand-fed daily using a honey-water solution. Eggs were typically deposited on the leaves and petioles of the host after one or two days. Once hatched, larvae were closely monitored. To negate cannibalism, second-instar larvae were moved to new plants so that no more than ten were on a single plant. Host plants were replaced regularly. Small pieces of pine bark were added to the soil surface of the potted plant during the final instar. Larvae would then crawl under the bark to pupate. Adults were transported to the release site in the afternoon following eclosion. The rearing unit was enclosed in a double barrier and isolated from other invertebrates in the collection. Instruments, as well as the floor, benches and other equipment were regularly disinfected.

5.4.3 Results

From 1998 to 2002, nearly 1700 adults were released at the Ohio reintroduction site. Since the cessation of captive breeding activities in 2002, the butterfly has expanded its range beyond the initial site and is now found throughout the 200 ha preserve. In addition, there has been a quantified large shift in population density from the original release site to another location 1000 m downwind. Recent efforts by the recovery team are focusing on the preparation of additional release sites and studying oviposition preferences of females *in situ*.

5.5 The barberry carpet moth, *Pareulype berberata*

5.5.1 Programme background

Previously widespread in Wales and England, as far north as Yorkshire, the UK population of the barberry carpet moth, *Pareulype bererata*, suffered a dramatic decline as a result of hedgerow loss and eradication of its once common food plant *Berberis vulgaris*. By 1987, the species was restricted to a single known site in Suffolk and was made a *Schedule 5* and Biodiversity Action Plan listed species. The British population was saved from imminent extinction by Paul Waring, who bred sufficient numbers from the remaining population to enable a concerted conservation initiative to be developed. This effort was initially led by the Joint Nature Conservancy Council, and then in 1991 was adopted by English Nature's Species Recovery Programme in partnership with a group of UK zoos. The breeding programme remit called for participating zoos to breed large numbers of late-instar moth larvae, together with their food plant, to be used to establish new populations in restored areas of the species' former UK range. The species is capable of producing two generations per year with moths emerging between April–June, and July–September, with the second generation of pupae over-wintering to emerge the following spring (Waring, 1990).

5.5.2 Management summary

Five participating zoos, Bristol, Dudley, Paignton, Penscynor and Whipsnade, along with a number of private individuals followed a simple breeding and rearing protocol. This was based on a combination of larvae reared on individually netted food plants and in a larger rearing units housing around 20 potted food plants. To reduce disease risk, the rearing areas were isolated from non-native invertebrate species. Other biological barrier measures included servicing the moths before other invertebrate species, wearing overalls and disposable gloves, and using a disinfectant foot dip. All equipment required for care of the moths remained within the rearing unit and a double door system reduced the risk of inadvertent escape of free flying adults.

5.5.3 Results

Increasingly successful breeding and rearing results were achieved by most participating institutions. Provision of animals for field release reached a peak in the year 2000 when a combined 147 moths emerged in the spring and produced a surplus of 4413 eggs and larvae of which 3793 larvae went to release projects. The season ended with approximately 1000 pupae being over-wintered at seven institutions in readiness for the 2001 season (Hughes, 2000). In recent years the breeding programme's emphasis has shifted to help improve understanding of the moth's autecology, especially egg-laying preferences, over-wintering and summer pupation requirements and adult flight behaviour. The establishment of new populations within the grounds of participating institutions has emerged as the most practical way of gathering these data. Accordingly, large-scale plantings of the moth's larval food plant, *B. vulgaris*, are currently underway to create suitable establishment habitats.

5.6 The American burying beetle, *Nicrophorus americanus*

5.6.1 Programme background

American burying beetles (ABBs) are the largest *Nicrophorus* spp. in the USA, measuring up to 37 mm. For successful reproduction ABBs require a vertebrate carcass (raging between 100 and 200 g), which is buried and prepared by both male and female for use as a food source for their larvae. The historic range of the ABB was eastern and central USA (35 states) and along the southern borders of Ontario, Quebec and Nova Scotia in Canada. A serious decline in this species was noticed in the late 1800s through the mid-1900s. Now the only naturally occurring population east of the Mississippi river is found on Block Island (BI) off the southern coast of Rhode Island, West of the Mississippi river. ABBs can still be found in eastern Oklahoma, Arkansas, eastern Kansas, central Nebraska, extreme southern South Dakota, and just recently were discovered in Texas. Reasons for the disappearance over 90% of the ABBs range may include loss of carcass-base in the necessary weight range for reproduction, such as the passenger pigeon *Ectopistes migratorus* and the greater prairie chicken *Tympanuchus cupido*. Habitat loss, alteration and fragmentation are causing a change in species composition resulting in greater competition for the carrion resources needed for reproduction. Other factors may include pesticides, disease, artificial lighting and electric bug zappers. The US Fish and Wildlife Service (USFWS) listed the ABB as endangered in 1989 and by 1991 had completed a recovery plan for the species (Raithel, 1991). The recovery plan called for the monitoring, managing and protection of existing populations, searches for additional populations and to implement a reintroduction plan using captive reared beetles.

5.6.2 Management summary

A pilot reintroduction and study was launched in 1990 and continued through 1993 on Penikese Island (PI), Massachusetts using beetles captive reared at Boston University (BU) by Andrea Kozol. The success of this pilot study led to a second reintroduction in 1994 on Nantucket Island (NI), Massachusetts. Roger Williams Park Zoo (RWPZ) was asked to participate in the recovery effort and received 19 male and 11 female beetles from BU that had been collected as larvae on BI. This colony was reared by RWPZ using the husbandry and breeding protocol developed at BU (A.J. Kozol, Concord, 1992, unpublished data). Beetles were maintained at 20–23°C with a 12-hour light cycle. Depending on the size of container used, 1–20 same sex sibling beetles were housed together on a moistened paper towel substrate. Newly emerged beetles are ravenous feeders and were fed heavily for the first 2 weeks (8–12 mealworms a day) after which feeding rates reduced to 6–8 mealworms a day. Breeding was carried out in 11-l plastic buckets filled with soil to about 5 cm from the top and covered with plexiglas lids. A pair of beetles was placed on the surface of the soil and given an optimal size rat or quail carcass.

5.6.3 Results

The NI reintroduction programme continued from 1994 to 2005 with RWPZ rearing and supplying USFWS with over 2500 beetles for release on NI. The status of this population continues to be regularly monitored. This programme has shown how zoos working in partnership with federal and local wildlife agencies can successfully meet the conservation breeding requirements of such species recovery initiatives (Amaral and Prospero, 1999). In addition to providing large numbers of animals for field release, the breeding programme for this species has allowed for the collection of data on husbandry and reproductive behaviours not easily observed in the wild (Wetzel, 1995). This effort has also led to the establishment of educational programmes providing public awareness of the ecosystem roles of insects and the importance of invertebrate conservation (Perrotti *et al.*, 2001).

5.7 The Frégate Island giant tenebrionid beetle, *Polposipus herculeanus*

5.7.1 Programme background

The Frégate Island giant tenebrionid beetle, *Polposipus herculeanus*, is a large, flightless beetle endemic to wooded habitat on Frégate Island in the Seychelles. The species has an IUCN Red List designation of 'Critically Endangered A2e' (Baillie *et al.*, 2004) on the basis of its extremely limited distribution and the accidental introduction of the brown rat, *Rattus norvegicus*, to the island in 1995 (Lucking and Lucking, 1997; Millet, 1999). In 1996, with the support of Frégate Island Private, Government of Seychelles, the Nature Protection Trust of Seychelles and Nature Seychelles, an *ex situ* population was established at the Zoological Society of London with 47 wild-caught founders, followed by an additional founder line of 20 animals in May 1999. The conservation remit was to establish a secure *ex situ* population and to provide as much life-history, reproductive and disease profile data as possible to inform *in situ* conservation management efforts.

5.7.2 Management summary

The management regime is comprehensively detailed in Ferguson and Pearce-Kelly (2004). The beetles were housed in large plastic tubs with a minimum 30 cm depth of soil substrate to allow larvae to burrow and pupate. A tree branch, secured vertically within each tub, allowed natural arboreal behaviour to be expressed and increased available surface area. Each tub accommodated between 50 and 100 beetles. Ambient night temperature was about 25°C and rose to approximately 28°C during the day, and relative humidity ranged between 65% and 75%. Natural spectrum fluorescent lights provided 12 h of daylight. The beetles' largely nocturnal behaviour could be studied using red spectrum lighting to which the beetles appear to be insensitive. Their diet consisted of a variety of fruit and vegetables, decaying leaf litter and wood. The beetles were normally kept as single generation populations.

5.7.3 Results

The Frégate beetles have proved to be a relatively straightforward species to maintain in culture with modest maintenance needs. The *ex situ* programme has realized its husbandry development remit with additional breeding groups successfully established in four other European zoos (Bristol, Artis, Riga and Poznan) culminating in a formalized EEP in 2002. A range of life-history, reproductive and health-related studies have helped clarify longevity, life-stage durations and generation length (Ferguson and Pearce-Kelly, 2005).

Standardized husbandry guidelines have been published (Ferguson and Pearce-Kelly, 2004) including protocols for taking biometric measurements, adult emergence and death records, as well as necropsy investigations.

Rat eradication has since been successfully achieved on the island and measures put into place to prevent future re-invasion (Shah, 2001). However, the beetle remains vulnerable due to its restricted range and potential *in situ* conservation options include possible translocations to other Seychelles islands, which may have been part of the species former range (Gerlach *et al.*, 1997). The discovery of an entomopathogenic fungal infection, *Metarhizium anisopliae* var. anisopliae (Elliot, 2003; Ferguson and Pearce-Kelly, 2004), in the *ex situ* population highlights the importance of health-screening protocols. Clarifying the significance and molecular stain source of the *Metarhizium* infection, including its potential presence in the *in situ* population, is a current conservation priority for informing *in situ* management decisions.

5.8 The Lord Howe Island stick insect, *Dryococelus australis*

5.8.1 Programme background

The Lord Howe Island stick insect (LHISI) was once common on Lord Howe Island, 700 km off the coast of New South Wales, Australia. It became extinct on Lord Howe Island a few years after rats were accidentally released in 1918 (Gurney, 1947), but was rediscovered in 2001 living on a small group of *Melaleuca* bushes on a rocky outcrop, called Ball's Pyramid, 25 km off Lord Howe Island. LHISIs were classified at the time as endangered under the New South Wales Threatened Species Conservation Act 1995 and presumed extinct in the IUCN Red Data List. A Draft Recovery Plan was developed by the New South Wales Department of Environment and Conservation (D. Priddel *et al.*, Sydney, 2002, unpublished data), and in 2003 two adult pairs were removed from Ball's Pyramid for captive breeding. One pair went to a private breeder in Sydney, the other pair to Melbourne Zoo. At that point almost nothing was known of their biology and ecology (Lea, 1916), except for observations made during collection. The remaining wild population is now thought to be less than 40 individuals living on a few bushes on the side of a cliff (Priddel *et al.*, 2003).

5.8.2 Management summary

LHISIs at Melbourne Zoo are kept under temperature and humidity regimes as close as possible to those of Lord Howe Island and are offered *Melaleuca*, as well as a number of other plant species. The original pair were intensively studied for the first month after arrival but, as the species is nocturnal, observations are now limited to health checks and inferences of behaviour. The eggs are buried in sand by the female and the nymphs emerge after 6–9 months. In order to collect as much data as possible, each egg is removed from the sand, weighed, measured and placed in a range of incubation media under different moisture regimes.

5.8.3 Results

At the time of going to press there are in excess of 5000 individuals, including around 100 adults and more than 1000 eggs. The LHISIs will remain in captivity until rats are eradicated from Lord Howe Island. This will be one of the most complex eradication programmes ever undertaken and will not take place for several years due to the necessity for studies on non-target species. The LHISI project illustrates two of the pitfalls of invertebrate conservation efforts: the first is the difficulty of working with a species about which nothing is known, particularly when the remaining wild population cannot be studied; the second is the lack of veterinary knowledge available when individual specimens become ill. It also illustrates that some invertebrate conservation programmes are closely analogous to vertebrate conservation programmes when the species, such as the LHISI, is high profile. This may have the disadvantage that the project can become mired in politics and bureaucracy, as many vertebrate programmes do. It also has the advantages that the project can attract as much public and media interest as any vertebrate programme and that the invertebrate species, as in this case, can act as a flagship for threat abatement programmes for a number of vertebrate and invertebrate species within the same habitat.

6 Discussion

Insects are an incredibly large and diverse group dominating earth's animal life (Wilson, 1987, 1992) and typify the assertion that habitat preservation alone represents the only realistic conservation option for the majority of endangered species. However, the insects also contain among their ranks some of the most technically feasible and cost-effective conservation breeding programme candidates that zoos and other conservation bodies can undertake. As the programme case studies section of this chapter illustrate, endangered insect species from a range of taxonomic orders can make excellent programme recipients with good chances of successful conservation outcome, providing best management practice is followed. The public awareness raising role that insect breeding programmes can engender is an additional significant conservation benefit, as is the wider conservation informing role that invertebrate programmes can provide, for both *ex situ* and *in situ* management contexts. Such programmes reflect latest thinking as to the role and value of modern zoos (Conway, 1995a,b; Balmford *et al.*, 1996; Miller *et al.*, 2004; WAZA, 2005). This suitability combined with increasingly sophisticated evaluation tools, including phylogenetic distinctiveness and taxa rarity (Redding and Mooers, 2006; Isaac *et al.*, 2007) helps address the 'overwhelming' species numbers issue.

At the 18th General Assembly of IUCN in Perth, Australia, a resolution was adopted urging zoos and butterfly houses to increase their participation in invertebrate conservation breeding and establishment programmes (IUCN, 1991). Over the intervening period, the value of developing such initiatives has been further demonstrated. Hopefully the international zoo community, museums, universities, governmental agencies and other like-minded organizations will increasingly realize their significant potential to help to conserve many of our planet's most remarkable animal species and direct their energies accordingly.

Acknowledgements

The authors gratefully acknowledge the following collaborating colleagues and organizations: Paul Atkin, Onnie Byers, Oliver Cheesman, John Cooper, SSC Conservation Breeding Specialist Group, Andrew Cunningham, Mike Edwards, English Nature, Amanda Ferguson, Frégate Island Private, Justin Gerlach, Richard Gibson, Sebastian Grant, Ian Hughes, Heather Koldewey, Daniel Koch, Land Care Research, Rob and Vicky Lucking, Donald MacFarlane, Bob Merz, Nature Seychelles, Lenka Nealova, New Zealand Department of Conservation, Romain Pizzi, John Pullin, David Priddel, Ann Pocknell, Matthew Robertson, Ilona Roma, Ratajsczak Radoslaw, Tony Sainsbury, David Sheppard, Rob Shappell, John Shaughnessy, Jane Stevens, Ian Stringer, US Fish and Wildlife Service, Craig Walker, Gerard Visser, Paul Waring, Chris West, Chris Winks, Wildlife Department of Seychelles Government and Brian Zimmerman.

This chapter is dedicated to the memory of the St Helena giant earwig, *Labidura herculeana* – one of many remarkable species that might still be with us had a conservation breeding initiative been attempted.

References

Amaral, M. and Prospero, M.L. (1999) One zoo, two islands, and a beetle. *Endangered Species Bulletin* 24.

Amin, R., Pearce-Kelly, P. and Wang, J. (2005) A demonstration tool for determining genetic relatedness in group managed species. Project completion report for European Association of Zoos and Aquaria. Unpublished.

Andrews, C. (1990) The management of aquatic invertebrate displays in zoos. In: Collins, N.M. (ed.) *Management and Welfare of Invertebrates in Captivity*. National Federation of Zoological Gardens, London, pp. 21–27.

Baillie, J.E.M., Hilton-Taylor, C. and Stuart, S.N. (eds) (2004) *IUCN Red List of Threatened Species. A Global Species Assessment*. IUCN, Gland, Switzerland and Cambridge.

Balmford, A., Mace, G.M. and Leader Williams, N. (1996) Designing the ark: setting priorities for captive breeding. *Conservation Biology* 10, 719–727.

Barrett, P. (1991) *Keeping Wetas in Captivity*. Wellington Zoological Gardens, Department of Conservation and New Zealand Entomological Society, New Zealand.

Brown, K.R. (1991) Conservation of neotropical environments: insects as indicators. In: Collins, N.M. and Thomas, J.A. (eds) *The Conservation of Insects and their Habitats, 15th Symposium of the Royal Entomological Society of London*. Academic Press, London, pp. 349–401.

Byers, O. and Seal, U.S. (2003) The conservation breeding specialist group: activities, core competencies and vision for the future. *International Zoo Yearbook* 38: 43–53.

Cheesman, O. (1999) Biodiversity and populations. *Aspects of Applied Biology*, 53: *Challenges in Applied Population Biology*. Association of Applied Biologists, Warwick, UK, pp. 213–218.

Cherrill, A.J. (1993) The conservation of Britain's wart biter bush-crickets. *British Wildlife* 5, 26–31.

Cherrill, A.J. and Brown, V.K. (1993) The life cycle and distribution of the wart-biter *Decticus verrucivorus* (L) (Orthoptera: Tettigoniidae) in a chalk grassland in southern England. *Biological Conservation* 53, 125–143.

Collins, N.M. (1986) *Butterfly Houses in Britain: The Conservation Implications*. IUCN, Cambridge.

Conway, W. (1995a) Wild and zoo animal management and habitat conservation. *Biodiversity and Conservation* 4, 573–594.

Conway, W. (1995b) The conservation park: a new zoo synthesis for a changed world. In: Wemmer, C.M. (ed.) *The Ark Evolving: Zoos and Aquariums in Transition*. Smithsonian Institution Press, Washington, DC, pp. 259–276.

Cooper, J.E. and Cunningham, A.A. (1991) Pathological investigation of captive invertebrates. *International Zoo Yearbook* 30, 137–143.

Cunningham, A.A. (1996) Disease risks of wildlife translocations. *Conservation Biology* 10, 347–648.

Cunningham, A.A. (1997) Invertebrate pathology: a developing and essential science for invertebrate conservation. *Memoirs of the Museum of Victoria* 56, 647–648.

Cunningham, A.A., Frank, M.J., Croft, P., Clarke, D. and Pearce-Kelly, P. (1997) Mortality of captive British wart-biter crickets: implications for reintroduction programs. *Journal of Wildlife Diseases* 33, 673–676.

Daniel, B.A., Molur, S. and Walker, S. (1998) *Report of the Workshop Conservation Assessment and Management Plan for Selected Soil Invertebrates of Southern India*. BCPP Endangered Species Project, Zoo Outreach Organisation, CBSG, Coimbatore, India.

Dirig, R. (1994) Historical notes on wild lupine and the Karner blue butterfly at the Albany Pine Bush, New York. In: Andow, D.A., Baker, R. and Lane, C. (eds) *Karner Blue Butterfly: A Symbol of a Vanishing Landscape*. Miscellaneous Publication 1984–1994. Minnesota Agricultural Experiment Station, University of Minnesota, Minnesota.

Edwards, M., Patmore, J.M. and Sheppard, D. (1996) The field cricket – preventing extinction. *British Wildlife* 8, 87–91.

Elliot, P. (2003) A case of systemic fungal infection (*Metarhizium anisopliae*) in a Frégate Island giant tenebrionid beetle (*Polposipus herculeanus*). Report for MSC. In: *Wild Animal Health*. Zoological Society of London, London.

English Nature (1995) Guidance notes for invertebrate translocations and introductions. *Species Conservation Handbook*. English Nature, Peterborough, UK.

Erhardt, A. and Thomas, J.A. (1991) Lepidoptera as indicators of change in the semi-natural grasslands of lowland and upland Europe. In: Collins, N.M. and Thomas, J.A. (eds) *The Conservation of Insects and their Habitats, 15th Symposium of the Royal Entomological Society of London*. Academic Press, London, pp. 213–236.

Ferguson, A.J. and Pearce-Kelly, P. (2004) *Husbandry Guidelines for the Fregate beetle Polposipus herculeanus*. The Federation of Zoological Gardens of Great Britain and Ireland, London.

Ferguson, A.J. and Pearce-Kelly, P. (2005) Captive Management of the Frégate Island Giant Tenebrionid Beetle *Polposipus herculeanus*. *Phelsuma*, Zoological Society of London, London.

Frankham, R., Ballou, J. and Briscoe, D. (2004) *Introduction to Conservation Genetics*. Cambridge University Press, Cambridge.

Fry, R. and Lonsdale, D. (eds) (1991) *Habitat Conservation for Insects – A Neglected Green Issue*. The Amateur Entomologists' Society, Middlesex, UK.

Gerlach, J. (1997) *Seychelles Red Data Book 1997*. The Nature Protection Trust of Seychelles, pp. 62–63.

Gerlach, J., Daszak, P. and Jones, C.G. (1997) On *Polposipus herculeanus* on Round Island, Mauritus. *Phelsuma* 5, 43–48.

Gibbs, G.W. (1998) *New Zealand Weta*. Reed Publishing, Auckland, New Zealand.

Gurney, A.B. (1947) Notes on some remarkable Australian walkingsticks, including a synopsis of the genus *Extatosoma* (Orthoptera: Phasmatidae). *Annals of the Entomological Society of America* 40, 373–396.

Howarth, F.G. and Ramsay, G.W. (1991) The conservation of island insects and their habitats. In: Collins, N.M. and Thomas, J.A. (eds) *The Conservation of Insects and their*

Habitats, 15th Symposium of the Royal Entomological Society of London. Academic Press, London, pp. 71–107.

Hughes, I. (2000) *Report on Barberry Carpet Moth* Pareulype berberata *Conservation Breeding Project for the Year 2000*. An English Nature Species Recovery Programme Report, English Nature, Peterborough, UK.

Hughes, D.G. and Bennett, P.M. (1991) Captive breeding and the conservation of invertebrates. *International Zoo Yearbook* 30, 45–51.

Hutchings, P.A. and Ponder, W.F. (1999) Workshop: criteria for assessing and conserving threatened invertebrates. In: Ponder, W. and Lunney, D. (eds) *The Other 99%. The Conservation and Biodiversity of Invertebrates*. Transactions of the Royal Zoological Society of New South Wales, Mosman, New South Wales, Australia, pp. 264–315.

Ingrish, S. (1994) Embryonic development of *Decticus verrucivorous*. *Entomologia Generalis* 10, 1–9.

Isaac, N.J.B., Turvey, S.T., Collen, B., Waterman, C. and Baillie, J.E.M. (2007) Mammals on the EDGE: Conservation priorities based on threat and phylogeny. *PLoS* ONE (in press).

IUCN (1987) *Translocation of Living Organisms: Introductions, Re-introductions, and Restocking*. IUCN Position Statement. IUCN, Gland, Switzerland.

IUCN (1990) *Programme for Conservation of Species and Genetic Resources*. Annex 5, IUCN, Gland, Switzerland.

IUCN (1991) *18th General Assembly Resolution and Recommendations*. IUCN, Gland, Switzerland.

JCCBI (1996) Insect re-establishment – a code of conservation practice. *Antenna* 10, 13–18.

Johns, P.M. (1998) The Gondwanaland weta: Family Anostostomatidae (formerly in Stenopelmatidae, Henicidae or Mimnermidae): nomenclatural problems, world checklist, New genera and species. *Journal of Orthoptera Research 6*, 125–138.

Jones, R., Pearce-Kelly, P. and Cheesman, O. (1999) Population biology and management considerations in invertebrate rearing programmes with particular reference to Orthoptera. In: *Aspects of Applied Biology, 53: Challenges in Applied Population Biology*. Association of Applied Biologists, Warwick, UK, pp. 219–226.

Lea, A.M. (1916) Notes on the Lord Howe Island phasmid, and on an associated longhorn beetle. *Proceedings of the Royal Society of South Australia* 40, 145–147.

Lees, D. (1989) Practical considerations and techniques in the captive breeding of insects for conservation purposes. *The Entomologist* 180, 77–96.

Lucking, R. and Lucking, V. (1997) Preliminary results of monthly surveys of *Polposipus herculeanus*. *Phelsuma* 5, 5.

Mallinson, J.J.C. (1995) Conservation breeding programmes: an important ingredient for species survival. *Biodiversity and Conservation* 4, 617–635.

McIntyre, M. (2001) The ecology of some large weta species in New Zealand. In: Field, L.H. (ed.) *The Biology of Wetas, King Crickets and their Allies*. CAB International, Wallingford, UK.

Miller, B. Conway, W., Reading, P.R., Wemmer, C., Wildt, D., Kleiman, D., Monfort, S., Rabonowitch, A., Armstrong, B. and Hutchings, M. (2004) Evaluating the conservation mission of zoos, aquariums, botanic gardens, and natural history museums. *Conservation Biology* 18, 86–93.

Millet, J. (1999) Giant tenebrionid beetle in danger. *Birdlife Seychelles Newsletter* 4, 12.

Morton, A.C. (1991a) Captive breeding of butterflies and moths: I Advances in equipment and techniques. *International Zoo Yearbook* 30, 80–89.

Morton, A.C. (1991b) Captive breeding of butterflies and moths: II. Conserving genetic variation and managing biodiversity. *International Zoo Yearbook* 30, 89–97.

New, T.R. (1995) *Introduction to Invertebrate Conservation Biology*. Oxford University Press, Oxford.

Pearce-Kelly, P. (1994) Invertebrate propagation and re-establishment programmes: the conservation potential for zoos and related institutions. In: Olney, P.J.S., Mace, G.M. and Feistner, A.T.C. (eds) *Creative Conservation: Interactive Management of Wild and Captive Animals*. Chapman & Hall, London.

Pearce-Kelly, P. (1997) Coping with pathogen infections in captive populations in British field and wart-biter crickets. *Veterinary Invertebrate Society (VIS) Newsletter* 12, 7–8.

Pearce-Kelly, P., Clarke, D., Robertson, M. and Andrews, C. (1991) The display, culture and conservation of invertebrates at London Zoo. *International Zoo Yearbook* 30, 21–30.

Pearce-Kelly, P., Jones, R., Clarke, D., Walker, C., Atkin, P. and Cunningham, A. (1998) The captive rearing of threatened Orthoptera: a comparison of the conservation potential and practical considerations of two species' breeding programmes at the Zoological Society of London. *Journal of Insect Conservation* 2, 201–210.

Perrotti, L., Prospero, L.M. and Spevak, E. (2001) *The Effort to Save the Endangered American Burying beetle on a Show-string Budget. Making Conservation Resources go Farther with Invertebrates*. Aquarium and Zoo Association Annual Conference Proceedings, Gland, Switzerland.

Pizzi, R. (2004) Disease diagnosis and control in *ex situ* terrestrial invertebrate conservation programmes. *European Association of Zoo and Wildlife Veterinarians 5th Scientific Meeting,* 19–23 May 2004, Ebeltoft, Denmark.

Priddel, D., Carlile, N., Humphrey, M., Fellenberg, S. and Hiscox, D. (2003) Rediscovery of the 'extinct' Lord Howe Island stick-insect (*Dryococelus australis*) (Montrouzier) (Phasmatodea) and recommendations for its conservation. *Biodiversity and Conservation* 12, 1391–1403.

Pullin, A. (1999) Priorities in terrestrial invertebrate conservation. In: *Aspects of Applied Biology*, 53: *Challenges in Applied Population Biology*. Association of Applied Biologists, Warwick, UK, pp. 33–40.

Pullin, A. (2004) *Conservation Biology*. Cambridge University Press, Cambridge.

Rabb, G.B. (1994) The changing roles of zoological parks in conserving biological diversity. *American Zoologist* 34, 159–164.

Raithel, C. (1991) *American Burying Beetle (Nicrophorus americanus) Recovery Plan*. US Fish and Wildlife Service, Region 5. Newton Corner, Massachusetts.

Redding, D.W. and Mooers, A.O. (2006) Incorporating evolutionary measures into conservation priorization. *Conservation Biology* 20, 1670–1678.

Rivers, C.F. (1990) The control of diseases and pests of insects. In: Collins, N.M. (ed.) *The Management and Welfare of Invertebrates in Captivity*. National Federation of Zoological Gardens, London, pp. 45–49.

Rivers, C.F. (1991) The control of diseases in insect cultures. *International Zoo Yearbook* 30, 131–137.

Robinson, M.H. (1991) Invertebrates: exhibiting the silent majority. *International Zoo Yearbook* 30, 1–7.

Samways, M.J. (1994) *Insect Conservation Biology*. Chapman & Hall, London.

Samways, M.J. (1999) Landscape triage for conserving insect diversity. Workshop: criteria for assessing and conserving threatened invertebrates. In: Ponder, W. and Lunney, D. (eds) *The Other 99%. The Conservation and Biodiversity of Invertebrates,* Transactions of the Royal Zoological Society of New South Wales, Mosman, New South Wales, Australia, pp. 269–273.

Samways, M.J. (2005) *Insect Diversity Conservation*. Cambridge University Press, Cambridge.

Seal, U.S., Foose, T.J. and Ellis, S. (1994) Conservation Assessment and Management Plans (CAMPS) and Global Captive Action Plans (GCAPS). In: Olney, P.J.S., Mace, G.M. and Feistner, A.T.C. (eds) *Creative Conservation: Interactive Management of Wild and Captive Animals*. Chapman & Hall, London.

Shah, N.J., Payet, R. and Lowe-Henri, K. (eds) (1997) *Seychelles National Biodiversity Strategy and Action Plan*. Government of Seychelles.

Shaughnessy, J.P. and Cheesman, O.D. (2005) Wart-biter bush-cricket (*Decticus verrucivorus*) 2004. CAB International Bioscience, Ref: U3050 (National Trust Contract), Wallingford, UK.

Sherley, G.H. (1998) *Threatened Weta Recovery Plan*. Threatened Species Recovery Plan #25. Science and Research Unit, Department of Conservation, Wellington, New Zealand.

Shirt, D.B. (1987) *British Red Data Books, 2: Insects*. Nature Conservancy Council, Peterborough, UK.

Sullivan, E., Phillips, G., Howorth, P., Magdich, M., Mason, T., Morgan, R., Snider, A. and Stevens, J. (2005) *Terresterial Invertebrate Taxon Advisory Group Regional Collection Plan 2005–2008*. Terrestrial Invertebrate Taxon Advisory Group, American Association of Zoos and Aquaria report.

Sutton, S.L. and Collins, N.M. (1989) Insects and Tropical Forest Conservation. In: Collins, N.M. and Thomas, J.A. (eds) *15th Symposium of the Royal Entomological Society of London*. Academic Press, London, pp. 405–424.

van Helsdingen, P.J., Willemse, L. and Speight, M.C.D. (eds) (1996) *Background Information on Invertebrates of the Habitats Directive and the Bern Convention*. Nature and Environment, No. 80. Council of Europe, Strasbourg, France.

Visser, G., Gibson, R., Koldewey, H., Zimmerman, B., Veltman, K. and Pearce-Kelly, P. (eds) (2005) *Lower Vertebrate and Invertebrate Taxon Advisory Group Manual for Regional Collection Planning*. August 2005, European Association of Zoos and Aquariums.

Waring, P. (1990) Conserving Britain's rarest moths. *British Wildlife* 1(5), 87–91.

WAZA (2005) *Building a Future for Wildlife – The World Zoo and Aquarium Conservation Strategy*. In: Olney, P.J.S. (ed.) WAZA Executive Office. 3012 Bern, Switzerland.

Wetzel, D.L. (1995) Husbandry and Conservation Initiatives for the Endangered American Burying Beetle at Roger Williams Park Zoo. *Invertebrates in Captivity Conference, Proceedings*, Sonoran Arthropod Studies Institute, Arizona, pp. 37–42.

Wilson, E.O. (1987) The little things that run the world: the importance and conservation of invertebrates. *Conservation Biology* 1, 344–346.

Wilson, E.O. (1992) *The Diversity of Life*. The Belknap Press of Harvard University Press, Cambridge, Massachusetts.

Wilson, A.C. and Stanley-Price, M.R. (1994) Reintroduction and a reason for captive breeding. In: Olney, P.J.S., Mace, G.M. and Feistner, A.T.C. (eds) *Creative Conservation: Interactive Management of Wild and Captive Animals*. Chapman & Hall, London.

Yen, A.L. (1993) The role of museums and zoos in influencing public attitudes towards invertebrate conservation. In: Gaston, K.J., New, T.R. and Samways, M.J. (eds) *Perspectives on Insect Conservation*. Intercept, Andover, Massachusetts, pp. 213–229.

4 What Have Red Lists Done for Us? The Values and Limitations of Protected Species Listing for Invertebrates

MARTIN S. WARREN, NIGEL BOURN, TOM BRERETON, RICHARD FOX, IAN MIDDLEBROOK AND MARK S. PARSONS

Butterfly Conservation, Manor Yard, East Lulworth, Dorset BH20 5QP, UK

Keywords: Red Lists, insects, conservation, Lepidoptera

1 Introduction

Red Lists have been an important tool in conservation ever since they were formalized by the International Union for the Conservation of Nature (IUCN) in 1963. Since then they have evolved rapidly and have been used within many individual countries, as well as to compile a global list of threatened species.

The two main aims of the IUCN Red Lists are:

1. To identify species threatened with extinction;
2. To promote their conservation.

Other more political aims stated by IUCN are to convey the scale and urgency of the problems facing biodiversity to the public and policy makers, and to motivate the global community to take action (www.iucn.org/themes/ssc/RedLists). In 2000, a single global Red List for animals and plants was published for the first time and contained 18,000 species assessments (Hilton-Taylor, 2000). This vast database is now available on searchable website www.iucnredlist.org. A more recent development has been to use the global Red List to provide a global index of the state of degeneration of certain taxa (Butchart *et al.*, 2005).

Red Lists have been used for invertebrates since their inception but the criteria have been widely criticized as being difficult to apply to this diverse group due to lack of precise data about their status (e.g. Sutherland, 2000; New and Sands, 2004). Moreover, only 70 insect extinctions have been documented in the last 600 years, despite predictions that the real figure should be near 40,000 (Dunn, 2005). Because of the lack of data, can the global Red List be meaningful if the criteria cannot be used to assess most invertebrates, which comprise over two-thirds of the world's described species?

This is a serious issue given that recent evidence from Britain shows that butterflies are declining faster than either birds or plants (Thomas *et al.*, 2004). Similarly, rapid declines have since been demonstrated amongst a far larger group of 337 common moths in Britain (Conrad *et al.*, 2004, in preparation), giving further weight to the view that the extinction crisis may be far worse than that estimated earlier. For invertebrates, the Red Listing process must be precautionary and initiate conservation action on the best available evidence, because a delay to gather conclusive data may be too late for many species (Samways, 2005).

A fundamental question is thus whether Red Lists are a sensible approach to identifying priorities amongst such a diverse and species-rich group as invertebrates, when the criteria for selection have been developed primarily for more well-known groups, such as mammals and birds.

In this chapter, we explore the use of Red Lists for invertebrates, using examples of Lepidoptera and other taxa in the UK and Europe to demonstrate their influence on conservation practice. We draw conclusions about the use of Red Lists for invertebrates and their potential for promoting their conservation.

2 IUCN Criteria

The initial IUCN criteria included six categories (Table 4.1): Extinct (Ex), Endangered (E), Vulnerable (V), Rare (R), Intermediate (I) and Insufficiently Known (K). These criteria were widely adopted by government and non-governmental organizations (NGOs) around the world and were used to classify species in many taxa, including invertebrates (see Sections 3 and 4). However, they were largely subjective and open to interpretation by different users, which led to problems in their use and credibility (e.g. Fitter and Fitter, 1987).

After many years of consultation, IUCN published a new set of criteria in 1994 that were designed to give a more objective and transparent method of assessing extinction threat (IUCN, 1994). Detailed guidance has since been developed both for use at global (IUCN, 2001, and updated on the Red List website) and regional levels (IUCN, 2003).

A notable feature of the new criteria is that they use assessments of species population trends, as well as rarity in order to assess extinction risk (Table 4.2). Thus, although these criteria are widely held to be a great improvement on the earlier ones, they present even more problems when assessing invertebrates,

Table 4.1. Old IUCN definitions and criteria used for insects in Britain. (From Shirt, 1987.)

RDB 1 – Endangered
RDB 2 – Vulnerable
RDB 3 – Rare (defined in UK as 15 or fewer 10 km grid squares)
RDB 4 – Out of danger
RDB K – Insufficiently known

Table 4.2. New IUCN criteria, based on IUCN (2001) as updated by IUCN (2005).

Criteria	Critically endangered (CE)	Endangered (E)	Vulnerable (V)
A. Population reduction	<80% in 10 years	<50% in 10 years	<20% in 10 years
B1. Extent of occurrence	EOO < 100 km^2	<5,000 km^2	<20,000 km^2
B2. Area of occurrence	AOO < 10 km^2 + two of: 1. Severely fragmented/ single location 2. Continuing decline 3. Extreme fluctuations	<500 km^2 + two of: 1. Severely fragmented/ <5 locations 2. Continuing decline 3. Extreme fluctuations	<2,000 km^2 + two of: 1. Severely fragmented/ <10 locations 2. Continuing decline 3. Extreme fluctuations
C. Population estimate	<250 individuals + strong decrease	<2,500 individuals + strong decrease	<10,000 individuals + strong decrease
D. Population estimate	<50 individuals	<250 individuals	<1,000 individuals
E. Probability of extinction	>50% within 10 years	>20% within 10 years	>10% within 10 years
Near threatened (NT)	**Least concern (LC)**	**Data deficient (DD)**	**Not evaluated (NE)**
Species close to qualifying as threatened	Species that do not meet criteria	Inadequate information	Not evaluated

because they need far more precise data on status and trends. For example, few invertebrate taxa have precise data on trends over the required 10-year period and even assessments of area of occupancy are likely to be very imprecise. The criteria recognize these problems to some extent and suggest the use of proxy data, such as the loss of habitat and extrapolation from smaller data-sets, provided this can be justified. However, these have yet to be used widely for invertebrates and lack of data is likely to remain an overriding problem for many groups.

Aside from the assessment problems caused by insufficient data, the application of the new IUCN criteria still involve sufficient subjectivity to have significant impacts on the resultant classification (e.g. Regan *et al.*, 2005). To mitigate the impact of subjectivity, those carrying out assessments have developed modified procedures (Keller *et al.*, 2005) and called for further revision of the IUCN criteria (Eaton *et al.*, 2005).

3 Invertebrate Red Lists Around the World

3.1 The IUCN global Red List

The latest IUCN global Red List covers assessments for 38,000 species of which almost half (15,503) were classified as threatened (IUCN, 2004). However, only 771 insects have so far been evaluated, of which 559 (73%) are classified as threatened. This is clearly a very partial assessment, comprising small samples of better known orders, such as Lepidoptera and Odonata, totally just 0.06% of described insect species. In contrast, the assessment includes all birds and amphibians and the majority of mammals.

As the footnotes to this analysis point out, the true percentage of threatened insect species lies somewhere between 0.06% and 73%. Moreover, as insects comprise such a large percentage of total described species, the number of globally threatened species has so far been vastly underestimated.

3.2 A brief overview of Red Lists for invertebrates around the world

The IUCN criteria have been used to produce lists of threatened invertebrates in at least 19 countries (based on an Internet search in September 2005), including the USA, Canada, South Africa, Australia, Russia and 12 European countries. Most of these assessments used the old Red List criteria, but some are now being updated using the new criteria. The old criteria have been used even more widely to assess better known invertebrate groups, such as butterflies, for which Red Lists have been produced for at least 36 European countries (van Swaay and Warren, 1999). A new Red List of threatened butterflies in Europe has also been produced, based on the new IUCN criteria, but adapting them for use with the type of data available (see Section 4).

4 The Use of Red Lists within Britain

4.1 British Red Data Books for invertebrates

The first Red Data Book (RDB) for Insects was published by the government agency the Nature Conservancy Council in 1987 (Shirt, 1987). It was compiled after almost a decade of detailed work by the RDB Criteria and Species Selection Committee, and a further 3 years by an RDB Production Committee. These Committees consulted a wide range of experts in different taxonomic groups, and called on data that had been gathered by the Biological Records Centre at the Centre for Ecology and Hydrology (CEH), Monks Wood.

The RDB for Insects listed 1786 species as threatened using the original IUCN criteria, 14.5% of the total insect fauna. This was followed by a RDB for other invertebrates, which covered 144 species (Bratton, 1991). In the absence of detailed criteria for each IUCN category, criteria were defined for use at the national level (Table 4.1). Although most species that fell into the Endangered and Vulnerable category were undoubtedly under great threat, over half of the species listed fell into category 3, Rare, which was defined as any species found in 15 or fewer 10 km grid squares. Although some of these may be threatened, many had always been highly restricted due to their specific ecological requirements.

Despite the enormous effort in compiling the two RDBs, their impact on conservation policy was somewhat limited. The presence of Red Listed invertebrates was included in the criteria for the designation of Sites of Special Scientific Interest (SSSI) (Nature Conservancy Council, 1989), but few sites were designated specifically for invertebrates. The majority of SSSIs were designated primarily on habitat grounds, with the aim of covering the best examples of each habitat within each 'area of search' within Britain. Perhaps the most important legacy of the RDBs was that they raised awareness of the huge number of insects under threat in Britain and the need to find out more about them. The need to identify important sites for RDB species also helped start the Invertebrate Site Register Project, which aimed to identify and document important sites for the conservation of invertebrates in Britain. The publications also provided a clear focus for the many invertebrate recording schemes that were run by volunteers and coordinated by the Biological Records Centre (see list in Shirt, 1987; Hawksworth, 2001).

4.2 The habitats versus species debate

A key question for conservationists at the time was does habitat conservation lead to effective species conservation? It was widely assumed that this was the case and that if you conserved the best habitats across the country, most species would also be conserved. However, most invertebrate ecologists knew that this argument did not follow, because many invertebrates have very demanding requirements that may or may not be met within these habitats, and a large number of species did not occur in the 'best' selection of habitats because these were chosen largely on botanical grounds.

Thus key invertebrate habitats, such as dead wood and river shingles were rarely included, and the designated sites were rarely managed sufficiently to sustain populations of the more specialized invertebrates (e.g. Fry and Lonsdale, 1991). Two classic examples are those of National Nature Reserves at Monks Wood and Castor Hanglands in Cambridgeshire, which have lost 11 and 14 species of butterfly, respectively (one-third of their totals), including most of the threatened species listed in the RDB and many other habitat specialists (Thomas, 1991). With hindsight, we now know that some of these losses may have been inevitable because the reserves are too small and isolated to maintain viable populations in the long term. However, many others were lost due to insufficient habitat management. For example, at Monks Wood the cessation of coppicing, which formerly provided open habitats for many butterflies, is seen as a major cause of the loss of specialist species. The wood still holds some important invertebrate populations, but it is a shadow of its former self. The sad part is that the same story has been repeated in hundreds of sites up and down Britain, which has resulted in major declines in butterflies and many other invertebrates (e.g. Thomas and Morris, 1994; Asher *et al.*, 2001; Hawksworth, 2001).

4.3 The UK biodiversity action plan

Following the signing of the Convention on the Conservation of Biodiversity at the Rio de Janeiro conference in 1992, the UK government initiated its own response: the UK Biodiversity Action Plan (UK BAP). In order to press the case for greater action, a group of NGOs came together to form the Biodiversity Challenge Group. This group published its own detailed plans of how the government could take concerted action to conserve dwindling wildlife and wild habitats (Wynne *et al.*, 1995). The Biodiversity Challenge Report took the pragmatic view that a combined habitat and species approach was necessary to conserve biodiversity, and to take targeted action for species most under threat, or for which the UK had particular global responsibility (e.g. endemics). The UK BAP took on board much of the rationale of this document, which led to the publication of a series of species action plans and habitat action plans (Department of the Environment, 1994, 1995).

To qualify as Priority Species (on the short list), they had to meet one of two main criteria:

1. Rapidly declining (>50% in the last 25 years);
2. Globally threatened.

Thus, although the list took into account some of the principles of the IUCN criteria, the intention was to produce a list of conservation priorities. Detailed species action plans were published for each priority species, including the following sections:

1. Current status;
2. Current factors causing decline;
3. Current action;
4. Action plan objectives and targets;
5. Proposed action with lead agencies (including sections on policy and legislation, site safeguard and management, species management and protection, advisory, future research and monitoring, communication and publicity).

The plans also established a cycle to review and modify plans at intervals of 3 years.

Although government agencies were identified for taking the lead on each individual action in the plans, a novel approach was to nominate a 'Lead Partner' for each species from the NGO community. Thus, the implementation of the plans was intended to be a partnership that involved numerous government departments, NGOs and volunteers, and the business community.

Building on the platform of the British RDBs, new data flowing from numerous recording schemes, and the expertise of Alan Stubbs, it was possible to compile a more relevant list of invertebrate priorities within the UK BAP (based on Wynne *et al.*, 1995). The initial UK BAP list contained over 300 priority species, over half of which were invertebrates from a wide range of taxa (Table 4.3). This is the first time that invertebrates had been recognized in such a prominent way within the UK and has led to concerted action in recent years. The following two sections give two examples.

Table 4.3. Invertebrates listed as Priority Species within the UK Biodiversity Action Plan. (From Department of the Environment, 1995.)

	No. spp.
Ants	4
Beetles	53
Bees	10
Butterflies	9
Crickets	4
Crustacea	3
Damselflies	1
Flies/mayflies	12
Molluscs	12
Moths	53
Worms	1
Total	164

4.4 Implementing plans for Lepidoptera

Within the UK BAP, Butterfly Conservation was appointed as Lead Partner (or Joint Lead Partner) for 61 priority species of Lepidoptera, comprising 9 butterflies and 52 moths. The butterfly list was taken from an assessment of priorities using the three axes of a 'conservation cube': (i) national status (populations and trends using the new IUCN criteria adapted for use with the distribution survey data available for butterflies); (ii) international importance; and (iii) European/global conservation status (Warren *et al.*, 1997). The listing of priority moths used similar criteria and trends, which were taken from an analysis of pre- and post-1960 records held by the National Scarce Moth Recording Scheme (Wynne *et al.*, 1995).

A series of UK-wide conservation projects have since been started under the Action for Butterflies and Action for Threatened Moths programmes funded in a large part by the statutory conservation agencies. This has led to positive action for all the species listed, involving a wide range of organizations at local and national level, as well as the involvement of many thousands of volunteers (Warren, 2002). Several of these projects have received high profile media coverage and ministerial involvement, which has helped to raise awareness of biodiversity loss and the plight of invertebrates in general. The need to objective information to identify conservation priorities and review progress has also been a major driver to develop a detailed recording scheme for butterflies (Asher *et al.*, 2001) and more recently one planned for moths (Fox *et al.*, 2005).

4.5 The Action for Invertebrates project

Nine of the priority species of invertebrate listed in the UK BAP, initially had no obvious organization to act as Lead Partner. To ensure that conservation effort was directed at these species, a consortium was formed in 2000 through members of the Biodiversity Challenge group, English Nature and Invertebrate Link. This led to a Project Officer being employed under the *Action for Invertebrates* project, which covers a diverse group of invertebrates including a freshwater Bryozoan, *Lophopus crystallinus*, a stonefly, *Brachyptera putata*, and several Coleopteran and other species. The project continues today with support from English Nature, the Royal Society for the Protection of Birds, Butterfly Conservation and Buglife (Middlebrook, 2000, 2002 and 2005).

One of the species that the project currently covers is the cranefly, *Lipsothrix nigristigma*, which breeds in coarse woody debris within woodland streams. When it was originally listed as a priority species, there were recent records from only two sites but, thanks to targeted survey work funded by the BAP process, knowledge of the species has grown substantially and 34 sites had been identified by 2004 (S. Hewitt and J. Parker, 2004; A. Godfrey, 2005, unpublished data). In addition to providing specific conservation advice at these sites, the project has acted as a spear-head for raising awareness of a whole suite of invertebrates associated with woody debris in rivers and streams (see Table 4.4).

There are similar examples of other BAP priority species spear-heading several other crucial invertebrate conservation issues, such as the importance of dead wood and veteran trees (Bowen, 2003). The focus on threatened invertebrates listed in the BAP has led to widespread media coverage, even for some unlikely species. The media seems to be particularly fascinated by the 'intrigue factor' for species, such as the Depressed River Mussel (*Pseudanodonta complanata*) and the New Forest Beetle (*Tachys edmondsi*). It has also stimulated some popular surveys, including one for the Stag Beetle (*Lucanus cervus*), which involved 1300 recorders (Smith, 2003).

Table 4.4. Species associated with dead, wet timber: a neglected habitat for invertebrates that has been highlighted by action for the threatened cranefly *Lipsothrix nigristigma*. (From Godfrey, 2003.)

Group	No. associated species
Molluscs	6
Crustaceans	4
Mayflies	4
Damselflies	2
Stoneflies	4
True bugs	1
Caddisflies	25
Beetles	21
True flies	79
Total	146

4.6 Revising the UK BAP lists

A significant difference between 'traditional red lists' and BAP listings is the strong commitment to review the relevance of the BAP priorities on a regular cycle. Within the BAP a review of the priority species and habitats was set in place on a 10-year cycle and was initiated late in 2004 for intended publication in 2005. A working group of Invertebrate Link, a UK wide gathering of invertebrate specialist and conservation organizations, which is co-chaired by Butterfly Conservation and the government's Joint Nature Conservation Committee, contracted Buglife to coordinate the invertebrate specialists of the UK. The review is a good example of how the BAP process has brought together statutory, NGOs and other expertise.

To date, over 40 specialist taxonomic coordinators have brought together the views of over 300 experts to develop the initial invertebrate species lists. The criteria were similar to those used in the original BAP (Department of Environment, 1995), but with much more flexibility to enable use of the best data-sets available. For example, if survey intervals produce decline rates for 35 years rather than the preferred 25 years these have been accepted provided the rate is high enough to meet the criteria when adjusted to take into account the longer recording period.

This approach has enabled the working group to recommend over 500 candidate species that meet the criteria for inclusion in the new list. In addition, there is an agreed second stage to the process where consideration will be given to the practical mechanisms necessary to deliver conservation action. The working group is currently developing this approach with a view to the government publishing the updated list in 2006. A large number of additional Lepidoptera spp. are included in the proposed list due to new data on their rate of decline. They also include for the first time a group of 71 widespread moth species that meet both IUCN vulnerable criteria and BAP criteria due to their rapid rate of decline (Conrad *et al.*, in preparation). It is proposed that these are grouped into a single plan for action, mainly to research the reasons for their decline, which are still a matter for conjecture.

Overall, it is expected that the number of all priority species (animals and plants) in the UK BAP will rise from ~300 to over 1500 at the forthcoming review, as a result of the rapid decline and better knowledge of many groups. To cope with this large number of additional species, there will have to be renewed effort to integrate species action plans within the relevant habitat action plans, a process which so far has been very patchy. This would also be a very healthy process as it would force better communication between entomologists and practitioners primarily concerned with habitat conservation, a process which has also been very partial in the past.

5 Red Lists and European Legislation

5.1 The Bern Convention

The first list of priority invertebrates to be produced at a European level was in 1988 when 71 arthropod species were added to Appendix II of Council of

Europe's Convention on the Conservation of European Wildlife and Natural Resources (commonly referred to as the Bern Convention). These included 51 species of insects, comprising 1 Mantodea, 16 Odonata, 2 Orthoptera, 8 Coleoptera and 24 Lepidoptera (21 butterflies and 3 moths).

The lists in the Bern Convention were based on a review by the IUCN Conservation Monitoring Centre, which incorporated recommendations of various expert groups and a review on information published in Red Lists across Europe (Wells *et al.*, 1983). Criteria for selection included that species must be threatened according to (old) IUCN criteria but restricted the selection to a 'moderate number in order that achievable conservation objectives could be set'. The selection aimed to cover representatives from as wide array of habitats as possible but restricted itself to species reasonably easy to identify. Although the Bern Convention is largely voluntary, the lists were used subsequently to develop stronger legislation in many countries across Europe and also by the European Union (EU) within the Habitats and Species Directive (see Section 5.2).

5.2 The EU Habitats and Species Directive

The EU Habitats and Species Directive (92/43/EEC) is one of the strongest pieces of wildlife legislation in Europe. It lists a wide range of priority species and habitats to be protected in all member states. A total of 85 arthropod species (including 54 insects) are listed in the Annexes, 59 (36 insects) of them are listed in Annexe II, which requires the designation and protection of important breeding areas as Special Areas of Conservation. It also requires member states to maintain these species at a 'Favourable Conservation Status' (defined as maintaining the species range within each member state). A total of 71 of these species are also listed in Annexe IV (including 46 insects some of which are also listed in Annexe II). This requires member states to provide strict protection for the species (see lists in van Helsdingen *et al.*, 1996).

Although most invertebrate zoologists agree that the lists in these two pieces of legislation are far from perfect, they were based on the lengthy deliberations of expert committees using the best information available at the time. They include many species that have since been confirmed as being highly threatened across Europe and have received much needed conservation action as a result.

One problem is the extremely long time lag between new information becoming available and any revision of the lists. Thus, the species on the Habitats and Species Directive are not likely to be reviewed in the near future despite them being based on information collected for the Bern Convention in the 1980s. There is an urgent need for revision of the lists now that far better information is available, e.g. within the RDB of European Butterflies. Despite these criticisms, the production of priority lists within EU legislation has focussed the minds of entomologists across Europe and has directly initiated some valuable survey work that will enable the construction of far more accurate lists in future (e.g. Grootaert *et al.*, 2001; see Section 6).

6 The European Butterfly Red List and its Use

6.1 The Red Data Book of European butterflies

The first comprehensive review of the status of butterflies in Europe was commissioned by the Council of Europe in 1997. The review was compiled using over 50 expert compilers who completed questionnaires covering 45 countries. The resulting RDB of European butterflies followed the new IUCN criteria as closely as possible but adapted them for use with the data currently available (van Swaay and Warren, 1999, 2006). A key adaptation was that the criteria for rates of change were applied over a 25-year period rather than the 10 years specified by the IUCN criteria, partly because this is a more sensible time frame for butterflies that often undergo large yearly fluctuations in population size and partly because it is easier for country compilers to provide a reliable assessment of trends over a longer period. Because the level of information varies considerably between countries, data quality was ranked from 1 to 4.

The review found that 71 (12%) of Europe's 576 butterfly species were threatened, many due to their rapid rate of decline across the continent. These included 19 of Europe's 189 endemic species, which were consequently classified as globally threatened. A further 43 species were classified as Near Threatened because of their rates of decline and many more were found to be declining across substantial parts of their range. Taken together, these results demonstrated a major crisis in Europe's butterfly fauna and a comprehensive action programme was recommended (van Swaay and Warren, 1999; Warren and van Swaay, 2006). The results are significant because they are the first comprehensive data available for an insect group across Europe, and suggest that similar serious declines are likely to be occurring in other insect groups. Many countries have since used the analysis to revise their own conservation priorities in a global and European context.

6.2 Prime butterfly areas of Europe

As a follow-up to the RDB of European butterflies, a project was instigated to identify some of the key areas (particularly those that could be targeted for priority conservation action) for 34 of the most threatened species. Using the same network of country experts, 433 sites were identified covering more than 21 million hectares, equivalent to 1.8% of the land area of Europe (van Swaay and Warren, 2003, 2006). The results highlighted that over half of these vitally important sites were not protected and that many of the target species were still declining within those that had been protected. The results are being incorporated into other initiatives to plan conservation strategies across Europe, such as the pan European and the Emerald Network being developed by the Council of Europe. This important development would not have been possible without the initial incentive and funding to produce the RDB.

6.3 Developing a European indicator for butterflies

The RDB of European butterflies has also been used to develop the first pan-European indicator for an insect group (van Swaay *et al.*, in press). The results show that, overall, butterflies have declined substantially across Europe at a rate of 11% reduction in distribution over the last 25 years. The distributions of 25 most 'generalist' species are declining only slowly (–1%) compared to specialist butterflies of grassland (–19%), wetlands (–15%) and forests (–14%). It should be noted that these losses have been calculated from changes in distribution and that population level losses are likely to have been even greater (Thomas and Abery, 1995; Warren *et al.*, 1997).

Equivalent pan-European data are currently not available for any other wildlife group apart from birds (Gregory *et al.*, 2005). Given the sensitivity of butterflies to environmental change (e.g. Warren *et al.*, 2001; Parmesan, 2003; Thomas, 2005), they are uniquely placed to provide a complementary indicator to birds to assess how Europe is performing against their target of halting biodiversity loss by 2010. Such a development has flowed naturally from the Red Listing process and the need to collate objective, quantitative data.

6.4 Formation of Butterfly Conservation Europe

The RDB for European Butterflies and Prime Butterfly Areas of Europe highlighted the need to take urgent and concerted action for butterflies at both the country and pan-European level. A new organization, Butterfly Conservation Europe, has since been formed as an umbrella group to coordinate and stimulate conservation action for Lepidoptera (see www.bc-europe.eu). It is hoped this will operate in a similar way to Birdlife International and Planta Europa, with partners in each country.

7 Conclusions

1. There are undoubtedly some serious limitations when applying the new IUCN criteria to invertebrates. The most serious of these are that there are too many species, many of them as yet not described, and too little data available to apply the very onerous and detailed new criteria. The biggest problem is undoubtedly in the tropics where other approaches, such as a focus on hotspots, need to be developed to identify priorities (e.g. Samways, 2005).
2. Nevertheless, the rationale behind the new IUCN criteria are sound and are still a valuable tool for focussing conservation action in countries with a better known invertebrate fauna.
3. The new IUCN criteria have been used to compile Red Lists in several countries around the world, at least for some better known invertebrate groups. The enormous interest in biological recording in many countries will enable better lists to be produced in future in many more countries and hopefully for more taxonomic groups.

4. Red Lists should be based on the precautionary principle, using the best data available to inform conservation action to address the serious loss of invertebrate biodiversity. A good example is the use of the principles of the new IUCN criteria to help identify conservation priorities within the UK BAP.

5. Red Lists have been compiled successfully for butterflies within Europe and have been extremely useful in raising awareness of the decline of this group and have stimulated much needed conservation action, as well as the formation of a new umbrella organization, Butterfly Conservation Europe.

6. Red Lists and BAP lists have identified species that have spear-headed crucial conservation issues for invertebrates, such as the importance of dead wood (and within that dead, wet wood, etc.).

7. The need for objective data to compile Red Lists has provided a welcome drive for detailed recording schemes on invertebrates, so that better information will be available in the future to target conservation action more closely.

8. More data are becoming available to help raise the profile of some groups, for example butterflies, which can help raise awareness of the general plight of invertebrates.

9. Red Lists and BAP priority lists are a very useful political tool for raising awareness of issues amongst policy makers, as well as the general public. People can relate more easily to species issues and the fascination of invertebrates can be a great advantage in gaining publicity for neglected species and neglected issues.

10. Red Lists should be regarded as just one way to help guide invertebrate conservation and we should use these lists as far as possible while recognizing their limitations, especially when applied to less known taxa in less known regions of the world. Priority Species lists are developed from Red Lists to perform a rather different function: to address urgent conservation problems following massive habitat degradation during the twentieth century. Although species and habitat loss still remain high, conservationists will continue to use the latter to prioritize their efforts.

Acknowledgements

We would like to thank the following for funding surveys and action programmes: Council of Europe, Dutch Ministry of Agriculture, Nature Management and Fisheries. Dutch Butterfly Conservation and Butterfly Conservation (UK) (European Red Lists and Prime Butterfly Areas); Countryside Council for Wales, English Nature and Scottish Natural Heritage (Action for Butterflies and Moths programmes); the Royal Society for the Protection of Birds, the Wildlife Trusts, Buglife – The Invertebrate Conservation Trust, and Butterfly Conservation (UK) (Action for Invertebrates Programme); and the Environment Agency, RSPB, and the Joint Nature Conservation Committee (review of invertebrates in the UK BAP).

References

Asher, J., Warren, M., Fox, R., Harding, P., Jeffcoate, G. and Jeffcoate, S. (2001) *The Millennium Atlas of Butterflies in Britain and Ireland*. Oxford University Press, Oxford.

Bowen, C.P. (ed.) (2003) *Proceedings of the Second Pan-European Conference on Saproxylic Beetles; Held at Royal Holloway, University of London, June 2002*. People's Trust for Endangered Species, London.

Bratton, J. (ed.) (1991) *British Red Data Books. 3: Invertebrates other than Insects*. Nature Conservancy Council, Peterborough, UK.

Butchart, S.H.M., Stattersfield, A.J., Baillie, J., Bennun, L.A., Stuart, S.N., Akçakaya, H.R., Hilton-Taylor, C. and Mace, G.M. (2005) Using Red List Indices to measure progress towards the 2010 target and beyond. *Philosophical Transactions of the Royal Society of London Series B – Biological Sciences* 360, 255–268.

Conrad, K.F., Woiwod, I.P., Parsons, M.S., Fox, R. and Warren, M.S. (2004) Long-term population trends in widespread British moths. *Journal of Insect Conservation* 8, 119–136.

Conrad, K.F., Warren, M.S., Fox, R., Parsons, M.S. and Woiwod, I.P. (2006) Rapid declines of common, widespread British moths provide evidence of a potential insect biodiversity crisis. *Biological Conservation* 132, 279–291.

Department of the Environment (1994) *Biodiversity: The UK Action Plan*. HMSO, London.

Department of the Environment (1995) *Biodiversity: The Steering Group Report*. HMSO, London.

Dunn, R.R. (2005) Modern insect extinctions, the neglected majority. *Conservation Biology* 19, 1030–1036.

Eaton, M.A., Gregory, R.D., Noble, D.G., Robinson, J.A., Hughes, J., Procter, D., Brown, A.F. and Gibbons, D.W. (2005) Regional IUCN Red Listing: the process as applied to birds in the United Kingdom. *Conservation Biology* 19, 1557–1570.

Fitter, R. and Fitter, M. (eds) (1987) *The Road to Extinction*. IUCN, Gland, Switzerland.

Fox, R., Spalding, A., Tunmore, M. and Parsons, M. (2005) Planning a new national macro-moth recording schemes. *British Journal of Entomology and Natural History* 18, 26–36.

Fry, R. and Lonsdale, D. (1991) *Habitat Conservation for Insects: A Neglected Green Issue*. Amateur Entolomogist's Society, Middlesex, UK.

Godfrey, A. (2003) A review of the invertebrate interest of coarse woody debris in England. *English Nature Research Reports* 513. English Nature, Peterborough, UK.

Gregory, R.D., van Strien, A., Vorisek, P., Gmelig Meyling, A.W., Noble, D.G., Foppen, R.P.B. and Gibbons, D.W. (2005) Developing indicators for European birds. *Philosophical Transactions of the Royal Society of London Series B – Biological Sciences* 360, 269–288.

Grootaert, P., Pollet, M. and Maes, D. (2001) A Red Data Book of empidid flies of Flanders (northern Belgium) (Diptera, Empididae S.l.): Constraints and possible use in nature conservation. *Journal of Insect Conservation* 5, 117–129.

Hawksworth, D.L. (ed.) (2001) *The Changing Wildlife of Great Britain and Ireland*. Taylor & Francis, London.

Hilton-Taylor, C. (2000) *2000 IUCN Red List of Threatened Species*. IUCN, Gland, Switzerland and Cambridge.

IUCN (1994) *IUCN Red List Categories*. Prepared by the IUCN Species Survival Commission. IUCN, Gland, Switzerland.

IUCN (2001) *IUCN Red List Categories and Criteria: Version 3.1*. IUCN Species Survival Commission. IUCN, Gland, Switzerland and Cambridge.

IUCN (2003) *Guidelines for Application of IUCN Red List Criteria at Regional Levels: Version 3.0*. IUCN Species Survival Commission. IUCN, Gland, Switzerland and Cambridge.

IUCN (2004) The 2004 IUCN Red List of threatened species: summary statistics. Available at: www.redlist.org/info/table/table1

IUCN (2005) Guidelines for the use of IUCN Red List categories and criteria, April 2005. Available at: www.iucn.org/themes/ssc/red-lists

Keller, V., Zbinden, N., Schmid, H. and Volet, B. (2005) A case study in applying the IUCN Regional Guidelines for National Red Lists and justifications for their modification. *Conservation Biology* 19, 1827–1834.

Middlebrook, I. (2000) Action for Invertebrates. *Biodiversity News* 12, 3.

Middlebrook, I. (2002) Action for invertebrates – a progress report. *Biodiversity News* 20, 8.

Middlebrook, I. (2005) Action for invertebrates – a progress report. *Biodiversity News* 32, 12–13.

Nature Conservancy Council (1989) *Guidelines for the Selection of Biological Sites of Special Scientific Interest*. Nature Conservancy Council, Peterborough, UK.

New, T.R. and Sands, D.P.A. (2004) Management of threatened insect species in Australia, with particular reference to butterflies. *Australian Journal of Entomology* 43, 258–270.

Parmesan, C. (2003) Butterflies as bioindicators for climate change effects. In: Boggs, C.L., Watt, W.B. and Ehrlich, P.R. (eds) *Butterflies – Ecology and Evolution taking Flight*. University of Chicago Press, Chicago, Illinois, pp. 541–560.

Regan, T.J., Burgman, M.A., McCarthy, M.A., Master, L.L., Keith, D.A., Mace, G.M. and Andelman, S.J. (2005) The consistency of extinction risk classification protocols. *Conservation Biology* 19, 1969–1977.

Samways, M.J. (2005) *Insect Diversity Conservation*. Cambridge University Press, Cambridge.

Shirt, D.B. (ed.) (1987) *British Red Data Books, 2: Insects*. Nature Conservancy Council, Peterborough, UK.

Smith, M.N. (2003) *National Stag Beetle Survey 2002*. People's Trust for Endangered Species, London.

Sutherland, W.J. (2000) *The Conservation Handbook: Research, Management and Policy*. Blackwell Science, Cambridge.

Thomas, C.D. and Abery, J.C.G. (1995) Estimating rates of butterfly decline from distribution maps: the effect of scale. *Biological Conservation* 73, 59–65.

Thomas, J.A. (1991) Rare species conservation: butterfly case studies. In: Spellerburg, I.F., Goldsmith, F.B. and Morris, M.G. (eds) *The Scientific Management of Temperate Communities for Conservation*. Blackwell Scientific, Oxford, pp. 149–98.

Thomas, J.A. (2005) Monitoring change in the abundance and distribution of insects using butterflies and other indicator groups. *Philosophical Transactions of the Royal Society of London Series B – Biological Sciences* 360, 339–357.

Thomas, J.A. and Morris, M.G. (1994) Patterns, mechanisms and rates of extinction among invertebrates in the UK. *Philosophical Transactions of the Royal Society of London Series B – Biological Sciences* 344, 47–54.

Thomas, J.A., Telfer, M.G., Roy, D.B., Preston, C., Greenwood, J.J.D., Asher, J., Fox, R., Clarke, R.T. and Lawton, J.H. (2004) Comparative losses of British butterflies, birds, and plants and the global extinction crisis. *Science* 303, 1879–1881.

van Helsdingen, P.J., Willemse, L. and Speight, M.C.D. (1996) *Background Information on Invertebrates of the Habitats Directive and Bern Convention*, Part I – *Crustacea, Coleoptera and Lepidoptera*. Nature and Environment Series No. 79, Council of Europe, Strasbourg, France.

van Swaay, C.A.M. and Warren, M.S. (1999) *Red Data Book of European Butterflies (Rhopalocera)*. Nature and Environment No. 99, Council of Europe, Strasbourg, France.

van Swaay, C.A.M. and Warren, M.S. (2003) *Prime Butterfly Areas in Europe: Priority Sites for Conservation*. National Reference Centre for Agriculture, Nature and Fisheries, Wageningen, The Netherlands.

van Swaay, C.A.M. and Warren, M.S. (2006) Prime butterfly areas in Europe: priority sites for conservation. *Journal of Insect Conservation,* 10(1), 5–11.

van Swaay, C.A.M., Warren, M.S. and Lois, G. (2006) Biotope use and trends of European butterflies. *Journal of Insect Conservation* 10, 189–209.

Warren, M.S. (2002) Species recovery work on butterflies and moths: an overview. In: Stone, D., Tither, J. and Lacey, P. (eds) *Species Recovery Programme: Proceedings of 10th Anniversary Conference, 2001*. English Nature, Peterborough, UK, pp. 25–30.

Warren, M.S. and van Swaay, C. (2005) Red Data Book of European butterflies – a pragmatic approach to target conservation action. In: Procter, D. and Harding, P.T. (eds) *Red Lists for Invertebrates: Their Application at Different Spatial Scales*. Report no. 367, Proceedings of *IN*Cardiff 2003. Joint Nature Conservation Committee, Peterborough, UK, pp. 61–70.

Warren, M.S., Barnett, L.K., Gibbons, D.W. and Avery, M.I. (1997) Assessing national conservation priorities: an improved red list of British butterflies. *Biological Conservation* 82, 317–328.

Warren, M.S., Hill, J.K., Thomas, J.A., Asher, J., Fox, R., Huntley, B., Roy, D.B., Telfer, M.G., Jeffcoate, S., Harding, P., Jeffcoate, G., Willis, S.G., Greatorex-Davies, J.N., Moss, D. and Thomas, C.D. (2001) Rapid responses of British butterflies to opposing forces of climate and habitat change. *Nature* 414, 65–69.

Wells, S.M., Pyle, R.M. and Collins, N.M. (1983) *The IUCN Invertebrate Red Data Book*. IUCN, Gland, Switzerland.

Wynne, G., Avery, M., Campbell, L., Gubbay, S., Hawkswell, S., Juniper, T., King, M., Newbery, P., Smart, J., Steel, C., Stones, T., Stubbs, A., Taylor, J., Tydeman, C. and Wynde, R. (1995) *Biodiversity Challenge*, Part 2. Royal Society for the Protection of Birds, Sandy, Bedfordshire, UK.

5 Species Conservation and Landscape Management: A Habitat Perspective

Roger L.H. Dennis,[1,2] Tim G. Shreeve[3] and David A. Sheppard[4]

[1]NERC Centre for Ecology and Hydrology, Monks Wood, Abbots Ripton, Huntingdon, Cambridgeshire PE28 2LS, UK; [2]Institute for Environment, Sustainability and Regeneration, Mellor Building, Staffordshire University, College Road, Stoke on Trent ST4 2DE, UK; [3]School of Life Sciences, Oxford Brookes University, Headington, Oxford OX3 0BP, UK; [4]Natural England, Northminster House, Northminster Road, Peterborough PE1 1UA, UK

1 Introduction

As long as there is a will for conservation and the resources for it, whatever these are (cash, volunteers, legacies, government-assisted schemes), there will be controversy about the direction conservation should take. Currently, there are three prominent issues vying for choice: (i) the species versus 'habitat' approach; (ii) the 'habitat' (= patch) versus entire landscape approach; and (iii) the single (= rare) species versus multispecies approach. Some choices, as the focus of attention, have already become redundant. For instance, the single patch (i.e. habitat) versus the multiple patch issue (i.e. single large or several small – SLOSS) has largely been resolved within metapopulation models and empirical findings in favour of multiple integrated patchworks (McCarthy and Lindenmayer, 1999; Ovaskainen, 2002; McCarthy *et al.*, 2005). Other choices or ploys fall within the compass of the three issues identified above (e.g. use of indicator taxa; the role of landscape heterogeneity; bias of attention to specialists or rare species versus generalists; resources for change; uniformitarianism versus catastrophism in management). Each of these issues has some independence. After all, there is a big difference between conserving for single versus multiple entities. But, perhaps what has not been realized with any degree of clarity is just how these different approaches are closely tied up with one another. They all depend on how habitat is envisaged and defined. To illustrate this, a useful starting point is to consider the difference between species and 'habitat' approaches in conservation, based largely on the work carried out on butterflies.

2 Species or 'Habitat' Approaches to Conservation?

There is a constant debate between the relative merits of species conservation and habitat conservation. Is it better to concentrate on individual species be they red kites, lady's slipper orchids or field crickets? Or, should we be looking after 'habitats', such as grassland or heathland with the implicit assumption that the latter will facilitate persistence of constituent and dependent species?

The debate exists because there is, currently, no clear distinction between the two approaches. Management for a target species inevitably involves whole sites or at least the appropriate parts of sites. The continued presence of the target species then becomes a measure of the success of the prescribed management for an assemblage of species. If the objective of management is the continued presence of a 'habitat' in its own right, the habitat is defined in species terms and the continued presence of a selected number of those species is used as a measure of management success. In the general understanding of the term habitat, these are simply two ways of looking at the same situation. But, are these approaches actually interchangeable and, if not, does it matter? Here, we explore a resource-based definition of habitat to demonstrate that there are important differences between them. We expand on this to illustrate what this means for the multispecies as opposed to a single species case, and argue that conservation cannot ignore the intervening ground between so-called habitat patches, the matrix. We also argue that conservation must move in emphasis from a focus on single species on prime patchworks to multispecies assemblages over the entire landscape and manage for change, not stasis.

To understand the distinctions between a species approach and a 'habitat' approach, we first need to understand what is meant by habitat. The habitat has long been treated as the basic unit for both theoretical developments and practical applications in the ecology and population biology of organisms (Elton, 1966; Southwood, 1977). However, developments have been frustrated by inconsistencies of definition and treatment of habitat (Hall *et al.*, 1997). Although descriptions typically refer to an identifiable locality or to the environment (e.g. topography, soils, vegetation types) and its subdivisions (i.e. microhabitats), practical guidance to the recognition of an organism's habitat has been lacking. Consequently, habitats have been described with a lack of precision (Rosenzweig, 1995). Habitat is most frequently regarded as being synonymous with a vegetation category or biotope. This is often entirely wrong as we explain below; vegetation associations can be described at a hierarchy of levels (Rodwell, 1991 et seq.) and species often extend over a number of distinct vegetation types regardless of scale (Dennis *et al.*, 2003, 2006).

In relation to the British Isles, confusion is increased by synonymizing of the habitat with vegetation assemblages (e.g. the Joint Nature Conservation Committee (JNCC) Phase 1 Habitat Classification; JNCC, 2003); in turn, the latter is increasingly equated with the national vegetation classification (NVC). The NVC scheme (Rodwell, 1991 et seq.) involves sampling a large number of 'representative' vegetation stands over a given time period; the resultant data are subject to multivariate techniques and lead to the grouping of stands to produce 'type' vegetation categories. Because these 'types' or categories represent the average composition of grouped stands they do not themselves represent actual vegetation assemblies, as in reality, patches of vegetation are classified to categories by their percentage similarity. A major drawback of this is that the original data were gathered over a set time period and categories may not be representative of potential future vegetation assemblages given the current and predicted future environmental changes and differential species' dynamics. Equating species' habitats to constructs is not necessarily the best practice, especially when some site management plans have the goal of changing current vegetation to average (ideal) categories of the NVC. A further serious problem is that a key plant for a phytophagous insect may not invariably be present in the NVC category.

2.1 Just what is a habitat?

So just what is a habitat? In all empirical and theoretical population studies habitat is implicitly or explicitly a bounded space (e.g. den Boer and Reddingius, 1996; Hanski and Gilpin, 1997). The fundamental problem with this is that it is often unclear what this space comprises, especially when it may be defined by the presence of a single resource, such as a host plant for a phytophagous insect, without reference to other essential resources. As a habitat is necessarily the location where an organism lives out its life cycle it should be possible to map the bounds of a habitat in terms of life-history requirements (Dennis and Shreeve, 1996; Dennis *et al.*, 2003). The approach we have taken is to regard species as requiring a set of resources and conditions in order to function; a convenient way of categorizing such resources for arthropods is under each stage of the life cycle. For example, an adult butterfly or moth would minimally require resources for egg laying, mate location, resting, roosting, feeding and predator escape. Other stages can be treated similarly and their resources mapped. The habitat is then the logical extension of this reasoning, defined by the intersection and union of these resources (Fig. 5.1; Dennis and Shreeve, 1996), the links being forged by flights of adults and movements of larvae. The resources required by each stage may be visualized as belonging to two groups, consumables (i.e. host plant parts, adult food) and utilities. The latter describe the conditions for existence and persistence, such as physical sites for various activities (e.g. sites for thermoregulation, mate location and pupation), and suitable conditions for development and activities (i.e. suitable local climates and microclimates), and enemy-free space. This latter group of resources, so well appreciated in bird and mammal ecology (e.g. Lahaye *et al.*, 1994; Lindenmayer, 2000), is often ignored in habitat definitions of arthropods. For example, the recent expansion of the butterfly *Hesperia comma* L. (Hesperiidae) in southern England adequately demonstrates that the conditions under which host plants can be exploited vary with temperature (Davies *et al.*, 2005). Early surveys, in cool conditions,

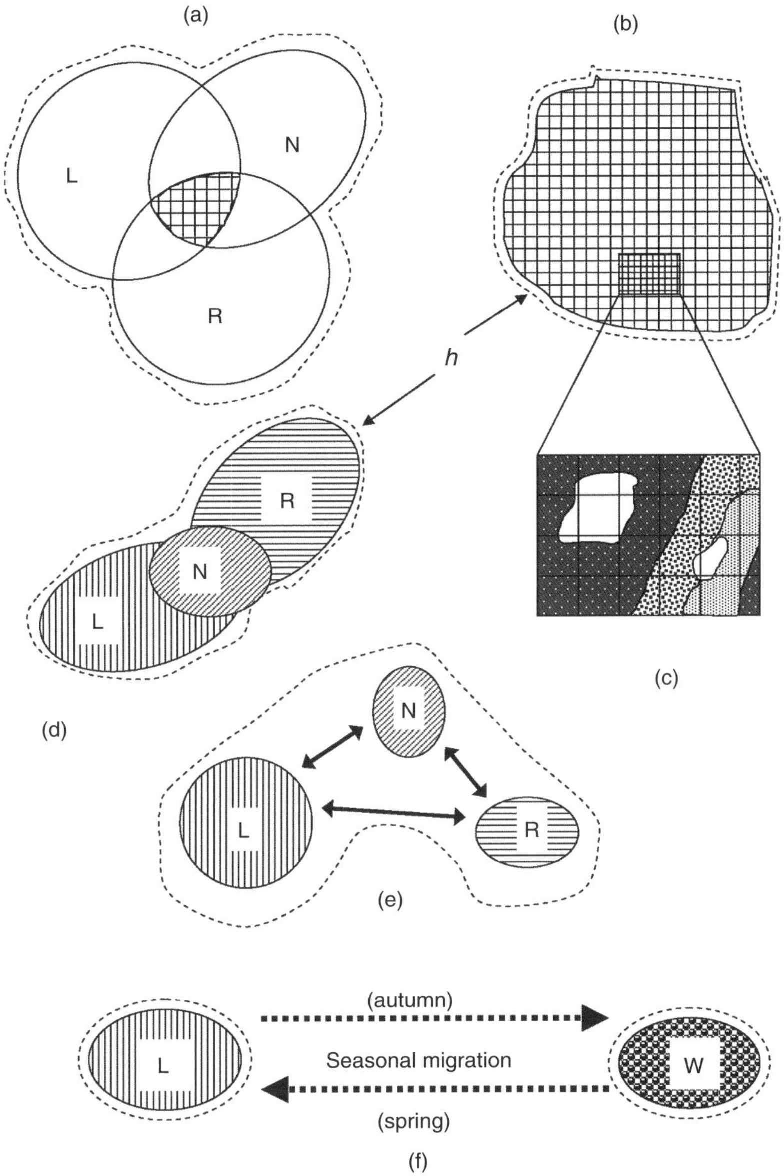

Fig. 5.1. The habitat model based on resource distributions and individual movements (see formal treatment in the text and in Dennis and Shreeve, 1996). For simplicity, resources are shown schematically as sets or envelopes; the elements of sets are arbitrary units of ground space based on fine-scale responses of individual butterflies, e.g. 1 m units, as illustrated by the grid in (c). A maximum of three resources are illustrated in each diagram: N, nectar resource; L, larval resource; R, roost sites; W, over wintering sites; *h*, habitat boundary. Resources are combined by daily search flights to give habitats by: (a) intersection and union; (b) equivalence and equality (e.g. *Cardamine pratensis* (L.) Hitt. is a host plant, nectar source and roost substrate for *Anthocharis cardamines*); (d) contiguous union; (e) disjointed union linked by back and forth daily flights; and (f) disjointed non-union linked by seasonal migration. Shading is used to distinguish resources. (c) illustrates a small part of (b) in which light shading is host plant, medium shading density is nectar, heavy shading is jointly nectar and host plant and white is neither but can be used for other activities, such as pupation, adult resting, etc. The habitat core is illustrated by cross shading. Distinctions are typically made between explorative (so-called trivial) flights within habitats and direct linear movements between habitats (Van Dyck and Baguette, 2005) but see Fig. 5.3.

produced a restricted set of relatively high temperature conditions (short turf on mainly south facing slopes) where females would lay eggs, and this set of conditions was used to define the habitat of the butterfly. Expansion of site occupancy was in part accompanied by the butterfly occupying areas earlier identified as non-habitat. These areas included longer turf and north and east facing aspects. A resource-based approach, in which microclimate is included, means that changing environmental conditions have transformed resources in the matrix to high-quality resource areas (= patch). It is not hard to envisage that further changes could transform some high-quality areas to lower-quality resource areas (= matrix). A functional definition of habitat is thus a practical solution to Hutchinson's concept of a hyperdimensional niche (Whittaker *et al.*, 1973); habitat describes real ground conditions (e.g. occupied space), whereas niche formulates biological space (vectors of influential agents).

A habitat, then, comprises the collection of resources required by, and accessible to, individuals of a species at a location. More formally, as a strict guide to practice in the field, habitat is the intersection and union of necessary complementary resources for an organism, linked by stage-to-stage movements of most individuals (>95%), to ensure an intrinsic rate of population increase ≥1. When defined in this way, habitats for most species will be discernible from the overlap or contiguity of resources and the movements of adults and larvae, searching for, visiting and returning to distinct resource zones (Fig. 5.1). However, the correct identification of a suite of resources with differing spatial attributes as an area that can potentially support a population can only be confidently determined with an understanding of the capacity of individual organisms to move within these areas. Empirical work has been done to test the capacity for determining habitat bounds in some butterflies (Vanreusel and Van Dyck, 2007; see Dennis *et al.*, 2006). For the majority of arthropod species there are two basic problems. First, there is virtually no information on their resource needs; second, there is a lack of information on their capacity to move between resource sets. Yet, obtaining this information is critical for the species approach and for understanding how they can persist in locations with particular resource distribution patterns. There are further problems of habitat identification associated with changing conditions. Most butterfly species have a greater capacity for movement when conditions are calm, solar radiation loads are high and temperatures warm (Dennis, 1993; Dennis and Bardell, 1996; Dennis and Sparks, 2006). Thus, an area with diffuse resources may be suitable under ideal conditions for activity, but not in marginal conditions. This behaviour may be matched by other arthropods, but there are likely to be important exceptions involving species (e.g. aphids, ballooning spiders) being transported to great distances by strong winds.

Defining a habitat on the basis of resource requirement and movement does not require that the resources either overlap or be compact (Fig. 5.1e and f). Admittedly, there are situations where, because of the ubiquitous and diffuse nature of resources and enormous movements involved, the habitat is extremely difficult to map. This situation is epitomized by pierid butterflies, such as *Pieris brassicae* L., *P. rapae* L. and *P. napi* L. (Dennis and Hardy, 2007), where search flight is demonstrated to occur in a wide range of what were thought to be thoroughly unpromising biotopes (Table 5.1). Seasonal movements create similar difficulties in determining habitat bounds (e.g. the butterfly *Gonepteryx rhamni* L., Pieridae) (Pollard and Hall, 1980). Other circumstances arise, where lifetime movements for whole cohorts of individuals are unique, and the geography of the organism changes seasonally on a vast scale with outbreaks in different regions (e.g. butterfly *Vanessa cardui* (L., Nymphalidae)) (Dennis, 1993). In these situations, it is questionable whether a habitat model applies at all, one reason why we focus on resources rather than on an abstraction, habitat, based on them.

Table 5.1. Frequency of direct flight, resource seeking and resource using behaviour of pierid butterfly species in biotopes within Greater Manchester, UK during summer 2005. (From Dennis and Hardy, 2007.)

Species	% Activity and host plant[a]	Garden	Urban	Pasture	Arable	Wood	Tall herb	Waste	Scrub	Total
Pieris brassicae	Direct flight	21.3	50.8	88.9	50.0	8.8	12.3	20.0	0.0	87
	Resource seeking	21.3	7.9	11.1	16.7	38.6	26.3	0.0	0.0	76
	Resource using	57.3	41.3	0.0	33.3	52.6	61.4	80.0	0.0	175
	Host plant	6.7	15.9	16.7	83.3	15.8	19.3	20.0	0.0	338
	Total[b]	75	63	18	6	57	114	5	0	338
Pieris rapae	Direct flight	38.0	50.0	30.0	9.7	6.8	5.0	6.5	29.6	89
	Resource seeking	24.0	27.6	0.0	45.2	34.1	28.9	26.1	37.0	144
	Resource using	38.0	22.4	70.0	45.2	59.1	66.2	67.4	33.3	263
	Host plant	8.9	13.8	30.0	100.0	29.6	45.8	69.6	85.2	496
	Total[b]	79	58	10	31	44	201	46	27	496
Pieris napi	Direct flight	4.6	40.0	14.3	0.0	1.5	3.4	0.0	0.0	13
	Resource seeking	25.0	0.0	21.4	0.0	40.3	35.1	0.0	25.0	103
	Resource using	70.4	60.0	64.3	0.0	58.2	61.5	0.0	75.0	192
	Host plant	2.3	20.2	7.1	0.0	31.3	51.2	0.0	50.0	308
	Total[b]	44	5	14	0	67	174	0	4	308

[a]For ease of comparison, figures provided are percentages except the totals for actual numbers.
[b]Summed totals for species: *P. brassicae* 338, *P. rapae* 496 and *P. napi* 308.

2.2 How does habitat relate to vegetation classes and biotope?

When habitat is regarded as being synonymous with a vegetation category or biotope (Fig. 5.2a), the outcome is that a habitat (or metapopulation patch) is mapped as if a vegetation unit or a biotope. Although this can arise, usually because of severe landscape fragmentation (e.g. an abandoned field corner cut off by road construction; Fig. 5.2b), it is often highly inappropriate (Dennis *et al.*, 2003, 2006). Arthropod species often extend over a number of distinct vegetation types regardless of scale, as well as being variably incident and abundant in the same vegetation type, again regardless of scale. This situation has recently been described for *Plebejus argus* L. (Lycaenidae) on the Great Ormes Head, a 3 × 2 km headland in North Wales (Dennis, 2004b; Dennis and Sparks, 2006); the butterfly not only occupies shorter turf areas (<15 cm) of calcareous grassland (NVC CG1 and CG2 categories; see Table 5.2) occupied by its host plants (*Helianthemum* spp. and *Lotus corniculatus* L.) in association with ants of the genus *Lasius* (Thomas, 1985; Thomas and Harrison, 1992), but also adjacent areas of scrub, which it uses for adult feeding (e.g. *Cotoneaster* spp., *Rubus* spp.), mate location, thermoregulation, daytime resting in cool, windy and cloudy conditions and roosting (Fig. 5.2c); these scrub areas also have small pockets of host plant that are used for egg laying.

The co-occurrence of species in a vegetation type or biotope in one location but lack of association within the same vegetation assembly in another location within the same climatic region (e.g. butterflies *P. argus* L., *G. rhamni* L., *Euphydryas aurinia* Rottemburg, Nymphalidae) indicates the inadequacy of defining habitats by vegetation or biotope alone. In the case of *P. argus* in North Wales it occurs on acid heath and mossland, but does not occupy similar areas over boulder clay on the Great Ormes Head adjacent to the calcareous grassland even where this has host plants used in the other biotopes, as well as host plants occurring on the calcareous grassland, such as *L. corniculatus* L. (Thomas, 1985). Despite this, there is continued adherence to regarding habitats as occurring in distinct vegetation patches and the treatment of such patches as being uniform in composition, all to the detriment of extinction risk assessment and management.

2.3 Are the species approach and 'habitat' approach interchangeable?

Having determined what a habitat comprises (resources and utilities) we can return to the question whether a species approach and a 'habitat' approach are interchangeable. In the case where a habitat is defined on the basis of linked resources they are interchangeable, whereas in the situation where a habitat is regarded as synonymous with a vegetation unit they are not. In a species approach to conservation, based on essential resources, this would require identifying and mapping the resources; such exploited resources, linked by individual movements, identifies the habitat and can be spread over a number of vegetation classes and biotopes (Fig. 5.2c). A habitat (= biotope or vegetation unit) approach would not necessarily include all the complementary

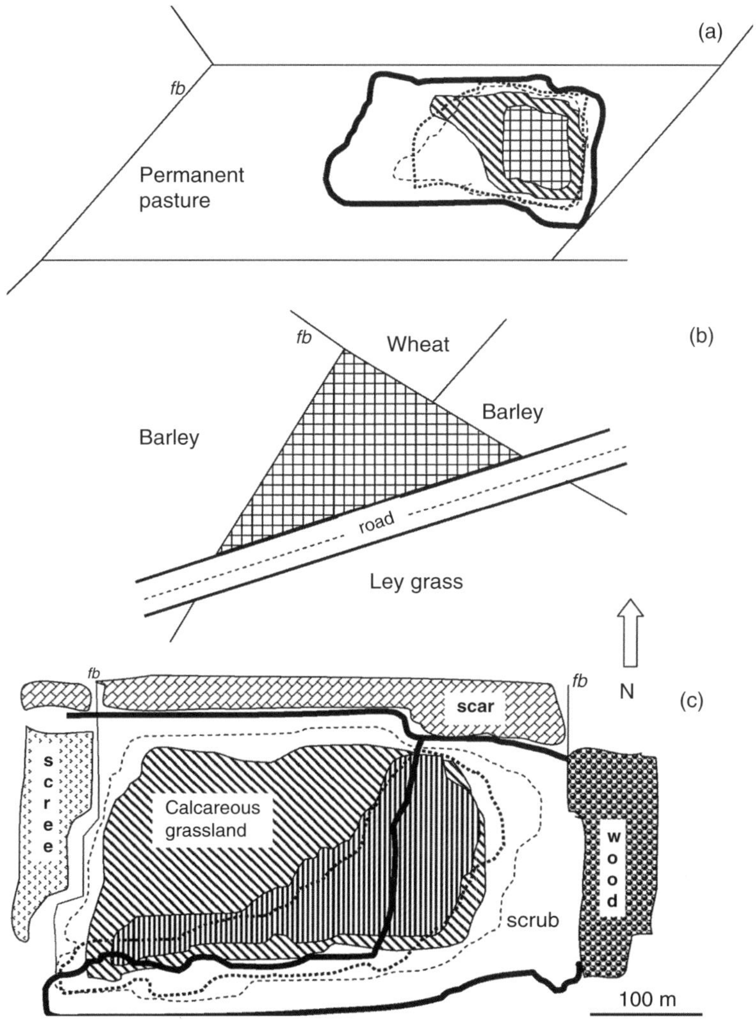

Fig. 5.2. Relationships between resource distributions, forming habitats for butterflies, vegetation units and field boundaries. (a) Resources coinciding with but occupying less area than both the vegetation unit (valley mire) and the field boundary; e.g. *Euphydryas aurinia* Rott., Nymphalidae. (b) Resources coinciding exactly with vegetation unit (tall herb grassland from abandoned arable land) and field boundary; e.g. *Maniola jurtina* L. Satyridae. (c) Resources overlapping vegetation units but within old field boundaries; e.g. *Plebejus argus* L. Lycaenidae. Lines and shading: thin lines marked *fb* = field boundaries; thick continuous lines = vegetation boundaries; diagonal shading = host plant; horizontal line shading = host plant suitable for egg laying and larval development in period of study; pecked line = area of nectar sources; dotted line = area of mate location, cross hatching, coincidence of suitable host plants, nectar sources and conditions for mate location.

resources for a target species and therefore not envelop the entire habitat for the species. It may, of course, include both resources and entire habitats for other species, whereas the precise demarcation of habitat in a species approach for any target species is less likely to coincide with habitat bounds of another species. These may, at first, appear to be fine distinctions as the two approaches will inevitably involve some overlap of ground. But, the difference is one of precision about resources, an understanding of habitat and ultimately the part played by the wider landscapes that becomes particularly relevant when multispecies conservation is considered.

2.4 New methodologies for delimiting habitats

When the focus is limited to single species, the first challenge in conservation is to find out what the insect's habitat really comprises. In most cases, despite the species being known for a couple of centuries, the habitat, in any meaningful sense, is still not known (Dennis *et al.*, 2003; Dennis, 2004a). An interesting example is provided by *Carabus intricatus* L. (Coleoptera: Carabidae). In the British Isles this Red Listed beetle only occurs in deciduous woodlands in steep-sided valleys, preferably running south-west, containing flowing water, having an annual rainfall in excess of 150 cm, atypically grazed by sheep and occupied by the tree-dwelling slug, *Limax marginatus* Müller. This is very precise information, but *C. intricatus* does not feed solely on *L. marginatus* or even follow the slime trails up and down the tree trunks, even though the beetle does spend a lot of time on the trunks and lower branches of the trees. Neither does the beetle or the slug appear to eat the sheep dung nor interact in any way with the sheep, but the beetle does not occur in woods from which grazing has been excluded. So, although it is now possible to define the type of sites that the beetle needs and the management that has to be in place, and conservation action has been put in place to secure these, it is still not possible to explain the resource requirements of this species (Boyce and Walters, 2001); behavioural links could usefully be sought with the modification of vegetation structure (e.g. dispersal). What the study has done is to identify surrogate markers for environmental conditions and resources used by the beetle and apply these to the proposed management.

When targeting single species, what we need are procedures for directly identifying habitats. Mapping of habitats is made easy where resources coincide and correspond to vegetation units. This is only likely to happen when biotopes have been so degraded (e.g. lowland northern Europe) that what is left comprises small parcels (patches) of semi-natural vegetation amidst intensively used farmland, patches that can be easily mapped (Fig. 5.2b). In most cases resources will be in parcels of different sizes and shapes, isolated from one another by, potentially, non-resource zones, a situation illustrated in Fig. 5.1e. In small areas it is possible to map the resources directly and to assess their use either by direct measurement of movements applying within site mark-release-recapture (MRR) techniques (Henderson, 2003) or by following individuals (e.g. Cant *et al.*, 2005).

As potentially suitable areas of biotope increase in size for a target organism, and thus the areas of potential resources for it, this direct approach becomes impracticable. In these situations, identification of habitats requires a two-tier process. First, smaller areas of study are required to identify a set of an organism's resources within vegetation zones and to determine the capacity for the species to move between resource outlets. Second, a broader scale-mapping programme of vegetation zones and resources within vegetation units is required. Mapping of habitats is then based on the conjunction of resources buffered with daily movements. Hence, it becomes possible to delineate functional habitat units or 'patches' that do not necessarily reflect physical patches or homogeneous zones in terms of vegetation. Moreover, functional units of invertebrate habitat may cover different vegetation types relating to different requirements (e.g. roosting in trees and foraging in nectar-rich grassland). The recognition of the spatial scale at which different resources form functional units depends on our understanding of the behavioural ecology of movements and hence of resource tracking by individuals. In view of both spatial and temporal variability in the occurrence of resources used, it is necessary that a full understanding of a species' habitat is based on autecological studies in different settings and in different conditions. For butterflies, this approach has so far only been carried out on one species *Callophrys rubi* (L.) Lycaenidae in Belgium's National Park Hoge Kempen (Vanreusel and Van Dyck, 2007; see Dennis *et al.*, 2006). Although this study did not apply all potential resources to habitat delineation, it indicates a suitable methodology for determining habitat based on resources and movements in line with a resource-based definition of habitats.

3 Focusing on Patches or an Entire Landscape Approach?

The issue about a species versus 'habitat' approach, whether the habitat is envisaged as resources or biotope, ignores crucial issues in conservation. One is that survival depends on extensive patchworks not just single patches – now well founded on metapopulation models and empirical testing of metapopulations (e.g. Thomas *et al.*, 1992; Hill *et al.*, 1996; Hanski and Gilpin, 1997; Mennechez *et al.*, 2003; Wilson *et al.*, 2002). A second is that single species or 'habitat' approaches often ignore the intervening matrix, the resources and structures in the wider landscape (Dennis *et al.*, 2003, 2006). This second issue is also ignored by metapopulation models with its focus on patches, as indicated below. The third issue is that a single species approach versus the 'habitat' approach confuses single as opposed to multispecies maintenance as the focus of management objectives.

It is important to appreciate that although explicit habitat delineation, to the extent that it can ever be adequate, may provide immediate local solutions for single species, it does not extend to establishing the impact of matrix components on that species and it has nothing to say, in terms of habitat, for other organisms, either arthropods or other taxa. For the former, we need to move to a resource-based view of the entire landscape and for the latter, we must necessarily find ways of measuring habitat suitability for whole assemblages and communities. The following sections deal with these points.

3.1 Metapopulations: how distinct are patch and matrix?

The focus on patchworks, patch and matrix, developed with metapopulation modelling (Gilpin and Hanski, 1991; Hanski and Gilpin, 1997; Ehrlich and Hanski, 2004). This is an extension of island biogeography to terrestrial situations in which dynamic homeostasis (equilibria) is envisaged between colonization and extinction within a patchwork of potential habitat units (Levins, 1969, 1970; Gotelli and Kelley, 1993). In these models population size is largely equated with patch size and isolation with distance across the matrix between patches. Increasingly, attention has been given to the quality of patches and matrix; the former has been demonstrated to influence population incidence, in some cases more than either patch size or isolation (Dennis and Eales, 1997, 1999; Thomas *et al.*, 2001; Matter *et al.*, 2003; Valimaki and Itamies, 2003), and the latter equally profoundly to influence transit of individuals (Dover *et al.*, 2000; Roland *et al.*, 2000; Ouin *et al.*, 2004). Quality has been regarded by the proponents of metapopulation models as being subsumed in patch area (Nieminen *et al.*, 2004), but this is clearly not the case, substantiated by firm examples (butterflies *Coenonympha tullia* Müller, Satyrinae; Dennis and Eales, 1997, 1999; *Parnassius* spp., Matter and Roland, 2002; Auckland *et al.*, 2004); patch area makes no reference to the composition and structure of resources comprising habitat patches nor to the connectivity amongst resources *within* patches (Dennis *et al.*, 2003, 2006). The emphasis has also been primarily on a single consumable resource, larval host plants. In exactly the same way, the matrix has been treated as if sea, without content or structure. This is one reason why traditional metapopulation models may reasonably refer to patchworks of habitat but not to more sophisticated topologies, such as networks of habitat.

To the extent that a patch (= habitat) is not a discrete, homogeneous entity, i.e. where there is virtual 1:1 correspondence between resources and a vegetation unit typically a host plant with a NVC unit (Rodwell, 1991 et seq.), there will be difficulty in distinguishing patch from matrix. Such situations are atypical of industrial farming systems (e.g. East Anglian cereal farmland, UK), which generate landscape simplification and severe fragmentation. Normally, where environmental conditions are described as 'semi-natural', the sheer difficulty in establishing rules for determining habitat patchworks is all too clear and commented on elsewhere (Dennis *et al.*, 2006); matrix is an extension of non-resource space *within* habitats (Fig. 5.1c). Metapopulation modellers and empiricists have to face up to two uncomfortable axioms. The greater the fraction of the complement of resources that make up habitats used to define them, inevitably the more resource types and elements will be found in the matrix. But, the fewer the resource types that are used to define habitat bounds, the more resources that should be included within the habitat space will be allocated to the matrix. Metapopulation modellers, whether they limit their definition of a habitat to a single resource or encompass the entire complement in the process, will find that they have resources defining a habitat for the organism dispersed throughout what they categorize as matrix. Suddenly, purely by changing our view of what is or not a habitat, the matrix becomes a zone of resources. Another truth is that the fewer the resources used to define a habitat patch the smaller it will be; simultaneously, the bigger becomes the matrix around it, and the more likely it contains resources. Thus, patch dimensions become truncated. The reason for these artefacts in studies is that arthropods, notably butterflies, which have been extensively studied, use different substrates and vegetation structures for different activities, particularly utilities compared to consumables (Dennis *et al.*, 2003).

There are also problems associated with scale and effort. Human observers have a tendency to filter out small items, simply because of inaccessibility or lack of resources for fine-scale surveys (Dennis *et al.*, 2006). Units or packages below a certain size tend to be ignored. But, this is what is distinctive about consumable resources, if not utilities, within the matrix: the resources are often in small, even tiny, pockets and they are disparate. Among these resources may be found host plants in the right condition for exploitation by a target species, but they are difficult to pick up on survey if only because of access, limitations of search time and numbers of surveyors compared to the area being covered. In plots of patch area against isolation, account is rarely made of small resource (host plant) patches <0.01 ha and certainly not patches of 1^{-04} ha (0.0001 ha or 1×1 m^2) (Thomas *et al.*, 1992; Hanski and Thomas, 1994; Lewis *et al.*, 1997; Baguette *et al.*, 2003). Yet, these can occur in matrix contexts for oligophagous species (e.g. *P. argus* on Great Ormes Head, North Wales; R. Dennis, personal observations) and are much more likely for species that exploit a wide range of larval host species (e.g. butterflies *Maniola jurtina* L.; *Pyronia tithonus* L.; Dennis, 2004a; *Pieris* spp.; Dennis and Hardy, 2007). In the case of many grass feeders we know very little about host plant preferences (but see *Pararge aegeria* L.; Shreeve, 1986; *Lasiommata megera* L., Nymphalidae; Dennis and Bramley, 1985) and even less about their other highly complex resource requirements and are thus more likely to misinterpret the role of matrix resources.

Small resource elements and items in the matrix are frequently regarded either as below the scale to which insects respond or, if used, as inflicting a cost, slowing movement and acting as sinks in reproduction (Pulliam, 1988). However, in the business of defining patch and matrix, very little attention is actually paid to what arthropods perceive and respond. Pertinent questions are: Do arthropods actually experience the environment polarized as patch and matrix or as landscape with variable resource distributions? How much does size matter when it comes to resource recognition and use? Empirical studies of just what arthropods do in landscapes can reveal how behaviour is related to biotopes, vegetation units and substrates. It is expected that, with a typical habitat (= vegetation patch) model, movements will be of two basic types: routine searching, sinuous flights and direct linear flights (Van Dyck and Baguette, 2005). Direct linear flights will dominate what is supposed to be the matrix. However, two studies on butterflies (e.g. *M. jurtina*; *P. tithonus*; Dennis, 2004a; *Pieris* spp.; Dennis and Hardy, 2007) reveal that they treat the matrix as comprising resources. In the matrix – defined either on the basis of traditionally accepted unsuitable biotopes or very low density of the target organism – they engage more in resource searching and resource using than they do in direct linear flights, typical of dispersal and escape (Table 5.1). However, pierid butterflies switch between direct linear flight and search flight in response to resource cues across a variety of substrate or vegetation surfaces (Fig. 5.3). Close attention to resource attributes indicates that movements are affected by more than just inter-patch distances, by at least five aspects of resource geography and timing (Fig. 5.4). This lies at the root of the occurrence of both types of movements observed in *Pieris* spp. across landscapes; frequent switches between the two types are expected (Fig. 5.3). The key is that consumer resources, if not utility resources, in the matrix will tend to be in smaller lots, scattered and different in composition (type) and structure (shape). Host plants occur in the matrix but are often so small that they are missed by human observers. But, they are found and used by arthropods even when they are not visible to the observer (e.g.

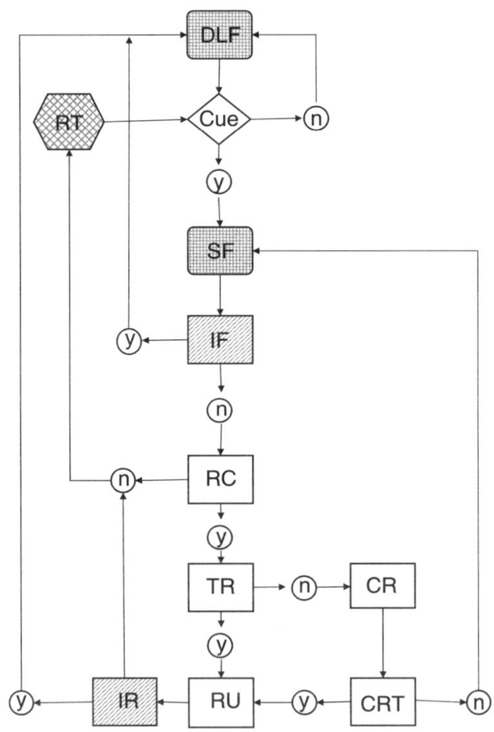

Fig. 5.3. Flowchart of suggested switches in flight behaviour in response to resource cues in *Pieris* butterflies within both habitats and matrix. Boxes: hexagon/diamond shading, RT = resource targeted; round cross-shaded squares: DLF = direct linear flight and SF = search flight; white squares: resource variables, TR = targeted resource, CR = complementary resource, CRT = complementary resource targeted, RU = resource use; diagonal shaded boxes: IF = interaction in transit (with butterfly or predator), IR = interaction on resource; round boxes: connectors, 'y' yes or 'n' no; diamond box, proximate 'cues' including visual and scent stimuli triggering switch in flight types.

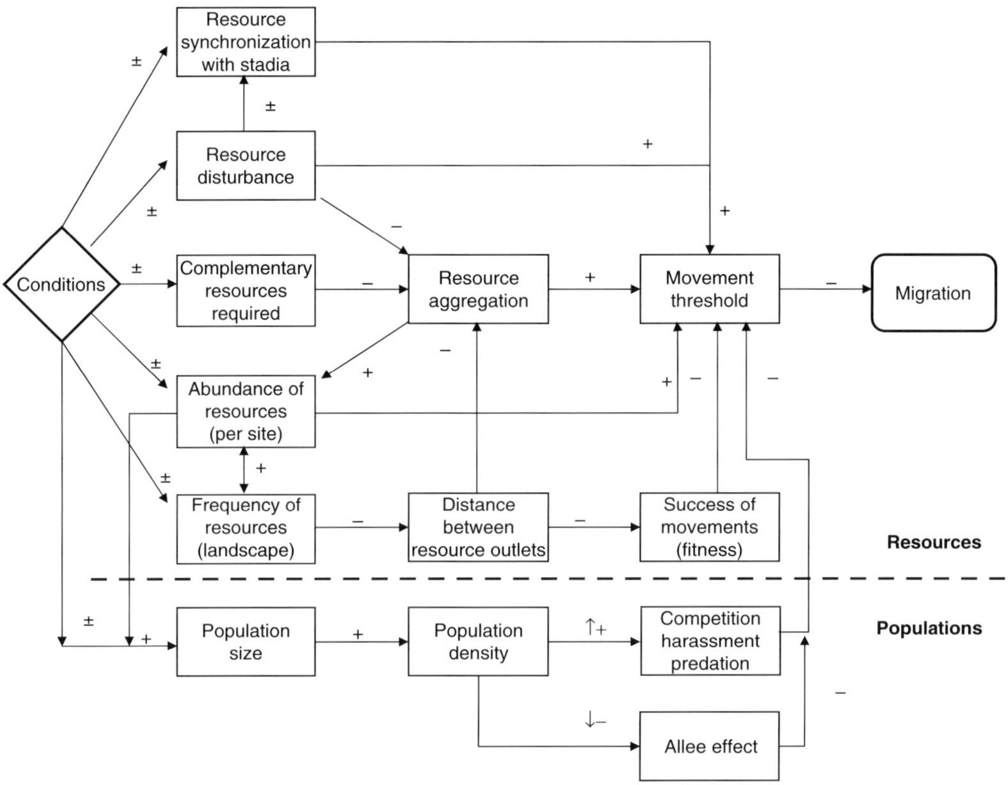

Fig. 5.4. Resource variables inducing movement and migration in butterflies. Initial conditions include a variety of agents (e.g. vegetation succession, human management, weather and climate). This has impacts on five basic attributes of resource distributions that, in turn, over different space–time frames influence the tendency to move over the landscape. Resource disturbance refers to vegetation changes (Grime, 1974) and host plant dynamics (i.e. generation time). Complementary resources refer to non-substitutable resource outlets (Dunning *et al.*, 1992). Some degree of dependence occurs among the resource variables (illustrated). No attempt is made to expand on individual (e.g. lifespan) and population influences on movement and migration distances in this simple process-response model other than to indicate a link to resources, nor on the direct influence of conditions (e.g. weather; Dennis and Bardell, 1996; Dennis and Sparks, 2006). In Baker's (1978) initiation factor model, the probability of an individual initiating migration depends on its migration threshold being exceeded. According to this model, both population density and environmental conditions interact to effect mobility, including changes to resources generated by individual resource use (e.g. the ideal free distribution; Calow, 1999).

host plants in *Pieris napi*; Courtney, 1988). There is a serious lack of data on interactions between behaviour and matrix components, yet such data are critical to understand the potential of the matrix and the functioning of species in industrial farming landscapes. There is an urgent need to know more about scales of resources used and distances over which resource elements can be sensed and the part played by vision and olfaction in resource tracking (Vane-Wright and Boppré, 1993; Cant *et al.*, 2005).

Patch and matrix bounds are further confounded by temporal changes (Wiens, 1996; Thomas and Kunin, 1999) and spatial (regional) variation. Just what appears to be a habitat patch changes on scales of seconds to decades. Those engaged in conservation practice are constantly faced with successional changes on sites, as well as changes in conditions induced by human activities (Sheppard, 2002; Offer *et al.*, 2003; Underhill-Day, 2005). Change is integral for sites and in the following sections we argue for conservation to be geared to managing dynamics. Change on fine timescales has important implications for recognizing just what resources are important for organisms; there are also inevitable implications for habitat recognition. The habitat space used by *P. argus* (Lycaenidae) on the Carboniferous limestone headland of the Great Orme (North Wales) oscillates upslope and across slope from the vicinity and shelter of scrub with changes in sunshine, temperatures and wind speeds; the warmer the conditions, the larger the area used, and the response is immediate on weather changes (Dennis and Sparks, 2006). Such changes are known for other invertebrate taxa (e.g. *Oedipoda caerulescens* L., Orthoptera, Acrididae; Maes *et al.*, 2006). Changes with weather and seasonal conditions have also been recorded in mate location surfaces and elements in *Inachis io* (Nymphalidae) at three different spatial scales: the landscape, the surface substrates and in relation to microfeature topography (Dennis, 2004c; Dennis and Sparks, 2005). Seasonal shifts are well known in vegetation and biotope occupancy in both sexes of *P. aegeria* (Shreeve, 1984, 1985, 1987) and for vegetation and host plant use by different generations of the butterfly *Polyommatus bellargus* Rottemburg (Lycaenidae) (Roy and Thomas, 2003). Distinction of habitat and matrix has also to contend with regional changes in apparent resource use, involving substantial shifts in biotope occupancy (e.g. *P. napi*; Dennis, 1977; *Anthocharis cardamines* L. Pieridae; Courtney and Duggan, 1983). Changes in resource use, including movements and resource tracking are under evolutionary change and not static (Dennis, 1977; Merckx and Van Dyck, 2002; Van Dyck and Baguette, 2005). All this frustrates the distinction of habitat and matrix for management based, as it is, on limited resources. At the very least it means that attention should be given to the matrix (surrounding landscape) and the resources it can usefully provide, as well as the obstacles it presents, for a target species.

3.2 New principles and practices for arthropod landscapes

The question is: How are we to proceed with a resource-based approach that takes in the complete landscape rather than a typical metapopulation patch-based approach that treats most of the landscape as sea? Stages in advance are not difficult to envisage. A first stage would be to measure resource attributes of the patches, as has been done (Dennis and Eales, 1997, 1999; J.A. Thomas *et al.*, 2001). A further stage would be to include measures on the matrix.

Adding measures on quality (composition and structure) to patches, to complement that of patch area, can be done through very simple extensions of explicit metapopulation modelling and field trials. The topology remains the same as that of isolated patchworks. A step-up, but retaining this patchwork model, can be accomplished by further parameterizing isolation based on the matrix components. Movement is no longer perceived as an isotropic, simple negative exponential or negative power function applied to the entire patchwork. Instead, it has direction, changes of direction thus increased length, and explicit values tied into explicit attributes of links (e.g. barriers, flyways, etc.) and resource effects. This approach is required even of organisms dependent on patchworks that have distinct patch boundaries, such as pond arthropods, plants and amphibians (Kirchner *et al.*, 2003; Briers and Biggs, 2005; Smith and Green, 2005). To achieve advances from this, we have to expand on the patchwork model to include a greater variety of target patches (polygons in geographic information system (GIS) language) or move to raster-based data approaches, which cover the entire surface (Longley *et al.*, 2001). A start can be made with broad statistical approaches, searching for factors that affect species diversity, and incidence and population size in species as has been done for butterflies (Dover, 1996, 1997; Dover *et al.*, 1997, 2000; Dover and Sparks, 2000). Some revelations have been disclosed in response to what may be considered trivial features, for instance, the impact of a tape drawn across a cereal field on dispersal (Dover and Fry, 2001). Any advance on these approaches necessarily must abandon the patchwork topology and consider new ones involving networks and interactions among points, line and surface phenomena (viz., edges, vertices, disconnected vertices and edges), as geographers did years ago for population interactions among humans (Haggett, 1965). It is not difficult to envisage modelling species landscapes in a raster context where the effect of landscape attributes are combined; each cell can be quantified in terms of production (natality), losses (mortality) and transfers to adjoining cells (immigration, emigration, transfer direction), all based on resources (Fry, 1995).

The next question is what kind of information do we require for these increasingly sophisticated landscape models? Habitats and matrix impact on species' life history strategies and population dynamics through three distinct aspects of their resource distributions: composition, physiognomy and connectivity (Dennis *et al.*, 2006; R. Dennis, unpublished data). Composition refers to the occurrence (or absence) of a specific resource or resource component (i.e. one or several nectar sources or host plants) and the variation in its make-up and context. Context relates to the conditions in which a resource may occur, all of which affects its quality and therefore exploitation by individuals in a population; for instance, one vital resource – a host plant, may occur in wetter or drier conditions, more basic or more acidic substrates, grow in the open or shade of trees and clear or overtopped by other vegetation. Composition also includes density, frequency and abundance of a resource, though the latter two attributes are probably better dealt with under physiognomy. Resources varying in composition occur for all life history stages or phases of activity and, at a particular site, specific resources may be: single or multiple (e.g. number of host plants or nectar plants used); restricted or unrestricted (e.g. limited to parts of host plant used or access to the whole of it); singular or transferable (i.e. a resource used by a single stage or several stages or phases); main or subsidiary (e.g. host plants as primary, secondary, unsuitable and novel (Wiklund, 1981)).

Physiognomy refers to the geography of a resource patch. Each resource patch or the array of patches can be described in terms of: location, both absolute in terms of coordinates (x, y, z) and relative to other resource patches; height (altitude or elevation); size, for instance, length and breadth (area); shape, from circular to linear; orientation; slope; fragmentation and comminution (the degree to which resource elements and individual resource patches are broken up by other resource or non-resource types; frequency of patches) and, contagion, whether random in distribution, over-dispersed or clustered (aggregated). As such, resources can be described as any other geographical feature and at different scales. Connectivity refers to the potential links between resource elements. Connectivity can be described in terms of: overlap, contiguity (contact, neighbourhoods) and isolation, and barriers and obstacles. Connectivity involves more than resource geography as it depends on the mobility of all life history stages.

Although we have some understanding of how these three aspects of resource distributions can influence the incidence and population status of target organisms they are rarely considered in any study, little or no consideration is given to how they change and impact in time, and they are ignored in the management of a target species. Two basic approaches have been applied to understand how 'resource' items in both habitat and the matrix influence butterfly biology. The first is feature-oriented, in which the focus of attention is the feature and insect behaviour is studied in response to it (Dover and Sparks, 1997; Dover *et al.*, 1997; Dover and Fry, 2001). The second is insect-oriented, in which behaviour is monitored in response to the surroundings – to biotopes, vegetation, surfaces and substrates used (Dennis, 2004a; Dennis and Hardy, 2007). Both are statistical approaches based on relative frequencies and actual occurrences. Both these approaches have successfully identified substantial responses to unexpected features and 'resources' and suggest that single species conservation is still highly dependent on autecological study. It is simply unacceptable that metapopulation studies continue to be based on host plant-based patchworks alone or ignore the matrix.

4 Single Species versus Multispecies Conservation

Although it is manifestly feasible to conserve (e.g. Field Cricket, *Gryllus campestris* L. Orthoptera: Gryllidae; Edwards *et al.*, 1995) or restore single species (e.g. large blue butterfly, *Maculinea arion* L., Lycaenidae; Thomas *et al.*, 1998), there are simply too many species and insufficient human resources to allocate to each the same attention. Worldwide there are estimated to be at least 8 million insect species and perhaps as many as 30 million (Erwin, 1982; May, 1990, 1992; Stork, 1993; Gaston and Hudson, 1994; Samways, 2005). Single species studies will not resolve current extinction rates of between 0.4% and 5% per century (Hambler and Speight, 2004; Thomas, 2005), let alone predictions of 15–37% extinction by 2050 with climate change (Thomas *et al.*, 2004).

Without denying the importance of a single species focus, current losses and predicted losses in arthropod diversity generate a greater urgency to find appropriate directions for conserving biodiversity. A large part of this process is to understand the place of a resource-based concept of habitat, the role of indicator species to help focus human resources, and the need to develop general principles for conservation given current and predicted future environmental changes.

4.1 Rarity and flagship species

It can be argued that as conservation is an entirely human-resourced percept, it follows that there are sound, logical reasons for conserving any single species. If there is a will to do so and resources are freely generated for the purpose – an immediate implication being that conserving the species carries no cost to an indifferent wider public – then there is no reason why action should be negated. There are much stronger reasons, and usually more public support, when the species is important to us economically or because of emotional ties (e.g. a flagship species, such as the spider *Dolomedes plantarius* Clerk, Pisauridae in the British Fenland) especially when coupled to human phylogenetic proximity (e.g. Great Apes). A major criterion in the selection of species for autecological investigations or targeted conservation management is usually a rarity or at least perceived rarity. The single species approach tends to be a rare species approach, actual or threatened. Over the last 15 years, the English Nature Species Recovery Programme (SRP) has been attempting to address the conservation issues affecting a selection of species from a variety of taxa, not just invertebrates. Species were selected for action based on a series of criteria but principally they were species whose survival in England was thought to be under severe threat. This criterion, understandable though it is, brings with it a fundamental problem in that such rare species are often difficult to find. Occasionally, this may actually be the reason for their rarity, but rare species are not that uncommon; they are not unexpected from frequency distributions (Odum, 1963). Rare species can become common with environmental changes much as common species can become rare. In fact, rarity is often defined in terms of relative abundance or range sizes (see Gaston, 1994) and each study often conveys its own definition. Gaston (1994) prefers the quartile definition, a cut-off of 25% (the first quartile of the frequency distribution of species abundance or range sizes); the unfortunate implication is that the more species there are in any taxon, the more rare species whose future there will be to contemplate. Some might prefer different definitions of rarity for each of spatial cover, abundance and biotope affiliation – these can be related but can also be distinct. A different kind of rarity is based on extinction risk, perceived threat measured in decline: becoming rarer or simply becoming rare; many such schemes exist (International Union for Conservation of Nature (IUCN), Davis *et al.*, 1986; see Munton, 1987). The point is, not which definition of rarity is appropriate, as each faces us with a formidable load of rarity, but how do we cater for rare species when there are apparently so many of them. Is it very different from dealing with biodiversity generally? The implication is obvious that the few rare species that do get attention or receive the resources are regarded as special to us in some way. More resources will be found, very likely, for a magnificent large blue butterfly than a Hymenopteran parasite (e.g. *Ichneumon eumerus* (Wesmael) (Hymenoptera: Ichneumonidae) a parasite of *Maculinea rebeli* Hir., Lycaenidae; Hochberg *et al.*, 1996) living on it, which can be rarer still.

So what of the many rare species that lack attention? Do we actually know much about them? Even in the UK after more than 200 years of published entomological research (Barnard, 1999), it is amazing how little is known about the British insect fauna. Most textbooks and a lot of detailed studies fail to provide the basic information (resources and environmental conditions) needed to implement conservation action. It is not unusual to read that the 'habitat' of a particular rare species is 'grassland, heathland, sand dunes and some urban post-industrial sites' – and that it is still considered to be rare. What is really meant is that the insect's habitat occurs in grassland, heathland, sand dunes and some urban post-industrial sites but that the actual resources, the habitat, remain unknown. It is perhaps no understatement that the habitats of most of our British insects are still not known – and this is the most thoroughly studied insect fauna in the world. It is on this foundation of ignorance that we begin the conservation of these species. Consequently, a single species approach is expensive since we first have to acquire adequate data on resources and conditions to implement a programme of management. Where this is feasible, the approaches discussed above become relevant. The inevitable cautionary note is that cutting corners by using surrogate resource data for patchworks augurs long-term failure (see below) and to manage biodiversity and multispecies systems different approaches are needed. A prerequisite is to understand habitat in a multispecies context.

4.2 The habitat in a multispecies context: one species' matrix is another species' habitat

As species have distinct resources, their habitats will be unique and their habitat bounds are likely to differ as a consequence, the more so as resources differ. Even for species sharing the same basic resource types, such as phytophagous insects exploiting the same host plant, there can be fundamental and striking differences in micro-resource requirements that affect distributions and habitat suitability (e.g. saproxylic Coleoptera: Cerambycidae and Diptera: Syrphidae; Fayt *et al.*, 2006). A classic cautionary tale is illustrated by the light conditions required for creating microstructures by larvae of the butterfly *Limenitis camilla* L. (Nymphalidae) on *Lonicera percylmenum*, very different from those required by the moth *Hemaris fuciformis* L. (Sphingidae) on the same plant (Fox, 2005). The probability of species having different habitat bounds increases when their resources fail to overlap or to intersect (discontinuous union). The structuring of resources and their connectivity can potentially affect congruence in habitat bounds as much as does composition (e.g. use of same host plant but in different settings; Gutiérrez *et al.*, 2001). The probability of species' habitat congruence is further reduced if their resources are associated with different vegetation units, as distinctions in vegetation units infer quantum shifts in resource types. One highly important generalization emerging from these percepts is the realization that one species' matrix may well contain another species' resources. As an axiom for conservation practice it may be somewhat less accurately but usefully restated to make the point as: one species' matrix is another species' habitat.

This can be tested. To provide an insight the example of the Great Ormes Head in North Wales (Cowley *et al.*, 2000, 2001) can be taken. This headland has already been used as a template for mapping the metapopulation patches of several butterfly and moth species (e.g. Thomas and Harrison, 1992; Lewis *et al.*, 1997; Gutiérrez *et al.*, 1999; León-Cortés *et al.*, 2000, 2003). Key amongst them is that of *P. argus* occupying calcicolous grassland (specifically parts of NVC categories CG1 and CG2; Stevens *et al.*, 1995; see Table 5.2). A table of other NVC categories and land use types illustrates that one butterfly species or another has key resources in virtually every other vegetation or substrate unit (Table 5.2), including bare rock, mining spoil, scree and cliffs (*Hipparchia semele* L. Nymphalidae; Dennis, 1977), and minor components, such as hedges, verges and surface excavations (e.g. *P. tithonus*, *M. jurtina*; R. Dennis, personal observations). Substrates thought to be of little inherent value none the less have a role: such are walls of buildings in built up areas used by nymphalids (e.g. *Vanessa atalanta* L. and *V. cardui* L.) for thermoregulation and territorial perches, and intensively (sheep) grazed pastures walled off as farmland used as breeding sites for other nymphalids (e.g. nettle patches for *Aglais urticae* L. and *I. io* L.; thistle patches for *V. cardui*). Even the most disturbed and eroded biotope, the grassland summit subject to severe human trampling, is used for mate location by hill topping species (Dennis and Dennis, 2006). The few vegetation and substrate units that could arguably be regarded as not forming parts of butterfly habitats are most certainly habitats for other organisms (e.g. a children's play area on the summit has clumps of *Marrubium vulgare* L. (Lamiaceae) for the monophagous plume moth *Wheeleria spilodactylus* (Curtis) (Pterophoridae) (Menéndez and Thomas, 2000)); the sea cliffs exposed to sea spray are invaluable nesting sites for a variety of sea birds (e.g. *Rissa tridactyla*, *Phalacrocorax carbo*) with their associated beetle, fly and flea faunas (A. Fowles, personal communication) as is dense scrub for insectivorous birds (e.g. stonechats), which is also used as daytime shelter by the moth *Idaea dilutaria* (Hübner) (Geometridae). As it is, ensembles of Lepidoptera sharing the same host plant in the calcareous heath (e.g. *Helianthemum* larval feeders, such as *P. argus*, *Aricia agestis*, *Adscita geryon* (Hübner) (Zygaenidae), and *L. corniculatus* larval feeders, such as *Erynnis tages*, *Polyommatus icarus* and *Zygaena filipendulae*; R.J. Wilson and C.D. Thomas, unpublished data; Gutiérrez *et al.*, 2001) have very different distributions, indicative of differences in resources. Recent research has demonstrated that essential resources have

Table 5.2. Vegetation and other substrates on the Great Ormes Head, North Wales, UK and breeding resources for butterflies occupying the Carboniferous limestone headland.

Vegetation and substrate class[a]	Butterfly species[c]
CG1 *Festuca ovina–Carlina vulgaris* grassland	*O. venata*, ***C. croceus***, ***P. argus***, ***A. agestis***, ***P. icarus***, ***A. aglaja***, ***L. megera***, ***H. semele***, *P. tithonus*, *M. jurtina*, *A. hyperantus*, ***C. pamphilus***
CG2 *Festuca–Avenula pratensis* grassland	*O. venata*, ***C. croceus***, ***P. argus***, ***A. agestis***, ***P. icarus***, ***A. aglaja***, ***L. megera***, ***H. semele***, ***P. tithonus***, ***M. jurtina***, *A. hyperantus*, ***C. pamphilus***
CG6 *Avenula pubescens* grassland	***T. sylvestris***, *O. venata*, *C. croceus*, *A. agestis*, *P. icarus*, *A. aglaja*, *L. megera*, *H. semele*, ***P. tithonus***, ***M. jurtina***, *A. hyperantus*, *C. pamphilus*
CG10 *Festuca ovina–Agrostis capillaris–Thymus praecox* grassland	*T. sylvestris*, *O. venata*, *C. croceus*, *P. icarus*, *A. aglaja*, *L. megera*, *H. semele*, *P. tithonus*, ***M. jurtina***, *A. hyperantus*, ***C. pamphilus***
U4 *Festuca ovina–Agrostis capillaris–Galium saxatile* grassland	*T. sylvestris*, *C. croceus*, *P. icarus*, ***V. cardui***, *A. aglaja*, *L. megera*, *H. semele*, *P. tithonus*, ***M. jurtina***, ***C. pamphilus***
MG1 *Arrhenatherum elatius* grassland	***T. sylvestris***, ***O. venata***, *C. croceus*, ***P. icarus***, *V. cardui*, *A. aglaja*, *L. megera*, *H. semele*, ***P. tithonus***, ***M. jurtina***, ***A. hyperantus***, *C. pamphilus*
MG6 *Lolium–Cynosurus* grassland (semi-improved grassland; cemetery)	*T. sylvestris*, *C. croceus*, ***L. phlaeas***, ***P. icarus***, *V. atalanta*, ***V. cardui***, *A. urticae*, *I. io*, *L. megera*, *H. semele*, *P. tithonus*, *M. jurtina*, *C. pamphilus*
MC4 *Brassica oleracea* maritime cliff-ledge community	***P. brassicae***, ***P. rapae***, ***P. napi***
MC8 *Festuca rubra–Armeria maritima* maritime grassland[b]	*P. aegeria*, *H. semele*, *P. tithonus*, ***M. jurtina***, *C. pamphilus*
MC9 *Holcus lanatus* maritime grassland[b]	*T. sylvestris*, *P. aegeria*, ***L. megera***, *H. semele*
M24 *Molinia–Cirsium dissectum* fen meadow[b]	***O. venata***, ***M. jurtina***
H8 *Calluna vulgaris–Ulex gallii* heath	*O. venata*, *A. agestis*, *P. icarus*, *A. aglaja*, *H. semele*, ***P. tithonus***, ***M. jurtina***, ***C. pamphilus***
Brachypodium sylvaticum grassland[b]	***O. venata***, *P. aegeria*, ***L. megera***, *A. hyperantus*
H10 *Calluna vulgaris–Erica cinerea* heath	*T. sylvestris*, *P. icarus*, *A. aglaja*, *L. megera*, *H. semele*, ***P. tithonus***, ***M. jurtina***, ***C. pamphilus***
'CGH' calcicolous grass heath	*T. sylvestris*, *C. croceus*, *P. argus*, *A. agestis*, *P. icarus*, *A. aglaja*, *L. megera*, *H. semele*, ***P. tithonus***, ***M. jurtina***, ***C. pamphilus***
U20 *Pteridium aquilinum–Galium saxatile* community (dense bracken)	***T. sylvestris***, ***O. venata***, ***A. aglaja***, *H. semele*, ***P. tithonus***, ***M. jurtina***, *C. pamphilus*
Scrub (*Ulex europaeus*, *Rubus* spp.)	***O. venata***, *L. phlaeas*, ***P. aegeria***, ***P. tithonus***, *M. jurtina*
Woodland	*P. c-album*, ***P. aegeria***
Exposed rock (cliffs, crags, pavement, erosion scars, scree, quarries, rock walls)	*P. napi*, *C. argiolus*, *V. atalanta*, ***L. megera***, ***H. semele***

Table 5.2. *Continued*

Vegetation and substrate class[a]	Butterfly species[c]
Amenity (improved) grassland (playing fields; intensely used farmland)	***V. atalanta*, *V. cardui*, *A. urticae*, *I. io*, *P. c-album***
Urban and gardens	***G. rhamni*, *P. brassicae*, *P. rapae*, *P. napi*, *A. cardamines*, *C. argiolus*, *V. atalanta*, *V. cardui*, *A. urticae*, *I. io*, *P. c-album*, *P. aegeria***, *H. semele*, ***P. tithonus***
Hedges, ditches, verges, tracks, paths, banks, springs	*T. sylvestris*, ***O. venata*, *P. napi*, *L. phlaeas***, *C. argiolus*, ***A. urticae*, *I. io*, *P. aegeria*, *L. megera***

[a]Mainly UK national vegetation classification (NVC) categories mapped for the headland by D.G. Guest and S.L.N. Smith in 1994 (Countryside Council for Wales, Bangor; www.ccw.gov.uk/) (Stevens *et al.*, 1995); note, some do not entirely match NVC classes (e.g. CG2, H8, *Brachypodium sylvaticum* grassland).
[b]Vegetation units covering small areas.
[c]Butterfly species recorded having host plants in >50% quadrats. Bold, suitable breeding biotope, most supported by observations of egg laying and occurrence of both sexes (R. Dennis, personal observations). Nectar and utility resources not disclosed but ensure wider use of vegetation and substrate on the headland than listed (see text).

even been omitted from metapopulation patchworks for at least one of these species (e.g. contiguous shrubs and bracken are essential roosts, mate location sites and thermoregulation sites for *P. argus*; Dennis, 2004b; Dennis and Sparks, 2006). These few examples do not begin to impress the full extent of the observation, that there is not an organismal empty space on the headland that can be dismissed as an empty set. The more species there are, the more the entire landscape becomes relevant for conservation. All this is grist to the argument for moving to a resource-based view of landscape: merging patch and matrix.

There are, however, situations in which this variation in resource geography is so dramatically reduced that there is some excuse for considering a part of a landscape as an empty set, but this discounts restoration of the matrix (see below). Current fragmentation of landscapes, with intensive agricultural practices, leads to smaller patchworks and inevitable homogenization of vegetation units. What tends to get left behind is not an unbiased sample of the original vegetation or substrates but that which is least valuable for human exploitation. As semi-natural vegetation units become reduced in size and homogenized, there is increased probability that species will share much the same habitat bounds. But, there is still an issue of whether they share the same fine-scale substrates within single vegetation units, a level below any of the most detailed mapping programmes (e.g. 10 × 10 m). Each substrate or vegetation subunit has its own dynamics and congruence in habitat boundaries is not synonymous with identity in resource use and lifespan. There is clearly much to test with the new resource-based habitat definition in a multispecies context. As part of this, there is an urgent need to develop techniques for identifying resource use in numerous species

(Dennis, 2004a), as well as generating principles of resource impact on species other than area and isolation.

4.3 The biodiversity crisis: indicators of what and how to manage?

Management for conservation is typically faced with two quite similar problems; species to manage over a site(s) or a site(s) to manage for species. The question is, as habitats for organisms are unique and costly and time-consuming to determine, how do we account for them in management of whole faunas, even of rarer fractions on single sites. Although databases on organisms' habitats are essential for long-term maintenance (Shreeve *et al.*, 2001, 2004) – monitoring by broad-scale biotope is not informative – there is simply insufficient time before action needs to be taken on many sites; these are real pressures, evidenced by loss rates. In fact, there are no simple and 'clean' approaches; not knowing what and how to manage one is forced back on 'quick and dirty' solutions. These tend to be taxon (species)-orientated and/or feature (resource)-orientated.

Before giving any advice about a site in a taxon approach, extensive survey is required to determine the current invertebrate interest of the site. This would involve the identification of all of the insects to species and allocation to their true habitats. This does not happen because there is never enough time and there are not enough skilled entomologists. There is not sufficient knowledge about most of the British species. The fallback position is to use indicator taxa but the question is 'indicator of what exactly' and then what makes for a good indicator? The problem faced in delineating species' habitats in a multispecies context is that habitat bounds are very likely to differ for each species. It is improbable that a single rare species can provide a focal marker or an indicator of habitat bounds for a wide variety of other species, simply because a rare species, by definition, will have some resource(s) that is restricted (Leibig's law) and that is not shared by a large target group. Even so, they may be useful as an indicator for a particular guild or community based on a vegetation unit as in the case of some species selected for SRP action (Stone *et al.*, 2002). The habitat view of this is that the degree to which taxa differ in resources (and especially in resource distributions) the less able is one to act as an indicator of the habitat for the other. In situations where rare species reflect on conditions that need to be managed for a scarce faunal component with similar resource types and threats, a rare species may provide suitable indication on sites.

However, it is difficult in conservation to measure success based on the presence of rarities. They are too easy to miss. It is much more meaningful to base a monitoring scheme, designed to assess the success of conservation actions, on the presence of an assemblage of species, which more broadly represents the features of interest on that site. Success in conservation can be assessed by the proportion of registered species found during subsequent monitoring surveys. The absence of parts of the assemblage and the presence of additional species gives some idea of changes to the insect fauna of the site. This approach can provide an early warning of things changing, which is more than can be deduced from the apparent absence of a rare species.

Recently, Webb and Lott (2006) of English Nature describe an ambitious, but absolutely essential, programme to develop a habitat-based invertebrate assemblage classification system – invertebrate species information system (ISIS) – along these lines. The objective is to produce an invertebrate assemblage system for English terrestrial and freshwater systems for assessing the quality and conditions of sites for conservation. This approach is based on the expertise of a large number of specialists (Table 5.3); it recognizes the sheer scale of multispecies conservation, and the need to fuse botanists and entomologists into working partnerships. Systems such as this one can make good use of life history strategies. Life history features, particularly those linked to conservation strategy review (CSR), identify differences in dynamism

Table 5.3. Examples of assemblages of arthropods identified for UK biotopes.
(a) Assemblages identified that characterize the insect fauna of selected biotopes.

Biotope	Reference
Lowland calcareous grassland	Alexander, 2003
Wet grassland	Drake, 1998
Grazing marshes	Drake, 2004
Inundated wetlands	Lott, 2003
Seepages	Boyce, 2002
Acid mires	Boyce, 2004
Peatlands	Coulson, 1988
Living and decaying timber	Alexander, 2002
Coarse woody debris	Godfrey, 2003
Coastal shingle	Shardlow, 2001
Coastal soft cliffs	Howe, 2003

(b) Phylogenetically limited assemblages characterizing selected biotopes.

Biotope	Taxon	Reference
Peatlands	Carabidae	Holmes *et al.*, 1993
Peatlands	Araneae	Coulson and Butterfield, 1986
Moorlands	Carabidae	Gardner *et al.*, 1997
Mires	Tipuluidae	Salmela and Ilmonen, 2005
Scrub	Araneae	Rushton, 1988
Scrub	Carabidae	Rushton *et al.*, 1990
Woodland	Carabidae	Fowles *et al.*, 1999
Upland grassland	Araneae	Coulson *et al.*, 1984
	Arameae	Coulson and Butterfield, 1986
	Araneae	Luff and Rushton, 1989
	Opiliones	Coulson *et al.*, 1984
	Carabidae	Luff and Rushton, 1989
Saltmarsh	Hemiptera	Denno, 1977
Heathland	Carabidae	Van Essen, 1994
Riverine sediments	Staphylinidae, Carabidae and other beetle families	Sadler, *et al.*, 2004

and vulnerability to extinction. This was initially determined for plants and has now been extended to one group of phytophagous arthropods, butterflies (Hodgson, 1993; Shreeve *et al.* 2001; Dennis *et al.*, 2004); Food and/or resource specialists, stress tolerators, can be more sensitive to human disturbance and fragmentation than generalists in these features (Kitihara *et al.*, 2000; Steffan-Dewenter and Tscharntke, 2000; Dennis *et al.*, 2004; Stefanescu *et al.*, 2005).

An alternative approach, which is likely to become increasingly associated with using assemblage indicators, has been to focus on site features (i.e. vegetation units, specific substrates, microclimates) forming part of insect habitats considered to be rich in species or significant in some way with respect to their invertebrate fauna. Earlier, such links have been influenced strongly by the knowledge and experience of the entomologists involved. In this approach, the site is dissected into component parts, which hold different invertebrate interests and require different management. A range of multivariate ordination techniques is, of course, now available (e.g. Ludwig and Reynolds, 1988; ter Braak and Smilauer, 2002) that takes the guesswork out of linking species with resources, substrates and structures (Dennis, 2004b; Eyre *et al.*, 2004). One cautionary note; a habitat view would suggest that there is need to find all the features crucial for a target group of insects; discovery of clusters of a species at one time in one place does not ensure the presence of other vital resources not identified at the time of survey.

5 Managing for Threat, Diversity and Environmental Dynamics

Moving from 'what' to manage to 'how' and 'where' to manage raises a number of recurring issues in conservation: taxonomic interests, site dimensions and potential, site heterogeneity and dynamics, conflicts of interest on sites and the role of the wider landscape around sites. Just what implications these issues have for conservation depends on how habitats and the matrix are viewed.

5.1 Reconciling past, current and future interests of sites

Current interests and future potential of locations is, in part, dependent on their landscape context and past history. One thing they probably all have in common is that it is unlikely that there are adequate data on habitats to provide management with unequivocal site design for specific taxa. As such, management has to fall back to a broader base, that of substrates and structures, vegetation units and biotopes.

Some sites may be regarded as important for invertebrates because of species that have been found there in the past. This informs us of the past structure of the site, how it was managed and past landscape context, but may be of little meaning for the future. This certainly seems to be the case for species subject to metapopulation dynamics that now have lost their surrounding patchworks and have become restricted to single sites (e.g. butterfly *E. aurini;* Fowles and Smith, 2006). Unless we can manage the whole landscape we have to accept that this particular past interest of a site is lost. This should not stop us from trying to create a suitable landscape in the long term for some future recolonization event (Joyce and Pullin, 2003), particularly where opportunities arise for repairing the matrix and creating patchworks and networks.

It is not desirable to create an outdoor museum where things never change, though such changes cause understandable anguish (e.g. Waring, 2001) and attempts are made to cater for changes in management of sites (Kirby, 1992). Conservation cannot advance by way of attempted preservation only if changes on sites are inevitable. Insects are so successful because they exploit change. All sites go through changes on a range of timescales and these are part of normal site dynamics; without them many species will be lost. Of particular importance to many arthropod species is the maintenance of botanical and structural variability in time and space. For example, the richness of many semi-natural grassland areas is the result of variability of treatments in the past when the sites were of economic value. However, varying agricultural economics, landholder requirements and variable environments resulted in different stocking rates, grazing intensities, grazing times and periods and animal mixes, often over short time periods, contributing to structural and compositional diversity (Smith, 1980). Likewise, woodland management was rarely uniform (Rackham, 1980) with woodland lots undergoing varying management in relation to economic cycles and local demand variation. Such diverse use introduced heterogeneity at both site and landscape scale. In the absence of details on resource dimensions for almost all species and the clear evidence that more species are declining than improving in status, despite decades of conservation management, it is probably the promotion of this spatial heterogeneity that will do most to maintain the maximum resources for suites of species, simply because it is a strategy that is most likely to maintain a diversity of (unquantified) resources.

Just what can be achieved on sites will depend much on site dimensions and ownership. Conflict of interest is not inevitably linked to site dimensions. With increasing size of site there is greater likelihood of variety in substrate and vegetation units and therefore greater potential in catering for diversity, among this rarity. On smaller sites there is greater likelihood of homogeneity in vegetation and substrate; there may also be fewer species to cater for and greater coincidence of habitat *bounds*. Part of the challenge of invertebrate conservation is to develop achievable objectives to sustain, enhance or create invertebrate interests and to ensure that these are taken into account in site management planning (Offer *et al.*, 2003). Far more work is required to test what is most suitable for species in different circumstances and it is likely to become prominent with the development of evidence-based conservation (Pullin and Knight, 2001; Sutherland *et al.*, 2004). This may require bolder methods than those applied in the past, including substitutions for catastrophic events, to achieve objectives. A classic example has been the creation of bare ground through severe disturbance regimes for field crickets at the instigation of staff at the Invertebrate Conservation Centre in London Zoo (Edwards *et al.*, 1995).

5.2 Sites and landscapes, opportunities for the future

Changes to agricultural subsidies in the European Union (EU) (Common Agricultural Policy – CAP – reforms), including Environmental Stewardship, Countryside Stewardship, Environmentally Sensitive Areas, Farm Woodland Premium Scheme, Hill Farming Allowance, Organic Farming Scheme and Woodland Grant Scheme in the UK, combined with the formulation of Biodiversity Action Plans, Natura 2000 site designations and other statutory conservation area declarations, are perceived as vehicles to achieve the EU objective of reversing the decline in biodiversity by 2010. Increasingly, means are being made available to address environmental problems within Europe. More than 300 different policy measures are implemented in the member countries of the Organization for Economic Cooperation and Development (OECD) addressing biodiversity and landscape protection (Herzog, 2005). In the EU, farmland covers about 50% of the land surface and the proportion under agri-environment schemes has risen from about 15% in 1998 to 27% in 2001 and continues to rise (European Commission, 2003). Current practice with available subsidy is to try to target it to specific locations and landscapes, in attempts to increase the size of, or buffer, existing prime sites, or to achieve regional targets of increasing the area of specific biotopes. Often, this involves the practice of attempting to replace past biotopes in specific locations, with the tacit (untested) assumption that past landscapes were best. However, landscapes have always changed and the key point for such an approach is at what time was it best? In the absence of specific and rigorous tests that targeted conservation to particular locations on the basis of history and location does more to promote biodiversity at the landscape scale than conservation measures in random locations, we advocate the promotion of within-landscape heterogeneity.

In promoting heterogeneity, it is unwise to ignore the potential of the general matrix for at least three reasons: first, because resources are present in the landscape matrix for species (Dover and Sparks, 2000; Dennis, 2004a) and can be promoted in the matrix; second, because we know so little of its importance for so many species; third, because species clearly search for resources even in what is regarded as unprofitable biotopes (Dennis and Hardy, 2007). In addition, it is evident that one species' matrix is another species' patch. Therefore, the importance of the general landscape cannot be overlooked. We, therefore, advocate paying as much (if not more) attention to the restoration of the matrix as to the preservation and enhancement of specific sites. Promotion of landscape heterogeneity is more likely to facilitate species persistence in the face of climate change than focusing on specific locations, simply because it is most likely to maximize resource diversity (consumables and utilities) and heterogeneity.

6 Summary

Alternatives in conservation, choices in management, are highly dependent on the definition of habitat. The implications of a resource-based definition of habitat are explored for species conservation and site management. Habitat is defined as the co-occurrence of essential resources within the exploratory range of individuals. As such, habitats can extend both over different vegetation types and over different physical structures, or comprise subsets of vegetation classes, dependent on the scale of movement of the organism. Thus, habitats are not synonymous with vegetation units. Practical advice is given as to how to recognize habitats in the field. The role of habitat is explored in relation to three strategies in conservation: the species versus 'habitat' approach, the 'habitat' (= patch) versus entire landscape approach and the single (= rare) species versus multispecies approach. The role of indicator organisms is examined in the context of habitat and landscape changes. Arguments are made for moving towards a focus on multispecies and whole landscape conservation. Key issues to this end are: the resources occurring in the matrix; search by species for resources within the matrix; the complexity and diversity of resources used by species and the interchangeability of habitat and matrix for different organisms. We advocate management for substrate and resource diversity and dynamics as an immediate remedy to the biodiversity crisis current and in prospect; we urge the development of new techniques for determining species' resource use, the development of new principles of resource impact on species' populations beyond patch area and isolation and the development of more sophisticated spatial models than those currently based on metapopulation patchworks.

Acknowledgements

The Species Recovery Programme projects have been a partnership of commitment between organizations and individuals, too many to list in total but all deserving of the greatest praise. Our grateful thanks to Robert J. Wilson for allowing us to cite his unpublished work on the Creuddyn Peninsula, North Wales, to Adrian Fowles for access to unpublished data and to Keith Alexander and an anonymous referee for their most helpful comments.

References

Alexander, K.N.A. (2002) *The Invertebrates of Living and Decaying Timber in Britain and Ireland – A Provisional Annotated Checklist*. English Nature Research Reports No. 467. English Nature, Peterborough, UK.

Alexander, K.N.A. (2003) *A Review of the Invertebrates Associated with Lowland Calcareous Grassland*. English Nature Research Reports No. 512. English Nature, Peterborough, UK.

Auckland, J.N., Debinski, D.M. and Clark, W.R. (2004) Survival, movement, and resource use of the butterfly *Parnassius clodius*. *Ecological Entomology* 29, 139–149.

Baguette, M., Mennechez, G., Petit, S. and Schtickzelle, N. (2003) Effect of habitat fragmentation on dispersal in the butterfly *Proclossiana eunomia*. *Comptes Rendus Biologies* 326 (Suppl. 1), S200–S209.

Baker, R.R. (1978) *The Evolutionary Ecology of Animal Migration*. Hodder & Stoughton, London.

Barnard, P.C. (1999) *Identifying British Insects and Arachnids: An Annotated Bibliography of Key Works*. Cambridge University Press, Cambridge.

Boyce, D.C. (2002) *A Review of Seepage Invertebrates in England.* English Nature Research Reports No. 452. English Nature, Peterborough, UK.

Boyce, D.C. (2004) *A Review of the Invertebrate Assemblage of Acid Mires*. English Nature Research Reports No. 592. English Nature, Peterborough, UK.

Boyce, D. and Walters, J. (2001) The conservation of the blue ground beetle in south-west England. *British Wildlife* 13, 101–108.

Briers, R.A. and Biggs, J. (2005) Spatial patterns in pond invertebrate communities: separating environmental and distance effects. *Aquatic Conservation – Marine and Freshwater Ecosystems* 15, 549–557.

Calow, P. (1999) *Blackwell's Concise Encyclopaedia of Ecology*. Blackwell Science, Oxford.

Cant, E.T., Smith, A.D., Reynolds, D.R. and Osborne, J.L. (2005) Tracking butterfly flight paths across the landscape with harmonic radar. *Proceedings of the Royal Society of London B–Biological Sciences* 272, 785–790.

Coulson, J.C. (1988) The structure and importance of invertebrate communities on peatlands and moorlands, and effects of environmental and management changes. In: Thompson, D.B.A. and Usher, M.B. (eds) *Ecological Change in the Uplands*. Blackwell, Oxford, pp. 365–380.

Coulson, J.C. and Butterfield, J.E.L. (1986) The spider communities of peat and upland grasslands in northern England. *Holarctic Ecology* 9, 229–239.

Coulson, J.C., Butterfield, J.E.L., Barrett, B.I.P. and Harrison, S. (1984) The spiders and harvestmen of some peat and upland grassland sites in Yorkshire. *Naturalist* 109, 103–110.

Courtney, S.P. (1988) Oviposition on peripheral hosts by dispersing *Pieris napi* (L.) (Pieridae). *Journal of Research on the Lepidoptera* 26, 58–63.

Courtney, S.P. and Duggan, A.E. (1983) The population biology of the orange-tip butterfly *Anthocharis cardamines* in Britain. *Ecological Entomology* 8, 271–281.

Cowley, M.J.R., Wilson, R.J., León-Cortés, J.L., Gutiérrez, D., Bulman, C.R. and Thomas, C.D. (2000) Habitat-based statistical models for predicting the spatial distribution of butterflies and day-flying moths in a fragmented landscape. *Journal of Applied Ecology* 37, 60–72.

Cowley, M.J.R., Thomas, C.D., Roy, D.B., Wilson, R.J., León-Cortés, J.L., Gutiérrez, D., Bulman, C.R., Quinn, R.M., Moss, D. and Gaston, K.J. (2001) Density-distribution relationships in British butterflies. I. The effect of mobility and spatial scale. *Journal of Animal Ecology* 70, 410–425.

Davis, S.D., Droop, S.J.M., Gregerson, P., Henson, L., Leon, C., Villa-Lobos, J.L., Synge, H. and Zantovska, J. (1986) *Plants in Danger: What Do We Know?* International Union for Conservation of Nature and Natural Resources, Cambridge, UK and Gland, Switzerland.

Davies, Z.G., Wilson, R.J., Brereton, T.M. and Thomas, C.D. (2005) The re-expansion and improving status of the silver-spotted skipper butterfly (*Hesperia comma*) in Britain: a metapopulation success. *Biological Conservation* 124, 189–198.

den Boer, P.J. and Reddingius, J. (1996) *Regulation and Stabilization Paradigms in Population Ecology*. Chapman & Hall, London.

Dennis, R.L.H. (1977) *The British Butterflies. Their Origin and Establishment.* E.W. Classey, Faringdon, UK.

Dennis, R.L.H. (1993) *Butterflies and Climate Change*. Manchester University Press, Manchester, UK.

Dennis, R.L.H. (2004a) Butterfly habitats, broad-scale biotope affiliations, and structural exploitation of vegetation at finer scales: the matrix revisited. *Ecological Entomology* 29, 744–752.

Dennis, R.L.H. (2004b) Just how important are structural elements as habitat components? Indications from a declining lycaenid butterfly with conservation status. *Journal of Insect Conservation* 8, 37–45.

Dennis, R.L.H. (2004c) Landform resources for territorial nettle-feeding nymphalid butterflies: biases at different spatial scales. *Animal Biodiversity and Conservation* 27, 37–45.

Dennis, R.L.H. and Bardell, P. (1996) The impact of extreme weather on Great Orme populations of *Hipparchia semele* (Linnaeus,

1758) and *Plebejus argus* (Linnaeus, 1758) (Papilionoidea: Satyrinae and Lycaenidae): hindsight, inference and lost opportunities. *Entomologist's Gazette* 47, 211–225.

Dennis, R.L.H. and Bramley, M.J. (1985) The influence of man and climate on dispersion patterns within a population of adult *Lasiommata megera* (L.) (Satyridae) at Brereton Heath, Cheshire. *Nota Lepidopterologica* 8, 309–324.

Dennis, R.L.H. and Dennis, M.P. (2006) Hill-topping in British butterflies: incidence and cues in a cool, windy climate? *Entomologist's Gazette* 57, 17–20.

Dennis, R.L.H. and Eales, H.T. (1997) Patch occupancy in *Coenonympha tullia* (Lepidoptera: Satyridae): habitat quality matters as much as patch size and isolation. *Journal of Insect Conservation* 1, 167–176.

Dennis, R.L.H. and Eales, H.T. (1999) Probability of patch site occupancy in *Coenonympha tullia* (Müller) (Lepidoptera: Satryinae) determined from geographical and ecological data. *Biological Conservation* 87, 295–301.

Dennis, R.L.H. and Hardy, P.B. (2007) Support for mending the matrix: resource seeking by butterflies is apparent non-resource zones. *Journal of Insect Conservation*, DOI 10.1007/s10841-006-9032-y.

Dennis, R.L.H. and Shreeve, T.G. (1996) *Butterflies on British and Irish Offshore Islands*. Gem Publishing, Wallingford, UK.

Dennis, R.L.H. and Sparks, T.H. (2005) Landscape resources for the territorial nymphalid butterfly *Inachis io*: microsite landform selection and behavioural responses to environmental conditions. *Journal of Insect Behaviour* 18, 725–742.

Dennis, R.L.H. and Sparks, T.H. (2006) When is a habitat not a habitat? Dramatic resource use changes under differing weather conditions for the butterfly *Plebejus argus*. *Biological Conservation* 129, 291–301.

Dennis, R.L.H., Shreeve, T.G. and Van Dyck, H. (2003) Towards a functional resource-based concept for habitat: a butterfly biology viewpoint. *Oikos* 102, 417–426.

Dennis, R.L.H., Hodgson, J.G., Grenyer, R., Shreeve, T.G. and Roy, D.B. (2004) Host plants and butterfly biology. Do host-plant strategies drive butterfly status? *Ecological Entomology* 29, 12–26.

Dennis, R.L.H., Shreeve, T.G. and Van Dyck, H. (2006) Habitats and resources: the need for a resource-based definition to conserve butterflies. *Biodiversity and Conservation* 15, 1943–1966.

Denno, R.F. (1977) Comparison of the assemblages of sap-feeding insects (Homoptera-Hemiptera) inhabiting two structurally different salt marsh grasses in the genus Spartina. *Environmental Entomology* 6, 359–372.

Dover, J.W. (1996) Factors affecting the distribution of satyrid butterflies on arable farmland. *Journal of Applied Ecology* 33, 723–734.

Dover, J.W. (1997) Conservation headlands: effects on butterfly distribution and behaviour. *Agriculture Ecosystems and Environment* 63, 31–49.

Dover, J.W. and Fry, G.L.A. (2001) Experimental simulation of some visual and physical components of a hedge and the effects on butterfly behaviour in an agricultural landscape. *Entomologia Experimentalis et Applicata* 100, 221–233.

Dover, J. and Sparks, T. (2000) A review of the ecology of butterflies in British hedgerows. *Journal of Environmental Management* 60, 51–63.

Dover, J.W., Sparks, T.H. and Greatorex-Davies, J.N. (1997) The importance of shelter for butterflies in open landscapes. *Journal of Insect Conservation* 1, 89–97.

Dover, J.W., Sparks, T., Clarke, S. Gobbett, K. and Glossop, S. (2000) Linear features and butterflies: the importance of green lanes. *Agriculture Ecosystems and Environment* 80, 227–242.

Drake, C.M. (1998) The important habitats and characteristic rare invertebrates of lowland wet grassland in England. In: Joyce, B.B and Wade, P.M. (eds) *European Wet Grasslands: Biodiversity, Management and Restoration*. Wiley, Chichester, UK, pp. 137–149.

Drake, C.M. (2004) *Grazing Marsh Assemblages and Site Classification using Invertebrates*. English Nature Research Reports No. 579. English Nature, Peterborough, UK.

Dunning, J.B., Danielson, B.J. and Pulliam, H.R. (1992) Ecological processes that affect populations in complex landscapes. *Oikos* 65, 169–175.

Edwards, M., Patmore, J.M. and Sheppard, D. (1995) The field cricket – preventing extinction. *British Wildlife* 8, 87–91.

Ehrlich, P.R. and Hanski, I. (2004) *On the Wings of Checkerspots: A Model System for Population Biology*. Oxford University Press, Oxford.

Elton, C.S. (1966) *The Pattern of Animal Communities*. Methuen, New York.

Erwin, T.L. (1982) Tropical forests: their richness in Coleoptera and other arthropod species. *Coleopterists' Bulletin* 36, 74–75.

European Commission (2003) *Agriculture and the Environment*. European Commission, Directorate General for Agriculture, Brussels.

Eyre, M.D., Luff, M.L. and Woodward, J.C. (2004) Beetles (Coleoptera) on brownfield sites in England: an important conservation resource? *Journal of Insect Conservation* 7, 223–231.

Fayt, P., Dufrêne, M., Branquart, E., Hastir, P., Pontégnie, C., Henin, J.-M. and Versteirt, V. (2006) Contrasting responses of saproxylic insects to focal habitat resources: the example of longhorn beetles and hoverflies in Belgian deciduous forests. *Journal of Insect Conservation* 10, 129–150.

Fowles, A.P. and Smith, R.G. (2006) Mapping the habitat quality of patch networks for the marsh fritillary *Euphydryas aurinia* (Rottemburg, 1775) (Lepidoptera, Nymphalidae) in Wales. *Journal of Insect Conservation* 10, 161–177.

Fowles, A.P., Alexander, K.N.A. and Key, R.S. (1999) The saproxylic quality index: evaluating wooded habitats for the conservation of dead-wood Coleoptera. *The Coleopterist* 8, 121–141.

Fox, B.W. (2005) The larva of the white admiral butterfly, *Limenitis camilla* (Linnaeus, 1764) – a master builder. *Entomologist's Gazette* 56, 225–236.

Fry, G. (1995) Landscape ecology of insect movement in arable ecosystems. In: Glen, D.M., Greaves, M.P. and Anderson, H.M. (eds) *Ecology and Integrated Farming Systems*. Wiley, Chichester, UK, pp. 177–202.

Gardner, S.M., Hartley, S.E., Davies, A. and Palmer, S.C.F. (1997) Carabid communities on heather moorlands in northeast Scotland: The consequences of grazing pressure for community diversity. *Biological Conservation* 81, 275–286.

Gaston, K.J. (1994) *Rarity*. Chapman & Hall, London.

Gaston, K.J. and Hudson, E. (1994) Regional patterns of diversity and estimates of global insect species richness. *Biodiversity and Conservation* 3, 493–500.

Gilpin, M. and Hanski, I. (eds) (1991) *Metapopulation Dynamics: Empirical and Theoretical Investigations*. Academic Press, London.

Godfrey, A. (2003) *A Review of the Invertebrate Interest of Coarse Woody Debris in England*. English Nature Research Reports No. 513. English Nature, Peterborough, UK.

Gotelli, N.J. and Kelley, W.G. (1993) A general model of metapopulation dynamics. *Oikos* 68, 36–44.

Grime, J.P. (1974) Vegetation classification by reference to strategies. *Nature* 250, 26–31.

Gutiérrez, D., Thomas, C.D., León-Cortés, J.L. (1999) Dispersal, distribution, patch network and metapopulation dynamics of the dingy skipper butterfly (*Erynnis tages*). *Oecologia* 121, 506–517.

Gutiérrez, D., León-Cortés, J.L., Menéndez, R., Wilson, R.J., Cowley, M.J.R. and Thomas, C.D. (2001) Metapopulations of four Lepidopteran herbivores on a single host plant, *Lotus corniculatus*. *Ecology* 82, 1371–1386.

Haggett, P. (1965) *Locational Analysis in Human Geography*. Edward Arnold, London.

Hall, L.S., Krausman, P.R. and Morrison, M.L. (1997) The habitat concept and a plea for standard terminology. *Wildlife Society Bulletin* 25, 173–182.

Hambler, C. and Speight, M.R. (2004) Extinction rates and butterflies. *Science* 305, 1563.

Hanski, I.A. and Gilpin, M.E. (eds.) (1997) *Metapopulation Biology. Ecology, Genetics, and Evolution*. Academic Press, London.

Hanski, I. and Thomas, C.D. (1994) Metapopulation dynamics. A spatially explicit model applied to butterflies. *Biological Conservation* 68, 167–180.

Henderson, P.A. (2003) *Practical Methods in Ecology*. Blackwell, Oxford.

Herzog, F. (2005) Agri-environment schemes as landscape experiments – preface. *Agriculture Ecosystems and Environments* 108, 175–177.

Hill, J.K., Thomas, C.D. and Lewis, O.T. (1996) Effects of habitat patch size and isolation on dispersal by *Hesperia comma* butterflies: implications for metapopulation structure. *Journal of Animal Ecology* 65, 725–735.

Hochberg, M.E., Elmes, G.W., Thomas, J.A. and Clarke, R.T. (1996) Mechanisms of local persistence in coupled host-parasitoid associations: the case model of *Maculinea rebeli* and *Ichneumon eumerus*. *Philosophical Transactions of the Royal Society of London Series B – Biological Sciences* 351, 1713–1724.

Hodgson, J.G. (1993) Commonness and rarity in British butterflies. *Journal of Applied Ecology* 30, 407–427.

Holmes, P.R., Fowles, A.P., Boyce, D.C. and Reed, D.K. (1993) The ground beetle (Coleoptera: Carabidae) fauna of Welsh peatland biotopes – species assemblages in relation to peatland habitats and management. *Biological Conservation* 65, 61–67.

Howe, M.A. (2003) Coastal soft cliffs and their importance for invertebrates. *British Wildlife* 14, 323–331.

JNCC (2003) *Handbook for phase 1 habitat survey: a technique for environmental audit: Field manual.* Joint Nature Conservation Committee, Peterborough, UK.

Joyce, D. and Pullin, A.S. (2003) Conservation implications of the distribution of genetic diversity at different scales: a case study using the Marsh Fritillary butterfly (*Euphydryas aurinia*). *Biological Conservation* 114, 453–461.

Kirby, P. (1992) *Habitat Management for Invertebrates: A Practical Handbook*. RSPB, Sandy, UK.

Kirchner, F., Ferdy, J.B., Andalo, C., Colas, B. and Moret, J. (2003) Role of corridors in plant dispersal: an example with the endangered *Ranunculus nodiflorus*. *Conservation Biology* 17, 401–410.

Kitihara, M., Sei, K. and Fujii, K. (2000) Patterns in the structure of grassland butterfly communities along a gradient of human disturbance: further analysis based on the generalist/specialist concept. *Population Ecology* 42, 135–144.

Lahaye, W.S., Gutierrez, R.J. and Akcakaya, H.R. (1994) Spotted owl metapopulation dynamics in southern California. *Journal of Animal Ecology* 63, 775–785.

León-Cortés, J.L., Cowley, M.J.R. and Thomas, C.D. (2000) The distribution and decline of a widespread butterfly *Lycaena phlaeas* in a pastoral landscape. *Ecological Entomology* 25, 285–294.

León-Cortés, J.L., Lennon, J.L. and Thomas, C.D. (2003) Ecological dynamics of extinct species in empty habitat networks. 2. The role of host plant dynamics. *Oikos* 102, 465–477.

Levins, R. (1969) Some demographic and genetic consequences of environmental heterogeneity for biological control. *Bulletin of the Entomological Society of America* 15, 237–240.

Levins, R. (1970) Extinction. In: Gerstenhaber, M. (ed.) *Some Mathematical Problems in Biology*. American Mathematical Society, Providence, Rhode Island, pp. 75–107.

Lewis, O.T., Thomas, C.D., Hill, J.K., Brookes, M.I., Crane, T.P.R., Graneau, Y.A., Mallet, J.L.B. and Rose, O.C. (1997) Three ways of assessing metapopulation structure in the butterfly *Plebejus argus*. *Ecological Entomology* 22, 283–293.

Lindenmayer, D.B. (2000) Factors at multiple scales affecting distribution patterns and their implications for animal conservation – Leadbeater's Possum as a case study. *Biodiversity and Conservation* 9, 15–35.

Longley, P.A., Goodchild, M.F., Maguire, D.J. and Rhind, D.W. (2001) *Geographical Information Systems and Science*. Wiley, New York.

Lott, D.A. (2003) *An Annotated List of Wetland Ground Beetles (Carabidae) and Rove Beetles (Staphylinidae) Found in the British Isles Including a Literature Review of their Ecology*. English Nature Research Reports No. 488. English Nature, Peterborough, UK.

Ludwig, J.A. and Reynolds, J.F. (1988) *Statistical Ecology. A Primer on Methods and Computing*. Wiley, New York.

Luff, M.L. and Rushton, S.P. (1989) The ground beetle and spider fauna of managed and unimproved upland pasture. *Agriculture, Ecosystems and Environment* 25, 195–205.

Maes, D., Ghesquiere, A., Logie, M. and Bonte, D. (2006) Habitat use and mobility of two threatened coastal dune insects: implications for conservation. *Journal of Insect Conservation* 10, 105–115.

Matter, S.F. and Roland, J. (2002) An experimental examination of the effects of habitat quality on the dispersal and local abundance of the butterfly *Parnassius smintheus*. *Ecological Entomology* 27, 308–316.

Matter, S.F., Roland, J., Keyghobadi, N. and Sabourin, K. (2003) The effects of isolation, habitat area and resources on the abundance, density and movement of the butterfly *Parnassius smintheus*. *American Midland Naturalist* 150, 26–36.

May, R.M. (1990) How many species? *Philosophical Transactions of the Royal Society of London Series B – Biological Sciences* 330, 293–304.

May, R.M. (1992) Past efforts and future prospects towards understanding how many species there are. In: Solbrig, O.T., van Emden, H.M. and van Oordt, P.G.W.J. (eds) *Biodiversity and Global Change*. Monograph 8, IUBS Press, Paris, pp. 71–81.

McCarthy, M.A. and Lindenmayer, D.B. (1999) Incorporating metapopulation dynamics of greater gliders into reserve design in disturbed landscapes. *Ecology* 80, 651–667.

McCarthy, M.A., Thompson, C.J. and Possingham, H.P. (2005) Theory for designing nature reserves for single species. *American Naturalist* 165, 250–257.

Menéndez, R. and Thomas, C.D. (2000) Metapopulation structure depends on spatial scale in the host-specific moth *Wheeleria spilodactylus* (Lepidoptera: Pterophoridae). *Journal of Animal Ecology* 69, 935–951.

Mennechez, G., Schtickzelle, N. and Baguette, M. (2003) Metapopulation dynamics of the bog fritillary: comparison of demographic parameters and dispersal between a continuous and a highly fragmented landscape. *Landscape Ecology* 18, 279–291.

Merckx, T. and Van Dyck, H. (2002) Interrelations between habitat use, behavior, and flight-related morphology in two co-occurring satyrine butterflies, *Maniola jurtina* and *Pyronia tithonus*. *Journal of Insect Behaviour* 15, 541–561.

Munton, P. (1987) Concepts of threat to the survival of species used in Red Data Books and similar compilations. In: Fitter, R. and Fitter, M. (eds) *The Road to Extinction. Problems of Categorising the Status of Taxa Threatened with Extinction*. IUCN, Gland, Switzerland, pp. 71–95.

Nieminen, M., Siljander, M. and Hanski, I. (2004) Structure and dynamics of Melitaea cinxia populations. In: Ehrlich, P.R. and Hanski, I. (eds) *On the Wings of Checkerspots: A Model System for Population Biology*. Oxford University Press, Oxford, pp. 63–91.

Odum, E.P. (1963) *Ecology*. Holt, Rinehart & Winston, New York.

Offer, D., Edwards, M. and Edgar, P. (2003) *Grazing Heathland: A Guide to Impact Assessment for Insects and Reptiles*. English Nature Research Reports No. 497. English Nature, Peterborough, UK.

Ouin, A., Aviron, S., Dover, J. and Burel, F. (2004) Complementation/supplementation of resources for butterflies in agricultural landscapes. *Agriculture Ecosystems and Environment* 103, 473–479.

Ovaskainen, O. (2002) Long-term persistence and the SLOSS problem. *Journal of Theoretical Biology* 218, 419–433.

Pollard, E. and Hall, M.L. (1980) Possible movement of *Gonepteryx rhamni* (L.) (Lepidoptera: Pieridae) between hibernating and breeding areas. *Entomologist's Gazette* 31, 217–220.

Pulliam, H.R. (1988) Sources, sinks and population regulation. *American Naturalist* 132, 652–661.

Pullin, A.S. and Knight, T.M. (2001) Effectiveness in conservation practice. Pointers from medicine and public health. *Conservation Biology* 15, 50–54.

Rackham, O. (1980) *Ancient Woodland: Its History, Vegetation and Uses in England*. Edward Arnold, London.

Rodwell, J.S. (1991 et seq.) *British Plant Communities*, Vols. 1–5. Cambridge University Press, Cambridge.

Roland, J., Keyghobadi, N. and Fownes, S. (2000) Alpine Parnassius butterfly dispersal. Effects of landscape and population size. *Ecology* 81, 1642–1653.

Rosenzweig, M.L. (1995) *Species Diversity in Space and Time*. Cambridge University Press, Cambridge.

Roy, D.B. and Thomas, J.A. (2003) Seasonal variation in the niche, habitat availability and population fluctuations of a bivoltine thermophilous insect near its range margin. *Oecologia* 134, 439–444.

Rushton, S.P. (1988) The effects of scrub management regimes on the spider fauna of chalk grassland, Castor Hanglands National Nature Reserve, Cambridgeshire, UK. *Biological Conservation* 46, 169–182.

Rushton, S.P., Eyre, M.D. and Luff. M.L. (1990) The effects of scrub management on the ground beetles of Oolitic limestone grassland at Castor Hanglands National Nature Reserve, Cambridgeshire, UK. *Biological Conservation* 51, 97–111.

Sadler, J.P., Bell, D. and Fowles, A. (2004) The hydroecological controls and conservation value of beetles on exposed riverine sediments in England and Wales. *Biological Conservation* 118, 41–56.

Salmela, J. and Ilmonen, J. (2005) Cranefly (Diptera: Tipuloidea) fauna of a boreal mire system in relation to mire trophic status: implications for conservation and bioassessment. *Journal of Insect Conservation* 9, 85–94.

Samways, M.J. (2005) *Insect Diversity Conservation*. Cambridge University Press, Cambridge.

Shardlow, M.E.A. (2001) A Review of the conservation importance of shingle habitats for invertebrates in the United Kingdom (UK). In: Packham, J.R., Randall, R.E., Barnes, R.S.K. and Neal, A. (eds) *Ecology and Geomorphology of Coastal Shingle*. Westbury Academic & Scientific Publishing, Otley, UK, pp. 355–379.

Sheppard, D.A. (2002) The importance of scrub and bare ground. In: Stone, D., Tither, J. and Lacey, P. (eds) *The Species Recovery Programme Proceedings of the 10th Anniversary Conference*. English Nature, Peterborough, UK, pp. 67–71.

Shreeve, T.G. (1984) Habitat selection, mate location and microclimatic constraints on the activity of the speckled wood butterfly *Pararge aegeria*. *Oikos* 42, 371–377.

Shreeve, T.G. (1985) The population biology of the speckled wood butterfly (*Pararge aegeria* L.) PhD thesis. Oxford Polytechnic, Oxford.

Shreeve, T.G. (1986) Egg-laying by the speckled wood butterfly (*Pararge aegeria*) – the role of female behaviour, host plant abundance and temperature. *Ecological Entomology* 11, 229–236.

Shreeve, T.G. (1987) The mate location behaviour of the speckled wood butterfly, Pararge aegeria, and the effect of phenotypic differences in hind-wing spotting. *Animal Behaviour* 35, 682–690.

Shreeve, T.G., Dennis, R.L.H., Roy, D.B. and Moss, D. (2001) An ecological classification of British butterflies: ecological attributes and biotope occupancy. *Journal of Insect Conservation* 5, 145–161.

Shreeve, T.G., Dennis, R.L.H., and Van Dyck, H. (2004) Resources, habitats and metapopulations – whither reality? *Oikos* 106, 404–408.

Smith, C.J. (1980) *Ecology of the English Chalk*. Academic Press, London.

Smith, M.A. and Green, D.M. (2005) Dispersal and the metapopulation paradigm in amphibian ecology and conservation: are all amphibian populations metapopulations? *Ecography* 28, 110–128.

Southwood, T.R.E. (1977) Habitat, the templet for ecological strategies. *Journal of Animal Ecology* 46, 337–365.

Stefanescu, C., Peñuelas, J. and Filella, I. (2005) Butterflies highlight the conservation value of hay meadows highly threatened by land-use changes in a protected Mediterranean area. *Biological Conservation* 126, 234–246.

Steffan-Dewenter, I. and Tscharntke, T. (2000) Butterfly community structure in fragmented habitats. *Ecology Letters* 3, 449–456.

Stevens, P.A., Bell, S.A., Brittain, S.A., Hughes, S. and Lowe, J.A.H. (1995) *Soil/Plant Interactions in Lowland Grasslands – Great Orme Study, Final Report*. CCW Contract Science Report No. 142. Countryside Council for Wales. Available at: www.ccw.gov.uk

Stone, D., Tither, J. and Lacey, P. (2002) *The Species Recovery Programme. Proceedings of the 10th Anniversary Conference 5–7 December 2001*. English Nature, Peterborough, UK.

Stork, N.E. (1993) How many species are there? *Biodiversity and Conservation* 2, 215–232.

Sutherland, W.J., Pullin, A.S., Dolman, P.M. and Knight, T.M. (2004) The need for evidence-based conservation. *Trends in Ecology and Evolution* 19, 305–308.

ter Braak, C.F.J. and Smilauer, P. (2002) *CANOCO Reference manual and CanoDraw for Windows User's guide: Software for Canonocial Community Ordination*. Version 4.5. Microcomputer Power, Ithaca, New York.

Thomas, C.D. (1985) The status and conservation of the butterfly *Plebejus argus* L. (Lepidoptera: Lycaenidae) in North West Britain. *Biological Conservation* 33, 29–51.

Thomas, C.D. and Harrison, S. (1992) Spatial dynamics of a patchily distributed butterfly species. *Journal of Animal Ecology* 61, 437–446.

Thomas, C.D and Kunin, W.E. (1999) The spatial structure of populations. *Journal of Animal Ecology* 68, 647–657.

Thomas, C.D., Warren, M.S. and Thomas, J.A. (1992) Distributions of occupied and vacant butterfly habitats in fragmented landscapes. *Oecologia* 92, 563–567.

Thomas, J.A. (2005) Monitoring change in the abundance and distribution of insects using butterflies and other indicator groups. *Philosophical Transactions of the Royal Society of London Series B – Biological Sciences* 360, 339–357.

Thomas, J.A., Clarke, R.T., Elmes, G.W. and Hochberg, M.E. (1998) Population dynamics in the genus *Maculinea* (Lepidoptera: Lycaenidae). In: Dempster, J.P. and McLean, I.F. (eds) *Insect Populations: Theory and Practice*. Symposium of the Royal Entomological Society, 19, Chapman & Hall, London, pp. 261–290.

Thomas, J.A., Bourn, N.A.D., Clarke, R.T., Stewart, K.E., Simcox, D.J., Pearman, G.S., Curtis, R. and Goodger, B. (2001). The quality and isolation of habitat patches both determine where butterflies persist in fragmented landscapes. *Proceedings of the Royal Society of London Series B – Biological Sciences* 268, 1791–1796.

Thomas, J.A., Telfer, M.G., Roy, D.B., Preston, C.D., Greenwood, J.J.D., Asher, J., Fox, R., Clarke, R.T. and Lawton, J.H. (2004) Comparative losses of British butterflies, birds and plants and the global extinction crisis. *Science* 303, 1879–1881.

Underhill-Day, J.C. (2005) *A Literature Review of Urban Effects on Lowland Heaths and Their Wildlife*. English Nature Research Reports No.623. English Nature, Peterborough, UK.

Valimaki, P. and Itamies, J. (2003) Migration of the clouded Apollo butterfly *Parnassius mnemosyne* in a network of suitable habitats – effects of patch characteristics. *Ecography* 26, 679–691.

Van Dyck, H. and Baguette, M. (2005) Dispersal behaviour in fragmented landscapes: routine or special movements? *Basic and Applied Ecology* 6, 535–545.

Van Essen, S.J. (1994) A method to evaluate the condition of heathlands by using catches of carabid beetles. *Series Entomologica (Dordrecht)* 51, 383–386.

Vane-Wright, R.I. and Boppré, M. (1993) Visual and chemical signaling in butterflies – functional and phylogenetic perspectives. *Philosophical Transactions of the Royal Society of London Series B – Biological Sciences* 340, 197–205.

Vanreusel, W. and Van Dyck, H. (2007) When functional habitat does not match vegetation types: a resource-based approach to map butterfly habitat. *Biological Conservation*.

Waring, P. (2001) Grazing and cutting as conservation management tools – the need for a cautious approach, with some examples of rare moths which have been adversely affected. *Entomologist's Record and Journal of Variation* 113, 193–200.

Webb, J.R. and Lott, D.A. (2006) The development of ISIS: a habitat-based invertebrate assemblage classification system for assessing conservation interest in England. *Journal of Insect Conservation* 10, 179–188.

Whittaker, R.H., Levin, S.A. and Root, R.B. (1973) Niche, habitat, and ecotope. *American Naturalist* 107, 321–338.

Wiens, J.A. (1996) Wildlife in patchy environments: metapopulations, mosaics, and management. In: McCullough, D.R. (ed.) *Metapopulations and Wildlife Conservation*. Island Press, Washington, DC, pp. 53–84.

Wiklund, C. (1981) Generalist vs specialist oviposition behavior in Papilio machaon (Lepidoptera) and functional-aspects on the hierarchy of oviposition preferences. *Oikos* 36, 163–170.

Wilson, R.J., Ellis, S., Baker, J.S., Lineham, M.E., Whitehead, R.W. and Thomas, C.D. (2002) Large-scale patterns of distribution and persistence at the range margins of a butterfly. *Ecology* 18, 3357–3368.

6 Implementing Ecological Networks for Conserving Insect and other Biodiversity

MICHAEL J. SAMWAYS

Department of Conservation Ecology and Entomology and Centre for Invasion Biology, University of Stellenbosch, Private Bag X1, Matieland 7602, South Africa

1 A *Sense of Time* and the Historical Context for Landscape Design

Conservation activities depend first on clearly defining the goals. Whatever those goals are, some form of habitat conservation is likely to be included, as habitat loss is the greatest of all threats to insects (Mawdsley and Stork, 1995). In turn, protecting and restoring habitats depends on managing whole landscapes, or countryside-wide management (Ricketts *et al.*, 2001). When visualizing the landscape to be conserved, it is essential to have a historical perspective, so as to conserve for *when*, as well as for *what* against a background of ever-changing landscapes. Although Lockwood (2001) has rightly argued for a sense of *place* in insect conservation, a parallel concept is also that of *time*.

Evidence from various taxa suggests that the current biodiversity crisis began thousands of years ago, and the current global extinction spasm has a historic precedent (Steadman, 1995; Burney *et al.*, 2001; Burney and Flannery, 2005). With increasing impact on the landscape, humans have placed many natural communities into disequilibrium. This means that current planning across the land mosaic must consider the past to appreciate what would actually be living in a particular area in the absence of humans (Buchwald and Svenning, 2005). But how far back in time should we go? From a global perspective, a reasonable time would be the Upper Quaternary, i.e. the last 130,000 years, and for which there are good records, at least for the northern hemisphere (Fig. 6.1). This timescale covers three major events: (i) the last interglacial (the Eemian, 130–110 kyr BP), in which the vegetation continued to evolve under the effect of climate changes, and indigenous vegetation response was in the absence of human impact; (ii) the glacial period (110–10 kyr BP), with alternating cold and temperate flushes with a cold maximum (25–15 kyr BP) and considerable changes in insect assemblages at any one location (Elias, 1994; Coope, 1995; Ponel *et al.*, 2003); and (iii) the current

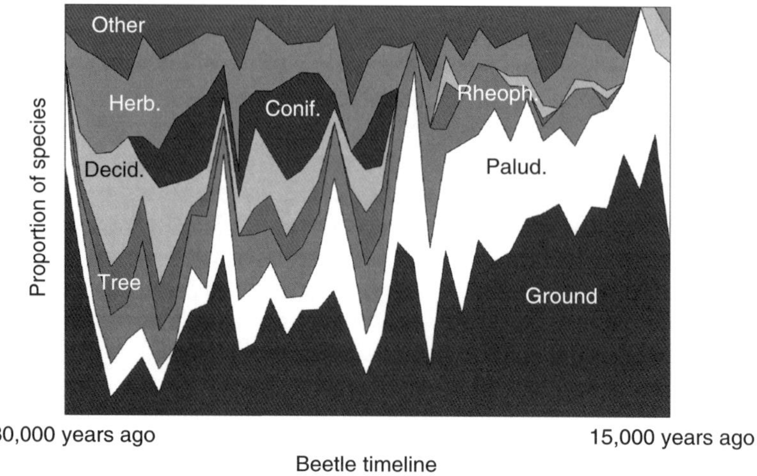

Fig. 6.1. Changes over time (130–15 kyr BP) of proportionate European beetle taxon richness in various habitat types: Ground = ground inhabiting, Palud. = paludal or marsh, Rheoph. = rheophilous or stream, Tree = tree canopy, Decid. = deciduous forest, Conif. = coniferous forest, Herb. = Herbs and grasses, Other = coprophagous, necrophagous, non-specialist and undetermined taxa. (From Ponel *et al.*, 2003.)

interglacial period (10 kyr BP–present) where primary evolutionary pressure moves from rapid climate swings and natural migratory dynamics to anthropogenic land mosaics.

These modern mosaics are harsh, and present an 'adapt (Stockwell *et al.*, 2003) or die' evolutionary backdrop. The gradual global climatic warming over the last 15,000 years (Fagan, 2004) is being accelerated in the industrial age, and this warming together with fragmentation is a 'deadly anthropogenic cocktail' for biodiversity (Travis, 2003), which appears to have been borne out by the decline of British butterflies (Warren *et al.*, 2001).

2 Comparative Challenges in the South Relative to the North

The conservation challenges in the northern hemisphere present some common challenges with the southern hemisphere and also some differences. Among the common challenges are habitat fragmentation and loss, as well as threats from invasive alien organisms. Differences include the disproportionately species-rich South, with its wide range of narrow endemics (Samways, 1995), compared to the relatively species-poor North and many species with widespread distributions (Niemelä, 1997). Indeed, most of the world's global hotspots are at low latitudes or in the South (Myers *et al.*, 2000). Furthermore, farther North, populations of geographically widespread species become increasingly genetically impoverished (Schmidtt and Seitz, 2001; Hewitt, 2003). Apparently, there are no comparable data for the southern hemisphere, a research challenge.

Two-thirds of the land surface is in the northern hemisphere. In the North, 39% of the surface is land, while in the South it is 19%. Of the southern landmass, 29% is Antarctica, poor in insect species. The rest of the South is climatically relatively moderate, with a rich insect fauna. This is partly beçause the Pleistocene in the South did not brush away the biota as it did in the North. At the highest elevations, there were only periglacial conditions, with temperatures in the Western Cape 5–6°C cooler 21–17 kyr ago than they are today. At the Holocene maximum, 8–5 kyr BP, conditions were warmer than today, with the Western Cape drier and the Kalahari wetter. The point is that over the last few tens of millennia, the North has presented a different ecological stage and evolutionary grist than the South. Arguably, the North need only consider conservation along the time line of the current interglacial, whereas the South must consider deeper time and its consequent large number of narrow-range endemics.

In terms of post-industrial anthropogenic impact on land mosaics, the South is rapidly catching up with the North, as population extinctions become more widespread and commonplace (African Wildlife, 2005). In addressing the conservation challenges, there is one major difference between the North and South. The land mosaic in the warm, temperate North has been largely pre-empted for use by humans, leaving little physical room for designing the landscape mosaic with new geometries (but see Jongman and Pungetti, 2004). Nevertheless, the South has not been immune from human impact, with the megafauna both in South Africa and Australia suffering considerably (Flannery, 1994). Intensification of the anthropogenic land mosaic in the North does not mean that there is no opportunity for landscape design; witness the agri-environment schemes (Kleijn and Sutherland, 2003), which involve both landscape design and wildlife-friendly methods (New, 2005). However, it is in the South that the opportunities are greatest in terms of conserving irreplaceable biota.

The aim of this chapter is to illustrate how some of the biodiversity challenges are being met with in the South, emphasizing landscape design based on land sparing, which is currently one of the greatest opportunities for conservationists (Mattison and Norris, 2005).

3 The Fundamental Underpinning of Topography, Fire and Megaherbivores

A major consideration in the southern hemisphere, with its long history without glaciations, is that topography plays a major role in driving biodiversity (Fjeldså and Lovett, 1997). Even microtopography can determine local invertebrate distributions (Greenslade, 1993). Various factors interplay, with cold-air drainage among them. Grasshopper assemblages are richer on tops of hills than in valleys, with the difference magnified on recently burnt hills compared with those clothed in grass material (Samways, 1990). The difference is also enhanced by grazing by megaherbivores, which have a greater

Fig. 6.2. Topography and grazing by wild megaherbivores play an important role in determining habitat and microhabitat heterogeneity, and thus suitable conditions for a wide range of insects in South Africa.

impact on the flatlands than on the hilltops (Gebeyehu and Samways, 2006a), although the hills do not have to be large for this to occur, with small hills having a major positive effect on local assemblages (Gebeyehu and Samways, 2006b) (Fig. 6.2). Additionally, the type of vegetation covering the hills also makes a difference, with butterflies hilltopping both when there is natural grassland or open-canopy alien eucalyptus, but not when there is dense natural forest or dense-canopy eucalyptus (Lawrence and Samways, 2002).

It is not so much the type of grazer, whether indigenous game or domestic livestock, that changes the African insect fauna, but the intensity of the impact (Rivers-Moore and Samways, 1996). Heavy grazing and trampling are impoverishing, whether from game or livestock (Samways and Kreuzinger, 2001), although impoverishment is usually from overstocking with livestock. When there is such overstocking, it is the abundance of insects, such as grasshoppers, which decline the most, not species richness (Fig. 6.3). During restoration, when natural game-stocking levels replace heavy livestock pressure, there is a return to the natural assemblage structure (Gebeyehu and Samways, 2002). Where there is very high elephant impact, which is continuous pressure, there is impoverishment of the local dragonfly fauna, favouring common habitat generalists.

What these results remind us is that when considering land sparing as a mitigation procedure, it is essential to consider the third or vertical dimension and its interaction with natural impacts, such as fire and megaherbivores. These natural disturbance factors have, after all, been major drivers of ecological processes on the African landscape for many millennia.

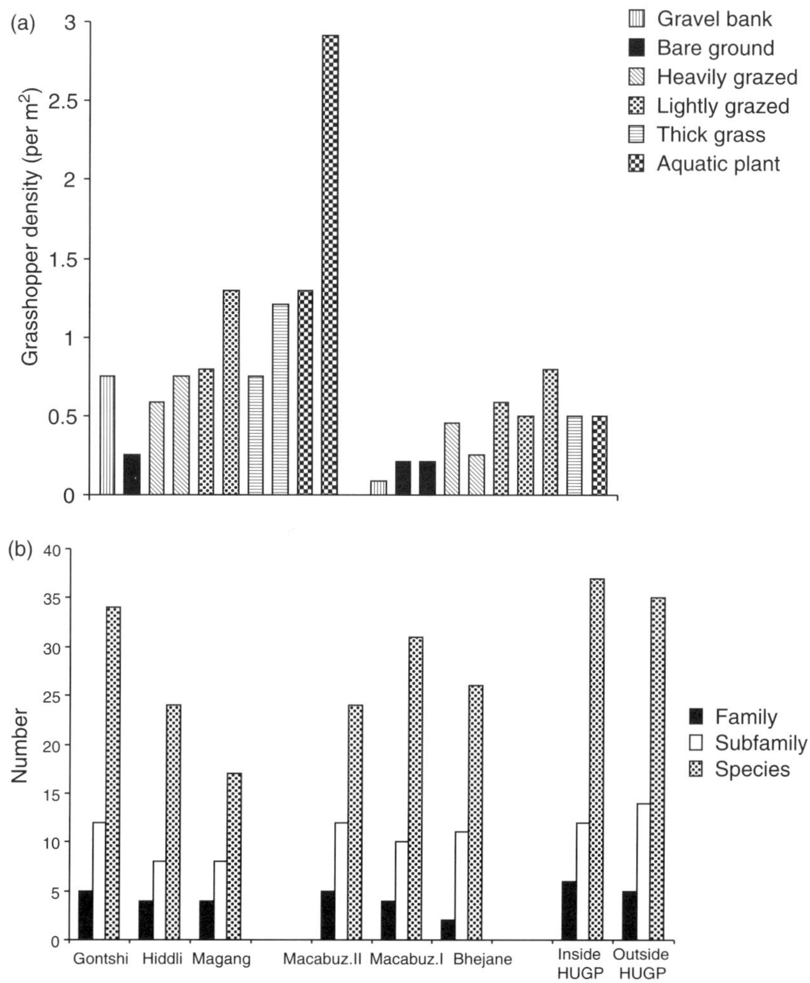

Fig. 6.3. (a) Grasshopper densities inside versus outside the Hluhluwe-Umfolozi Game Park (HUGP) at 20 sites associated with waterholes. Densities were lower outside where large numbers of cattle replaced game animals. (b) Abundance of families, subfamilies and species of grasshoppers at three waterholes inside and three outside HUGP (and combined on the right hand side), which show that species richness was overall the same inside versus outside the reserve. (Redrawn from Samways and Kreuzinger, 2001, with permission from Springer.)

4 Significance of Landscape Linkages

Ensuring connectivity between habitat patches is crucial for long-term maintenance of biodiversity (Bennett, 1999). A linear landscape element, broadly termed linkage or corridor, may be for animal movement. It can also function as a habitat when resources are available to fulfil the life functions of a species. When the whole biological community is considered, a linkage is more than a binary phenomenon (movement versus habitat). It becomes a spectrum, as more and more species are considered. However, with an increasing number of species, the linkage also becomes a differential filter (Ingham and Samways, 1996), such that certain organisms may or may not move the whole length of the linkage. When they do move along it, the journey may be within one lifetime (making it a true movement corridor), or it may be over several lifetimes, making it a genetic linkage. Even among individuals within a species there may be difference in movement capability (Wood and Samways, 1991; Denno *et al.*, 1996).

Some insects are extraordinarily vagile, with founder populations appearing in far away places (Thornton, 1996). The converse is that, in some sedentary insects there is nevertheless some long-distance gene spread, enough to maintain population viability (Peterson, 1996).

Linkages in the urban landscape, often called greenways by planners, have become a major feature in landscape design (Smith and Hellmund, 1993; Rosenberg *et al.*, 1997; Bennett, 1999; Jongman and Pungetti, 2004). Linkages are also being considered in the agricultural context (Burel and Baudry, 1995). Although there has been much discussion on the theoretical value of (Forman, 1995) and etymological confusion surrounding (Hess and Fischer, 2001) 'corridors', there has been little reporting on practical successes of linkages, particularly for connecting nodes of quality remnant habitat for ingenous and endemic biodiversity (but see Hilty *et al.*, 2006). In response, a very large-scale, multiple array of remnant landscape linkages have been successfully deployed to conserve biodiversity in an agroeconomic context in South Africa (Fig. 6.4).

5 Implementing Ecological Networks

Parts of South Africa have been described as biodiversity hotspots of global note (Myers *et al.*, 2000). In addition, there are high levels of endemism among much of its biota which, in many cases, is also threatened (African Wildlife, 2005). Furthermore, some reserve areas have been designated World Heritage Sites (e.g. Table Mountain and the Greater St Lucia Wetlands Park). Against this background of great biotic wealth, there has been increasing impact from agricultural development, which is particularly extensive in the 'Cinderella biome', grassland (Neke and Du Plessis, 2004). Among this development has been the large-scale expansion of plantation forestry that today accounts for about 1.5 million hectares within the country. The forestry industry has been criticized for impacting upon biodiversity (Armstrong *et al.*, 1998) with the industry responding to the criticism (Pott, 1997).

Some of the dispute has arisen as a result of confusion of spatial scales. At the scale of patch, specifically a plantation forestry patch, there may be an adverse impact (Donnelly and Giliomee, 1985; Sinclair and New, 2004), although it depends very strongly on which type of plantation has been established (Samways *et al.*, 1996; Lawrence and Samways, 2002). Yet plantation managers are operating at the larger scale of landscape, not at the smaller scale of patch. They are also asking how we should grow trees while maintaining biotic capital. This does not ignore regional planning and large-scale

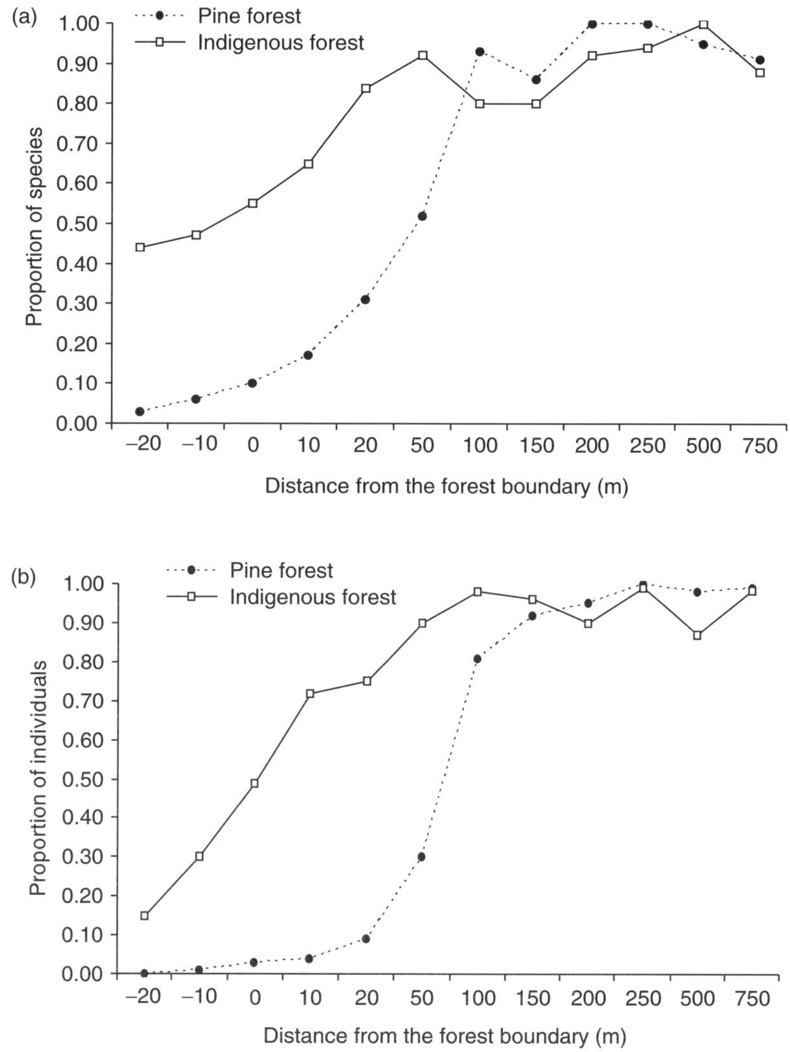

Fig. 6.4. Responses of African butterflies to the boundaries along the edge of natural grassland linkages. The boundary is designated as '0'. Both proportion of species (a) and proportion of individuals (b) penetrate the natural forest boundaries to a much greater extent than into alien pine-tree boundaries (negative values of the graph). In other words, pines are a much harder boundary than natural forest. Furthermore, the edge effect of the alien pines into the grassland has a much more depressing effect on both abundance and species richness than natural forest (positive values on the graph). (Redrawn from Pryke and Samways, 2001, with permission from Elsevier.)

processes but requires that the multitude of landscape-scale patterns and functions make up the regional conservation perspective. The outcome from the dispute has been to leave a network of remnant grassland linkages and nodes, which form an ecological (and evolutionary) network between the plantation stands, with this land sparing equal to about a third of the total land surface, and with proclaimed reserves additive upon this third (Fig. 6.4). This is a major undertaking because these networks are being implemented at a sub-regional spatial scale and not merely at one or few locations.

The design of these linkages must also consider the third, or vertical, dimension (topography) over the two-dimensional geometrical design. Topographic considerations do not focus only on biodiversity per se, but also on land-form effects and how they affect ecosystem patterns and processes (Swanson *et al.*, 1988). Of particular concern have been hydrological processes, where riparian linkages and wetlands are maintained as intact as possible. This inevitably produces conditions for a wide range of biodiversity, at various elevations and land-surface structures, such as rocky summits of hills. Behavioural activities, such as natural insect hilltopping, are also encouraged by these networks. Indeed, there is a vast array of subtle and often complex biotic interactions (Mevi-Schütz and Erhardt, 2005), which must be accommodated when designing such ecological networks. Among these interactions is the effect of megaherbivores, such as eland (*Taurotragus oryx*) at higher elevations and elephant (*Loxodonta africana*) at lower ones. These large animals are part of the interactive landscape, and necessary for encouraging a range of plants and insects that require particular types and level of disturbance.

6 Biodiversity Value of the Ecological Networks

6.1 Linkages

At the smaller spatial scale, the interface between plantation patches and remnant, indigenous grasslands has a fuzzy effect on insects at the edge (Samways and Moore, 1991). Alien pine trees, for example, can influence grasshopper species richness and abundance 30 m, and occasionally more, into the natural grassland matrix. Indeed, pines have a much harder edge than do natural forests. This was shown for both butterfly abundance and species richness (Pryke and Samways, 2001) (Fig. 6.5). Less than 10% of grassland species penetrated 10 m into the pine patch, whereas 45% of the species penetrated natural forest. Furthermore, the effect of the pine trees on these butterflies extended into the grassland as it did for grasshoppers. The effect was significant up to 50 m from the pine edge, whereas it was negligible relative to the natural forest edge. This supports the recommendations of Fry and Lonsdale (1991) and Kirby (1992), that softening edges is beneficial for most insect species.

Management at the landscape spatial scale using linkages also takes into account temporal factors. A linkage that acts as a movement corridor enables those mobile organisms to locate suitable habitats to complete their life histories. In the case of South African butterflies, only the most mobile, generalist species entered the narrow corridors (less than 50 m wide), when they flew 13 times faster than they did in wide (greater than 250 m) linkages. The slowest movements were recorded in the widest linkages, which was due to the butterflies using these linkages as habitats (stopping to nectar, drink, sunbask and rest). Interestingly, butterflies also flew faster through linkages that

Fig. 6.5. An ecological network in South Africa designed to optimize agroforestry while at the same time maintaining indigenous biodiversity within a global hotspot. Although there is loss of biodiversity at the spatial scale of the pine stand, there is maintenance of quality biodiversity at the larger spatial scale of the landscape.

were highly disturbed, had few nectar flowers, a high density of alien plants, short grasses and high impact from cattle. However, linkage width did not significantly affect movement speeds of many migrant species. Likewise, some of the sedentary and local endemic species were unaffected by linkage width, generally flying less than 200 m in both intermediate, as well as wide linkages. These specialists rarely entered linkages less than 250 m wide, and when they did, they only spent a short time in them.

Preferably, linkages must also be habitats per se, enabling completion of life histories. In other words, linkages can only be considered successful in the long term if they are a network of habitats (ecological network) that are resistant and/or are resilient enough in the face of environmental fluctuations to permit and promote long-term survival of all the local, indigenous species.

What evidence is there that the ecological networks being established across South Africa have the potential for critical conservation? The first results, on adult butterflies, were surprisingly positive (Pryke and Samways, 2001, 2003). Linkages more than 250 m wide were actual habitats within the network. Butterflies penetrated deep into the ecological network, far from the outside indigenous grassland. Specialists were limited to the least disturbed sites, both inside and outside the network, and shared some common traits, such as being small, sedentary and having specific habitat requirements, such as grassy slopes, hilltops and tall grasses. The important point is that the high quality linkages, even deep within the network, supported the rare, specialized and endemic species.

Good quality grassland corridors were rich in species, although three common widespread species were not recorded within them. Two of these at least seem to prefer very wide open spaces, larger even than the major corridors. This illustrates that large, open nodes are also required to supplement these corridors. This is the case, for example, for the Karkloof blue, *Orachrysops ariadne*, a Red Listed species, which has very specific habitat requirements, including particular slopes, a very special host plant and an ant mutualist, all within a specific type of grassland. Indeed, this species now benefits from a large set-aside node, which is carefully managed within the ecological networks (Lu and Samways, 2002).

Surprisingly, these ecological networks also supported three other species that were not present in the natural nodes in the immediate vicinity. One of these species, *Alaena amazoula*, is a localized national endemic, emphasizing that these networks can have add-on effects and act as important natural reserves in their own right.

The most important influence on the corridors detracting from their effectiveness was disturbance, especially from domestic cattle. The effects of their disturbance could be measured by changes in plant compositional and structural diversity. Major disturbance had a highly impoverishing effect, particularly upon abundance. Preliminary results on caterpillars (indicators of residency of butterflies), grasshoppers (herbivore functional group) and flower-inhabiting arthropods (pollinators, flower-eaters, seed-eaters, predators) (Bullock and Samways, 2005) are also suggesting that the width of linkage and its interiorness are not critical for ensuring movement throughout the network. Nevertheless, to retain species and functional biodiversity in the long term, it is essential to have wide corridors and nodes, as well as adjacent natural reserves (Fig. 6.6).

Flower–arthropod associations remained intact whatever the character of the corridor, so long as the flowers were present. The associations were only lost when the plants were lost, not when the plants were present, but under some disturbance pressure. As a result of these studies, it is now possible to develop design guidelines for such ecological networks (Fig. 6.7).

6.2 Dealing with linkages as human conduits

Although these corridors, which amount to about a third of the whole local land surface, mitigate the effects of the pine afforestation, they are, in places, also conduits receiving intense human impact. This means that implementation of these ecological networks must also consider human social and commercial activities, as well as the biology of the organisms. Another way of viewing this is that these corridors are not necessarily conservation end points, but rather, they are a conservation-enabling strategy. The wider the linkage, and the more natural they are, the more they become nodes, and thus linkages intergrade into nodes. Such large linkages and nodes can accommodate the high impact of a limited number of vehicles, which, although locally intensive, constitute only a small amount of the total area. More difficult to address are thoroughfares for domestic cattle, the impact of which is greatest where closest to their enclosures. As the effect of many

Fig. 6.6. This ecological network is adjacent to a natural reserve (1) and includes considerable habitat heterogeneity (2). Linkages (3) and associated nodes (4) are an insurance for all the subtle, unrecorded biotic interactions that take place and need to be conserved across the landscape. This craggy hilltop (3) is essential as a thermal refugium and hilltopping site for insects, as well as a special habitat for many plants and insects. Management involves activities, such as encouragement of grazing by indigenous game, as well as limited number of domestic livestock (5). Ideally, there should be reduction of contrast between the impoverishing pine stands and the natural grassland (6). This can only be achieved in these networks by having wide (>250 m) linkages (3), as well as nodes (4) and adjacent reserves (1). At the population level, the aim is to maintain the metapopulation trio of large patch (habitat) size, good patch quality and reduced patch isolation (7).

cattle is impoverishing upon the local plant and insect diversity, this impact must be viewed in the same way as a pine patch, subject to the same triage, and considered as a deficit for biodiversity.

6.3 Riparian corridors

These corridors must also function as retainers of hydrological processes. This means that many of the linkages have roads or are riparian corridors. Studies on dragonflies (Kinvig and Samways, 2000) have indicated that these riparian zones are maintaining quality aquatic diversity, as measured by the presence of localized, endemic species. This however, presumes that invasive alien plants, which radically alter the structural diversity of the riparian

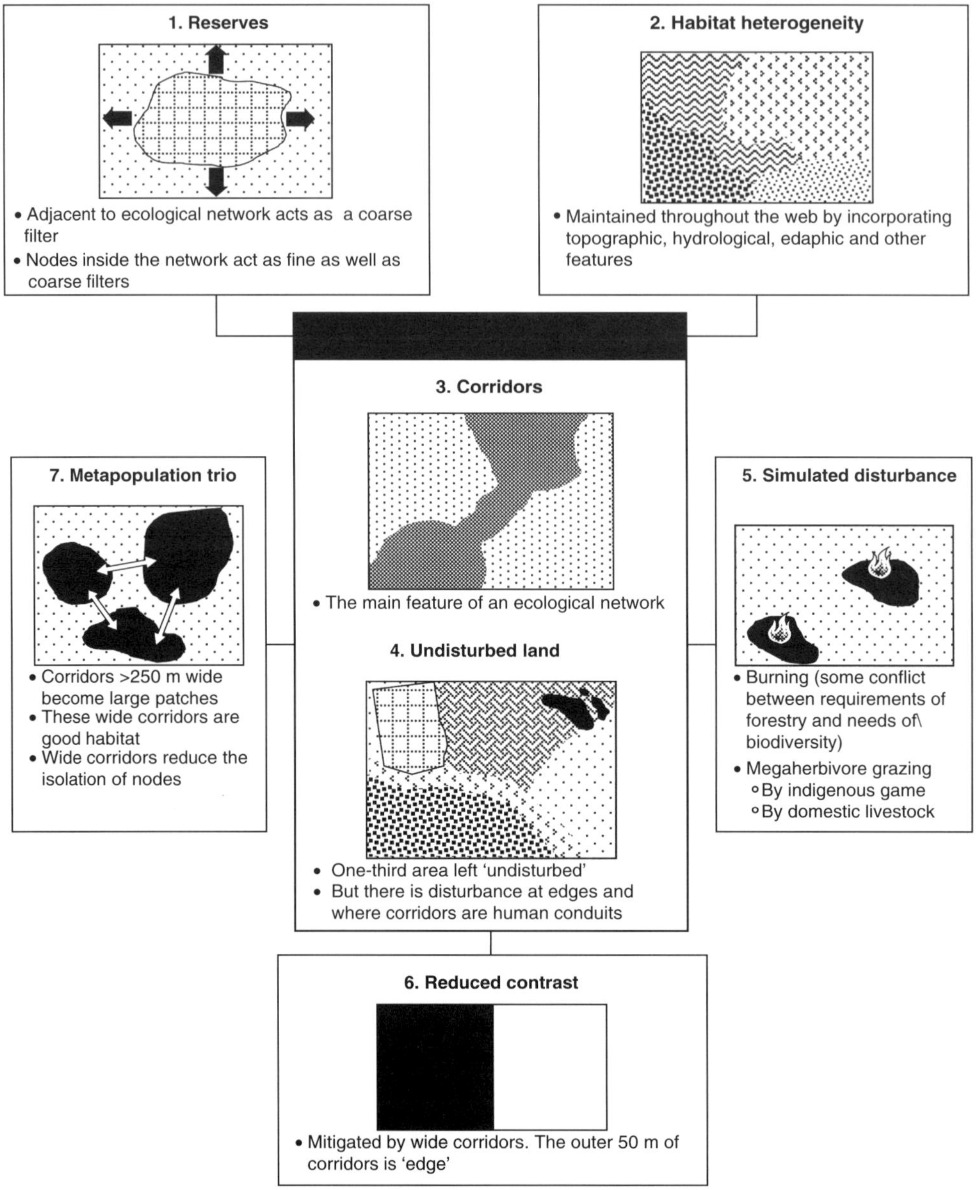

Fig. 6.7. A summary of the design elements of ecological networks. The success of these networks hinges first on the first two key premises of maintaining undisturbed land (nodes) (4) and instigating linkages (3). These in turn, link with adjacent reserve areas (1). Throughout the reserve areas, nodes and linkages, the aim is always to maintain quality heterogeneity (2). The linkages are subject to edge effects because contrast between afforested stands and grassland remnants is great (6). This edge effect is mitigated by wide corridors (>250 m) that have central areas away from the pines, which are natural habitats. Within the ecological network, natural disturbance is simulated (5). Running through all the above landscape approaches is the golden thread of the metapopulation trio of large patch (habitat) size, good patch quality and reduced patch isolation, which is a function of good quality, wide corridors (7).

zone, are removed. Indeed, preliminary results indicate that removal of these aliens leads to a remarkably fast recovery of rare, endemic and other odonate species (Samways *et al.*, 2005).

6.4 Emergent properties of ecological networks

The context and contrast of adjoining landscape elements (Wiens *et al.*, 1993) can result in emergent properties. Little evidence is yet available, but the ecotones between plantation trees and natural grasslands do appear to be favoured habitats for some invertebrates, e.g. scorpions (Ingham and Samways, 1996). The boundaries are also significant in that they attract certain vertebrates, including the threatened oribi antelope (*Ourebia ourebi*), which shelters in the plantation yet grazes in the linkages (R. Pott, Pietermaritzburg, 2001, personal communication).

7 What are the Disadvantages of Ecological Networks?

As the linkages are conduits of human activity, and the interface between the plantation and grassland is effectively an area of disturbance, it inevitably leads to the encouragement of ruderal species, including alien invasive plants, such as bugweed *Solanum mauritianum* and bramble *Rubus cuneifolius*. These plants must be removed or at least contained because they can be highly threatening even in major nodes (Lu and Samways, 2002).

The sophistication of mechanized agroforestry and the concentration of biodiversity must, in turn, not overlook the needs of the local people, their cultural background and their way of life. Cattle are a major component in the equation, whereas it is well known that some disturbance is a great encourager of plant diversity (Grime, 1998). Plant structural, compositional and functional diversity in turn enhances invertebrate diversity (Koricheva *et al.*, 2000). This means that some cattle grazing, in the absence of former indigenous megaherbivores, is desirable, particularly when it simulates the positive impacts of wild game (Gebeyehu and Samways, 2002). However, when grazing and trampling are too severe, whether from game or domestic animals, it can greatly reduce invertebrate diversity (Rivers-Moore and Samways, 1996). The converse is that certain species are adapted to heavy trampling and grazing, which can be encouraged yet contained by having marshy waterholes that localize severe impacts (Samways and Kreuzinger, 2001).

If grazing and fire are excluded from the landscape, the plant community proceeds along a seral path to wooded conditions, which impoverishes the indigenous grass-feeding anthropod guild (Chambers and Samways, 1998). When the conservation goal is to maintain the natural fire-dependent grassland community, it is necessary to artificially induce fire. As firebreaks must be employed to protect the plantation trees, the ideal compromise is to optimize fire management for plantation forestry and for biodiversity conservation at the same time (Teie, 2005).

8 The Way Forward

The pivotal question is whether these ecological networks are sustainable in the long term. This would be particularly so in times of climatic adversity, whether from severe drought or severe flood, particularly in this El Niño-prone area. This is why it is so important to obtain baseline data against which future assessments can be made. Additionally, there needs to be a comparative yardstick that gives a measure to the 'quality' of the biodiversity from one site to the next, as well as over time. Ideally, such a measure should encompass qualities, such as ecosystem health (Rapport *et al.*, 1998) and ecological integrity, which could be construed as functional and compositional diversity, respectively. Such a measure must be relatively easy to use, give repeatable results, be a fair surrogate for biodiversity and must be sensitive enough to measure changes or differences. One problem is that biodiversity has often not been honestly brokered, with great emphasis on vertebrates and plants, and little emphasis on invertebrates, which are often the webmasters of the ecosystem (Coleman and Hendrix, 2000). Once we have a measure, we can see how well we are doing in terms of sustainable forestry, and know whether or not it is smart forestry.

Acknowledgements

Financial support was provided by World Wildlife Fund (WWF), South Africa, National Research Foundation, Mondi Forests and South African Pulp and Paper Industries (SAPPI) Ltd.

References

African Wildlife (2005) Disappearing Species. *African Wildlife* 59 (Special Issue).

Armstrong, A.J., Benn, G., Bowland, A.E., Goodman, P.S., Johnson, D.N., Maddock, A. and Scott-Shaw, C.R. (1998) Plantation forestry in South Africa and its impact on biodiversity. *South African Forestry Journal* 182, 59–65.

Bennett, A.F. (1999) *Linkages in the Landscape: The Role of Corridors in Wildlife Conservation*. IUCN, Gland, Switzerland and Cambridge.

Buchwald, E. and Svenning, J.-C. (2005) Premature rejection of a controversial hypothesis? Primeval forest openness revisited. *Journal of Ecology* 93, 168–177.

Bullock, W.L. and Samways, M.J. (2005) Conservation of flower–arthropod interactions in remnant grassland linkages among pine afforestation. *Biodiversity and Conservation* 14, 3093–3103.

Burel, F. and Baudry, J. (1995) Farming landscapes and insects. In: Glen, D.M., Greaves, M.P. and Anderson, H.M. (eds) *Ecology and Integrated Farming Systems*. Wiley, Chichester, UK, pp. 203–220.

Burney, D.A. and Flannery, T.F. (2005) Fifty millennia of catastrophic extinctions after human contact. *Trends in Ecology and Evolution* 20, 395–401.

Burney, D.A., James, H.F., Burney, L.P., Olson, S.L., Kikuchi, W., Wagner, W.L., Burney, M., McCloskey, D., Kikuchi, D., Grady, F.V., Gage, R. II and Nishek, R. (2001) Fossil evidence for a diverse biota from Kuaa'i and its transformation since human arrival. *Ecological Monographs* 71, 615–641.

Chambers, B.Q. and Samways, M.J. (1998) Grasshopper response to a 40-year experimental burning and mowing regime, with recommendations for invertebrate conservation management. *Biodiversity and Conservation* 7, 985–1012.

Coleman, D.C. and Hendrix, P.F. (eds) (2000) *Invertebrates as Webmasters in Ecosystems*. CAB International, Wallingford, UK.

Coope, E.R. (1995) Insect faunas in ice age environments: why so little extinction? In: Lawton, J.H. and May, R.M. (eds) *Extinction Rates*. Oxford University Press, Oxford, pp. 55–74.

Denno, R.F., Roderick, G.K., Peterson, M.A., Huberty, A.F., Döbel, H.G., Eubanks, M.D., Losey, J.E. and Langellotto, G.A. (1996) Habitat persistence underlies intraspecific variation in dispersal strategies of planthoppers. *Ecological Monographs* 66, 389–408.

Donnelly, D. and Giliomee, J.H. (1985) Community structure of epigaeic ants in a pine plantation and in newly burnt fynbos. *Journal of the Entomological Society of Southern Africa* 48, 259–265.

Elias, S.A. (1994) *Quaternary Insects and their Environments*. Smithsonian Institution Press, Washington, DC.

Fagan, B. (2004) *The Long Summer: How Climate Changed Civilization*. Granta, London.

Fjeldså, J. and Lovett, J. (1997) Geographical patterns of old and young species in African forest biota: the significance of montane areas as evolutionary centres. *Biodiversity and Conservation* 6, 325–346.

Flannery, T.F. (1994) *Future Eaters*. Braziller, New York.

Forman, R.T.T. (1995) *Land Mosaics: The Ecology of Landscapes and Regions*. Cambridge University Press, Cambridge.

Fry, R. and Lonsdale, D. (eds) (1991) *Habitat Conservation – A Neglected Green Issue*. The Amateur Entomologists' Society, Middlesex, UK.

Gebeyehu, S. and Samways, M.J. (2002) Grasshopper assemblage response to a restored national park (Mountain Zebra National Park, South Africa). *Biodiversity and Conservation* 11, 283–304.

Gebeyehu, S. and Samways, M.J. (2006a) Conservation refugium value of a large mesa for grasshoppers in South Africa. *Biodiversity and Conservation* 15, 717–734.

Gebeyehu, S. and Samways, M.J. (2006b) Topographic heterogeneity plays a crucial role for grasshopper diversity in a southern African megabiodiversity hotspot. *Biodiversity and Conservation* 15, 231–244.

Greenslade, P. (1993) Australian native steppe-type landscapes: neglected areas for invertebrate conservation in Australia. In: Gaston, K.J., New, T.R. and Samways, M.J. (eds) *Perspectives on Insect Conservation*. Intercept, Andover, UK, pp. 51–73.

Grime, J.P. (1998) Benefits of plant diversity to ecosystems: immediate, filter and founder effects. *Journal of Ecology* 86, 901–910.

Hilty, J.A., Lidicker, Jr., W.Z. and Merenlender, A.M. (2006) *Corridor Ecology*. Island Press, Washington, DC.

Hess, G.R. and Fischer, R.A. (2001) Communicating clearly about conservation corridors. *Landscape and Urban Planning* 55, 195–208.

Hewitt, G. (2003) Ice ages, species distributions and evolution. In: Rothschild, L.J. and Lister, A.M. (eds) *Evolution on Planet Earth*. Academic Press, Amsterdam, The Netherlands, pp. 339–361.

Ingham, D.S. and Samways, M.J. (1996) Application of fragmentation and variegation models to epigaeic invertebrates in South Africa. *Conservation Biology* 10, 1353.

Jongman, R. and Pungetti, G. (eds) (2004) *Ecological Networks: Concept, Design, Implementation*. Cambridge University Press, Cambridge.

Kinvig, R. and Samways, M.J. (2000) Conserving dragonflies (Odonata) along streams running through commercial forestry. *Odonatologica* 29, 195–208.

Kirby, P. (1992) *Habitat Management for Invertebrates: A Practical Handbook*. Royal Society for the Protection of Birds, Sandy, Bedforshire, UK.

Kleijn, D. and Sutherland, W.J. (2003) How effective are European agri-environment schemes in conserving and promoting biodiversity? *Journal of Applied Ecology* 40, 947–969.

Koricheva, J., Mulder, C.P.H., Schmid, B., Joshi, J. and Huss-Danell, K. (2000) Numerical responses of different trophic groups of invertebrates to manipulations of plant diversity in grasslands. *Oecologica* 125, 271–282.

Lawrence, J.M. and Samways, M.J. (2002) Influence of hilltop vegetation type on an African butterfly assemblage and its conservation. *Biodiversity and Conservation* 11, 1163–1171.

Lockwood, J.A. (2001) The ethics of 'Classical' biological control and the value of Place. In: Lockwood, J.A., Howarth, F.G. and Purcell, M.F. (eds) *Balancing Nature: Assessing the Impact of Importing Non-native Biological Control Agents (an International Perspective)*. Entomological Society of America, Lanham, Massachusetts, pp. 100–109.

Lu, S.-S. and Samways, M.J. (2002) Conservation management recommendations for the Karkloof blue butterfly *Orachyrsops ariadne* (Lepidoptera: Lycaenidae). *African Entomology* 10, 149–159.

Mattison, E.H.A. and Norris, K. (2005) Bridging the gaps between agricultural policy, land use and biodiversity. *Trends in Ecology and Evolution* 20, 610–616.

Mawdsley, N.A. and Stork, N.E. (1995) Species extinctions in insects: ecological and biogeographical considerations. In: Harrington, R. and Stork, N.E. (eds) *Insects in a Changing Environment*. Academic Press, London, pp. 321–369.

Mevi-Schütz, J. and Erhardt, A. (2005) Amino acids in nectar enhance butterfly fecundity: a long-awaited link. *The American Naturalist* 165, 411–419.

Myers, N., Mittermeier, R.A., Mittermeier, C.G., Da Fonseca, G.A.B. and Kent, J. (2000) Biodiversity hotspots for conservation priorities. *Nature* 403, 853–858.

Neke, K.S. and Du Plessis, M.A. (2004) The threat of transformation: quantifying the vulnerability of grasslands in South Africa. *Conservation Biology* 18, 466–477.

New, T.R. (2005) *Invertebrate Conservation and Agricultural Systems*. Cambridge University Press, Cambridge.

Niemelä, J. (1997) Invertebrates and boreal forest management. *Conservation Biology* 11, 601–610.

Peterson, M.A. (1996) Long-distance gene flow in the sedentary butterfly, *Euphilotes enoptes* (Lepidoptera: Lycaenidae). *Evolution* 50, 1990–1999.

Ponel, P., Orgeas, J., Samways, M.J., Andrieu-Ponel, V., de Beaulieu, J.-L., Reille, M., Roche, P. and Tatoni, T. (2003) 110,000 years of Quaternary beetle diversity change. *Biodiversity and Conservation* 12, 2077–2089.

Pott, R.McC. (1997) Plantation forestry in South Africa and its impact on biodiversity and water. *South African Forestry Journal* 180, 45–48.

Pryke, S.R. and Samways, M.J. (2001) Width of grassland linkages for the conservation of butterflies in South African afforested areas. *Biological Conservation* 101, 85–96.

Pryke, S.R. and Samways, M.J. (2003) Quality of remnant indigenous grassland linkages for adult butterflies (Lepidoptera) in an afforested African landscape. *Biodiversity and Conservation* 12, 1985–2004.

Rapport, D., Costanza, R., Epstein, P.R., Gaudet, C. and Levins, R. (1998) *Ecosystem Health*. Blackwells, Oxford.

Ricketts, T.H., Daily, G.C., Ehrlich, P.R. and Fay, J.P. (2001) Countryside biogeography of moths in a fragmented landscape: biodiversity in native and agricultural habitats. *Conservation Biology* 15, 378–388.

Rivers-Moore, N.A. and Samways, M.J. (1996) Game and cattle trampling, and impacts of human dwellings on arthropods at a game park boundary. *Biodiversity and Conservation* 5, 1545–1556.

Rosenberg, D.K., Noon, B.R. and Meslow, E.C. (1997) Biological corridors: form, function and efficiency. *BioScience* 47, 677–687.

Samways, M.J. (1990) Land forms and winter habitat refugia in the conservation of montane grasshoppers in southern Africa. *Conservation Biology* 4, 375–382.

Samways, M.J. (1995) Southern hemisphere insects: their variety and the environmental pressures upon them. In: Harrington, R. and Stork, N.E. (eds) *Insects in a Changing Environment*. Academic Press, London, pp. 297–320.

Samways, M.J. and Kreuzinger, K. (2001) Vegetation, ungulate and grasshopper interactions inside vs outside an African savanna game park. *Biodiversity and Conservation* 10, 1963–1981.

Samways, M.J. and Moore, S.D. (1991) Influence of exotic conifer patches on grasshopper (Orthoptera) assemblages in a grassland matrix at a recreational resort, Natal, South Africa. *Biological Conservation* 57, 205–219.

Samways, M.J., Caldwell, P.M. and Osborn, R.M. (1996) Ground-living invertebrate assemblages in native, planted and invasive vegetation in South Africa. *Agriculture, Ecosystems and Environment* 59, 19–32.

Samways, M.J., Taylor, S. and Tarboton, W. (2005) Extinction reprieve following alien removal. *Conservation Biology* 19, 1329–1330.

Schmidtt, T. and Seitz, A. (2001) Allozyme variation in *Polyommatus coridon* (Lepidoptera: Lycaenidae): identification of ice-age refugia and reconstruction of post-glacial expansion. *Journal of Biogeography* 28, 1129–1136.

Sinclair, J.E. and New, T.R. (2004) Pine populations in south eastern Australia support highly impoverished ant assemblages (Hymenoptera: Formicidae). *Journal of Insect Conservation* 8, 277–286.

Smith, D.S. and Hellmund, P.C. (eds) (1993) *Ecology of Greenways: Design and Function of Linear Conservation Areas*. University of Minnesota Press, Minneapolis, Minnesota.

Steadman, D.W. (1995) Prehistoric extinctions of Pacific island birds: biodiversity meets zooarchaeology. *Science* 267, 1123–1131.

Stockwell, C.A., Hendry, A.P. and Keunison, M.J. (2003) Contemporary evolution meets conservation biology. *Trends in Ecology and Evolution* 18, 94–101.

Swanson, F.J., Kratz, T.K., Caine, N. and Woodmansee, R.G. (1988) Landform effects and ecosystem patterns and processes. *BioScience* 38, 92–98.

Teie, W.C. (2005) *Fire Manager's Handbook on Veld and Forest Fires*. Southern African Institute of Forestry, Menlo Park, California.

Thornton, I. (1996) *Krakatau: The Destruction and Reassembly of an Island Ecosystem*. Harvard University Press, Cambridge, Massachussetts.

Travis, J.M.J. (2003) Climate change and habitat destruction: a deadly anthropogenic cocktail. *Proceedings of the Royal Society of London Series B – Biological Sciences* 270, 467–473.

Warren, M.S., Hill, J.K., Thomas, J.A., Asher, J., Fox, R., Huntley, B., Roy, D.B., Telfer, M.G., Jeffcoate, S., Harding, P., Jeffcoate, G., Willis, S.G., Greatorex-Davies, J.N., Moss, D. and Thomas, C.D. (2001) Rapid responses of British butterflies to opposing forces of climate and habitat change. *Nature* 414, 65–69.

Wiens, J.A., Stenseth, N.C., van Horne, B. and Ims, R.A. (1993) Ecological mechanisms and landscape ecology. *Oikos* 66, 369–380.

Wood, P.A. and Samways, M.J. (1991) Landscape element pattern and continuity of butterfly flight paths in an ecologically landscaped botanic garden, Natal, South Africa. *Biological Conservation* 58, 149–166.

7 Insects and Bioindication: Theory and Progress

MELODIE A. MCGEOCH

Centre for Invasion Biology, Department of Conservation Ecology and Entomology, University of Stellenbosch, Private Bag X1, Matieland 7602, South Africa

1 Introduction

Bioindication has experienced renewed interest over the last decade, precipitated largely by the 1992 Convention on Biological Diversity (CBD) (Glowka *et al.*, 1994) and the 2010 target to reduce the rate of global biodiversity loss (United Nations, 2002; UNEP, 2003). The relative dearth of information on biodiversity, particularly in species-rich parts of the world, combined with rapid rates of human-induced species loss remains a significant challenge to conservation. This challenge can only be effectively met with efficient approaches to gather maximal information with minimal resource requirements. Bioindicators, that both readily reflect and represent the state of the environment, provide such a tool (Table 7.1). Bioindication has become an essential component of conservation strategies aimed at addressing a wide array of biodiversity threats. The use of selected, suitable species, or species groups, to reflect some component of their environment or biodiversity context is far from new, but has recently undergone critical evaluation in an attempt to establish bioindication as an effective, additional tool for addressing the biodiversity crisis (McGeoch, 1998; Caro and O'Doherty, 1999; Hilty and Merenlender, 2000; Duelli and Obrist, 2003; Niemi and McDonald, 2004).

Insects in particular have been flagged as promising bioindicators for over two and a half decades because of their significant contribution to global species richness, biomass and ecological function, as well as their responsiveness and extensive life history and behavioural diversity (Lenhard and Witter, 1977; Majer, 1983; Brown, 1991; Erhardt and Thomas, 1991; Holloway and Stork, 1991; Rosenberg and Resh, 1993; Chessman, 1995; Luff and Woiwod, 1995; Davis *et al.*, 2001; Lu and Samways, 2002; Balvanera *et al.*, 2005). However, insects have also been integral in recent efforts to improve the rigour of bioindication and enhance its value to biodiversity conservation (the term 'insect' is used for convenience throughout this chapter to more

Table 7.1. Definitions and categories of bioindication.

	Definition	Examples and related terms
Bioindicator	A species or group of species that readily: reflects the abiotic or biotic state of an environment; represents the impact of environmental change on a habitat, community or ecosystem or is indicative of the diversity of a subset of taxa, or of wholesale diversity, within an area	
Bioindication system	A tool to extract biological information from an ecosystem and to use this information for making scientifically based management decisions (van Straalen and Krivolutsy, 1996)	
Policy indicator	Indicates the success of policy implementation, or the requirement for intervention, in bringing about one or more conservation objectives. This may include bioindicators	Examples include: state of the environment indicators, the living planet index, indicators of sustainability
Three categories of bioindication		
1. Environmental indicator	A species or group of species that responds predictably, in ways that are readily observed and quantified, to environmental disturbance or to a change in environmental state	Related terms: Sentinel, exploiter, bioassay, accumulator, biomarker
2. Ecological indicator	A species or group of species that demonstrates the effects of environmental change (such as habitat alteration, fragmentation and climate change) on biota or biotic systems	
3. Biodiversity indicator	A biodiversity indicator is a group of taxa (e.g. genus, tribe, family or order, or a selected group of species from a range of higher taxa), or functional group, the diversity of which reflects some measure of the diversity (e.g. character richness, species richness, level of endemism) of other higher taxa in a habitat or set of habitats	Related terms: Surrogate, umbrella, flagship, focal species or taxon

broadly include a range of freshwater and terrestrial arthropod taxa). This chapter examines current thinking and recent advances in the field of bioindication and the position and role of insects within it. It does not aim to provide a comprehensive review, but rather to highlight key concepts and areas of significant progress. Topics that remain important to the field, but that are extensively dealt with elsewhere, include the selection criteria and characteristics of effective indicators (Brown, 1991; Holloway and Stork, 1991; Kremen, 1992; Hammond, 1994; McGeoch, 1998; Andersen, 1999; Buchs, 2003a) and the selection and development of bioindicators in particular environments, such as agriculture (e.g. Bailey *et al.*, 1999; Enami *et al.*, 1999; Fauvel, 1999; Marc *et al.*, 1999; Paoletti, 1999; Buchs, 2003b; Fox and MacDonald, 2003; Hoffmann and Greef, 2003; Sauberer *et al.*, 2004; Zalidis *et al.*, 2004; Halberg *et al.*, 2005) and forestry (e.g. Thor, 1998; Ferris and Humphrey, 1999; Jonsson and Jonsell, 1999; Lindenmayer, 1999; Simberloff, 1999; Gustafsson, 2000; Nilsson *et al.*, 2001; Rempel *et al.*, 2004; Schulze *et al.*, 2004; Dudley *et al.*, 2005). The chapter begins with a brief overview of the field, and synthesis of the taxa, environments and forms of bioindication that appear in recent literature. The methodological process by which bioindication systems are developed is summarized, and significant technical developments in ecological bioindication and progress in biodiversity indication highlighted. Finally, the chapter discusses the bioindication science–policy interface, and, in conclusion, provides a perspective on the theory of bioindication.

2 Defining the Field

The terminology associated with bioindication is varied and extensive, and its use and interpretation often inconsistent between fields (compare, e.g. ecotoxicology, community ecology and environmental policy), as well as between geographic regions (Hammond, 1994; McGeoch, 1998; Muller *et al.*, 2000; Duelli and Obrist, 2003; Niemi and McDonald, 2004). Bioindicators are, however, distinguished from abiotic indicators, such as soil quality, temperature and landscape structure (Humphrey *et al.*, 1999). They are also distinct from composite or indirect indicators, such as land-use metrics, environmental diversity or policy indicators (Table 7.1; Levy *et al.*, 2000; Faith, 2003; Hietala-Koivu *et al.*, 2004).

Three forms of bioindication have emerged that correspond to the three main applications of bioindicators (Fig. 7.1, Table 7.1). Although these categories are not always mutually exclusive, they do successfully distinguish the vast majority of bioindication studies, including those using insects and other taxa. The distinction is also useful because the categories have very different objectives, and subsequently different approaches, methods and necessary conditions that the bioindicator should fulfill (see McGeoch, 1998). Furthermore, the targets of indication in the categories differ in the degree to which they can be accurately quantified and thus predicted. For example, the target of environmental indication is generally some readily measured abiotic characteristic of the environment (such as pH, heavy metal concentration), whereas in ecological bioindication the environmental state of interest may be some, less accurately quantified or more complex variable, such as habitat disturbance or climate change. The variables to be indicated thus tend to have different statistical properties, levels of natural variability and levels of susceptibility to measurement error that will influence the performance of the bioindicator.

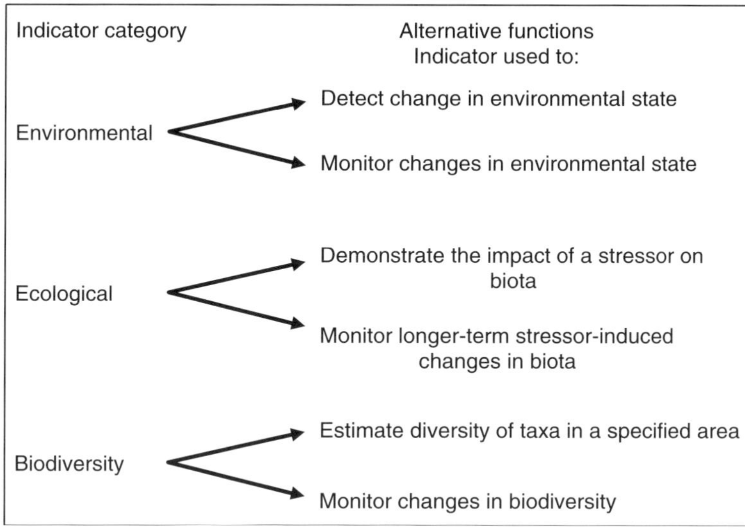

Fig. 7.1. The function of bioindicators in each category of bioindication. (Redrawn with permission from McGeoch, 1998.)

3 Objectives, Environments and Taxa

Although insects have long been promoted as bioindicators, their value remains contested; this is well illustrated by the following viewpoints:

> The wealth of existing, documented information on the relationship between invertebrates and habitat parameters, means that they offer great potential as indicators of biodiversity. In addition to being well-studied, invertebrates may be sampled using established, standardized methods, and expertise is widely available.
>
> (Ferris and Humphrey, 1999)

> Insects and other microfauna…are of limited use in terrestrial systems because of the cost of sampling and processing and because there is limited acceptance by resource managers, politicians, and the general public.
>
> (Niemi and McDonald, 2004)

Andersen and Majer (2004) also recognize the constraints to widespread adoption of insects as bioindicators by land managers, because of the sampling effort and taxonomic expertise that is commonly required. However, given the demand for the development of bioindication systems, and the several distinct advantages and long history that insects have as bioindicators, their continued use in bioindication systems seems both essential and inevitable (Dobson, 2005). None the less, bioindication clearly extends beyond the use of insects, and valuable insights are to be gained by examining the development of the subject and the position of insects within it. Furthermore, insects are increasingly used along with other taxa, as well as non-taxonomic indicators to achieve indication objectives (Watson, 2005), and thus should not be considered divorced from developments in the broader field.

A summary of recent literature[1] on bioindication demonstrates the dominance of environmental bioindication studies. Environmental bioindicators are generally at a more advanced stage of development than the other two categories, particularly freshwater monitoring schemes involving macroinvertebrates, e.g. the River Invertebrate Prediction Classification System used in the UK to monitor the pollution status of water courses (Wright *et al.*, 2000) and the South African Scoring System (SASS) (Chutter, 1972; Hodkinson, 2005; Revenga *et al.*, 2005). By comparison, the studies on biodiversity indicators remain surprisingly few (Fig. 7.2a). Insect (freshwater and terrestrial) publications are dominated by ecological, followed by biodiversity indication, illustrating the relative importance of these categories in insect bioindication (Fig. 7.2a). The volume of research in above-ground terrestrial environments is fairly similar to that conducted in aquatic (marine and freshwater) environments, whereas there are comparatively few soil-based studies (based on the search-terms used here) (Fig. 7.2b). The latter is surprising considering that bioindication systems for soils are comparatively well-established (van Straalen and Krivolutsky, 1996; van Straalen and Verhoef, 1997; Cortet *et al.*, 1999; Viard *et al.*, 2004; Parisi *et al.*, 2005), but perhaps reflects the more advanced status of this field. With the obvious exclusion of the marine environment, the distribution of insect studies across environments is similar to that for all studies, although there has been comparatively more work in above-ground terrestrial environments (Fig. 7.2b).

From a taxonomic perspective the literature is dominated by studies on plants (particularly lichens as pollution indicators) and invertebrates (including insects), together constituting over 65% of all publications (Fig. 7.3a). Amongst Arthropoda, the hexapods encompass the vast majority of stud-

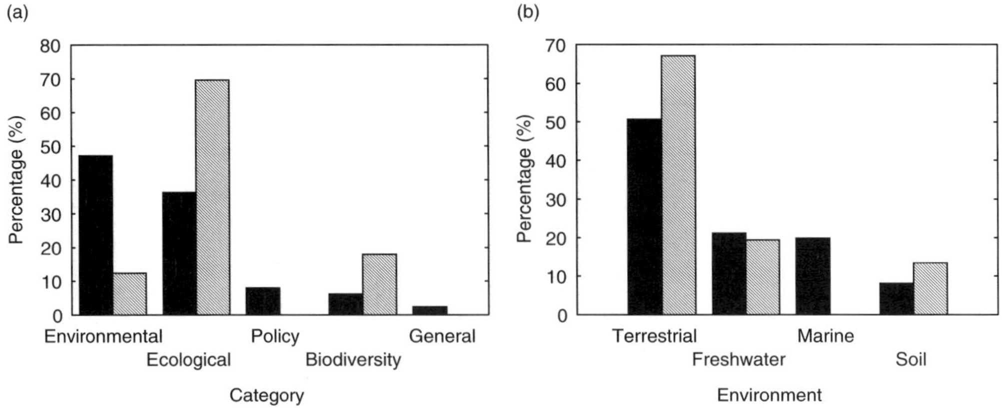

Fig. 7.2. Frequency of bioindicator publications: (a) on different forms of bioindication; and (b) conducted in different environments. Solid bars are for all bioindication publications (n = 2311 (a), 2088 (b)), whereas hashed bars are for arthropod bioindication publications only (n = 283) (see Endnote).

ies (Fig. 7.3b), with the Coleoptera and Hymenoptera (Fig. 7.3c), and ants, ground beetles and bloodworms (chironomid larvae) (Fig. 7.3d), most frequently represented. The Coleoptera, especially ground, tiger and dung beetles, are certainly well recognized as ecological bioindicators and have also been tested in biodiversity assessments (Pearson and Cassola, 1992; Pearson and Carroll, 1998; van Jaarsveld *et al.*, 1998). Dung beetles have been extensively used in studies as indicators of disturbance and habitat quality, particularly in the tropics and subtropics (Spector and Forsyth, 1998; Van Rensburg *et al.*, 1999; Davis *et al.*, 2001, 2004; Halffter and Arellano, 2002; Avendano-Mendoza *et al.*, 2005). Ground beetles have been applied in similar contexts, although studies are geographically skewed to higher latitudes (Kromp, 1999; Paoletti *et al.*, 1999; Magura *et al.*, 2000; Niemelä *et al.*, 2000a; Cole *et al.*,

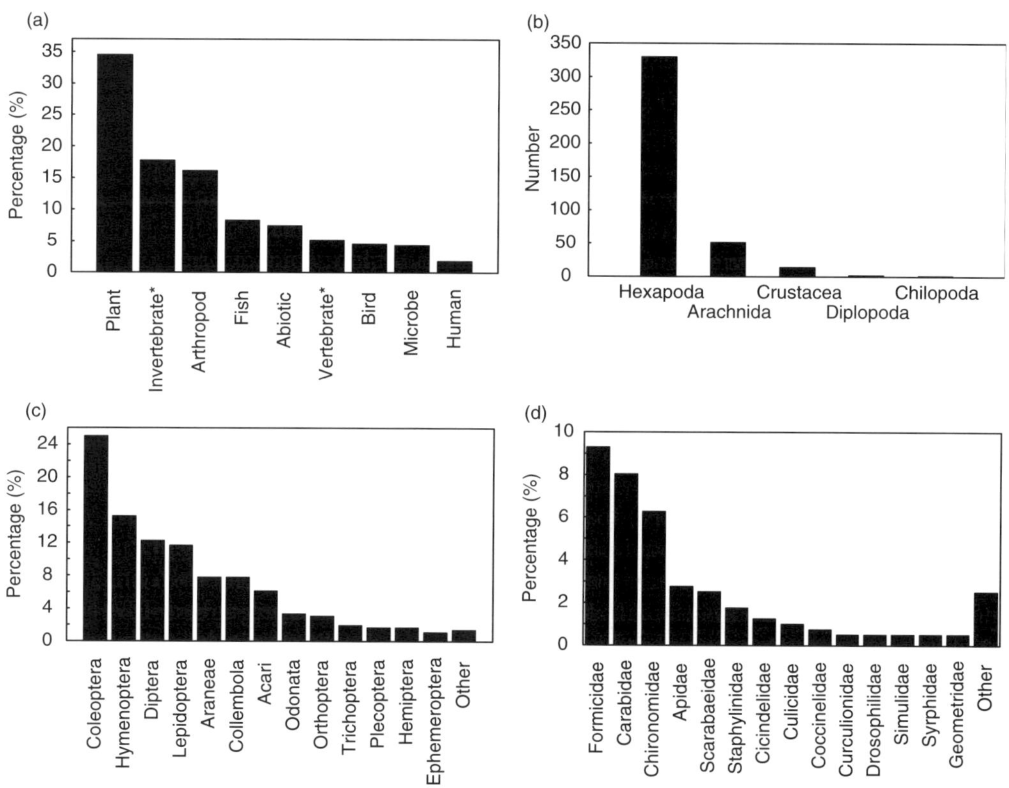

Fig. 7.3. Frequency of bioindication publications involving different taxa. (a) *Invertebrate category excludes terrestrial and freshwater arthropods and *vertebrate category excludes fish, birds and humans (n = 2061). The abiotic category includes studies that do not use species information as the indicator. Microbes include, for example, bacteria, protozoa and dinoflagellates. Publications including: (b) taxa in terrestrial and freshwater arthropod classes (n = 287); (c) arthropod orders ('Other' includes Isoptera, Dermaptera, Mantodea and Thysanoptera); and (d) insect families (n = 160, 'Other' includes Buprestidae, Cerambicidae, Chrysomelidae, Dytiscidae, Lucanidae, Tenebrionidae, Sarcophagidae, Braconidae, Chalcidoidea and Chrysopidae).

2002; Allegro and Sciaky, 2003; Buchs, 2003a). The use of Hymenoptera in bioindication studies includes largely ants, but also honeybees (particularly as environmental indicators of pollutant levels in agroecosystems (Celli and Maccagnani, 2003)) and other apidoid communities (Tscharntke *et al.*, 1998; Brown and Albrecht, 2001; Gayubo *et al.*, 2005). Ants have been strongly promoted as bioindicators, mostly of land use and restoration, because of their high diversity and functional importance, especially in the southern hemisphere (Brown, 1991; Andersen, 1997; King *et al.*, 1998; de Bruyn, 1999; Osborn *et al.*, 1999; Alonso, 2000; Armbrecht and Ulloa-Chacon, 2003; Matlock and de la Cruz, 2003; Andersen *et al.*, 2004; Parr *et al.*, 2004; van Hamburg *et al.*, 2004; Netshilaphala *et al.*, 2005). Ants are also amongst the insect ecological bioindicators most extensively adopted by land managers (Kaspari and Majer, 2000; Andersen *et al.*, 2002; Andersen and Majer, 2004) (see also Table 4 in Buchs, 2003a). A fairly novel application with apparent potential for future development is the use of invasive insect taxa, often Hymenoptera, in bioindication and monitoring (Kevan, 1999; Chapman and Bourke, 2001; Cook, 2003; Revenga *et al.*, 2005). Blood worms are commonly used as both environmental indicators of freshwater pollution (Pinder and Morley, 1995; Hamalainen, 1999; Orendt, 1999; Meregalli *et al.*, 2000; de Bisthoven *et al.*, 2005) and of habitat quality (Brodersen and Lindegaard, 1999; Milakovic *et al.*, 2001; Brodersen and Anderson, 2002).

The frequency of taxa in bioindication studies is, however, little different to the relative number of described species in each group, at least at the order level (Fig. 7.4). A clear exception is the Hemiptera that is under-represented in

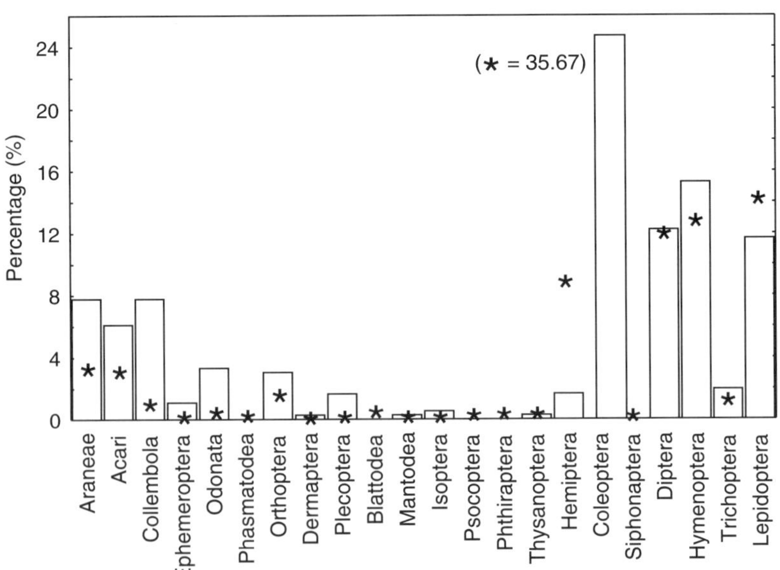

Fig. 7.4. The percentage of studies in which taxa appear in the literature (bars) compared with the percentage of described species in the same taxon (*). Described species percentages calculated from data in Gaston (1991) and Grimaldi and Engel (2005).

bioindicator studies. Although Auchenorrhyncha communities are considered good potential bioindicators (Duelli and Obrist, 2003; Nickel and Hildebrandt, 2003), the level of taxonomic knowledge of the group proves an obstacle to its use in many instances (Buchs, 2003a). By contrast, the spiders, mites and springtails are comparatively over-represented for their taxonomic diversity (Fig. 7.4). Mites and springtails are mostly used in agricultural and soil environments as indicators of habitat quality and contamination (Behan-Pelletier, 1999; Alvarez *et al.*, 2001; Zaitsev and van Straalen, 2001; Ponge *et al.*, 2003; Geissen and Kampichler, 2004; Sousa *et al.*, 2004), as are spiders (Gravesen, 2000; Wheater *et al.*, 2000; Horvath *et al.*, 2001; Gibb and Hochuli, 2002; Woinarski *et al.*, 2002; Cardoso *et al.*, 2004a). Reasons for the frequency distribution of studies across taxa (Fig. 7.4) generally include the proportional species richness of the groups, but also the selection of taxonomically manageable or better-known groups, and taxa that are conspicuous, abundant and readily sampled or quantified (Brown, 1991; Buchs, 2003a).

4 The Methodology of Bioindication

Between 1998 and 2000, at least three independent reviews of bioindication appeared in the literature (McGeoch, 1998; Caro and O'Doherty, 1999; Hilty and Merenlender, 2000). These marked the recognition of both the importance of bioindication and the need for a critical evaluation of the field. Although all three reviews agree on the advantages of insects as bioindicators, they highlight the requirement for clear objective setting, improved scientific rigour and the development of methods to facilitate progress in the field. The last decade has seen significant advances along these lines, as well as the widespread recognition of the importance of a sound, rigorous scientific approach to bioindication (Noss, 1990; Murtaugh, 1996; Lindenmayer, 1999; Niemelä, 2000; Mac Nally and Fleishman, 2002; McGeoch, 2002; Gregory *et al.*, 2005).

4.1 Quantification and predictability

Bockstaller and Girardin (2003) outline a number of levels of validation necessary for indicators. The first of these, design validation, which takes place in the absence of data, is based on expert opinion and is equivalent to the use of a priori bioindicator selection criteria (such as ease of sampling, cost-effectiveness and taxonomic knowledge: see McGeoch, 1998) to minimize the risk of the rejection of the putative bioindicator subsequent to substantial investment in its testing. The essence of bioindication is the predictability of the relationship between the bioindicator and the environmental parameter (EP) of interest. The critical criteria for any bioindicator thus remain the presence of a strong, significant and robust relationship between it and, for example, the concentration of a pollutant, habitat quality or the biodiversity of a particular taxon or area (Fig. 7.5). The empirical and statistical measurement of predictability first includes the identification of a significant relationship between the putative bioindicator and the EP, as well as the extent to which the variability in the EP is explained by the relationship (Fig. 7.5). Statistical significance is a necessary but not sufficient condition for bioindicators, whereas the higher the explanatory power of the relationship (the stronger the relationship) the more useful the bioindicator is likely to be (Fig. 7.5). For example, in an evaluation of the potential of trap-nesting bees and wasps as ecological bioindicators of habitat quality, the relationship between species richness of trap-nesters and other groups of bees and wasps was both significant and strong ($P < 0.001$, $r^2 = 81.6\%$) (Tscharntke *et al.*, 1998). However, although natural enemy species richness decreased significantly with increasing patch isolation ($P = 0.003$), the relationship was weak ($r^2 = 20\%$) (Tscharntke *et al.*, 1998) and therefore not suitable as a basis for bioindication. This first stage in the development of a bioindicator, i.e. establishing the 'relationship' (Fig. 7.5), forms the basis of a large component of the literature on bioindication.

Second, robustness is the degree to which the relationship between the putative bioindicator and the EP remains constant within the spatial and temporal context of inference, i.e. of the bioindication objective (Fig. 7.5). This stage of bioindication development is related to the 'output validation' process

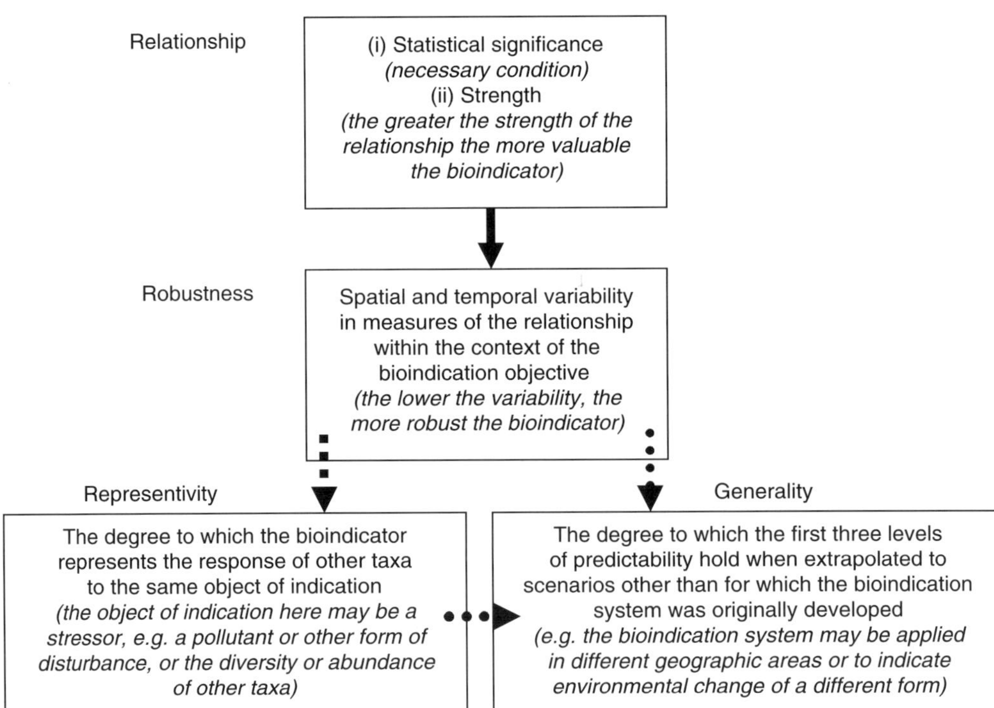

Fig. 7.5. The predictability hierarchy in bioindication. As a bioindicator moves along the hierarchy of levels of predictability it gains value for biodiversity assessment and conservation management.

described by Bockstaller and Girardin (2003). It essentially involves the use of independent data to test the consistency or repeatability of bioindicator performance (i.e. to test the hypothesis) and to quantify the degree of confidence with which the bioindicator may be used (McGeoch *et al.*, 2002). Thomson *et al.* (2005) provide an excellent example of establishing the robustness of biodiversity indication models. Models of bird and butterfly species richness were built from inventories conducted over an 8-year period, and then the performance of selected models were tested (using a more recent data-set collected over a shorter period in the same region) by comparing observed and predicted species richness values. Rösch *et al.* (2001) provide a further example by testing the repeatability of a moth assemblage as a bioindicator of habitat quality in urban areas. The study was repeated 3 years after the initial establishment of the relationship between moth species richness and habitat quality, using both the same and independent sampling sites. Although few such examples are to be found in the literature (but see, e.g. King *et al.*, 1998; Campbell *et al.*, 2000; Fleishman *et al.*, 2001; Hogg *et al.*, 2001; Geissen and Kampichler, 2004), the existence of significant, strong and robust relationships are the minimum necessary criteria in most instances for a system to warrant the 'bioindicator' tag. Bioindicators may be selected and applied without undergoing such rigorous assessment of their predictability. However, in the absence of such an assessment, no measured degree of confidence may be placed in the information provided by the bioindicator. In this case, the bioindicator is used in the hope that it reflects some unmeasured component of the environment. Indeed, this may be the only avenue possible under data-deficient scenarios, where measuring anything is better than measuring nothing in the hope that over time the purported bioindicator and its relationship with its environment will be better understood.

Thereafter, two additional criteria that the bioindicator may meet are representivity and generality (Fig. 7.5), and the requirement for them to do so is at least partly dependent on the bioindication category and objective. For example, the cadmium concentration in honeybees (Conti and Botre, 2001) is inclusive as an environmental bioindicator, and no relationship need exist between this bioindicator and heavy metal concentration in other taxa, i.e. representivity is not a requirement. However, it would be valuable, for example, to assess how representative the positive relationship between restorative grazing and Lepidoptera abundance is of the abundance responses of other insect taxa (e.g. Poyry *et al.*, 2005). Again, few studies have established the representivity of bioindicators (see Majer, 1983; Andersen *et al.*, 2004; Schulze *et al.*, 2004).

Finally, if a particular relationship between a bioindicator and a stressor is found to hold in a domain other than that for which it was first identified, e.g. in a different habitat, nature reserve, geographic region, then it may be said to have generality (achieving 'universal laws' or 'predictive theory' as outlined by Murray, 2000) (Fig. 7.5). The 'GLOBENET' initiative provides a unique example of an approach to establish generality in an ecological bioindication system. This initiative aims to assess and compare landscape changes across urban development gradients on a global scale, using a single group of invertebrates, i.e. ground beetles, and common methodology in a similar landscape, i.e. urban mosaics (Niemelä *et al.*, 2000b). The outcome of the programme to date has shown some success, with generality in the effect of urbanization on the composition of ground beetle assemblages found across several cities and continents (Ishitani *et al.*, 2003). Approaches, such as GLOBENET, present underexploited, yet potentially powerful opportunities for developing general bioindicators. However, because few such case studies exist, it is not possible to estimate under which circumstances or how commonly generality is likely to be found. None the less, the criterion of generality is not a necessary requirement for the successful local application of a bioindication system.

4.2 Indicator values

In many instances, in both environmental and ecological bioindication, the objective of bioindication involves identifying species that are both sensitive to environmental quality and conspicuously responsive to a change in that quality. The response is generally quantified using measures of species abundance and distribution. The process of developing a bioindicator in this context involves the quantitative identification of sensitive and suitable species from an assemblage of potential taxa. Important considerations include: (i) separating stochastic abundance fluctuations from those associated with the environmental change of interest; (ii) being able to associate a quantitative indicator value and associated level of significance with the bioindicator species; (iii) using several, or a 'basket' of, species to improve the reliability of the bioindicator (Hammond, 1994; Duelli and Obrist, 2003; Maes and Van Dyck, 2005); and (iv) ensuring that species selected are readily sampled and quantified, and likely to remain so.

The most significant methodological advance in this area is the *Indicator Value* (*IndVal*) method developed by Dufrêne and Legendre (1997). This method combines measurements of the degree of specificity of a species to an ecological state, e.g. a habitat type, and its fidelity (or frequency of occurrence) within that state, and was first applied to an assemblage of 189 ground beetle species across 69 localities and nine habitats in Belgium. The method is in fact based on species-sample matrices of the sort commonly compiled for insect assemblages. Species with a high specificity and high fidelity within an environmental state will have a high indicator value for that state (Dufrêne and Legendre, 1997) (Fig. 7.6). High fidelity (frequency of occurrence) of a species across sample sites is generally associated with a large abundance of individuals (Brown, 1984; Gaston *et al.*, 1997). Both these characteristics facilitate sampling and monitoring, an important requirement for a useful bioindicator (Kremen *et al.*, 1994).

The *IndVal* method permits the identification of both 'characteristic' species (i.e. with high specificity and fidelity to a state and thus a high % *IndVal*), and 'detector' species (i.e. species that span a range of ecological states and have intermediate specificity). McGeoch *et al.* (2002) demonstrate this using dung beetles as indicators of habitat conversion from closed canopy forest to

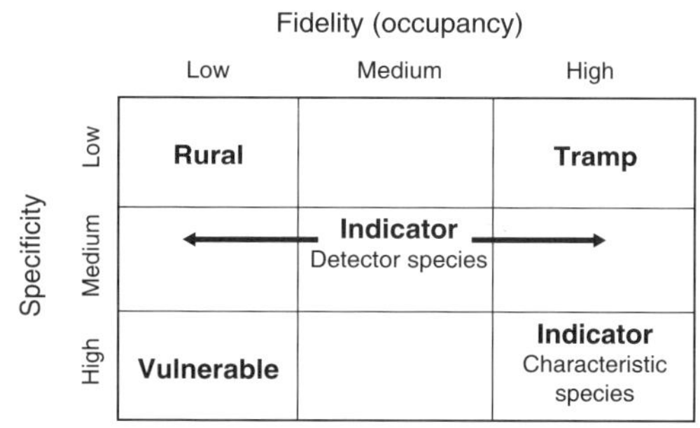

Fig. 7.6. Species characterized by a combination of their degree of environmental specificity and fidelity, and classified on this basis as either indicators (characteristic or detector species), tramp, rural or vulnerable species. (Redrawn with permission from McGeoch *et al.*, 2002.)

open, mixed woodland. Detector species may be more useful indicators of the direction of change than highly specific (characteristic) species restricted to a single state (Fig. 7.6). This is because the abundances (and thus the fidelity) of characteristic species may decline rapidly under changing environmental conditions to the point where they are regarded as vulnerable (Fig. 7.6). These species will become increasingly difficult to sample (Fig. 7.6), and may disappear rapidly with no further value for monitoring thereafter. Characteristic indicator species also provide no information on the direction of ecological change (although changes in their abundance may remain useful for monitoring within the habitat to which they are specific), because they are highly specific and thus restricted to a single ecological state (Fig. 7.6). By contrast, species with moderate specificity levels (detector species, Fig. 7.6), may thus be more useful for monitoring change. Bioindication using insects in aquatic and soil systems makes use of species such as these that have a range of preferences for different environmental states (e.g. van Straalen and Verhoef, 1997; Mouillot *et al.*, 2002), but this distinction has less commonly been made in above-ground terrestrial bioindication.

The *IndVal* method has several advantages over other indicator measures used for ecological bioindication (McGeoch and Chown, 1998). For example, the *IndVal* is calculated independently for each species, and there is complete flexibility with regard to the state (site, sample or habitat) categorization on which the *IndVal* measures are based (McGeoch and Chown, 1998). However, although habitat specificity is a comparatively inflexible species-specific trait (Southwood, 1988; Blackburn and Gaston, 2005), the abundances of species (and thus their fidelity) are likely to vary as a consequence of stochastic, seasonal, as well as disturbance factors. Insect abundances may also be higher under disturbed than undisturbed condition, as shown for dung beetles in coffee plantations versus cloud forest in Mexico (Pineda *et al.*, 2005). Abundance will thus not have a straightforward relationship with the EP of interest, resulting in potential problems with the interpretation of the *IndVal* (Hiddink, 2005). The sensitivity of the *IndVal* to such changes will thus ultimately determine its usefulness for bioindication.

Indeed, a comprehensive understanding of the behaviour and properties of indicator measures and indices in bioindication is critical (e.g. Chovanec and Waringer, 2001; Allegro and Sciaky, 2003; Garcia-Criado *et al.*, 2005). This must include an understanding of the formal relationship between index components, such as the fidelity and specificity components of the *IndVal* index. Failure to examine the properties of indicator measures or indices prior to their application can result in the misinterpretation of outcome values, the failure to recognize complex changes in the relationships between system components on which the aggregate measure is based, compounding of biases as a consequence of uncertainty or high variability in the constituent components and loss of information (Cousins, 1991; Gaston, 1996; Eiswerth and Haney, 2001; Niemi and McDonald, 2004; Loh *et al.*, 2005). Using a dung beetle assemblage, McGeoch *et al.* (2002) showed that species with significant, high *IndVal* tended to remain so when tested in different locations and at a different time (Fig. 7.7a). Although the fidelity component of *IndVal* is sensitive to species abundance fluctuations (Fig. 7.7b), the fidelity value

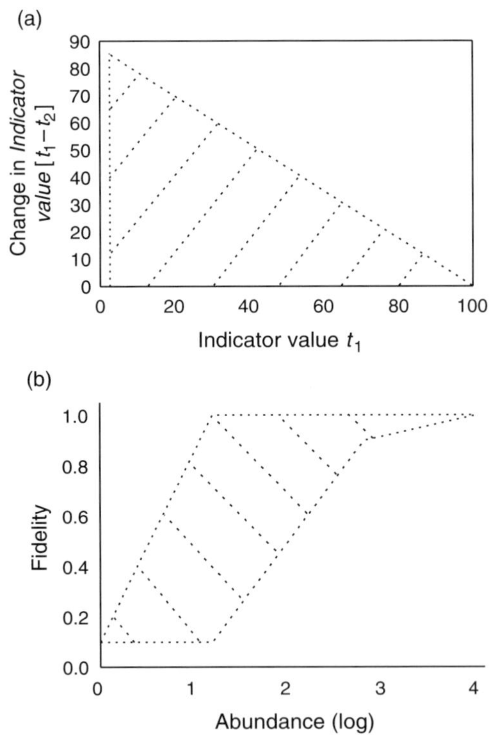

Fig. 7.7. Properties of the *Indicator Value* (*IndVal*) index of Dufrêne and Legendre (1997), stylized from McGeoch *et al.* (2002). (a) Relationship between the percentage indicator values of an assemblage of dung beetle species in one season and the change in this percentage 2 years later. (b) Relationship between the fidelity component (that lies between 0.0 and 1.0) of the *IndVal* and the logarithmically transformed abundance of species in the assemblage.

is calculated from relative, rather than absolute differences in the frequency of occurrence of a species across habitats. As a result, if the abundance of a species changes in a similar direction across environmental states of interest this may not affect a change in its fidelity value. Furthermore, the logistic nature of the relationship between fidelity and abundance (as well as the fact that abundance is logarithmic in the relationship) means that a substantial abundance change (over 1 order of magnitude) may not result in any change in fidelity (Fig. 7.7b). Properties such as these make the *IndVal* method a particularly effective tool for ecological bioindication. More widespread application of the method is, however, necessary to establish the generality of its properties.

5 Biodiversity Assessment and Monitoring

5.1 Assessment

Biodiversity indication is the youngest of the three categories of bioindication (Table 7.1), but has since the early 1990s received most attention. This has been driven by the urgent need to prioritize land areas for protection, given high rates of human-induced habitat destruction and species loss (Brown, 1991). Because our knowledge of the taxonomy and distribution of, particularly invertebrate, species is poor, comprehensive prioritization assessments are not possible. The only alternative is thus to use some surrogate of biodiversity as the basis for decision making, to both overcome the taxonomic impediment and save the time and expense required for comprehensive biodiversity surveys. Although abiotic or compound environmental correlates of biodiversity are often more practicable, and thus more likely to be adopted, these surrogate measures have been insufficiently tested and in some cases shown to be inadequate (Araujo *et al.*, 2001; Brooks *et al.*, 2004; Bonn and Gaston, 2005). Biodiversity indication has thus primarily concerned the use of individual species (e.g. Andelman and Fagan, 2000; Fleishman *et al.*, 2000), the species richness of target taxa (e.g. Kremen, 1994; Brehm and Fiedler, 2003; Araujo *et al.*, 2004; Grand *et al.*, 2004), higher taxon richness (e.g. Balmford *et al.*, 1996, 2000; Baldi, 2003; Cardoso *et al.*, 2004b), functional groups (e.g. Andersen, 1995; Horner-Devine *et al.*, 2003) and levels of rarity, endemism or threat (e.g. Broberg, 1999; Gustafsson, 2000; Orme *et al.*, 2005) to estimate a generally broader component of biodiversity.

However, despite the range of approaches and numerous case studies, the search for biodiversity indicators has met with limited success (Gaston and Blackburn, 1995; Gaston, 1996; Prendergast, 1997; Lindenmayer *et al.*, 2002a,b; Wilsey *et al.*, 2005). The preponderance of evidence demonstrates, at best, weak relationships between elements of biodiversity and proposed taxon-based surrogates thereof (e.g. Prendergast *et al.*, 1993; Humphrey *et al.*, 1999; Heino, 2001; Juutinen and Monkkonen, 2004; Kati *et al.*, 2004; Orme *et al.*, 2005; see also Hughes *et al.*, 2000; Kerr *et al.*, 2000; Moritz *et al.*, 2001; Garson *et al.*, 2002; Baldi, 2003). For example, the use of environmental rather than taxon-based variables as indicators has also yielded mixed results (e.g. Araujo *et al.*, 2001; Ricketts *et al.*, 2002; Ekschmitt *et al.*, 2003; Faith, 2003; Lassau and Hochuli, 2004; Bonn and Gaston, 2005; Dobson, 2005). None the less, particular approaches have proved more promising than others. For example, within-taxon surrogates generally perform better than across-taxon surrogates (Fleishman *et al.*, 2000, 2001), lower-taxonomic levels (e.g. genera) tend to predict species richness better than higher taxonomic levels (albeit scale-dependent) (e.g. Balmford *et al.*, 2000; Grelle, 2002; La Ferla *et al.*, 2002; Cardoso *et al.*, 2004b) and relationships show a tendency to be stronger at course than fine scales (Grand *et al.*, 2004; Sarkar *et al.*, 2005).

Despite the growing acceptance that different aspects of biodiversity (taxonomic, biogeographic and threat status) are often weakly correlated (Orme *et al.*, 2005), a novel recent approach does appear particularly promising. This involves the use of presence–absence data for selected species to model assemblage species richness (Mac Nally and Fleishman, 2002). The method was developed by modelling the species richness of butterflies in the central Great Basin (USA) as a function of the occurrence of a subset of selected, putative indicator species (Mac Nally and Fleishman, 2002, 2004). Widespread and rare species were excluded from the initial putative indicator species set on the basis that neither group is likely to serve as an effective indicator of spatial variation in biodiversity (see also rationale in McGeoch *et al.*, 2002). The best set of indicator species (explanatory variables) was then determined using a combination of information criterion and analysis of deviance approaches (Mc Cullagh and Nelder, 1989; Hooper *et al.*, 2002). A subset of four to five (<10%) of the butterfly species was found to explain a significant proportion of the variation in species richness (77–88%) (Mac Nally and Fleishman, 2002). Not only was the explanatory power of the models strong, but they also proved robust when tested using a formal validation process, including a test of the models using a spatially and temporally independent data-set (Mac Nally and Fleishman, 2004). Interestingly, the emergent indicator species represented the full range of flight phenologies in the assemblage, and also encompassed diverse (both taxonomic and growth form) larval host plants (Mac Nally and Fleishman, 2002). The approach worked equally well for birds (Fleishman *et al.*, 2005). Furthermore, a model incorporating the occurrence patterns of six butterfly species predicted 82% of the deviance in combined butterfly and bird species richness (over 130 species) (Fleishman *et al.*, 2005). Finally, these models are also particularly valuable because they were developed and tested at a scale at which conservation management decisions take place (McGeoch, 1998; Mac Nally and Fleishman, 2004). The species-occupancy modelling approach thus certainly warrants continued exploration, in different geographic regions and with different taxa, to establish its generality.

5.2 Monitoring

Biodiversity indicators are used not only to estimate broadscale biodiversity, but also to monitor biodiversity change over time and assess progress towards conservation targets (Fig. 7.1). For example, the biodiversity targets of the CBD 2010 include a detailed understanding of the rates of biodiversity change by 2010 (UNEP, 2003). However, options for achieving such understanding within the time frame are limited, and must necessarily rely heavily on both historical information and on the use of selected taxa, or biodiversity indicators, to represent wholescale biodiversity. A comprehensive discussion of the approaches and steps to achieving this target is presented in the recent discussion meeting issue *Beyond Extinction Rates: Monitoring Wild Nature for the 2010 Target* (Buckland *et al.*, 2005). As pointed out by Dobson (2005), to achieve the 2010 target, regular sampling of the selected species is required, preferably using a globally comparable method, and at least three data points are necessary for each species, community and location.

This raises at least two issues relevant to the application of biodiversity indicators, and the use of insects in monitoring schemes. First, a dichotomy immediately arises between those regions with long-term data-sets, where monitoring systems are already in place, and those without (Brown, 1991; Revenga *et al.*, 2005; Thomas, 2005). Second, the taxa used will be biased towards those for which most data are available, and in several regions are thus less likely to include insects than, e.g. birds and plants (Dobson, 2005; Gregory *et al.*, 2005; Loh *et al.*, 2005; Lughadha *et al.*, 2005; Thomas, 2005). Therefore, assessment for the 2010 targets will largely depend on existing data and current programmes, such as the four complementary schemes for assessing changes in butterfly biodiversity in the UK, i.e. Red Data Books on species conservation status, multiscale atlases and mapping schemes to monitor changes in species distributions, transects that generate population time series data and occasional surveys that quantify population characteristics of selected species across their range (Thomas, 2005). Other well-developed monitoring schemes for insects include those for freshwater macroinvertebrates (Dallas and Day, 1993; Wright *et al.*, 1993; Revenga *et al.*, 2005; Thomas, 2005); see also Conrad *et al.* (Chapter 9, this volume). Importantly, the extensive data that have been generated by monitoring schemes hold significant potential for developing methodologies and testing bioindicators (e.g. Buckland *et al.*, 2005; Thomas, 2005).

In regions without extensive, good-quality baseline data or established survey schemes, options for 2010 are more limited. Here, unvalidated biodiversity indicators and the use of any available data, as well as the establishment of new survey schemes will be required. Indeed, although few scientifically rigorous systems exist for monitoring changes in insect biodiversity, and those that do are neither geographically or taxonomically representative, or in most cases situated in high biodiversity regions, substantial experience exists to support the development of such schemes elsewhere. Thomas (2005) is of the view that this is not only possible, but that monitoring programmes similar to the extensive, robust system in place for butterflies in the UK, should be developed and tested for Odonata, as well as certain groups of Diptera and Hymenoptera, albeit with a view beyond 2010.

Therefore, while options to achieve the 2010 targets are limited, given good leadership and financial support (Thomas, 2005), there is enormous potential for the future development of monitoring programmes involving insects, as the few highly successful existing schemes demonstrate. However, successful biodiversity assessment and monitoring must necessarily lie in the employment of all available information, tools and approaches, including abiotic information, and not only selected biodiversity indicators (Bonn *et al.*, 2002; Faith, 2003; Bonn and Gaston, 2005). Comprehensive, good quality species data will always remain most valuable for assessing biodiversity, as well as setting and monitoring conservation targets (Brooks *et al.*, 2004).

6 The Challenge

The field of bioindication is over a century old, with sustained activity and increasing interest in it apparent since the 1960s (Niemi and McDonald, 2004). However, the field today remains driven by the same fundamental questions:

1. What is the scope of bioindication, and what are the criteria for bioindicators?
2. What are appropriate, robust and effective techniques for the identification and application of bioindicators and bioindication systems?
3. Is it broadly possible to identify effective bioindicators, i.e. do appropriate, simple, predictable relationships exist between taxa and their environments?
4. What is the relationship between the science of bioindication and the application of bioindicators in conservation management, policy development, implementation and monitoring?

As outlined earlier, the field has moved well beyond identifying and defining the scope of bioindication, and the last decade has seen substantial methodological progress. However, with a handful of significant exceptions and albeit to some extent demand driven, bioindicators and bioindication systems do not exist for the vast majority of environmental problems, ecosystems, geographic regions or aspects of biodiversity. We have not yet established how frequently, how broadly and under which circumstances it is possible to establish bioindicators and develop robust bioindication systems. Of the recent publications on insect bioindicators (i.e. insect database, see Endnote), 4.8% develop or test bioindication methods, 76.3% establish relationships that may form the basis of bioindication (Fig. 7.5), only 6.6% establish the robustness of bioindicator relationships (Fig. 7.5) and 12.2% of the publications provide reviews or overviews of the field. The low percentage of studies that establish the robustness of the putative bioindicator is a consequence of both the absence of strong, significant relationships, but also failure to pursue potential bioindicators that are identified. The mean number of citations per review in the 2003 and 2004 volumes of *Annual Review of Ecology, Evolution, and Systematics* is 167.52 (± 47.78, n = 46). This provides an estimate of 0.006 (± 0.002) reviews per paper, or 0.6%, in the fields of ecology, evolution and systematics. Within the insect bioindicator literature there are ~0.14 reviews per paper, which is 20-fold greater than other fields. Therefore, there is apparently a higher ratio of debate to empirical support in insect bioindication than may be expected.

Available evidence suggests that the success of bioindication is dependent on several factors, including the scale at which bioindication is undertaken. Bioindication may be conducted in the context of ecological or earth-system processes that operate from fine scale, short-term events to global, broadscale processes and evolutionary timescales (see Fig. 7.2 in McGeoch, 1998). Patterns are well known to be scale-dependent and often more predictable at broad than at fine scales (Wiens, 1989; McGeoch, 1998; Huston, 1999; Hamer and Hill, 2000). Other determinants of the success or otherwise of bioindication include: (i) the category of bioindication (environmental bioindicators have generally been more successful than ecological or biodiversity indicators) (McGeoch, 1998); (ii) the level of organization involved (e.g. heavy metal concentrations in animal tissue is more predictable than community structure) (Noss, 1990; Lawton, 1999; Murray, 2000); (iii) the environment in which bioindication is conducted (it is generally acknowledged, for example, that aquatic systems tend to be less complex and variable than terrestrial systems) (Steele, 1991); (iv) the method used (as demonstrated earlier for biodiversity indicators); and (v) the taxon considered (as a consequence of the diversity, knowledge of and responsiveness of different taxonomic groups) (Landries *et al.*, 1988; Holloway and Stork, 1991). A priori information on taxon responsiveness may, however, be unreliable. For example, the four dominant dung and carrion beetles in a montane cloud forest and shade coffee plantation landscape in Mexico differed significantly between habitat types (Arellano *et al.*, 2005), whereas in a rainforest–agroforestry system in Indonesia, the five dominant dung beetle species were unresponsive to habitat type (Shahabuddin *et al.*, 2005). However, there remain too few comprehensive case studies in the field, and generalizations regarding the nature of successful bioindicators are in many cases premature.

Finally, bioindication is an applied science and as such the adoption of bioindicators by end-users, i.e. conservation planners, land managers and policy makers, is the ultimate measure of its success. Increasing attention is being paid to the need to make bioindicators policy-relevant (Nicholson and Fryer, 2002; Failing and Gregory, 2003; Niemi and McDonald, 2004). Indeed, the last of Bockstaller and Girardin's (2003) three indicator validation stages is 'end-use validation', or establishing the usefulness of an indicator as a benchmark for decision making. Some of the challenges associated with the translation of bioindication science into policy are reflected by the following statements: 'we cannot simply talk about monitoring birds and butterflies to most policy-makers – it has little or no chance of working' (Watson, 2005); 'the practices we call "mistakes" make good sense from the standpoint of doing careful science but can lead to trouble, and surprising failures in the implementation of biodiversity initiatives' (Failing and Gregory, 2003) versus a bioindicator will never 'be a freeze-dried, talking bug on a stick, i.e. the simple, standardized and easily applied measuring rod, asked for by regulatory authorities' (van Straalen, 1998), and administrations are 'spoiled by the easy handling of abiotic indicators' (Buchs, 2003a).

A recent review of the criteria used by global conservation organizations to select, prioritize and monitor conservation areas, reflects the need to develop effective bioindicators (Gordon *et al.*, 2005). Criteria that have been used to assess the 'usefulness' of scientific assessments, such as bioindicators, include: (i) the demand for simple, user-friendly, cost-effective protocols that may be applied by non-specialists; (ii) outputs that are readily interpretable and easy to communicate; (iii) technically accurate results; (iv) quantification of the uncertainties involved; (v) provision of summary indicators or indices; (vi) presentation of alternative viewpoints and involvement of experts from a range of stakeholder groups; and (vii) a process that is transparent and incorporates institutional, local and indigenous knowledge (Andersen, 1999; Failing and Gregory, 2003; Loh *et al.*, 2005; Watson, 2005). Clearly some of these criteria fall within the ambit of the science of bioindication, whereas others are relevant to policy indicators that are often composite, encompassing several indicators, including economic, social and sustainable development indicators, and may or may not incorporate bioindicators (Table 7.1) (Bella *et al.*, 1994; Levy *et al.*, 2000; Muller *et al.*, 2000; Soberon *et al.*, 2000; Burger and Gochfeld, 2001; Osinski *et al.*, 2003). In the latter case, decision making generally involves several interested parties, and trade-offs between multiple, complex alternatives (Failing and Gregory, 2003). The information provided by bioindicators will thus form only a part of the information on which decisions are based when applying policy indicators (although see, e.g. Gregory *et al.*, 2005; Loh *et al.*, 2005). None the less, Watson (2005) urges that assessments of biodiversity change must proceed to illustrate the impact of such change on issues that people care about, such as livelihoods, health, security and well-being. In a system using low-mobility butterflies (amongst other) as indicators of conservation value, Noe *et al.* (2005) show how the goals and ideas of organic farmers on the conservation of wildlife quality differ from those of biologists. However, the involvement of these organic farmers in the bioindication process positively influenced their perception of biodiversity (Noe *et al.*, 2005).

In the context of bioindication (as defined in this chapter), the challenge to scientists is to thus help bridge the science–policy divide by developing bioindicators that optimize feasibility, cost, information content, relevance and simplicity of application and interpretation (Rempel *et al.*, 2004). However, the range of utilitarian demands for bioindicators to comply with policy and management requirements is not possible, or even desirable, in all instances. There is inevitably a trade-off between efficiency of application and the power of the bioindicator to reflect systems and changing processes. As a consequence, a change in the philosophy of end-users may be necessary (Lindenmayer, 1999; Buchs, 2003a). Reliability and validity cannot be sacrificed for convenience, and the exclusion of bioindicators from indicator sets is preferable to the inclusion of simple, inadequately validated or poorly understood indicators.

7 Conclusions: A Theory of Bioindication

The search for broad, repeatable patterns and the development of theory should be the major goal of biological disciplines, where theory is defined as 'empirically based mechanistic explanation of pattern in nature' (Price, 1991; Price *et al.*, 1995). The state of bioindication is now at the point where the framework for developing a theory of bioindication has been well established. The process of theoretical development in bioindication may be considered to include: (i) delineation of objectives and the empirical collection of facts supporting the identity of species responsive, or related, to the EP of interest; (ii) the generation of hypotheses regarding these responses or relationships; (iii) independent testing of these hypotheses and acceptance or rejection of putative bioindicators; and finally (iv) further development of selected bioindicators to facilitate their use and maximize their suitability for conservation management and policy. The current primary weakness in this framework, at least for insect bioindicators, is the dearth of studies that have established robust bioindicators, and the narrow set of bioindication scenarios and geographic regions addressed by those that have. Insects have contributed substantially to the development of new methods for bioindication, and patterns are beginning to emerge of those insect taxa best suited to bioindication with different objectives, in different environments and geographical regions. However, in spite of an enormous groundswell flagging the importance of bioindication and the potential of insect bioindicators, only a handful of rigorous, fully developed insect bioindication systems have been realized. Perhaps there is a dichotomy between the desired role of indicators and realistic constraints on that role. Alternatively, perhaps the incentive and demand for insect bioindicators have not been sufficiently great. Optimistically, the field has perhaps merely required time to mature, develop methods and establish sufficient direction, and the next decade will see a proliferation of robust insect bioindication systems, as well as their widespread adoption in policy and management.

Acknowledgements

I thank M.J. Samways and K.J. Gaston for discussion, A.E. Hugo for research assistance and S.L. Chown and two reviewers for constructive comments on the chapter. This work was supported by the National Research Foundation of South Africa (GUN 2053618).

Endnote

[1] The Science Citation Index on Web of Science was searched for entries for the period 1998 to March 2005 using the following keywords: bioindicat*, indicator species, surrogate and biodiversity, ecological indicat*, environmental indicat* and biodiversity indicat*. The year 1998 was chosen because of the appearance of several reviews

around this time that summarized the field up to this date (see text). References were categorized based on the environment in which the work was conducted, as well as the form of bioindication (according to definitions in Table 7.1). Two subsets of the main database were then constructed including only publications involving terrestrial and freshwater Arthropoda (hereafter the 'insect database') and the second all other studies. Taxonomic frequency distributions were compiled using only those references in which the taxon was specifically mentioned in the title, abstract or keywords. After deletion of inappropriate references the total remaining was 2311, of which 287 constituted the insect database.

References

Allegro, G. and Sciaky, R. (2003) Assessing the potential role of ground beetles (Coleoptera, Carabidae) as bioindicators in poplar stands, with a newly proposed ecological index (FAI). *Forest Ecology and Management* 175, 275–284.

Alonso, L.E. (2000) Ants as indicators of diversity. In: Agosti, D., Majer, J.D., Alonso L.E. and Schultz, T.R. (eds) *Ants: Standard Methods for Measuring and Monitoring Biodiversity*. Smithsonian Institution Press, Washington, DC, pp. 80–88.

Alvarez, T., Frampton, G.K. and Goulson, D. (2001) Epigeic Collembola in winter wheat under organic, integrated and conventional farm management regimes. *Agriculture Ecosystems and Environment* 83, 95–110.

Andelman, S.J. and Fagan, W.F. (2000) Umbrellas and flagships: efficient conservation surrogates or expensive mistakes? *Proceedings of the National Academy of Sciences of the United States of America* 97, 5954–5959.

Andersen, A.N. (1995) A classification of Australian ant communities, based on functional groups which parallel plant life-forms in relation to stress and disturbance. *Journal of Biogeography* 22, 15–29.

Andersen, A.N. (1997) Using ants as bioindicators: multiscale issues in ant community ecology. *Conservation Ecology* 1, 8–21.

Andersen, A.N. (1999) My bioindicator or yours? Making the selection. *Journal of Insect Conservation* 3, 61–64.

Andersen, A.N. and Majer, J.D. (2004) Ants show the way down under: invertebrates as bioindicators in land management. *Frontiers in Ecology and the Environment* 2, 291–298.

Andersen, A.N., Hoffmann, B.D., Muller, W.J. and Griffiths, A.D. (2002) Using ants as bioindicators in land management: simplifying assessment of ant community responses. *Journal of Applied Ecology* 39, 8–17.

Andersen, A.N., Fisher, A., Hoffmann, B.D., Read, J.L. and Richards, R. (2004) Use of terrestrial invertebrates for biodiversity monitoring in Australian rangelands, with particular reference to ants. *Austral Ecology* 29, 87–92.

Araujo, M.B., Humphries, C.J., Densham, P.J., Lampinen, R., Hagemeijer, W.J.M., Mitchell-Jones, A.J. and Gasc, J.P. (2001) Would environmental diversity be a good surrogate for species diversity? *Ecography* 24, 103–110.

Araujo, M.B., Densham, P.J. and Williams, P.H. (2004) Representing species in reserves from patterns of assemblage diversity. *Journal of Biogeography* 31, 1037–1050.

Arellano, L., Favila, M.E. and Huerta, C. (2005) Diversity of dung and carrion beetles in a disturbed Mexican tropical montane cloud forest and on shade-coffee plantations. *Biodiversity and Conservation* 14, 601–615.

Armbrecht, I. and Ulloa-Chacon, P. (2003) The little fire ant *Wasmannia auropunctata* (Roger) (Hymenoptera: Formicidae) as a diversity indicator of ants in tropical dry forest fragments of Colombia. *Environmental Entomology* 32, 542–547.

Avendano-Mendoza, C., Moron-Rios, A., Cano, E. and Leon-Cortes, J. (2005) Dung beetle community (Coleoptera: Scarabaeidae:

Scarabaeinae) in a tropical landscape at the Lachua Region, Guatemala. *Biodiversity and Conservation* 14, 801–822.

Bailey, A.P., Rehman, T., Park, J., Keatinge, J.D.H. and Tranter, R.B. (1999) Towards a method for the economic evaluation of environmental indicators for UK integrated arable farming systems. *Agriculture Ecosystems and Environment* 72, 145–158.

Baldi, A. (2003) Using higher taxa as surrogates of species richness: a study based on 3700 Coleoptera, Diptera, and Acari species in Central-Hungarian reserves. *Basic and Applied Ecology* 4, 589–593.

Balmford, A., Green, M.J.B. and Murray, M.G. (1996) Using higher-taxon richness as a surrogate for species richness. I. Regional tests. *Proceedings of the Royal Society of London Series B – Biological Sciences* 263, 1571–1575.

Balmford, A., Lyon, A.J.E. and Lang, R.M. (2000) Testing the higher-taxon approach to conservation planning in a megadiverse group: the macrofungi. *Biological Conservation* 93, 209–217.

Balvanera, P., Kremen, C. and Martiz-Ramos, M. (2005) Applying community structure analysis to ecosystem function: examples from pollination and carbon storage. *Ecological Applications* 15, 360–375.

Behan-Pelletier, V.M. (1999) Oribatid mite biodiversity in agroecosystems: role for bioindication. *Agriculture Ecosystems and Environment* 74, 411–423.

Bella, D.A., Jacobs, R. and Li, H. (1994) Ecological indicators of global climate change: a research framework. *Environmental Management* 18, 489–500.

Blackburn, T.M. and Gaston, K.J. (2005) Biological invasions and the loss of birds on islands: insights into the idiosyncrasies of extinction. In: Sax, D.F., Stachowicz, J.J. and Gaines, S.D. (eds) *Species Invasions: Insights into Ecology, Evolution, and Biogeography*. Sinauer, Sunderland, Massachusetts, pp. 85–110.

Bockstaller, C. and Girardin, P. (2003) How to validate environmental indicators. *Agricultural Systems* 76, 639–653.

Bonn, A. and Gaston, K.J. (2005) Capturing biodiversity: selecting priority areas for conservation using different criteria. *Biodiversity and Conservation* 14, 1083–1100.

Bonn, A., Rodrigues, A.S.L. and Gaston, K.J. (2002) Threatened and endemic species: are they good indicators of patterns of biodiversity on a national scale? *Ecology Letters* 5, 733–741.

Brehm, G. and Fiedler, K. (2003) Faunal composition of geometrid moths changes with altitude in an Andean montane rain forest. *Journal of Biogeography* 30, 431–440.

Broberg, L.E. (1999) Will management of vulnerable species protect biodiversity? *Journal of Forestry* 97, 12–18.

Brodersen, K.P. and Anderson, N.J. (2002) Distribution of chironomids (Diptera) in low arctic West Greenland lakes: trophic conditions, temperature and environmental reconstruction. *Freshwater Biology* 47, 1137–1157.

Brodersen, K.P. and Lindegaard, C. (1999) Classification, assessment and trophic reconstruction of Danish lakes using chironomids. *Freshwater Biology* 42, 143–157.

Brooks, T.M., da Fonseca, G.A.B. and Rodrigues, A.S.L. (2004) Protected areas and species. *Conservation Biology* 18, 616–618.

Brown, J.C. and Albrecht, C. (2001) The effect of tropical deforestation on stingless bees of the genus Melipona (Insecta: Hymenoptera: Apidae: Meliponini) in central Rondonia, Brazil. *Journal of Biogeography* 28, 623–634.

Brown, J.H. (1984) On the relationship between abundance and distribution of species. *The American Naturalist* 124, 255–279.

Brown, K.S. (1991) Conservation of neotropical environments: insects as indicators. In: Collins, N.M. and Thomas, J.A. (eds) *The Conservation of Insects and their Habitats*. Academic Press, London, pp. 349–404.

Buchs, W. (2003a) Biodiversity and agri-environmental indicators – general scopes and skills with special reference to the habitat level. *Agriculture Ecosystems and Environment* 98, 35–78.

Buchs, W. (ed.) (2003b) Biotic indicators for biodiversity and sustainable agriculture – introduction and background. *Agriculture, Ecosystems and Environment* 98, 1–16.

Buckland, S.T., Magurran, A.E., Green, R.E., Fewster, R.M. (2005) Monitoring change in biodiversity through composite indices. *Philosophical Transactions of the Royal Society of London Series B – Biological Sciences* 360, 243–254.

Burger, J. and Gochfeld, M. (2001) On developing bioindicators for human and ecological health. *Environmental Monitoring and Assessment* 66, 23–46.

Campbell, P.J., Brown, K.C., Harrison, E.G., Bakker, F., Barrett, K.L., Candolfi, M.P., Canez, V., Dinter, A., Lewis, G., Mead-Briggs, M., Miles, M., Neumann, P., Romijn, K., Schmuck, R., Shires, S., Ufer, A. and Waltersdorfer, A. (2000) A hazard quotient approach for assessing the risk to non-target arthropods from plant protection products under 91/414/EEC: hazard quotient trigger value proposal and validation. *Anzeiger Fur Schadlingskunde-Journal of Pest Science* 73, 117–124.

Cardoso, P., Silva, I., de Oliveira, N.G. and Serrano, A.R.M. (2004a) Higher taxa surrogates of spider (Araneae) diversity and their efficiency in conservation. *Biological Conservation* 117, 453–459.

Cardoso, P., Silva, I., de Oliveira, N.G. and Serrano, A.R.M. (2004b) Indicator taxa of spider (Araneae) diversity and their efficiency in conservation. *Biological Conservation* 120, 517–524.

Caro, T.M. and O'Doherty, G. (1999) On the use of surrogate species in conservation biology. *Conservation Biology* 13, 805–814.

Celli, G. and Maccagnani, B. (2003) Honey bees as bioindicators of environmental pollution. *Bulletin of Insectology* 56, 137–139.

Chapman, R.E. and Bourke, A.F.G. (2001) The influence of sociality on the conservation biology of social insects. *Ecology Letters* 4, 650–662.

Chessman, B.C. (1995) Rapid assessment of rivers using macroinvertebrates: a procedure based on habitat-specific sampling, family level identification and a biotic index. *Australian Journal of Ecology* 20, 122–129.

Chovanec, A. and Waringer, J. (2001) Ecological integrity of river-floodplain systems-assessment by dragonfly surveys (Insecta: Odonata). *Regulated Rivers-Research and Management* 17, 493–507.

Chutter, F.M. (1972) An empirical biotic index of the quality of water in South African streams and rivers. *Water Research* 6, 19–30.

Cole, L.J., McCracken, D.I., Dennis, P., Downie, I.S., Griffin, A.L., Foster, G.N., Murphy, K.J. and Waterhouse, T. (2002) Relationships between agricultural management and ecological groups of ground beetles (Coleoptera: Carabidae) on Scottish farmland. *Agriculture Ecosystems and Environment* 93, 323–336.

Conti, M.E. and Botre, F. (2001) Honeybees and their products as potential bioindicators of heavy metals contamination. *Environmental Monitoring and Assessment* 69, 267–282.

Cook, J.L. (2003) Conservation of biodiversity in an area impacted by the red imported fire ant, *Solenopsis invicta* (Hymenoptera: Formicidae). *Biodiversity and Conservation* 12, 187–195.

Cortet, J., Gomot-De Vauflery, A., Poinsot-Balaguer, N., Gomot, L., Texier, C. and Cluzeau, D. (1999) The use of invertebrate soil fauna in monitoring pollutant effects. *European Journal of Soil Biology* 35, 115–134.

Cousins, S.H. (1991) Species diversity measurement: choosing the right index. *Trends in Ecology and Evolution* 6, 190–192.

Dallas, H.F. and Day, J.A. (1993) *The Effect of Water Quality Variables on Riverine Ecosystems: A Review*. Technical Report Series No. TT 61/93. Water Research Commission, South Africa.

Davis, A.J., Holloway, J.D., Huijbregts, H., Krikken, J., Kirk-Spriggs, A.H. and Sutton, S.L. (2001) Dung beetles as indicators of change in the forests of northern Borneo. *Journal of Applied Ecology* 38, 593–616.

Davis, A.L.V., Scholtz, C.H., Dooley, P.W., Bharm, N. and Kryger, U. (2004) Scarabaeine dung beetles as indicators of biodiversity, habitat transformation and pest control chemicals in agro-ecosystems. *South African Journal of Science* 100, 415–424.

de Bisthoven, L.J., Gerhardt, A. and Soares, A. (2005) Chironomidae larvae as bioindicators of an acid mine drainage in Portugal. *Hydrobiologia* 532, 181–191.

de Bruyn, L.A.L. (1999) Ants as bioindicators of soil function in rural environments.

Agriculture Ecosystems and Environment 74, 425–441.

Dobson, A. (2005) Monitoring global rates of biodiversity change: challenges that arise in meeting the convention on Biological Diversity (CBD) 2010 goals. *Philosophical Transactions of the Royal Society of London Series B – Biological Sciences* 360, 229–241.

Dudley, N., Baldock, D., Nasi, R. and Stolton, S. (2005) Measuring biodiversity and sustainable management in forests and agricultural landscapes. *Philosophical Transactions of the Royal Society of London Series B – Biological Sciences* 360, 457–470.

Duelli, P. and Obrist, M.K. (2003) Biodiversity indicators: the choice of values and measures. *Agriculture Ecosystems and Environment* 98, 87–98.

Dufrêne, M. and Legendre, P. (1997) Species assemblages and indicator species: the need for a flexible asymmetrical approach. *Ecological Monographs* 67, 345–366.

Eiswerth, M.E. and Haney, J.C. (2001) Maximizing conserved biodiversity: why ecosystem indicators and thresholds matter. *Ecological Economics* 38, 259–274.

Ekschmitt, K., Stierhof, T., Dauber, J., Kreimes, K. and Wolters, V. (2003) On the quality of soil biodiversity indicators: abiotic and biotic parameters as predictors of soil faunal richness at different spatial scales. *Agriculture Ecosystems and Environment* 98, 273–283.

Enami, Y., Shiraishi, H. and Nakamura, Y. (1999) Use of soil animals as bioindicators of various kinds of soil management in northern Japan. *Jarq-Japan Agricultural Research Quarterly* 33, 85–89.

Erhardt, A. and Thomas, J.A. (1991) Lepidoptera as indicators of change in semi-natural grasslands of lowland and upland Europe. In: Collins, N.M. and Thomas, J.A. (eds) *The Conservation of Insects and their Habitats*. Academic Press, London, pp. 213–236.

Failing, L. and Gregory, R. (2003) Ten common mistakes in designing biodiversity indicators for forest policy. *Journal of Environmental Management* 68, 121–132.

Faith, D.P. (2003) Environmental diversity (ED) as surrogate information for species-level biodiversity. *Ecography* 26, 374–379.

Fauvel, G. (1999) Diversity of Heteroptera in agroecosystems: role of sustainability and bioindication. *Agriculture Ecosystems and Environment* 74, 275–303.

Ferris, R. and Humphrey, J.W. (1999) A review of potential biodiversity indicators for application in British forests. *Forestry* 72, 313–328.

Fleishman, E., Murphy, D.D. and Brussard, P.E. (2000) A new method for selection of umbrella species for conservation planning. *Ecological Applications* 10, 569–579.

Fleishman, E., Blair, R.B. and Murphy, D.D. (2001) Empirical validation of a method for umbrella species selection. *Ecological Applications* 11, 1489–1501.

Fleishman, E., Thomson, J.R., Mac Nally, R., Murphy, D.D. and Fay, J.P. (2005) Using indicator species to predict species richness of multiple taxonomic groups. *Conservation Biology* 19, 1125–1137.

Fox, C.A. and MacDonald, K.B. (2003) Challenges related to soil biodiversity research in agroecosystems – issues within the context of scale of observation. *Canadian Journal of Soil Science* 83, 231–244.

Garcia-Criado, F., Becares, E., Fernandez-Alaez, C. and Fernandez-Alaez, M. (2005) Plant-associated invertebrates and ecological quality in some Mediterranean shallow lakes: implications for the application of the EC Water Framework Directive. *Aquatic Conservation-Marine and Freshwater Ecosystems* 15, 31–50.

Garson, J., Aggarwal, A. and Sarkar, S. (2002) Birds as surrogates for biodiversity: an analysis of a data set from southern Quebec. *Journal of Biosciences* 27, 347–360.

Gaston, K.J. (1991) The magnitude of global insect species richness. *Conservation Biology* 5, 283–296.

Gaston, K.J. (1996) *Biodiversity: A Biology of Numbers and Difference*. Blackwell Science, Oxford.

Gaston, K.J. and Blackburn, T.M. (1995) Mapping biodiversity using surrogates for species richness: macro-scales and New World birds. *Proceedings of the Royal Society of London Series B – Biological Sciences* 262, 335–341.

Gaston, K.J., Blackburn, T.M. and Lawton, J.H. (1997) Interspecific abundance-range size relationships: an appraisal of mechanisms. *Journal of Animal Ecology* 66, 579–601.

Gayubo, S.F., Gonzalez, J.A., Asis, J.D. and Tormos, J. (2005) Conservation of European environments: the Spheciformes wasps as biodiversity indicators (Hymenoptera: Apoidea: Ampulicidae, Sphecidae and Crabronidae). *Journal of Natural History* 39, 2705–2714.

Geissen, V. and Kampichler, C. (2004) Limits to the bioindication potential of Collembola in environmental impact analysis: a case study of forest soil-liming and fertilization. *Biology and Fertility of Soils* 39, 383–390.

Gibb, H. and Hochuli, D.F. (2002) Habitat fragmentation in an urban environment: large and small fragments support different arthropod assemblages. *Biological Conservation* 106, 91–100.

Glowka, L., Burnhenne-Guilmin, F., Synge, H., McNeely, J.A. and Gundling, L. (1994) *A Guide to the Convention on Biological Diversity*. IUCN, Gland, Switzerland.

Gordon, E.A., Franco, O.E. and Tyrrell, M.L. (2005) Protecting biodiversity: a guide to criteria used by global conservation organizations. Report no. 6. Yale School of Forestry and Environmental Sciences, New Haven, Connecticut.

Grand, J., Buonaccorsi, J., Cushman, S.A., Griffin, C.R. and Neel, M.C. (2004) A multiscale landscape approach to predicting bird and moth rarity hotspots, in a threatened pitch pine-scrub oak community. *Conservation Biology* 18, 1063–1077.

Gravesen, E. (2000) Spiders (Araneae) and other invertebrate groups as ecological indicators in wetland areas. *Ekologia-Bratislava* 19, 39–42.

Gregory, R.D., van Strien, A., Vorisek, P., Meyling, A.W.G., Noble, D.G., Foppen, R.P.B. and Gibbons, D.W. (2005) Developing indicators for European birds. *Philosophical Transactions of the Royal Society of London Series B – Biological Sciences* 360, 269–288.

Grelle, C.E.V. (2002) Is higher-taxon analysis an useful surrogate of species richness in studies of Neotropical mammal diversity? *Biological Conservation* 108, 101–106.

Grimaldi, D. and Engel, M.S. (2005) *Evolution of the Insects*. Cambridge University Press, Cambridge.

Gustafsson, L. (2000) Red-listed species and indicators: vascular plants in woodland key habitats and surrounding production forests in Sweden. *Biological Conservation* 92, 35–43.

Halberg, N., Verschuur, G. and Goodlass, G. (2005) Farm level environmental indicators; are they useful? An overview of green accounting systems for European farms. *Agriculture Ecosystems and Environment* 105, 195–212.

Halffter, G. and Arellano, L. (2002) Response of dung beetle diversity to human-induced changes in a tropical landscape. *Biotropica* 34, 144–154.

Hamalainen, H. (1999) Critical appraisal of the indexes of chironomid larval deformities and their use in bioindication. *Annales Zoologici Fennici* 36, 179–186.

Hamer, K.C. and Hill, J.K. (2000) Scale-dependent effects of habitat disturbance on species richness in tropical forests. *Conservation Biology* 14, 1435–1440.

Hammond, P.M. (1994) Practical approaches to the estimation of the extent of biodiversity in speciose groups. *Philosophical Transactions of the Royal Society of London Series B – Biological Sciences* 345, 119–136.

Heino, J. (2001) Regional gradient analysis of freshwater biota: do similar biogeographic patterns exist among multiple taxonomic groups? *Journal of Biogeography* 28, 69–76.

Hiddink, J.G. (2005) Implications of Liebig's law of the minimum for the use of ecological indicators based on abundance. *Ecography* 28, 264–271.

Hietala-Koivu, R., Jarvenpaa, T. and Helenius, J. (2004) Value of semi-natural areas as biodiversity indicators in agricultural landscapes. *Agriculture Ecosystems and Environment* 101, 9–19.

Hilty, J. and Merenlender, A. (2000) Faunal indicator taxa selection for monitoring ecosystem health. *Biological Conservation* 92, 185–197.

Hodkinson, I.D. (2005) Terrestrial and aquatic invertebrates as bioindicators for environmental monitoring, with particlar reference to mountain ecosystems. *Environmental Management* 35, 649–666.

Hoffmann, J. and Greef, J.M. (2003) Mosaic indicators – theoretical approach for the development of indicators for species diversity in agricultural landscapes. *Agriculture Ecosystems and Environment* 98, 387–394.

Hogg, I.D., Eadie, J.M., Williams, D.D. and Turner, D. (2001) Evaluating fluctuating asymmetry in a stream-dwelling insect as an indicator of low-level thermal stress: a large-scale field experiment. *Journal of Applied Ecology* 38, 1326–1339.

Holloway, J.D. and Stork, N.E. (1991) The dimensions of biodiversity: the use of invertebrates as indicators of human impact. In: Hawksworth, D.L. (ed.) *The Biodiversity of Microrganisms and Invertebrates: Its Role in Sustainable Agriculture*. CAB International, Wallingford, UK, pp. 37–61.

Hooper, J.N.A., Kennedy, J.A. and Quinn, R.J. (2002) Biodiversity 'hotspots', patterns of richness and endemism, and taxonomic affinities of tropical Australian sponges (Porifera). *Biodiversity and Conservation* 11, 851–885.

Horner-Devine, M.C., Daily, G.C., Ehrlich, P.R. and Boggs, C.L. (2003) Countryside biogeography of tropical butterflies. *Conservation Biology* 17, 168–177.

Horvath, R., Magura, T. and Szinetar, C. (2001) Effects of immission load on spiders living on black pine. *Biodiversity and Conservation* 10, 1531–1542.

Hughes, J.B., Daily, G.C. and Ehrlich, P.R. (2000) Conservation of insect diversity: a habitat approach. *Conservation Biology* 14, 1788–1797.

Humphrey, J.W., Hawes, C., Peace, A.J., Ferris-Kaan, R. and Jukes, M.R. (1999) Relationships between insect diversity and habitat characteristics in plantation forests. *Forest Ecology and Management* 113, 11–21.

Huston, M.A. (1999) Local processes and regional patterns: appropriate scales for understanding variation in the diversity of plants and animals. *Oikos* 86, 393–401.

Ishitani, M., Kotze, D.J. and Niemelä, J. (2003) Changes in carabid beetle assemblages across an urban-rural gradient in Japan. *Ecography* 26, 481–489.

Jonsson, B.G. and Jonsell, M. (1999) Exploring potential biodiversity indicators in boreal forests. *Biodiversity and Conservation* 8, 1417–1433.

Juutinen, A. and Monkkonen, M. (2004) Testing alternative indicators for biodiversity conservation in old-growth boreal forests: ecology and economics. *Ecological Economics* 50, 35–48.

Kaspari, M. and Majer, J.D. (2000) Using ants to monitor environmental change. In: Agosti, D., Majer, J.D., Alonso, L.E. and Schultz, T.R. (eds) *Ants: Standard Methods for Measuring and Monitoring Biodiversity*. Smithsonian Institution Press, Washington, DC, pp. 89–98.

Kati, V., Devillers, P., Dufrene, M., Legakis, A., Vokou, D. and Lebrun, P. (2004) Testing the value of six taxonomic groups as biodiversity indicators at a local scale. *Conservation Biology* 18, 667–675.

Kerr, J.T., Sugar, A. and Packer, L. (2000) Indicator taxa, rapid biodiversity assessment, and nestedness in an endangered ecosystem. *Conservation Biology* 14, 1726–1734.

Kevan, P.G. (1999) Pollinators as bioindicators of the state of the environment: species, activity and diversity. *Agriculture Ecosystems and Environment* 74, 373–393.

King, J.R., Andersen, A.N. and Cutter, A.D. (1998) Ants as bioindicators of habitat disturbance: validation of the functional group model for Australia's humid tropics. *Biodiversity and Conservation* 7, 1627–1638.

Kremen, C. (1992) Assessing the indicator properties of species assemblages for natural areas monitoring. *Ecological Applications* 2, 203–217.

Kremen, C. (1994) Biological inventory using target taxa: a case study of the butterflies of Madagascar. *Ecological Applications* 4, 407–422.

Kremen, C., Merenlander, A.M. and Murphy, D.D. (1994) Ecological monitoring: a vital need for integrated conservation and development programs in the tropics. *Conservation Biology* 8, 388–397.

Kromp, B. (1999) Carabid beetles in sustainable agriculture: a review on pest control efficacy, cultivation impacts and enhancement. *Agriculture Ecosystems and Environment* 74, 187–228.

La Ferla, B., Taplin, J., Ockwell, D. and Lovett, J.C. (2002) Continental scale patterns of biodiversity: can higher taxa accurately predict African plant distributions? *Botanical Journal of the Linnean Society* 138, 225–235.

Landries, P.B., Vierner, J. and Thomas, J.W. (1988) Ecological uses of vertebrate indicator species: a critique. *Conservation Biology* 2, 316–329.

Lassau, S.A. and Hochuli, D.F. (2004) Effects of habitat complexity on ant assemblages. *Ecography* 27, 157–164.

Lawton, J.H. (1999) Are there general laws in ecology? *Oikos* 84, 177–192.

Lenhard, S.C. and Witter, J.A. (1977) Insects as biological indicators of environmental change. *ESA Bulletin* 23, 191–192.

Levy, J.K., Hipel, K.W. and Kilgour, D.M. (2000) Using environmental indicators to quantify the robustness of policy alternatives to uncertainty. *Ecological Modelling* 130, 79–86.

Lindenmayer, D.B. (1999) Future directions for biodiversity conservation in managed forests: indicator species, impact studies and monitoring programs. *Forest Ecology and Management* 115, 277–287.

Lindenmayer, D.B., Cunningham, R.B., Donnelly, C.F. and Lesslie, R. (2002a) On the use of landscape surrogates as ecological indicators in fragmented forests. *Forest Ecology and Management* 159, 203–216.

Lindenmayer, D.B., Manning, A.D., Smith, P.L., Possingham, H.P., Fischer, J., Oliver, I. and McCarthy, M.A. (2002b) The focal-species approach and landscape restoration: a critique. *Conservation Biology* 16, 338–345.

Loh, J., Green, R.E., Ricketts, T.H., Lamoureux, J., Jenkins, M.A., Kapos, V. and Randers, J. (2005) The living planet index: using species population time series to track trends in biodiversity. *Philosophical Transactions of the Royal Society of London Series B – Biological Sciences* 360, 289–295.

Lu, S.S. and Samways, M.J. (2002) Behavioural ecology of the Karkloof blue butterfly *Orachrysops ariadne* (Lepidoptera: Lycaenidae) relevant to its conservation. *African Entomology* 10, 137–147.

Luff, M.L. and Woiwod, I.P. (1995) Insects as indicators of land-use change: a European perspective, focusing on moths and ground beetles. In: Harrington, R. and Stork, N.E. (eds) *Insects in a Changing Environment*. Academic Press, London, pp. 400–424.

Lughadha, E.N., Baillie, J., Barthlott, W., Brummitt, N.J., Cheek, M.R., Farjon, A., Govaerts, R., Hardwick, K.A., Hilton-Taylor, C., Meagher, T.R., Moat, J., Mutke, J., Paton, A.J., Pleasants, L.J., Savolainen, V., Schatz, G.E., Smith, P.J., Turner, I., Wyse-Jackson, P. and Crane, P.R. (2005) Measuring the fate of plant diversity: towards a foundation for future monitoring and opportunities for urgent action *Philosophical Transactions of the Royal Society of London Series B – Biological Sciences* 260, 269–288.

Mac Nally, R. and Fleishman, E. (2002) Using 'indicator' species to model species richness: model development and predictions. *Ecological Applications* 12, 79–92.

Mac Nally, R. and Fleishman, E. (2004) A successful predictive model of species richness based on indicator species. *Conservation Biology* 18, 646–654.

Maes, D. and Van Dyck, H. (2005) Habitat quality and biodiversity indicator performances of a threatened butterfly versus a multispecies group for wet heathlands in Belgium. *Biological Conservation* 123, 177–187.

Magura, T., Tothmeresz, B. and Bordan, Z. (2000) Effects of nature management practice on carabid assemblages (Coleoptera: Carabidae) in a non-native plantation. *Biological Conservation* 93, 95–102.

Majer, J.D. (1983) Bio-indicators of minesite rehabilitation, land use and land conservation. *Environmental Management* 7, 375–383.

Marc, P., Canard, A. and Ysnel, F. (1999) Spiders (Araneae) useful for pest limitation and bioindication. *Agriculture Ecosystems and Environment* 74, 229–273.

Matlock, R.B. and de la Cruz, R. (2003) Ants as indicators of pesticide impacts in banana. *Environmental Entomology* 32, 816–829.

Mc Cullagh, P. and Nelder, J.A. (1989) *Generalized Linear Models*. Chapman & Hall, London.

McGeoch, M.A. (1998) The selection, testing and application of terrestrial insects as bioindicators. *Biological Reviews* 73, 181–201.

McGeoch, M.A. (2002) Bioindicators. In: El-Shaarawi, A.H. and Piergorsch, W.W. (eds) *Encyclopedia of Environmetrics*. Wiley, Chichester, UK, pp. 186–189.

McGeoch, M.A. and Chown, S.L. (1998) Scaling up the value of bioindicators. *Trends in Ecology and Evolution* 13, 46–47.

McGeoch, M.A., Van Rensburg, B.J. and Botes, A. (2002) The verification and application of bioindicators: a case study of dung beetles in a savanna ecosystem. *Journal of Applied Ecology* 39, 661–672.

Meregalli, G., Vermeulen, A.C. and Ollevier, F. (2000) The use of chironomid deformation in an in situ test for sediment toxicity. *Ecotoxicology and Environmental Safety* 47, 231–238.

Milakovic, B., Carleton, T.J. and Jefferies, R.L. (2001) Changes in midge (Diptera: Chironomidae) populations of sub-arctic supratidal vernal ponds in response to goose foraging. *Ecoscience* 8, 58–67.

Moritz, C., Richardson, K.S., Ferrier, S., Monteith, G.B., Stanisic, J., Williams, S.E. and Whiffin, T. (2001) Biogeographical concordance and efficiency of taxon indicators for establishing conservation priority in a tropical rainforest biota. *Proceedings of the Royal Society of London Series B – Biological Sciences* 268, 1875–1881.

Mouillot, D., Culioli, J.M. and Chi, T.D. (2002) Indicator species analysis as a test of non-random distribution of species in the context of marine protected areas. *Environmental Conservation* 29, 385–390.

Muller, F., Hoffmann-Kroll, R. and Wiggering, H. (2000) Indicating ecosystem integrity – theoretical concepts and environmental requirements. *Ecological Modelling* 130, 13–23.

Murray, B.G. (2000) Universal laws and predictive theory in ecology and evolution. *Oikos* 89, 403–408.

Murtaugh, P.A. (1996) The statistical evaluation of ecological indicators. *Ecological Applications* 6, 132–139.

Netshilaphala, N.M., Milton, S.J. and Robertson, H.G. (2005) Response of an ant community to mining in the arid Namaqualand coast, South Africa. *African Entomology* 13, 162–167.

Nicholson, M. and Fryer, R. (2002) Developing effective environmental indicators – does a new dog need old tricks? *Marine Pollution Bulletin* 45, 53–61.

Nickel, H. and Hildebrandt, J. (2003) Auchenorrhyncha communities as indicators of disturbance in grasslands (Insecta, Hemiptera) – case study from the Elbe flood plains (northern Germany). *Agriculture Ecosystems and Environment* 98, 183–199.

Niemelä, J. (2000) Biodiversity monitoring for decision-making. *Annales Zoologici Fennici* 37, 307–317.

Niemelä, J., Kotze, D.J., Venn, S., Penev, L., Stoyanov, I., Spence, J.R., Hartley, D. and Montes de Oca, E. (2000a) Carabid beetle assemblages (Coleoptera: Carabidae) across urban-rural gradients: an international comparison. *Landscape Ecology*, 17, 387–401.

Niemelä, J., Kotze, J., Ashworth, A., Brandmayr, P., Desender, K., New, T.R., Penev, L., Samways, M.J. and Spence, J.R. (2000b) The search for common anthropogenic impacts on biodiversity: a global network. *Journal of Insect Conservation* 4, 3–9.

Niemi, G.J. and McDonald, M.E. (2004) Application of ecological indicators. *Annual Review of Ecology Evolution and Systematics* 35, 89–111.

Nilsson, S.G., Hedin, J. and Niklasson, M. (2001) Biodiversity and its assessment in boreal and nemoral forests. *Scandinavian Journal of Forest Research*, 10–26.

Noe, E., Halberg, N. and Reddersen, J. (2005) Indicators of biodiversity and conservational wildlife quality on Danish organic farms for use in farm management: a multidisciplinary approach to indicator development and testing. *Journal of Agricultural and Environmental Ethics* 18, 383–414.

Noss, R.F. (1990) Indicators for monitoring biodiversity: a hierarchical approach. *Conservation Biology* 4, 355–364.

Orendt, C. (1999) Chironomids as bioindicators in acidified streams: a contribution to the acidity tolerance of chironomid species with a classification in sensitivity classes. *International Review of Hydrobiology* 84, 439–449.

Orme, C.D.L., Davies, R.G., Burgess, M., Eigenbrod, F., Pickup, N., Olson, V.A.,

Webster, A.J., Ding, T., Rasmussen, P.C., Ridgley, R.S., Stattersfield, A.J., Bennett, P.M., Blackburn, T.M., Gaston, K.J. and Owens, I.P.F. (2005) Global hotspots of species richness are not congruent with endemism or threat. *Nature* 436, 1016–1019.

Osborn, F., Goitia, W., Cabrera, M. and Jaffe, K. (1999) Ants, plants and butterflies as diversity indicators: comparisons between strata at six forest sites in Venezuela. *Studies on Neotropical Fauna and Environment* 34, 59–64.

Osinski, E., Kantelhardt, J. and Heissenhuber, A. (2003) Economic perspectives of using indicators. *Agriculture Ecosystems and Environment* 98, 477–482.

Paoletti, M.G. (1999) Using bioindicators based on biodiversity to assess landscape sustainability. *Agriculture Ecosystems and Environment* 74, 1–18.

Paoletti, M.G., Hu, D.X., Marc, P., Huang, N.X., Wu, W.L., Han, C.R., He, J.H. and Cai, L.W. (1999) Arthropods as bioindicators in agroecosystems of Jiang Han Plain, Qianjiang City, Hubei China. *Critical Reviews in Plant Sciences* 18, 457–465.

Parisi, V., Menta, C., Gardi, C., Jacomini, C. and Mozzanica, E. (2005) Microarthropod communities as a tool to assess soil quality and biodiversity: a new approach in Italy. *Agriculture Ecosystems and Environment* 105, 323–333.

Parr, C.L., Robertson, H.G., Biggs, H.C. and Chown, S.L. (2004) Response of African savanna ants to long-term fire regimes. *Journal of Applied Ecology* 41, 630–642.

Pearson, D.L. and Carroll, S.S. (1998) Global patterns of species richness: spatial models for conservation planning using bioindicator and precipitation data. *Conservation Biology* 12, 809–821.

Pearson, D.L. and Cassola, F. (1992) Worldwide species richness patterns of tiger beetles (Coleoptera: Cicindelidae): indicator taxon for biodiversity and conservation studies. *Conservation Biology* 6, 376–391.

Pinder, L.V.C. and Morley, D.J. (1995) Chironomidae as indicators of water quality – with a comparison of the chironomid faunas of a series of contrasting Cumbrian tarns. In: Harrington, R. and Stork, N.E. (eds) *Insects in a Changing Environment.* Academic Press, London, pp. 270–297.

Pineda, E., Moreno, C.A., Escobar, F. and Halffter, G. (2005) Frog, bat and dung beetle diversity in the cloud forest and coffee agroecosystems of Veracruz, Mexico. *Conservation Biology* 19, 400–410.

Ponge, J.F., Gillet, S., Dubs, F., Fedoroff, E., Haese, L., Sousa, J.P. and Lavelle, P. (2003) Collembolan communities as bioindicators of land use intensification. *Soil Biology and Biochemistry* 35, 813–826.

Poyry, J., Lindgren, S., Salminen, J. and Kuussaari, M. (2005) Responses of butterfly and moth species to restored cattle grazing in semi-natural grasslands. *Biological Conservation* 122, 465–478.

Prendergast, J.R. (1997) Species richness covariance in higher taxa: empirical tests of the biodiversity indicator concept. *Ecography* 20, 210–216.

Prendergast, J.R., Quinn, R.M., Lawton, J.H., Eversham, B.C. and Gibbons, D.W. (1993) Rare species, the coincidence of diversity hotspots and conservation strategies. *Nature* 365, 335–337.

Price, P.W. (1991) Darwinian methodology and the theory of insect herbivore population dynamics. *Annals of the Entomological Society of America* 48, 464–473.

Price, P.W., Craig, T.P. and Roininen, H. (1995) Working towards theory on galling sawfly population dynamics. In: Cappuccino, N. and Price, P.W. (eds) *Population Dynamics: New Approaches and Synthesis*. Academic Press, New York, pp. 321–338.

Rempel, R.S., Andison, D.W. and Hannon, S.J. (2004) Guiding principles for developing an indicator and monitoring framework. *Forestry Chronicle* 80, 82–90.

Revenga, C., Campbell, I., Abell, R., de Villiers, P. and Bryer, M. (2005) Prospects for monitoring freshwater ecosystems towards the 2010 targets. *Philosophical Transactions of the Royal Society of London Series B – Biological Sciences* 360, 397–413.

Ricketts, T.H., Daily, G.C. and Ehrlich, P.R. (2002) Does butterfly diversity predict moth diversity? Testing a popular indicator taxon at local scales. *Biological Conservation* 103, 361–370.

Rösch, M., Chown, S.L. and McGeoch, M.A. (2001) Testing a bioindicator assemblage: gall-inhabiting moths and urbanization. *African Entomology* 9, 85–94.

Rosenberg, D.M. and Resh, V. (1993) *Freshwater Biomonitoring and Benthic Macroinvertebrates*. Chapman & Hall, New York.

Sarkar, S., Justus, J., Fuller, T., Kelley, C., Garson, J. and Mayfield, M. (2005) Effectiveness of environmental surrogates for the selection of conservation area networks. *Conservation Biology* 19, 815–825.

Sauberer, N., Zulka, K.P., Abensperg-Traun, M., Berg, H.M., Bieringer, G., Milasowszky, N., Moser, D., Plutzar, C., Pollheimer, M., Storch, C., Trostl, R., Zechmeister, H. and Grabherr, G. (2004) Surrogate taxa for biodiversity in agricultural landscapes of eastern Austria. *Biological Conservation* 117, 181–190.

Schulze, C.H., Waltert, M., Kessler, P.J.A., Pitopang, R., Shahabuddin, Veddeler, D., Muhlenberg, M., Gradstein, S.R., Leuschner, C., Steffan-Dewenter, I. and Tscharntke, T. (2004) Biodiversity indicator groups of tropical land-use systems: Comparing plants, birds, and insects. *Ecological Applications* 14, 1321–1333.

Shahabuddin, G., Shulze, C.H. and Tscharntke, T. (2005) Changes of dung beetle communities from rainforests towards agroforestry systems and annual cultures in Sulawesi (Indonesia). *Biodiversity and Conservation* 14, 863–877.

Simberloff, D. (1999) The role of science in the preservation of forest biodiversity. *Forest Ecology and Management* 115, 101–111.

Soberon, J., Rodriguez, P. and Vazquez-Dominguez, E. (2000) Implications of the hierarchical structure of biodiversity for the development of ecological indicators of sustainable use. *Ambio* 29, 136–142.

Sousa, J.P., da Gama, M.M., Pinto, P., Keating, A., Calhoa, F., Lemos, M., Castro, C., Luz, T., Leitao, P. and Dias, S. (2004) Effects of land-use on Collembola diversity patterns in a Mediterranean landscape. *Pedobiologia* 48, 609–622.

Southwood, T.R.E. (1988) Tactics, strategies and templets. *Oikos* 52, 3–18.

Spector, S. and Forsyth, A.B. (1998) Indicator taxa for biodiversity assessment in the vanishing tropics. In: Mace, G.M., Balmford, A. and Ginsberg, J.R. (eds) *Conservation in a Changing World*. Cambridge University Press, Cambridge, pp. 181–209.

Steele, J.H. (1991) Can ecological theory cross the land–sea boundary? *Journal of Theoretical Biology* 153, 425–436.

Thomas, J.A. (2005) Monitoring change in the abundance and distribution of insects using butterflies and other indicator groups. *Philosophical Transactions of the Royal Society of London Series B – Biological Sciences* 360, 339–357.

Thomson, J.R., Fleishman, E., Nally, R.M. and Dobkin, D.S. (2005) Influence of the temporal resolution of data on the success of indicator species models of species richness across multiple taxonomic groups. *Biological Conservation* 124, 503–518.

Thor, G. (1998) Red-listed lichens in Sweden: habitats, threats, protection, and indicator value in boreal coniferous forests. *Biodiversity and Conservation* 7, 59–72.

Tscharntke, T., Gathmann, A. and Steffan-Dewenter, I. (1998) Bioindication using trap-nesting bees and wasps and their natural enemies: community structure and interactions. *Journal of Applied Ecology* 35, 708–719.

UNEP (2003) Proposed Biodiversity Indicators Relevant to the 2010 Target. UNEP/CBD/SBSTTA/9/INF/26. Montreal.

United Nations (2002) Report of the World Summit on Sustainable Development. A/CONF.199/20. Johannesburg, South Africa.

van Hamburg, H., Anderson, A.N., Meyer, W.J. and Robertson, H.G. (2004) Ant community development on rehabilitated ash dams in the South African Highveld. *Restoration Ecology* 12, 552–558.

van Jaarsveld, A.S., Freitag, S., Chown, S.L., Muller, C., Koch, S., Hull, H., Bellamy, C., Kruger, M., Endrody-Younga, S., Mansell, M.W. and Scholtz, C.H. (1998) Biodiversity assessment and conservation strategies. *Science* 279, 2106–2108.

Van Rensburg, B.J., McGeoch, M.A., Chown, S.L. and van Jaarsveld, A.S. (1999) Conservation of heterogeneity among dung beetles in the Maputaland Centre of Endemism, South Africa. *Biological Conservation* 88, 145–153.

van Straalen, N.M. (1998) Evaluation of bioindicator systems derived from soil arthropod communities. *Applied Soil Ecology* 9, 429–437.

van Straalen, N.M. and Krivolutsky, D.A. (1996) *Bioindicator Systems for Soil Pollution*. Kluwer Academic, Dordrecht, The Netherlands.

van Straalen, N.M. and Verhoef, H.A. (1997) The development of a bioindicator system for soil acidity based on arthropod pH preferences. *Journal of Applied Ecology* 34, 217–232.

Viard, B., Maul, A. and Pihan, J.C. (2004) Standard use conditions of terrestrial gastropods in active biomonitoring of soil contamination. *Journal of Environmental Monitoring* 6, 103–107.

Watson, R.T. (2005) Turning science into policy: challenges and experiences from the science-policy interface. *Philosophical Transactions of the Royal Society of London Series B – Biological Sciences* 360, 471–477.

Wheater, C.P., Cullen, W.R. and Bell, J.R. (2000) Spider communities as tools in monitoring reclaimed limestone quarry landforms. *Landscape Ecology* 15, 401–406.

Wiens, J.A. (1989) Spatial scaling in ecology. *Functional Ecology* 3, 385–397.

Wilsey, B.J., Chalcraft, D.R., Bowles, C.M. and Willig, M.R. (2005) Relationships among indices suggest that richness is an incomplete surrogate for grassland biodiversity. *Ecology* 86, 1178–1184.

Woinarski, J.C.Z., Andersen, A.N., Churchill, T.B. and Ash, A.J. (2002) Response of ant and terrestrial spider assemblages to pastoral and military land use, and to landscape position, in a tropical savanna woodland in northern Australia. *Austral Ecology* 27, 324–333.

Wright, J.F., Furse, M.T., Armitage, P.D. and Moss, D. (1993) New procedures for identifying running-water sites subject to environmental stress and for evaluating sites for conservation, based on the macroinvertebrate fauna. *Archives Hydrobiology* 127, 319–326.

Wright, J.F., Sutcliffe, D.W. and Furse, M.T. (2000) *Assessing the Biological Quality of Fresh Waters*. The Freshwater Biological Association, Ambleside, pp. 400.

Zaitsev, A.S. and van Straalen, N.M. (2001) Species diversity and metal accumulation in oribatid mites (Acari, Oribatida) of forests affected by a metallurgical plant. *Pedobiologia* 45, 467–479.

Zalidis, G.C., Tsiafouli, M.A., Takavakoglou, V., Bilas, G. and Misopolinos, N. (2004) Selecting agri-environmental indicators to facilitate monitoring and assessment of EU agri-environmental measures effectiveness. *Journal of Environmental Management* 70, 315–321.

8 Insect Populations in Fragmented Habitats

Ilkka Hanski[1] and Juha Pöyry[2]

[1]*Department of Biological and Environmental Sciences, University of Helsinki, PO Box 65, FIN-00014, Finland;* [2]*Finnish Environment Institute, PO Box 140, Helsinki, FIN-00251, Finland*

1 Introduction

Changes in human land use have driven and continue to drive much of the change in biodiversity locally, nationally and globally. The dominant trend is reduced area of habitats with high biodiversity and increased area of habitats with low biodiversity. Apart from lost area and general degradation of habitat quality, the remaining area of species-rich habitats has become increasingly fragmented. By increasing fragmentation we mean that a continuous habitat splits into two or more distinct fragments; that existing fragments become smaller; and that they become increasingly isolated from the rest of the habitat. In all these situations, habitat fragmentation is typically accompanied by further loss of the pooled area of habitat. None the less, it is possible, and appropriate, to ask questions about fragmentation per se, about the influence of the spatial configuration of a given amount of habitat on the abundance, distribution and viability of populations. This is highly relevant for practical conservation, which often deals with choices, such as exactly where habitat should be protected, managed or restored. To answer such questions about the spatial configuration of habitat, we need to employ relevant concepts and methods. In this chapter, we focus on highly fragmented landscapes, in which only a small fraction of the total landscape area belongs to the habitat type or types of interest, and in which the habitat that exists occurs in relatively small and discrete patches. Many insects with specific habitat requirements occur in landscapes that are highly fragmented for natural reasons or because human land use has turned a continuous habitat into a highly fragmented one.

Landscape ecologists have tackled the task of describing the spatial configuration of habitats with a diverse array of measures (Turner *et al.*, 2001), and they typically aim at characterizing the structure of entire landscapes. In contrast, population and metapopulation ecologists tend to approach the same task from the perspective of individual habitat fragments, and derive landscape-level measures from well-defined patch-level measures, including information on how the patches are connected to each other from the perspective of the species of interest (Moilanen and Hanski, 2001). If the ultimate aim is to improve our understanding of the dynamics of particular species in fragmented landscapes, the advantage of the latter approach is that it is based on well-established population processes. This assumes, of course, that the focal habitat can be meaningfully delineated, but knowing the habitat requirements of the species is in any case a prerequisite for any approach to spatial population ecology (Hanski, 2005). Unfortunately, it is often far from trivial to accumulate this basic information, as the account of the demise of the large blue butterfly (*Maculinea arion*) from the UK so well exemplifies (Thomas, 1980).

Viewed from the perspective of a particular habitat fragment, there are two fundamental population processes to consider. First, the performance of the local population in that fragment (if there happens to be a local population): the average size of the population, its temporal variability, emigration rate and expected life-time. Second, how well the fragment (whether occupied or not) is connected to the surrounding local populations in the fragmented landscape. Connectivity is best defined in terms of the expected rate of immigration (number of individuals arriving per unit time) to the focal fragment. This is captured in the following formula:

$$S_i = A_i\zeta_{im}\Sigma_{j\neq i}\exp(-\alpha d_{ij})p_j A_j\zeta_{em} \qquad (8.1)$$

where S_i is the connectivity of patch i, A_i the area of patch i, p_j the incidence (probability of occurrence of a population) in patch j, d_{ij} the distance between patches i and j (simple Euclidian distance or some other distance), $1/\alpha$ the average migration distance of the species, and ζ_{im} and ζ_{em} two parameters describing the scaling of immigration and emigration rates with patch area (these are often assumed to equal 0 and 1, respectively). This formula describes the expected rate of immigration to patch i on the assumption that the sizes of the source populations (occupying patches other than patch i) are proportional to the areas of the respective patches. If information on actual population sizes is available, this knowledge may be used to replace the surrogate A_j in Eq. 8.1, though in this case possible patch area-dependent emigration rate should be modelled separately. For further discussion of Eq. 8.1 and how it can be calculated in practice, see Hanski (1994, 1999) and Moilanen and Nieminen (2002). The important point is that Eq. 8.1 makes good biological sense, and it is generally preferable to other measures, such as the nearest-neighbour distance, for the purpose of characterizing the influence of connectivity and hence fragmentation on habitat occupancy, local abundance and so forth.

In general, we expect habitat fragmentation to have substantial and greater effects in insect populations than, say, in birds, another taxon that has been much studied in the literature. The reasons include the small body size of insects, which allows local populations to become established even in small fragments of habitat. At the same time, small local populations necessarily have high or relatively high rate of extinction, which means that the large-scale and long-term persistence of species cannot be understood at the scale of single habitat patches and the respective local populations. Extinctions can be compensated for by the establishment of new local populations by dispersing individuals; hence, questions about migration and, for example, the scaling of emigration and immigration rates with habitat patch area, as in Eq. 8.1, are important. However, though there are these reasons to assume that habitat fragmentation often plays a big role in the occurrence and dynamics of insects and other species in fragmented landscapes, researchers have not reached a consensus on the general importance of fragmentation. What is agreed by everybody is that habitat loss is the number one threat to populations of insects and most other organisms, for the reason that smaller area of habitat means smaller and less viable populations (e.g. Sala *et al.*, 2000; Hanski, 2005, though see Thomas *et al.*, 2004, who advocate an even stronger role for climate change). But whether habitat fragmentation independent of habitat loss matters is less obvious, and there are researchers (e.g. Fahrig, 1997, 2001, 2003) who have argued that persistence of species in fragmented landscapes primarily depends on the pooled area of habitat. The critical issue is whether population connectivity makes a difference to the distribution and spatial dynamics of species in fragmented landscapes. In other words, does the spatial location of a particular patch of habitat in relation to the existing local populations in the landscape make a difference to what happens in that patch?

In this chapter, we first review a selection of empirical studies on the influence of habitat fragmentation on insect populations, and ask specifically how one should properly compare the relative importance of landscape structure, typically habitat patch area and connectivity, and habitat quality. Contrary to our general expectation, many empirical studies on insects have failed to demonstrate a significant effect of fragmentation on the occurrence or abundance of populations. We discuss a number of reasons why empirical studies may fail to document significant fragmentation effects. Focusing on situations in which habitat fragmentation is of importance, we briefly outline the key predictions of metapopulation theory about the consequences of habitat fragmentation for metapopulation viability, illustrating some key concepts with the results of a long-term study on the Glanville fritillary butterfly (*Melitaea cinxia*) (Hanski, 1999; Nieminen *et al.*, 2004). From here we move on to examine the role of changing landscape structure on the dynamics of metapopulations and possible evolutionary responses to habitat fragmentation in insect populations. We conclude by giving an example of how conservation efforts can be misguided by ignoring the consequences of habitat fragmentation. We also discuss attempts to merge spatial population dynamics into the design of reserves in fragmented landscapes.

2 Local versus Regional Factors Influencing Habitat Occupancy and Population Processes

Ecologists and entomologists working on insects and other taxa have spent much time and effort in sorting out the relative contributions of local factors, typically related to habitat quality, and regional factors, typically related to habitat fragmentation, in influencing habitat occupancy and related population processes in their study systems. Important as it is to know what really matters in particular cases, not least for conservation and management, one should also ask about the generality of the results of such studies. Each empirical study is necessarily based on a limited number of habitat patches and environmental variables for one or more species. Exactly which patches are included in a study is obviously critical. Including more patches of very low quality will most likely increase the 'significance' of habitat quality in explaining occupancy; adding tiny patches, which an ecologist might not usually consider as patches supporting local populations, would increase the significance of patch area; and including some very isolated patches might do the same for the significance of connectivity and hence habitat fragmentation. The point is that there is no general answer, and there cannot be one. Answers for specific species and landscapes can be helpful for the management of those very species and landscapes, but one should not be misled to assume that ten studies demonstrating the 'importance' of habitat quality have somehow demonstrated the general unimportance of the spatial configuration of habitat for the dynamics of species in fragmented landscapes.

With the above caveat in mind, we conducted our own analysis of the relative importance of local versus regional factors. We examined a set of 38 articles published between 1993 and 2005 in major ecology journals; the articles are listed at the end of the References. Articles were selected by the subjective criterion of being substantial contributions to the field. Apart from tabulating what the authors concluded about the relative importance of local versus regional factors, we also attempted to determine in what kind of situations fragmentation effects have been found to explain habitat occupancy and population processes. The clear majority of the papers ($n = 33$) were solely or primarily concerned with butterflies and moths, whereas other insect orders received attention in only individual studies. The majority of the papers ($n = 22$) dealt with individual species, and the most common response variable was habitat patch occupancy, though population density, migration rate and colonization events also received some attention. With one exception, papers on insect communities ($n = 16$) used species richness as the response variable.

All but two of the papers reported observational studies and employed some type of general linear model (seven ANOVAs, two ANCOVAs, ten multiple linear regressions) or a generalized linear model (14 multiple logistic regressions, two Poisson regressions) in data analysis, applied in a stepwise fashion. Four of the papers that used logistic regression attempted to determine the relative importance of groups of variables by using partial regression methods, but only two papers applied actual partitioning methods, about which we say more at the end of this section.

Based on these papers, we built a data matrix with the following three binary response variables: ISOLATION (1 = isolation effect observed), REGIONAL (1 = regional factors deemed more important than local ones) and UNDETERMINED (1 = local and regional factors considered equally important). The lat-

Table 8.1. Number of studies that showed a significant positive or negative effect for the following three binary response variables: ISOLATION (1 = isolation effect observed); REGIONAL (1 = regional factors deemed more important than local ones); and UNDETERMINED (1 = local and regional factors considered equally important). REGIONAL and UNDETERMINED are dummy variables constructed to describe the relative importance of local versus regional factors. Percentages are shown in parentheses.

Variable	Positive observation	Negative observation	Not studied	Total[a]
ISOLATION	27 (71%)	8 (21%)	3 (8%)	38
REGIONAL	12 (33%)	24 (67%)	0 (0%)	36
UNDETERMINED	6 (17%)	30 (83%)	0 (0%)	36

[a]Two papers tested only the significance of isolation, whereas 36 papers included a comparison between local and regional factors.

ter two are dummy variables constructed to describe the relative importance of local versus regional factors. Due to the relatively small sample of articles, we did not attempt to distinguish between different types of response variables (occupancy, density, species richness, etc.). The results showed that there was a significant isolation effect in 71% of the studies, that regional factors were more important than local ones in 33% of the studies, whereas in 17% of the studies the authors concluded that regional and local factors were equally important (Table 8.1). Thus, habitat fragmentation and other regional factors often have a significant effect on insect populations.

Do any generalities exist in the relative contributions of local versus regional factors in influencing habitat occupancy and related processes? To answer this question, we defined seven explanatory variables, described in Table 8.2. We built generalized linear models with binomial error structure to study the relationships between the three response variables in Table 8.1 and the seven explanatory variables in Table 8.2. We started with a full model including all variables and allowed backward deletion and forward selection with $p = 0.1$. The results firstly indicate that the isolation effect was significantly related to the scope of the study: an isolation effect was more likely to be observed in single-species than in multispecies studies (Table 8.2). Furthermore, in three out of the seven studies in which the isolation effect was observed in a multispecies community, the test had actually been applied to a subset of all the species in the community. Secondly, studies that were scored as 'landscape ecological' and that typically involved description of the extent of particular habitat types within a buffer zone from the focal fragment were likely to find regional factors more important than local factors. This result probably reflects a bias in studies involving a large number of factors: those factors that the researcher is most interested in are likely to turn out to be of importance. Thirdly, and somewhat surprisingly in view of the first result, the relative strengths of local versus regional factors were likely to be considered equal when the study involved a single species rather than several species or a community of species. This result is none the less possible, because the response variables ISOLATION and UNDETERMINED were

Table 8.2. Binomial general linear models (GLMs) for factors potentially explaining the incidence of significant isolation effect, stronger impact of regional over local factors (REGIONAL), and equal importance of local and regional factors (UNDETERMINED) in 38 published studies on insects (Table 8.1). Model coefficients with standard errors (SE) and statistical significance (χ^2 test) are given.

Explanatory variable[a]	ISOLATION (n = 35)		REGIONAL (n = 36)		UNDETERMINED (n = 36)	
	Coeff.	p value	Coeff.	p value	Coeff.	p value
SCOPE	−2.15 ± 0.92	0.019			−2.24 ± 1.22	0.066
LANDECO			1.61 ± 0.85	0.058		
HYPOTHESIS					2.14 ± 1.22	0.079

[a]Explanatory variables: (i) measure of connectivity (nearest neighbour, buffer or Eq. 8.1); (ii) type of study (observational or experimental); (iii) binary variable indicating whether the study used a landscape ecological description of landscape structure (LANDECO; typically some description of habitat complexity within a buffer zone); (iv) taxonomic scope of the study (SCOPE; single species or multispecies); (v) metapopulation approach versus a more general comparison between local and regional factors (HYPOTHESIS; metapopulation studies focused on testing habitat patch area and connectivity effects against local quality); (vi) spatial scale of the study (in km);and (vii) year of publication (1993–2005).

derived independently, and hence studies that did not detect differences between the relative importance of local and regional factors included those in which a significant isolation effect was reported. Studies that compared habitat patch area and isolation effects with habitat quality effects in the metapopulation framework were more likely to find a difference between local and regional factors than studies conducted without the metapopulation framework. The latter results were only marginally significant (Table 8.2), however, and a larger sample of studies would be needed to confirm, or reject, these generalizations.

2.1 Partitioning methods for assessing fragmentation effects

Variation partitioning and hierarchical partitioning are statistical methods that allow unequivocal decomposition of variation in a response variable among explanatory variables or groups of variables (Chevan and Sutherland, 1991; Borcard *et al.*, 1992). These methods thus provide better understanding of the relative importance of different explanatory variables than traditional stepwise regression models (MacNally, 2000). A shortcoming of the latter methods is that in the case of multicollinear data-sets, variables with higher statistical significance tend to receive more attention than other variables that may be ecologically more meaningful (Graham, 2003; Heikkinen *et al.*, 2005).

We illustrate here with an example on the clouded apollo butterfly (*Parnassius mnemosyne*) how variation in habitat occupancy and abundance can be decomposed into independent and joint effects of three groups of variables (Heikkinen *et al.*, 2005). The data on butterfly occupancy and abundance and on the explanatory variables were collected in 1999 from 2408 grid squares 50 × 50 m in size and located along a river valley in south-west Finland. The three groups of explanatory variables were: (i) larval and adult resources in each grid cell, including the abundance of the sole larval host plant *Corydalis solida* and nectar sources; (ii) habitat quantity and connectivity, consisting of the coverage of the four main habitat types (semi-natural grassland, agricultural field, deciduous forest and coniferous forest) in each grid cell, and connectivity calculated for the breeding habitat (semi-natural grassland) using Eq. 8.1 but ignoring habitat occupancy (thus the p_j values were set to 1); and (iii) microclimate, including radiation and average wind speed in each grid cell (details in Luoto *et al.*, 2001). We also included in the analysis an autocovariate for the response variable, describing the number of butterflies in the surrounding grid cells and calculated with Eq. 8.1 including habitat occupancy (p_j).

The variation partitioning method (Borcard *et al.*, 1992) was used to decompose variation in grid occupancy among the three groups of explanatory variables. Variation in the occupancy data was partitioned using a series of partial binomial generalized linear models. Quadratic terms of the predictors were included to take into account curvilinear relationships between butterfly occupancy and the predictor variables. Partitioning among three environmental matrices results in eight fractions of variance (Liu, 1997; Anderson and Gribble, 1998). Variation in the abundance data was decomposed in a similar manner, but using only the 349 grid cells in which *P. mnemosyne* was present as well as a series of partial regressions with redundancy analysis. The largest variance fractions in the occupancy data were the independent effects of the habitat quantity variables (26.4%; Fig. 8.1a), the joint effect of habitat quantity and resources (17.3%) and the joint effect of all three groups of predictors (9.8%). The independent effects of resources and microclimate were small though still statistically significant. Fitting the autocovariate as an additional variable to the final model resulted in a statistically significant ($p < 0.001$) deviance change, accounting for 3.0% of the deviance in butterfly occupancy. In the results for abundance data, the independent effect of habitat quantity variables (9.2%) and the joint effect between them and the resource variables (4.3%) were the largest fractions (Fig. 8.1b). The independent effects of resources and microclimate were higher than in the corresponding results for habitat occupancy.

The hierarchical partitioning method (Chevan and Sutherland, 1991) considers all possible models in a hierarchical multivariate regression setting. This method involves calculation of the increase in the fit of all models, including a particular predictor, compared with the respective models without that variable, and averaging the improvement in the fit across all possible models with the focal predictor. Thus, hierarchical partitioning provides an estimate for each explanatory variable of the variance fractions that are independent and joint with all other variables (Chevan and Sutherland, 1991; MacNally, 2000). Hierarchical partitioning was conducted using the *hier.part package* (MacNally and Walsh, 2004). A drawback of the current implementation of this package is that it assumes monotonic relationships between

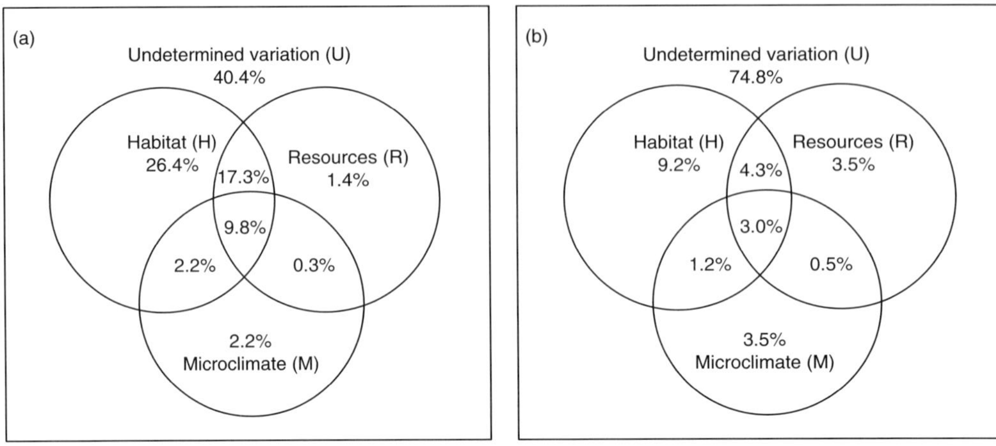

Fig. 8.1. Variation partitioning for (a) habitat occupancy and (b) local abundance of the butterfly *Parnassius mnemosyne*. Percentage of the explained variation is indicated for each fraction. Statistical models for habitat occupancy were built as generalized linear models with binomial errors and the significance of the variables (linear and non-linear effects) in each group was tested with an *F* ratio test. Abundance data were analysed using a series of partial regressions with redundancy analysis. (From Heikkinen *et al.*, 2005.)

the response and predictor variables. Some predictor variables were transformed to improve the linearity of their relationships with butterfly variables (Heikkinen *et al.*, 2005).

In the occupancy data, the independent effects of all variables were statistically significant, although some made only a small contribution (Fig. 8.2a). Consistent with variation partitioning, cover of semi-natural grassland and habitat connectivity made the highest independent contributions, and the independent contributions of the autocovariate and larval host plant abundance were also high. The negative joint contribution of radiation indicates that the majority of the relationships with other predictors are suppressive rather than additive (Chevan and Sutherland, 1991). In the abundance data, cover of semi-natural grassland made the largest independent contribution, followed by the autocovariate and nectar plant abundance (Fig. 8.2b). Independent effects of all predictors were statistically significant but a considerable part of the total variation was accounted for by their joint effects (Fig. 8.2b).

In summary, the independent effect of habitat quantity variables (habitat area and connectivity) accounted for the largest fraction of the variation in the clouded apollo habitat occupancy and abundance, though habitat connectivity made a major contribution for habitat occupancy only. Perhaps not surprisingly, the independent effects of resources and microclimate were greater for abundance than occupancy. A considerable amount of variation in the butterfly data was accounted for by the joint effects of the predictors and may thus be causally related to two or all three groups of variables. Abundance of the butterfly in the surroundings of the focal grid cell (the

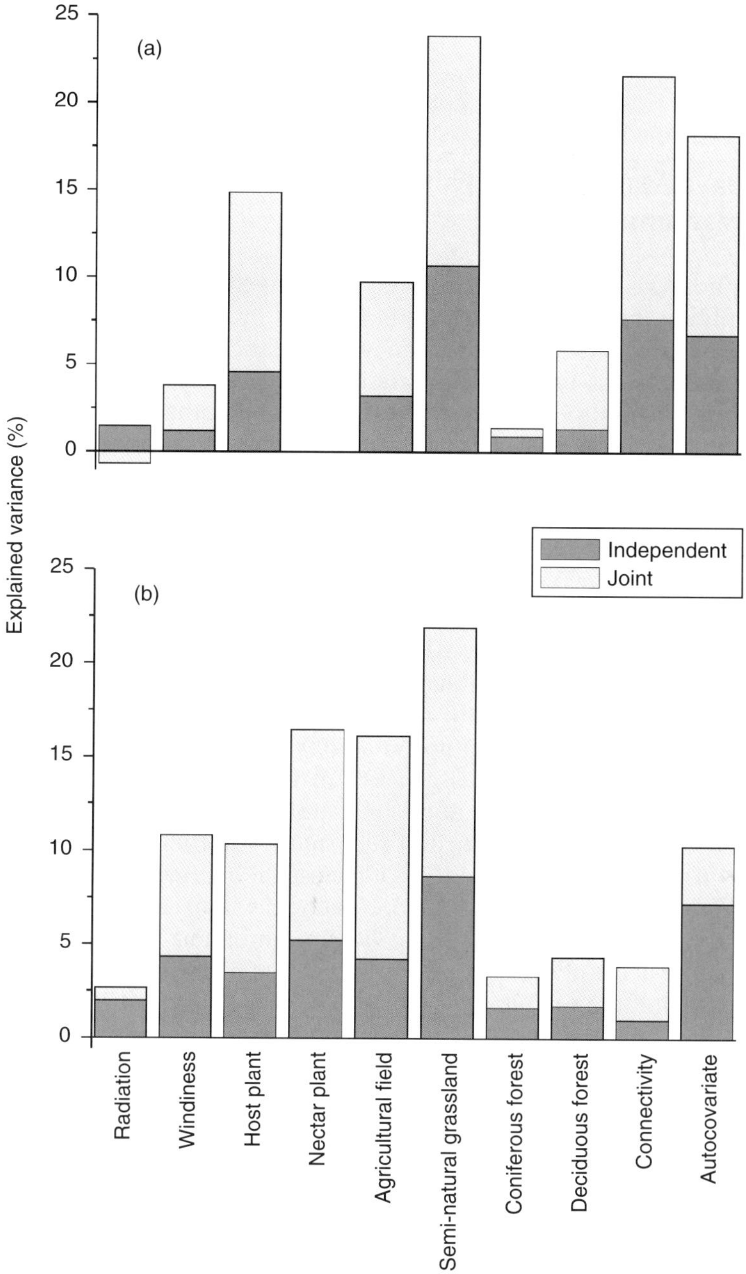

Fig. 8.2. The independent and joint contributions (as percentages of the total variance explained) of the predictor variables for (a) habitat occupancy and (b) local abundance of *Parnassius mnemosyne*, estimated with hierarchical partitioning. Habitat occupancy was analysed using binomial logistic regression and local abundance using linear regression. (From Heikkinen *et al.*, 2005.)

autocovariate) had a significant effect in all analyses, independently of the effects of other predictors. This result points to the role of migration in influencing habitat occupancy and local abundance.

3 Why Have Many Studies Failed to Detect Any Effect of Habitat Fragmentation?

Our analysis in Section 2 indicated that two-thirds of insect studies that have examined the role of connectivity (habitat fragmentation) have found a significant effect on habitat occupancy, abundance and other variables that were analysed. We do not, however, claim that our sample of studies is necessarily very representative, and we acknowledge that other researchers have found in their analyses of published studies a lower incidence of fragmentation effects (Fahrig, 2003). Below, we discuss a number of reasons why empirical studies might fail to demonstrate a significant effect of fragmentation.

- *Habitat wrongly defined*. Many insects are extreme habitat specialists; hence, erroneous knowledge of their habitat requirements may obscure the influence of population connectivity on their distribution and abundance. Some species require two different habitat types to complete their development. The archetypal example is many frogs, which require ponds for breeding but also an appropriate terrestrial summer habitat in the surroundings. Pope *et al.* (2000) report an example on the northern leopard frog (*Rana pipiens*), in which a buffer measure of connectivity to surrounding ponds did not explain the density of the frog in the focal pond when tested separately, but had a statistically significant effect when the availability of terrestrial habitat was also included in the model. A comparable situation occurs in some insect species, for instance in the apollo butterfly (*P. apollo*; Brommer and Fred, 1999) and bees and other aculeate wasps (e.g. Potts *et al.*, 2005), which have distinct breeding and foraging habitats.
- *Small spatial scale of the study*. If a study is conducted at such a small spatial scale that individuals of the focal species easily move from any habitat patch to any other within the study area, there is no reason to expect that fragmentation would matter greatly. This may explain why Fleishman *et al.* (2002) found no effect of connectivity on patch occupancy in the butterfly *Speyeria nokomis*. In this case the study area was small (4 km across) in comparison with the flight capacity of *S. nokomis*. Whether the spatial scale is small naturally depends on the scale of movements, which differs greatly among insect species. At one extreme, many species of aphids and other small-bodied insects disperse passively high up in the air for long distances, and their movements are not greatly limited by distance, though the direction of movements is evidently influenced by air currents with which the insects move (e.g. Mikkola, 1986; Nieminen *et al.*, 2000).
- *Poor measure of connectivity*. It is unfortunate that most empirical studies continue to use a simplistic measure of connectivity, distance to the

nearest population or, even worse, distance to the nearest habitat patch regardless of whether there is a population in that patch or not. These measures have limited biological justification and they can be expected to lack statistical power and bias results. In a meta-analysis of published papers, Moilanen and Nieminen (2002) found that studies using the nearest-neighbour distance as a measure of connectivity were less likely to report a significant effect of connectivity than studies employing the better-justified connectivity measure defined by Eq. 8.1.

- *Landscape not severely fragmented*. If most of the landscape represents the focal habitat and hence the degree of fragmentation is small, there is no reason to assume that an empirical study would uncover a statistically significant effect of fragmentation. In other words, we do not expect (major) fragmentation effects unless the landscape is (highly) fragmented (e.g. Shahabuddin *et al*., 2000; Collinge *et al*., 2003).
- *Mixture of dissimilar species studied*. Often the effect of connectivity is analysed with a large assemblage of species, not all of which may be specific to the habitat type investigated. Ubiquitous occurrence of generalist species may swamp any signal that might be present due to restricted occurrence of specialist species (e.g. Summerville and Crist, 2004). In a study attempting to distinguish between local and regional factors in lepidopteran communities in Finnish semi-natural grasslands, a significant effect of isolation was observed for the species in decline but not for the species whose populations were stable (J. Pöyry *et al*., unpublished data). The results in Table 8.2 show that fragmentation effects have been detected more frequently in single-species than in multispecies studies.
- *Changing environments*. The current occurrence of a species in a landscape may to a large extent reflect environmental conditions, including the degree of fragmentation that occurred at some time in the past. Hence any measure of connectivity calculated for the present landscape may be misleading. Such spurious lack of connectivity effect can be expected to occur especially in long-living organisms, such as perennial plants in forest fragments (Eriksson and Ehrlén, 2001) and in grasslands (Helm *et al*., 2006), where the past (50–100 years ago) habitat connectivity has been shown to explain the current species richness of vascular plants (Lindborg and Eriksson, 2004). We discuss an insect example in Section 5.
- *Use of inadequate statistical methods*. Studies investigating the effects of habitat fragmentation and other environmental factors on habitat occupancy and abundance have generally applied stepwise multiple regression or comparable models. We pointed out in Section 2 that partitioning methods (e.g. Chevan and Sutherland, 1991; Borcard *et al*., 1992; MacNally, 2000) represent a step forward, as with these methods it is possible to unequivocally distinguish between the independent and joint effects of various factors and groups of factors.

The list of circumstances under which we might not expect significant connectivity effects is so long and comprehensive that it may appear to leave very few cases where connectivity matters. However, this would be a rash conclusion. There are countless numbers of insect and other species that occur in highly fragmented landscapes and for which habitat fragmentation most likely is a real issue. Unfortunately, conducting studies on uncommon species at large spatial scales is difficult, and hence the existing sample of studies is biased away from situations in which connectivity can be expected to matter.

4 Metapopulations in Fragmented Habitats: Theoretical Predictions and Empirical Observations

Metapopulation theory for fragmented landscapes is concerned with the occurrence and dynamics of species in networks of discrete habitat patches, which are often so small that the respective local populations have a significant risk of extinction. In this situation, the long-term persistence of species, and any aspect of their ecology, genetics and evolution, cannot be properly understood by examining isolated local populations only but instead one has to investigate networks of local populations (Hanski, 1999; Hanski and Gaggiotti, 2004). The well-studied metapopulation of *M. cinxia* in the Åland Islands in south-west Finland (Hanski, 1999; Nieminen *et al.*, 2004) has become a helpful model system for metapopulation studies. Below, we briefly review key empirical results for the Glanville fritillary to illustrate essential theoretical concepts and model predictions. In the later sections of this chapter, we return to the same species in the context of changing landscapes and evolutionary responses of metapopulations to these changes. There are no reasons to assume that the Glanville fritillary metapopulation in Åland would be special in any other way than that a large amount of research has been conducted on it since 1991 (reviewed in Hanski, 1999, and in many chapters in Ehrlich and Hanski, 2004).

The habitat for the Glanville fritillary in Åland consists of dry meadows with at least one of the two larval host plant species *Plantago lanceolata* and *Veronica spicata* present. There are altogether 600 ha of such meadows, but fragmented into 4000 distinct patches. The average area of individual meadows is only 0.15 ha, and a meadow has on average 23 other meadows within a distance of 1 km, which is the usual migration range of the butterfly (Hanski *et al.*, 2000; Ovaskainen, 2004). Local populations inhabiting individual meadows are small and prone to extinction for many reasons (Hanski, 1998, 2003), whereas the probability of recolonization declines markedly with increasing isolation of a currently unoccupied meadow from the existing local populations in the surroundings. Figure 8.3 shows how the annual extinction and recolonization probabilities depend on patch area and connectivity.

The spatially realistic metapopulation theory assumes the kind of relationships between landscape structure and population processes depicted in Fig. 8.3 and incorporates them into dynamic models, which predict the occurrence of species in networks of habitat patches (for an introduction to the theory, see Hanski, 2001, 2005, and Ovaskainen and Hanski, 2004). The

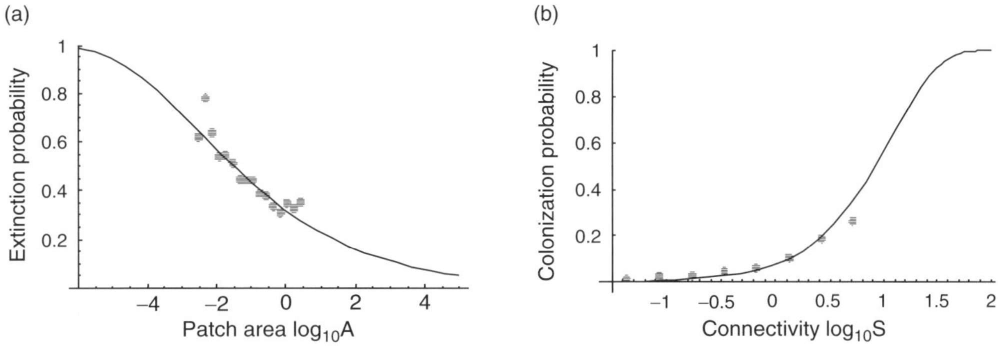

Fig. 8.3. (a) The dependence of extinction probability on patch area and (b) the dependence of recolonization probability on connectivity in the Glanville fritillary. Dots represent average values for classes of patch area and connectivity, based on data collected for nine successive generations in a network of 4000 habitat patches. Lines give maximum likelihood estimates based on the entire data-set. (From Ovaskainen and Hanski, 2004.)

models allow one to make predictions about the influence of the actual landscape structures – how much is habitat and what is the spatial configuration of that habitat – on the occurrence of the species. With decreasing amount and increasing fragmentation of habitat, a point is reached at which the viability of the entire metapopulation is lost. At this limit, called the extinction threshold, the colonization rate does not suffice to compensate for the losses due to local extinctions. Figure 8.4 gives an example of the Glanville fritillary, in which the observed habitat occupancy is related to an appropriate

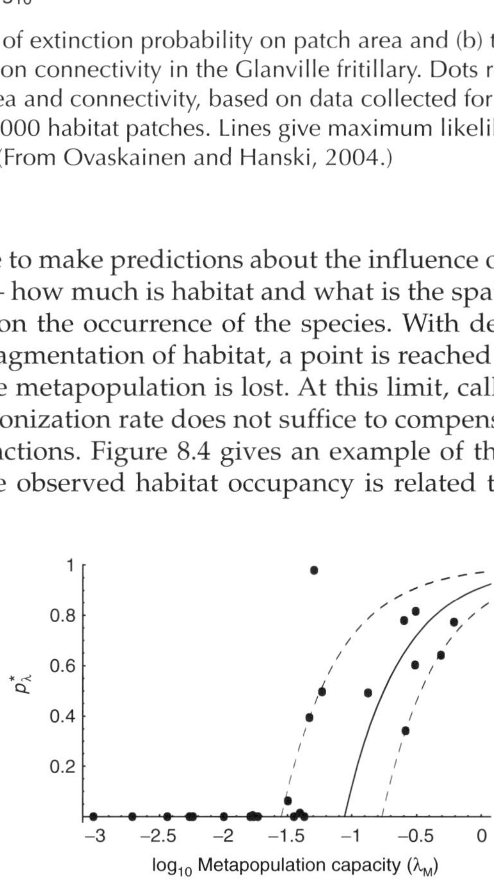

Fig. 8.4. Plot of metapopulation size (p^*_λ) against the logarithm of metapopulation capacity (λ_M) in 25 real habitat patch networks that are potentially occupied by the Glanville fritillary in the Åland Islands. The value of p^*_λ was calculated based on patch areas, spatial locations and the occurrence of the butterfly in the patches in 1993. For each network with $p^*_\lambda > 0.3$, the threshold value for persistence was calculated using the formula $\delta = \lambda_M (1 - p^*_\lambda)$. The continuous line is based on the average of the estimated p_λ values; the broken lines give the minimum and maximum estimates omitting the two networks yielding the most extreme values. (From Hanski and Ovaskainen, 2000.)

measure of landscape structure, the metapopulation capacity (Hanski and Ovaskainen, 2000). This measure summarizes in a single number the influence of the amount and fragmentation of habitat on metapopulation occurrence. Figure 8.4 provides strong evidence for an extinction threshold, as well as demonstrates a good fit of a metapopulation model to empirical data. Thomas and Hanski (2004) discuss other butterfly examples.

The example in Fig. 8.4 suggests that a metapopulation living in a landscape with small metapopulation capacity has a high risk of metapopulation extinction. The model used in Fig. 8.4 is deterministic and predicts metapopulation persistence when the extinction threshold is exceeded, but in reality metapopulations located just above the extinction threshold have a substantial risk of extinction for stochastic reasons. On the other hand, if the respective patch network is located close to existing metapopulations in nearby networks, the species may recolonize a network from which a metapopulation went extinct. Just like in the case of recolonization of individual habitat fragments, the probability of recolonization of an entire patch network is expected to increase with increasing connectivity. Figure 8.5 shows data for 9 years and tens of semi-independent patch networks (Hanski *et al.*, 1996) in the Åland Islands. If the above argument about network-wide extinctions and recolonizations is correct, we would expect that the incidence of network occupancy increases with increasing metapopulation capacity and network connectivity. This is indeed the case (Fig. 8.5), demonstrating that the occur-

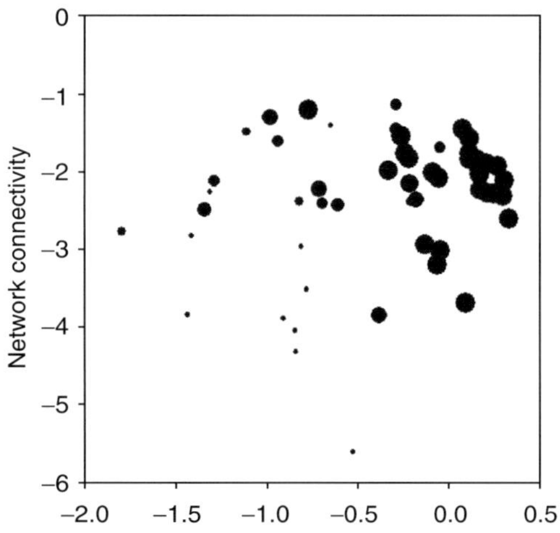

Fig. 8.5. Occupancy of habitat patch networks by the Glanville fritillary in the Åland Islands. Each point represents a separate semi-independent patch network. The number of years that a particular network was occupied out of 9 years is shown as a function of the metapopulation capacity of the network (a measure of network 'size') and its connectivity to metapopulations in the surrounding networks. The size of the dot is proportional to the number of years the network was occupied. (From Thomas and Hanski, 2004.)

rence of entire metapopulations is influenced by similar area and connectivity effects as the occurrence of local populations within metapopulations.

5 Changing Environments

Much of the metapopulation theory developed in the 1990s is concerned with static landscapes. Local populations were expected to go extinct for stochastic reasons, not because the habitat itself would have turned unsuitable. In the applications of the theory to landscapes that had become fragmented by man, the implicit assumption was that species would respond to changes in landscape structure so quickly that they would always occur close to a stochastic equilibrium with respect to the current landscape structure. This assumption may be warranted for species with fast changes in their population sizes and hence species responding quickly to environmental changes and living in slowly changing landscapes. But clearly there is no reason to assume that all species and landscapes would fit this description, especially when we consider larger spatial scales at which species' responses are necessarily slower.

There are also other issues that influence how bad or good the static landscape assumption is for particular species and landscapes. Figure 8.6 gives an example for the Glanville fritillary. This example assumes an empirically observed habitat loss and fragmentation over a period of 20 years (Fig. 8.6a and b), as well as a more hypothetical further loss of habitat over the next 20 years (Fig. 8.6c). The conclusions about the response of the butterfly metapopulation to changes in landscape structure are based on a model parameterized with empirical data; hence, we consider that Fig. 8.6 represents a realistic example. The bottom line is that how closely the metapopulation tracks the changing environment very much depends on how common the species is in the altered landscape. If the species is still common following a reduction in the amount of habitat and an increase in fragmentation, the metapopulation is predicted to track closely changing landscape structure (Fig. 8.6b). In contrast, if the viability of the metapopulation is threatened by habitat loss and fragmentation, and the metapopulation occurs close to the extinction threshold following environmental change, the transient time is predicted to be very long (Fig. 8.6c). This conclusion is supported by general theory (Fig. 8.6d; Ovaskainen and Hanski, 2002). The message for conservation is an important one, and grim. It is exactly those species about which we are most concerned – threatened species close to their extinction threshold – that exhibit the longest transient time in their response to environmental change, and for which we are therefore most likely to underestimate the threat posed by past habitat loss and fragmentation.

There is a rapidly expanding literature on the influence of climate change on the distribution and abundance of insect species. The pioneering study by Parmesan (1996) in California demonstrated a northward range shift in the checkerspot butterfly *Euphydryas editha* over 100 years, and subsequent work in Europe has confirmed the generality of the pattern (Hill *et al.*, 1999, 2002;

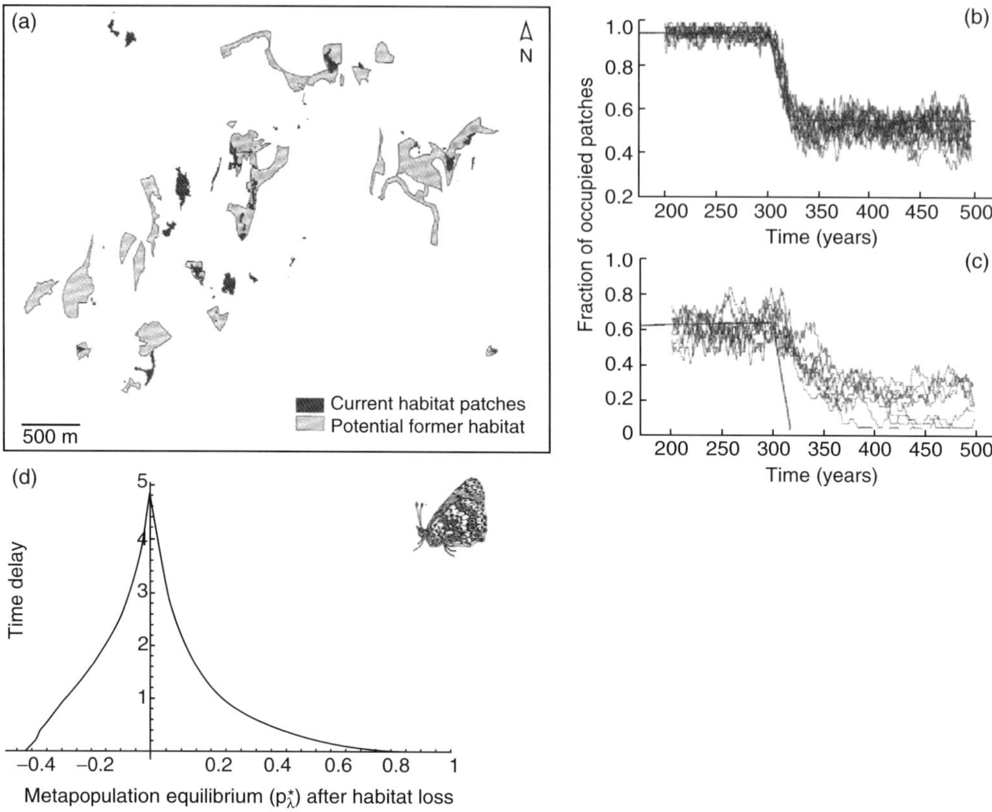

Fig. 8.6. The map (a) shows changing landscape structure in one part of the Åland Islands. Grey areas indicate the partly overgrown patches that were suitable habitat for the Glanville fritillary in 1973 but not 20 years later. Black areas were suitable in 1993. (b) Modelling results giving the metapopulation response to habitat loss in (a). Before and after the 20-year period when habitat was lost, the amount of habitat is assumed to stay constant. The thick line is the predicted quasi-equilibrium metapopulation size corresponding to the current structure of the landscape (amount of habitat and its fragmentation). The ten thin lines show the model-predicted trajectories of metapopulation size before, during and following the observed reduction in habitat area. (c) Similar results for a scenario of further loss of 50% of the area in each of the remaining patches in 1993. The equilibrium now moves to metapopulation extinction, but the actual predicted change in metapopulation size shows a long transient time. (d) The length of the transient time in metapopulation response (vertical axis) to a change in landscape structure. The horizontal axis gives metapopulation size following the change in landscape structure. Note that the transient time is especially long when the metapopulation occurs close to the extinction threshold, as in panel (c). (Panels a–c from Hanski *et al.*, 1996; panel d from Ovaskainen and Hanski, 2002.)

Parmesan *et al.*, 1999). However, species may move their ranges only if there is enough habitat in the landscape, and it is not too fragmented. In a very significant contribution on British butterflies, Warren *et al.* (2001) demonstrated this interaction between landscape structure and climate change-induced range shift: generalist species that have a large amount of suitable habitat did show the expected range shift, but specialist species dependent on presently scarce and highly fragmented habitat did not show it. For this reason, the more specialized species are expected to suffer more than the generalists in the course of climate warming. An additional twist is due to the possibility of behavioral changes in species in response to climate change. Thomas *et al.* (2001) describe an example of the silver-spotted skipper *Hesperia comma* in southern England. With increasing temperature, females are able to use a wider range of microhabitats for oviposition and thereby have a larger total area of habitat at the landscape level, which has facilitated their colonization and increased the current distribution (Thomas *et al.*, 2001). Change in microhabitat use may involve an evolutionary response as well as a behavioral response. In Section 6, we discuss possible evolutionary responses of species to habitat loss and fragmentation.

6 Evolutionary Responses to Habitat Fragmentation

Habitat loss and fragmentation have been so fast in the past decades, and these processes are so fast at present, that one would not expect evolutionary changes in species to greatly affect the overall impact of environmental change on biodiversity: there is little time for evolutionary changes. But at the same time, and just because the change in the natural environments is so great, it would be surprising to observe no evolutionary changes at all in the species. Exactly what we should expect is not entirely clear on the basis of current knowledge. Some consider that biologists tend to underestimate how fast evolutionary changes may take place. Reznick and Ghalambor (2001), Stockwell *et al.* (2003) and Thompson (2005), among others, have reviewed the evidence for contemporary evolution, meaning adaptive evolutionary changes that can be observed in decades or less than a few hundred years. There is indeed a rapidly growing list of more or less convincing examples. On the other hand, other researchers are more impressed by the apparent lack of rapid microevolution in natural populations, for which many possible reasons have been suggested (Merilä *et al.*, 2001), including biased estimates of heritability, fluctuating selection, selection on environmental deviations and correlated traits, evolutionary response masked by changing environment and simply lack of statistical power in field studies. Adaptive contemporary evolution has been mostly documented in response to anthropogenic environmental changes (Reznick and Ghalambor, 2001). The best and most numerous examples involve heavy metal tolerance, air pollution tolerance, insecticide resistance, herbicide resistance and industrial melanism (Reznick and Ghalambor, 2001). One class of rapid evolution that is well established and relevant for habitat loss-related conservation issues is changes taking place in captive populations. Captive populations are often seen as a means of improving the chances of long-term survival of species that have little current habitat but for which more habitat could exist in the future and into which individuals could be released from captive populations. Unfortunately, contemporary evolution may be especially common in captive populations, for which reason individuals raised in captivity may often do poorly when released in the wild (Stockwell and Weeks, 1999; Lynch and O'Hely, 2001). There are, of course, additional reasons why such introductions might fail.

In their review of contemporary evolution, Stockwell *et al.* (2003) cite three studies as providing examples of evolutionary response to habitat degradation and fragmentation, but these studies deal with processes such as heavy metal tolerance rather than habitat loss and fragmentation. The most likely evolutionary responses to habitat loss and fragmentation involve migration and movement behaviour. Changes in such traits are likely to be associated with correlated changes and trade-offs in other traits, such as fecundity. In metapopulations consisting of small extinction-prone local populations, such as the Glanville fritillary metapopulation in the Åland Islands, some migration is clearly necessary for long-term persistence. On the other hand, 'too much' migration may elevate mortality during migration so greatly and may lead to such an excessive loss of time for reproduction that persistence is again compromised (Comins *et al.*, 1980; Hanski and Zhang, 1993; Olivieri and Gouyon, 1997). Although natural selection does not operate to produce the optimal migration rate for the long-term survival of species or metapopulations (Comins *et al.*, 1980), it is none the less possible that an evolutionary change in migration rate following habitat loss and fragmentation might reduce the extinction risk of a metapopulation (Leimar and Nordberg, 1997). Kotiaho *et al.* (2005) found that threatened butterfly species in Finland had the most limited migration capacity, whereas Thomas (2000) reported that species with intermediate migration capacity have fared worst in the UK. In neither case is the explanation likely to be the cost of migration, but rather some correlation with, for example, habitat availability for different kinds of species.

But what is the likely change in migration rate in response to habitat loss and fragmentation? There are so many different selective forces affecting the evolution of migration rate (Ronce *et al.*, 2001; Ronce and Olivieri, 2004) that there is no simple answer to this question. For instance, habitat loss and fragmentation increase mortality during migration, because it becomes more difficult for migrants to locate another fragment of habitat, which should select for reduced migration (van Valen, 1971). Growing genetic relatedness of individuals in increasingly isolated local populations should select for increased migration (Hamilton and May, 1977; Gandon and Rousset, 1999), and so should the opportunity to recolonize habitat patches that have become unoccupied following local extinction, and more generally the chance to move to a low-density population (Gadgil, 1971; Roff, 1975). Given the multitude of often opposing selection pressures, it is perhaps not surprising that researchers have come up with conflicting suggestions as to what might be the net effect of habitat fragmentation on the evolution of migration rate. Thus, Dempster (1991) expected evolution to reduce migration rate in butterflies living in increasingly fragmented habitats (see also Thomas *et al.*, 1998; Hill *et al.*, 1999), whereas Hanski (1999) suggested that fragmentation would generally select for increased migration rate. The matter cannot be settled without having a means of considering all the major selective forces, and their interactions, at the same time. This cannot be done without employing appropriate models.

Heino and Hanski (2001) constructed a spatially realistic evolutionary model to investigate the evolution of migration rate in fragmented habitats, using the Glanville fritillary butterfly as an example. The model parameters were estimated with independent data whenever possible, whereas for the remaining parameters values were selected so that the model produced realistic short-term and long-term metapopulation dynamics. Reassuringly, when migration rate was allowed to evolve in the model, it settled to a value close to the empirically observed one (Heino and Hanski, 2001). This analysis suggested that the dominant selective forces were mortality and time lost during migration and the opportunity to establish new local populations in currently unoccupied patches. With increasing fragmentation, the predicted migration rate first declined due to increased cost of migration, but with further fragmentation migration rate increased when an increasing number of habitat patches was available for recolonization.

More detailed theoretical and empirical studies of the same butterfly metapopulation have revealed how the average migration rate of butterflies in particular local populations depends on their age and population dynamic connectivity to other populations (Hanski *et al.*, 2004). Migration rate is predicted and was observed to be higher in new than in old populations, apparently because new populations are likely to be established by exceptionally mobile individuals and because heritability of migration-related traits is generally high (Roff and Fairbairn, 1991). Among the new populations, migration rate increased with decreasing connectivity (increasing isolation), whereas among old populations the opposite was both predicted and observed. Reduced migration rate in old isolated populations is largely due to emigration of the more mobile individuals away from the population and limited immigration due to great isolation. These results satisfactorily resolve the two opposing verbal predictions that have been put forward about the impact of increasing habitat fragmentation on migration rate. Dempster (1991) emphasized increasing emigration losses and expected migration rate to reduce with fragmentation, increasing isolation of habitat patches; in the Glanville fritillary, this was observed for old populations. Hanski (1999) was primarily thinking of improved colonization opportunities with increasing fragmentation, hence expecting increased migration rate with increasing fragmentation, which was observed for new populations. Because both effects operate simultaneously in a metapopulation, one has to use a model to work out the overall consequences of fragmentation. In the case of the Glanville fritillary, and assuming realistic parameter values for this species and its natural landscape, the overall effect has been increasing migration rate with increasing fragmentation (Heino and Hanski, 2001; Hanski *et al.*, 2004), though the quantitative result will depend on the details, for instance on the spatial configuration of the landscape.

To return to the question of whether evolutionary changes may make a difference to the long-term survival of species in changing environments,

Heino and Hanski (2001) showed with the model described above that an evolutionary rescue is theoretically possible: natural selection may change migration rate to such an extent that a metapopulation will persist in a landscape in which it would go extinct without the evolutionary change. However, the calculations also indicated that in practice such a rescue is unlikely in the Glanville fritillary, largely because a change in migration rate has both positive and negative consequences for population sizes, and hence cannot much compensate for habitat loss and fragmentation. Only when the contrast is between relatively uniform and highly fragmented habitats is the level of migration likely to make a truly significant difference for population persistence, a point to which we will return in Section 6.1. The example of the silver-spotted skipper suggests that evolutionary changes may have more substantial consequences for population dynamics when habitat selection rather than migration rate is affected. None the less, conservationists can hardly count on evolution to solve the extinction crisis caused by habitat loss and fragmentation.

6.1 Species living in naturally fragmented versus newly fragmented landscapes

Fragmented landscapes harbour fragmented populations, whether the landscape is fragmented naturally or because of anthropogenic habitat loss and fragmentation. But the important difference is that species living in naturally fragmented landscapes have become adapted to living in such conditions – otherwise many of them would be extinct – whereas many species now found in newly fragmented landscapes may be well adapted to more continuous habitats. Species living in naturally fragmented habitats often exhibit a high rate of migration and related adaptations, such as density-dependent emigration and flight polymorphism (Roff, 1994; Denno *et al.*, 1996; Dingle, 1996). Indeed, a primary problem for the survival of many species in newly fragmented landscapes is likely to be insufficient migration and colonization rates, which were not selected for in the past as environmental conditions in their previously more continuous habitats favoured traits other than high level of mobility. Limited migration propensity compounded with often strict habitat selection explains why many forest species are especially sensitive to habitat fragmentation. One example from boreal forests is old-growth fungi (Penttilä *et al.*, 2006). In the tropics the situation is even worse, as most forest-living vertebrates as well as invertebrates appear unable to cross even narrow gaps of open habitat.

7 Conclusions

We emphasize the great practical significance of researchers' conclusions about the effects of habitat fragmentation on the viability of populations and metapopulations. If the influence of connectivity and hence fragmentation is ignored in situations where it truly matters, managers may end up implementing inadequate and wasteful conservation measures. Often it makes good sense to protect networks of such sites rather than a single larger but isolated site, but the favourable solution is always a compromise. Below, we first describe an extreme example in which the conservation effort is spread too thinly in space with no regard for the harmful effects of excessive fragmentation. Second, we outline an approach that can be used to find out the favourable compromise.

The example relates to the legislation and practices adopted in forestry in Finland and elsewhere in northern Europe in the 1990s. Forestry has become very intensive in the boreal forest zone in Europe. For instance, of the 10 million hectares of forests in southern Finland, only 1% remains in a state that can be called natural or semi-natural, and as a consequence, nearly 20% of the more than 20,000 forest species, most of which are insects, is nationally extinct, threatened or near-threatened (Siitonen and Hanski, 2004). There is thus an obvious and urgent need to do something to slow down and ultimately halt the decline of forest biodiversity. To this effect, the current legislation in Finland stipulates that special 'woodland key habitats' should be left intact while the rest of the forest stand is cut down. The key habitats are patches of habitat that clearly stand out from the rest of the forested land. Protection of key habitats is expected to preserve forest biodiversity, because many woodland key habitats represent habitat types that are appropriate for many threatened species (Annila, 1998). The snag here is that to qualify as a key habitat in the sense of this legislation, a habitat patch has to be well-delimited and very small; a larger area of the same habitat type would not count as a key habitat. The average size of the key habitats in Finland is only 0.5 ha and their density is around $0.6/km^2$ (Yrjönen, 2004). Applying these figures to the 10 million hectares of forest in southern Finland makes 30,000 ha, which would comprise several respectable forest reserves if located in a few pieces. But they are not: they are fragmented into tens of thousands of micropatches. We have many reasons to be very sceptical about the possibility that such a sparse network consisting of tiny habitat patches with massive edge effects (Aune *et al.*, 2005) would preserve metapopulations of endangered species. In a thorough study covering an area of 278 km^2 in southern Finland, Pykälä (2004) resurveyed the occurrence of 190 small local populations of 15 endangered lichen species. Many of these populations were found in key habitats, but during a 10-year period 40% of the local populations had disappeared, often due to outright habitat degradation, and practically no new local populations had been established. The 190 populations became established during previous decades when the forest landscape was generally more favourable for the species. Unfortunately, though the effectiveness of key habitats in preserving forest biodiversity is not supported by theory or by empirical results, they continue to be used as an argument as to why no more substantial measures to protect forest biodiversity are needed. Naturally, the reasons for devising the woodland key habitats as a major conservation instrument were political rather than ecological in the first place (Hanski, 2005).

There is an extensive literature in conservation biology on what is called the reserve site selection problem (Margules *et al.*, 1994; Pressey, 1994; Margules and Pressey, 2000). Given a particular set of sites with a particular set of species in each, and a certain amount of resources to protect only a subset of these sites, which subset should be protected to maximize the number of species protected? Hundreds of papers have addressed the many biological and technical issues that need to be considered while answering this and related questions (for reviews see Pressey, 1999; Cabeza and Moilanen, 2001; Cabeza *et al.*, 2004). Most of this work has been stimulated by the design of hypothetical reserve networks at continental or other very large spatial scales, but there is no reason why the same methods could not be applied at the landscape level. In fact, the latter applications are more helpful than the continental ones, because there are more landscapes than continents, and because there are generally more conservation and management choices to be made at small rather than large scales. The framework is flexible, and can address questions such as improving the habitat quality of particular sites, the value of restoring new habitat where it did not occur before, consequences of land use changes and so forth. To do all this properly at the landscape level, there is, however, the requirement of coupling reserve selection algorithms with metapopulation models, and asking about the long-term persistence of species in landscapes rather than about the current occurrence of species, which has been the focus of most reserve selection algorithms so far. At the landscape level, the occurrence of species is spatially dynamic due to small or relatively small population sizes, and this should be taken into account in the design of reserves.

A case study by Moilanen and Cabeza (2002) on landscape-level reserve selection for the butterfly *M. diamina* provides an excellent example. They examined a landscape of 20 × 30 km in area, where there were 125 patches of habitat of higher or lower quality for the butterfly (meadows). They combined a parameterized metapopulation model with an optimization algorithm to answer the question of which subset of the sites should be selected to maximize the long-term persistence of the species, given that each site has a cost and the amount of resources (money) available to acquire and manage the sites is limited. One may easily include spatially variable habitat quality in this model by allowing the performance of local populations to depend on habitat quality. Apart from showing that it is possible to provide a rigorous answer to the above question, Moilanen and Cabeza (2002) demonstrated how the optimal selection of sites may radically depend on the amount of resources available for conservation. The apparent drawback of this approach is that a large amount of ecological knowledge is needed. For many practical applications, and especially those involving communities of species rather than single species, simplifications are necessary. But the best available ecological knowledge should be used, whatever approach to conservation and management is adopted. One important advantage of models is that we need to make clear how good our knowledge actually is, and models can be helpful in allowing us to assess the sensitivity of management recommendations to inadequate knowledge.

References

Anderson, M.J. and Gribble, N.A. (1998) Partitioning the variation among spatial, temporal and environmental components in a multivariate data set. *Australian Journal of Ecology* 23, 158–167.

Annila, E. (1998) Uusittujen metsänkäsittelymenetelmien vaikutus uhanalaisiin lajeihin. In: Annila, E. (ed.) *Monimuotoinen metsä. Metsäntutkimuslaitoksen tiedonantoja* 705, 197–222.

Aune, K., Jonsson, B.G. and Moen, J. (2005) Isolation and edge effects among woodland key habitats in Sweden: is forest policy promoting fragmentation? *Biological Conservation* 124, 89–95.

Borcard, D., Legendre, P. and Drapeau, P. (1992) Partialling out the spatial component of ecological variation. *Ecology* 73, 1045–1055.

Brommer, J.E. and Fred, M.S. (1999) Movement of the apollo butterfly *Parnassius apollo* related to host plant and nectar plant patches. *Ecological Entomology* 24, 125–131.

Cabeza, M. and Moilanen, A. (2001) Design of reserve networks and the persistence of biodiversity. *Trends in Ecology and Evolution* 16, 242–248.

Cabeza, M., Moilanen, A. and Possingham, H.P. (2004) Metapopulation dynamics and reserve network design. In: Hanski, I. and Gaggiotti, O.E. (eds) *Ecology, Genetics, and Evolution of Metapopulations*. Elsevier Academic Press, Amsterdam, The Netherlands, pp. 541–564.

Chevan, A. and Sutherland, M. (1991) Hierarchical partioning. *The American Statistician* 45, 90–96.

Collinge, S.K., Prudic, K.L. and Oliver, J.C. (2003) Effects of local habitat characteristics and landscape context on grassland butterfly diversity. *Conservation Biology* 17, 178–187.

Comins, H.N., Hamilton, W.D. and May, R.M. (1980) Evolutionary stable dispersal strategies. *Journal of Theoretical Biology* 82, 205–230.

Dempster, J.P. (1991) Fragmentation, isolation and mobility of insect populations. In: Collins, N.M. and Thomas, J.A. (eds) *Conservation of Insects and Their Habitats*. Academic Press, London, pp. 143–154.

Denno, R.F., Roderick, G.K., Peterson, M.A., Huberty, A.F., Dobel, H.G., Eubanks, M.D., Losey, J.E. and Langellotto, G.A. (1996) Habitat persistence underlies intraspecific variation in the dispersal strategies of plant hoppers. *Ecological Monographs* 66, 389–408.

Dingle, H. (1996) *Migration: The Biology of Life on the Move*. Oxford University Press, Oxford.

Ehrlich, P. and Hanski, I. (2004) *On the Wings of Checkerspots: A Model System for Population Biology*. Oxford University Press, New York.

Eriksson, O. and Ehrlén, J. (2001) Landscape fragmentation and the viability of plant populations. In: Silvertown, J. and Antonovics, J. (eds) *Integrating Ecology and Evolution in Spatial Context*. Blackwell, Oxford, pp. 157–175.

Fahrig, L. (1997) Relative effects of habitat loss and fragmentation on population extinction. *Journal of Wildlife Management* 61, 603–610.

Fahrig, L. (2001) How much habitat is enough? *Biological Conservation* 100, 65–74.

Fahrig, L. (2003) Effects of habitat fragmentation on biodiversity. *Annual Review of Ecology, Evolution, and Systematics* 34, 487–515.

Fleishman, E., Ray, C., Sjogren-Gulve, P., Boggs, C.L. and Murphy, D.D. (2002) Assessing the roles of patch quality, area, and isolation in predicting metapopulation dynamics. *Conservation Biology* 16, 706–716.

Gadgil, M. (1971) Dispersal: population consequences and evolution. *Ecology* 52, 253–261.

Gandon, S. and Rousset, F. (1999) Evolution of stepping-stone dispersal rates. *Proceedings of the Royal Society of London Series B – Biological Sciences* 266, 2507–2513.

Graham, M.H. (2003) Confronting multicollinearity in ecological multiple regression. *Ecology* 84, 2809–2815.

Hamilton, W.D. and May, R.M. (1977) Dispersal in stable habitats. *Nature* 269, 578–581.

Hanski, I. (1994) A practical model of metapopulation dynamics. *Journal of Animal Ecology* 63, 151–162.

Hanski, I. (1998) Metapopulation dynamics. *Nature* 396, 41–49.

Hanski, I. (1999) *Metapopulation Ecology*. Oxford University Press, New York.

Hanski, I. (2001) Spatially realistic theory of metapopulation ecology. *Naturwissenschaften* 88, 372–381.

Hanski, I. (2003) Biology of extinctions in butterfly metapopulations. In: Boggs, C., Watt, W. and Ehrlich, P. (eds) *Butterflies: Ecology and Evolution Taking Flight*. University of Chicago Press, Chicago, Illinois, pp. 577–602.

Hanski, I. (2005) *The Shrinking World: Ecological Consequences of Habitat Loss*. International Ecology Institute, Oldendorf and Luhe, Germany.

Hanski, I. and Gaggiotti, O.E. (eds) (2004) *Ecology, Genetics, and Evolution of Metapopulations*. Elsevier Academic Press, Amsterdam, The Netherlands.

Hanski, I. and Ovaskainen, O. (2000) The metapopulation capacity of a fragmented landscape. *Nature* 404, 755–758.

Hanski, I. and Zhang, D.-Y. (1993) Migration, metapopulation dynamics and fugitive co-existence. *Journal of Theoretical Biology* 163, 491–504.

Hanski, I., Moilanen, A. and Gyllenberg, M. (1996) Minimum viable population size. *American Naturalist* 147, 527–541.

Hanski, I., Alho, J. and Moilanen, A. (2000) Estimating the parameters of survival and migration of individuals in metapopulations. *Ecology* 81, 239–251.

Hanski, I., Erälahti, C., Kankare, M., Ovaskainen, O. and Sirén, H. (2004) Variation in migration rate among individuals maintained by landscape structure. *Ecology Letters* 7, 958–966.

Heikkinen, R.K., Luoto, M., Kuussaari, M. and Pöyry, J. (2005) New insights into butterfly–environment relationships using partitioning methods. *Proceedings of the Royal Society of London Series B – Biological Sciences* 272, 2203–2210.

Heino, M. and Hanski, I. (2001) Evolution of migration rate in a spatially realistic metapopulation model. *American Naturalist* 157, 495–511.

Helm, A., Hanski, I. and Pärtel, M. (2006) Slow response of plant species richness to habitat loss and fragmentation. *Ecology Letters* 9, 72–77.

Hill, J.K., Thomas, C.D. and Huntley, B. (1999) Climate and habitat availability determine 20th century changes in a butterfly's range margin. *Proceedings of the Royal Society of London Series B – Biological Sciences* 266, 1197–1206.

Hill, J.K., Thomas, C.D., Fox, R., Telfer, M.G., Willis, S.G., Asher, J. and Huntley, B. (2002) Responses of butterflies to twentieth century climate warming: implications for future ranges. *Proceedings of the Royal Society of London Series B – Biological Sciences* 269, 2163–2171.

Kotiaho, J.S., Kaitala, V., Komonen, A. and Päivinen, J. (2005) Predicting the risk of extinction from shared ecological characteristics. *Proceedings of the National Academy of Sciences of the United States of America* 102, 1963–1967.

Leimar, O. and Nordberg, U. (1997) Metapopulation extinction and genetic variation in dispersal-related traits. *Oikos* 80, 448–458.

Lindborg, R. and Eriksson, O. (2004) Historical landscape connectivity affects present plant species diversity. *Ecology* 85, 1840–1845.

Liu, Q. (1997) Variation partitioning by partial redundancy analysis. *Environmetrics* 8, 75–85.

Luoto, M., Kuussaari, M., Rita, H., Salminen, J. and von Bonsdorff, T. (2001) Determinants of distribution and abundance in the clouded apollo butterfly: a landscape ecological approach. *Ecography* 24, 601–617.

Lynch, M. and O'Hely, M. (2001) Captive breeding and the genetic fitness of natural populations. *Conservation Genetics* 2, 363–378.

MacNally, R. (2000) Regression and model-building in conservation biology, biogeography and ecology: the distinction between – and reconciliation of – 'predictive' and explanatory models. *Biodiversity and Conservation* 9, 655–671.

MacNally, R. and Walsh, C.J. (2004) Hierarchical partitioning public-domain software. *Biodiversity and Conservation* 13, 659–660.

Margules, C.R. and Pressey, R.L. (2000) Systematic conservation planning. *Nature* 405, 243–253.

Margules, C.R., Nicholls, A.O. and Usher, M.B. (1994) Apparent species turnover, probability of extinction and the selection of nature reserves: a case study of the Ingleborough limestone pavements. *Conservation Biology* 8, 398–409.

Merilä, J., Kruuk, L.E.B. and Sheldon, B.C. (2001) Explaining stasis: microevolutionary studies of natural populations. In: Hendry, A. and Kinnison, M. (eds) *Contemporary Microevolution: Rate, Pattern and Process. Genetica* 112/113, pp. 199–222.

Mikkola, K. (1986) Direction of insect migration in relation to wind. In: Danthanarayana, W. (ed.) *Insect Flight: Dispersal and Migration*. Springer-Verlag, Berlin, Heidelberg, pp. 152–171.

Moilanen, A. and Cabeza, M. (2002) Single-species dynamic site selection. *Ecological Applications* 12, 913–926.

Moilanen, A. and Hanski, I. (2001) On the use of connectivity measures in spatial ecology. *Oikos* 95, 147–151.

Moilanen, A. and Nieminen, M. (2002) Simple connectivity measures for spatial ecology. *Ecology* 83, 1131–1145.

Nieminen, M., Leskinen, M. and Helenius, J. (2000) Doppler radar detection of exceptional mass-migration of aphids into Finland. *International Journal of Biometeorology* 44, 172–181.

Nieminen, M., Siljander, M. and Hanski, I. (2004) Structure and dynamics of *Melitaea cinxia* metapopulations. In: Ehrlich, P.R. and Hanski, I. (eds) *On the Wings of Checkerspots: A Model System for Population Biology*. Oxford University Press, New York, pp. 63–91.

Olivieri, I. and Gouyon, P.-H. (1997) Evolution of migration rate and other traits: the metapopulation effect. In: Hanski, I.A. and Gilpin, M.E. (eds) *Metapopulation Biology*. Academic Press, San Diego, California, pp. 293–324.

Ovaskainen, O. (2004) Habitat-specific movement parameters estimated using mark–recapture data and a diffusion model. *Ecology* 85, 242–257.

Ovaskainen, O. and Hanski, I. (2002) Transient dynamics in metapopulation response to perturbation. *Theoretical Population Biology* 61, 285–295.

Ovaskainen, O. and Hanski, I. (2004) Metapopulation dynamics in highly fragmented landscapes. In: Hanski, I. and Gaggiotti, O.E. (eds) *Ecology, Genetics, and Evolution in Metapopulations*. Elsevier Academic Press, Amsterdam, The Netherlands, pp. 73–104.

Parmesan, C. (1996) Climate and species' range. *Nature* 382, 765–766.

Parmesan, C., Ryrholm, N., Stefanescu, C., Hill, J.K., Thomas, C.D., Descimon, H., Huntley, B., Kaila, L., Kullberg, J., Tammaru, T., Tennent, W.J., Thomas, J.A. and Warren, M. (1999) Poleward shifts in geographical ranges of butterfly species associated with regional warming. *Nature* 399, 579–583.

Penttilä, R., Lindgren, M., Miettinen, O., Rita, H. and Hanski, I. (2006) Consequences of forest fragmentation for polyporous fungi at two spatial scales. *Oikos* 114, 225–240.

Pope, S.E., Fahrig, L. and Merriam, H.G. (2000) Landscape complemetation and metapopulation effects on leopard frog populations. *Ecology* 81, 2498–2508.

Potts, S.G., Vulliamy, B., Roberts, S., O'Toole, C., Dafni, A., Ne'eman, G. and Willmer, P. (2005) Role of nesting resources in organising diverse bee communities in a Mediterranean landscape. *Ecological Entomology* 30, 78–85.

Pressey, R.L. (1994) *Ad hoc* reservations: forward of backward steps in developing representative reserve systems? *Conservation Biology* 8, 662–668.

Pressey, R.L. (1999) Systematic conservation planning for the real world. *Parks* 9, 1–6.

Pykälä, J. (2004) Effects of new forestry practices on rare epiphytic macrolichens. *Conservation Biology* 18, 831–838.

Reznick, D.N. and Ghalambor, C.K. (2001) The population ecology of contemporary adaptations: what empirical studies reveal about the conditions that promote adaptive evolution. *Genetica* 112–113, 183–198.

Roff, D.A. (1975) Population stability and the evolution of dispersal in a heterogeneous environment. *Oecologia* 19, 217–237.

Roff, D.A. (1994) Habitat persistence and the evolution of wing dimorphism in insects. *American Naturalist* 144, 772–798.

Roff, D.A. and Fairbairn, D.J. (1991) Wing dimorphisms and the evolution of migratory polymorphisms among the Insecta. *American Zoologist* 31, 243–251.

Ronce, O. and Olivieri, I. (2004) Life history evolution in metapopulations. In: Hanski, I. and Gaggiotti, O.E. (eds) *Ecology, Genetics, and Evolution of Metapopulations*. Elsevier Academic Press, Amsterdam, The Netherlands, pp. 227–257.

Ronce, O., Olivieri, I., Clobert, J. and Danchin, E. (2001) Perspectives on the study of dispersal evolution. In: Clobert, J., Danchin, E., Dhondt, A.A. and Nichols, J.D. (eds) *Dispersal*. Oxford University Press, Oxford, pp. 341–357.

Sala, O.E., Chapin, F.S.I., Armesto, J.J., Berlow, E., Bloomfield, J., Dirzo, R., Huber-Sanwald, E., Huenneke, L.F., Jackson, R.B., Kinzig, A., Leemans, R., Lodge, D.M., Mooney, H.A., Oesterheld, M., Poff, N.L., Sykes, M.T., Walker, B.H., Walker, M. and Wall, D.H. (2000) Global biodiversity scenarios for the year 2100. *Science* 287, 1770–1774.

Shahabuddin, G., Herzner, G.A., Aponte, C. and Gomez, M.D. (2000) Persistence of a frugivorous butterfly species in Venezuelan forest fragments: the role of movement and habitat quality. *Biodiversity and Conservation* 9, 1623–1641.

Siitonen, J. and Hanski, I. (2004) Metsälajiston ekologia ja monimuotoisuus. In: Siitonen, J. and Hanski, I. (eds) *Metsän kätköissä, Suomen metsäluonnon monimuotoisuus*. Edita, Helsinki, pp. 76–105.

Stockwell, C.A. and Weeks, S.C. (1999) Translocations and rapid evolutionary responses in recently established populations of western mosquitofish (*Gambusia affinis*). *Animal Conservation* 2, 103–110.

Stockwell, C.A., Hendry, A.P. and Kinnison, M.T. (2003) Contemporary evolution meets conservation biology. *Trends in Ecology and Evolution* 18, 94–101.

Summerville, K.S. and Crist, T.O. (2004) Contrasting effects of habitat quantity and quality on moth communities in fragmented landscapes. *Ecography* 27, 3–12.

Thomas, C.D. (2000) Dispersal and extinction in fragmented landscapes. *Proceedings of the Royal Society of London Series B – Biological Sciences* 267, 139–145.

Thomas, C.D. and Hanski, I. (2004) Metapopulation dynamics in changing environments: butterfly responses to habitat and climate change. In: Hanski, I. and Gaggiotti, O.E. (eds) *Ecology, Genetics, and Evolution of Metapopulations*. Elsevier Academic Press, Amsterdam, The Netherlands, pp. 489–514.

Thomas, C.D., Hill, J.K. and Lewis, O.T. (1998) Evolutionary consequences of habitat fragmentation in a localised butterfly. *Journal of Animal Ecology* 67, 485–497.

Thomas, C.D., Bodsworth, E.J., Wilson, R.J., Simmons, A.D., Davies, Z.G., Musche, M. and Conradt, L. (2001) Ecological and evolutionary processes at expanding range margins. *Nature* 411, 577–581.

Thomas, C.D., Cameron, A., Green, R.E., Bakkenes, M., Beaumont, L.J., Collingham, Y.C., Erasmus, B.F.N., De Siqueira, M.F., Grainger, A., Hannah, L., Hughes, L., Huntley, B., Van Jaarsveld, A.S., Midgley, G.F., Miles, L., Ortega-Huerta, M.A., Peterson, A.T., Phillips, O.L. and Williams, S.E. (2004) Extinction risk from climate change. *Nature* 427, 145–148.

Thomas, J.A. (1980) Why did the large blue become extinct in Britain. *Oryx* 15, 243–247.

Thompson, J.N. (2005) *The Geographic Mosaic of Coevolution*. University of Chicago Press, Chicago, Illinois.

Turner, M.G., Gardner, R.H. and O'Neill, R.V. (2001) *Landscape Ecology in Theory and Practice*. Springer, New York.

van Valen, L. (1971) Group selection and the evolution of dispersal. *Evolution* 25, 591–598.

Warren, M.S., Hill, J.K., Thomas, J.A., Asher, J., Fox, R., Huntley, B., Roy, D.B., Telfer, M.G., Jeffcoate, S., Harding, P., Jeffcoate, G., Willis, S.G., Greatorex-Davies, J.N., Moss, D. and Thomas, C.D. (2001) Rapid responses of British butterflies to opposing forces of climate and habitat change. *Nature* 414, 65–69.

Yrjönen, K. (2004) Metsälain erityisen tärkeät elinympäristöt. MMM:n julkaisuja 9/2004, Vammala, Finland.

List of 38 articles used in the analysis reported in Tables 8.1 and 8.2

Anthes, N., Fartmann, T., Hermann, G. and Kaule, G. (2003) Combining larval habitat quality and metapopulation structure – the key for successful management of pre-alpine *Euphydryas aurinia* colonies. *Journal of Insect Conservation* 7, 175–185.

Bergman, K.-O., Askling, J., Ekberg, O., Ignell, H., Wahlman, H. and Milberg, P. (2004) Landscape effects on butterfly assemblages in an agricultural region. *Ecography* 27, 619–628.

Bonte, D., Lens, L., Maelfait, J.-P., Hoffmann, M. and Kuijken, E. (2003) Patch quality and connectivity influence spatial dynamics in a dune wolfspider. *Oecologia* 135, 227–233.

Brommer, J.E. and Fred, M.S. (1999) Movement of the apollo butterfly *Parnassius apollo* related to host plant and nectar plant patches. *Ecological Entomology* 24, 125–131.

Collinge, S.K. (2000) Effects of grassland fragmentation on insect species loss, colonization, and movement patterns. *Ecology* 81, 2211–2226.

Collinge, S.K., Prudic, K.L. and Oliver, J.C. (2003) Effects of local habitat characteristics and landscape context on grassland butterfly diversity. *Conservation Biology* 17, 178–187.

Dauber, J., Hirsch, M., Simmering, D., Waldhardt, R., Otte, A. and Wolters, V. (2003) Landscape structure as an indicator of biodiversity: matrix effects on species richness. *Agriculture Ecosystems and Environment* 98, 321–329.

Dennis, R.L.H. and Eales, H.T. (1997) Patch occupancy in *Coenonympha tullia* (Müller, 1764) (Lepidoptera: Satyrinae): habitat quality matters as much as patch size and isolation. *Journal of Insect Conservation* 1, 167–176.

Dennis, R.L.H. and Eales, H.T. (1999) Probability of site occupancy in the large heath butterfly *Coenonympha tullia* determined from geographical and ecological data. *Biological Conservation* 87, 295–301.

Doak, P. (2000a) Habitat patchiness and the distribution, abundance, and population dynamics of an insect herbivore. *Ecology* 81, 1842–1857.

Doak, P. (2000b) Population consequences of restricted dispersal for an insect herbivore in a subdivided habitat. *Ecology* 81, 1828–1841.

Fleishman, E., Ray, C., Sjögren-Gulve, P., Boggs, C.L. and Murphy, D.D. (2002) Assessing the roles of patch quality, area, and isolation in predicting metapopulation dynamics. *Conservation Biology* 16, 706–716.

Grand, J. and Mello, M.J. (2004) A multi-scale analysis of species–environment relationships: rare moths in a pitch pine–scrub oak (*Pinus rigida–Quercus ilicifolia*) community. *Biological Conservation* 119, 495–506.

Hanski, I., Pöyry, J., Pakkala, T. and Kuussaari, M. (1995) Multiple equilibria in metapopulation dynamics. *Nature* 377, 618–621.

Heikkinen, R.K., Luoto, M., Kuussaari, M. and Pöyry, J. (2005) New insights into butterfly–environment relationships using partitioning methods. *Proceedings of the Royal Society of London Series B – Biological Sciences* 272, 2203–2210.

Hill, J.K., Thomas, C.D. and Lewis, O.T. (1996) Effects of habitat patch size and isolation on dispersal by *Hesperia comma* butterflies: implications for metapopulation structure. *Journal of Animal Ecology* 65, 725–735.

James, M., Gilbert, F. and Zalat, S. (2003) Thyme and isolation for the Sinai baton blue butterfly (*Pseudophilotes sinaicus*). *Oecologia* 134, 445–453.

Krauss, J., Steffan-Dewenter, I. and Tscharntke, T. (2003a) How does landscape context contribute to effects of habitat fragmentation on diversity and population density of butterflies? *Journal of Biogeography* 30, 889–900.

Krauss, J., Steffan-Dewenter, I. and Tscharntke, T. (2003b) Local species immigration, extinction, and turnover of butterflies in relation to habitat area and habitat isolation. *Oecologia* 137, 591–602.

Krauss, J., Steffan-Dewenter, I. and Tscharntke, T. (2004) Landscape occupancy and local population size depends on host plant distribution in the butterfly *Cupido minimus*. *Biological Conservation* 120, 355–361.

Kuussaari, M., Nieminen, M. and Hanski, I. (1996) An experimental study of migration in the Glanville fritillary butterfly *Melitaea cinxia*. *Journal of Animal Ecology* 65, 791–801.

Matter, S.F., Roland, J., Keyghobadi, N. and Sabourin, K. (2003) The effects of isolation, habitat area and resources on the abundance, density and movement of the butterfly *Parnassius smintheus*. *American Midland Naturalist* 150, 26–36.

Moilanen, A. and Hanski, I. (1998) Metapopulation dynamics: effects of habitat quality and landscape structure. *Ecology* 79, 2503–2515.

Ricketts, T.H., Daily, G.C., Ehrlich, P.R. and Fay, J.P. (2001) Countryside biogeography of moths in a fragmented landscape: biodiversity in native and agricultural habitats. *Conservation Biology* 15, 378–388.

Roland, J., Keyghobadi, N. and Fownes, S. (2000) Alpine *Parnassius* butterfly dispersal: effects of landscape and population size. *Ecology* 81, 1642–1653.

Schmidt, M.H., Roschewitz, I., Thies, C. and Tscharntke, T. (2005) Differential effects of landscape and management on diversity and density of ground-dwelling farmland spiders. *Journal of Applied Ecology* 42, 281–287.

Shahabuddin, G., Herzner, G.A., Aponte, C. and Gomez, M.D. (2000) Persistence of a frugivorous butterfly species in Venezuelan forest fragments: the role of movement and habitat quality. *Biodiversity and Conservation* 9, 1623–1641.

Steffan-Dewenter, I. and Tscharntke, T. (2000) Butterfly community structure in fragmented habitats. *Ecology Letters* 3, 449–456.

Stoner, K.J.L. and Joern, A. (2004) Landscape vs. local habitat scale influences to insect communities from tallgrass prairie remnants. *Ecological Applications* 14, 1306–1320.

Summerville, K.S. and Crist, T.O. (2004) Contrasting effects of habitat quantity and quality on moth communities in fragmented landscapes. *Ecography* 27, 3–12.

Summerville, K.S., Steichen, R.M. and Lewis, M.N. (2005) Restoring Lepidopteran communities to oak savannas: contrasting influences of habitat quantity and quality. *Restoration Ecology* 13, 120–128.

Thomas, C.D. and Jones, T.M. (1993) Partial recovery of a skipper butterfly (*Hesperia comma*) from population refuges: lessons for conservation in a fragmented landscape. *Journal of Animal Ecology* 62, 472–481.

Thomas, J.A., Bourn, N.A.D., Clarke, R.T., Stewart, K.E., Simcox, D.J., Pearman, G.S., Curtis, R. and Goodger, B. (2001) The quality and isolation of habitat patches both determine where butterflies persist in fragmented landscapes. *Proceedings of the Royal Society of London Series B – Biological Sciences* 268, 1791–1796.

Välimäki, P. and Itämies, J. (2003) Migration of the clouded apollo butterfly *Parnassius mnemosyne* in a network of suitable habitats – effects of patch characteristics. *Ecography* 26, 679–691.

WallisDeVries, M.F. (2004) A quantitative conservation approach for the endangered butterfly *Maculinea alcon*. *Conservation Biology* 18, 489–499.

Weibull, A.-C., Bengtsson, J. and Nohlgren, E. (2000) Diversity of butterflies in the agricultural landscape: the role of farming system and landscape heterogeneity. *Ecography* 23, 743–750.

Weibull, A.-C., Östman, Ö. and Granqvist, Å. (2003) Species richness in agroecosystems: the effect of landscape, habitat and farm managem. *Biodiversity and Conservation* 12, 1335–1355.

Wettstein, W. and Schmid, B. (1999) Conservation of arthropod diversity in montane wetlands: effect of altitude, habitat quality and habitat fragmentation on butterflies and grasshoppers. *Journal of Applied Ecology* 36, 363–373.

9 Monitoring Biodiversity: Measuring Long-term Changes in Insect Abundance

KELVIN F. CONRAD,[1,3] RICHARD FOX[2] AND IAN P. WOIWOD[1]

[1]*Plant and Invertebrate Ecology Division, Rothamsted Research, Harpenden, Hertfordshire AL5 2JQ, UK;* [2]*Butterfly Conservation, Manor Yard, East Lulworth, Dorset BH20 5QP, UK;* [3]*Current address: Department of Biology, Trent University, Peterborough, Ontario, K9J 7B8, Canada*

1 Introduction

In addition to preserving threatened individual species, maintaining biodiversity is a predominant focus of modern conservation biology (Dobson, 2005). Measuring changes in abundance for species-centric conservation usually involves small populations in limited areas, often associated with specific resources. Monitoring abundance of threatened species is of key importance to evaluating the progress of conservation activities (Hellmann *et al.*, 2003).

In conserving biodiversity, measuring changes in abundance of multiple species is more demanding (Balmford *et al.*, 2005; Buckland *et al.*, 2005). Biodiversity indices, comprising long-term trend data for many species, are valuable tools for biodiversity monitoring (e.g. Gregory *et al.*, 2005; Loh *et al.*, 2005). However, the development of such indices for insects, which make up the majority of terrestrial species, is hindered by the rarity of long-term abundance trends across multiple species (Loh *et al.*, 2005).

The term 'monitoring' has been used to describe a variety of activities, but for the purposes of this chapter, we consider monitoring to be repeated sampling of populations for detecting patterns in the variation of abundance over time. We discuss the goals of long-term monitoring of biodiversity, the criteria for data required, and examine some of the long-term data-sets of insect abundance that can be used for biodiversity monitoring.

1.1 Monitoring

In ecology, evolution and conservation, which all have historical components, a temporal perspective is critical and can only be obtained from regular monitoring over long periods (Taylor, 1989; Woiwod and Harrington, 1994). Such long-term monitoring provides vital information for the development of effective biodiversity conservation strategies. For example, status and trend information enable proactive prioritization of the most threatened species and biotopes for new initiatives, whereas monitoring outcomes of existing conservation policies and practice enables retrospective assessment of efficacy, whether for small-scale habitat management on a particular site or for national-level policy, such as agri-environment schemes. Long-term studies of populations, communities and ecosystems are also vital for understanding ecological processes (Ehrlich *et al.*, 1975; Callahan, 1984; Hellmann *et al.*, 2003) and, therefore, for improving predictions of future impacts of climatic and other environmental change.

For many years monitoring was regarded rather pejoratively, as more or less mindless accumulation of additional data points (Sykes and Lane, 1997) or blind data gathering (Noss, 1990). Krebs (1991) suggested that monitoring of animal populations was 'ecologically banal' unless accompanied by experimental manipulation to help understand population variation, and Yoccoz *et al.* (2001) emphasized a priori hypotheses for the design of monitoring programmes, rather than retrospective analyses associating observed patterns. However, such views have been strongly contested by Taylor (1989, 1991), who considered them inimical to successful long-term studies. In either case, it is still essential to recognize, even during the establishment of a long-term monitoring programme, that additional value may be derived from the data collected by using them for purposes other than those of the original programme (Woiwod and Harrington, 1994).

Consistent monitoring means that subtle trends can be detected before they are noticed by a casual observer. Long-term data are essential to see beyond short-term fluctuations. The greater the period and amplitude of short-term variation in a time series, the longer a time series needs to be to overcome short-term fluctuations (Woiwod and Harrington, 1994).

To be successful, long-term monitoring must be based on a standardized collection method, established and followed from the outset. Monitoring must be repeated at intervals appropriate to detection of the changes of interest. The unique nature of the time-series data generated by long-term monitoring, in terms of its occurrence in time and space, means that missing data cannot be made up for, replaced or duplicated at a later time or another place (Woiwod and Harrington, 1994).

The measurement technique used must be reliable; it should be easy to execute, not too demanding of time or effort, and it should be simple and inexpensive to maintain (Woiwod, 1991; Woiwod and Hanski, 1992; Woiwod and Harrington, 1994). Excessive complexity, time, effort or expense in executing a monitoring programme can reduce its uptake and 'momentum' and result in an early end to the monitoring period, either because the expense and energy investment cannot be justified, or because the people doing the monitoring cannot manage the commitment and effort to continue it.

1.2 Measuring insect biodiversity

Biodiversity, to a considerable extent, has always been defined by the way it has been measured (Peet, 1974). The measurement of biodiversity has been the subject of many discussions (e.g. Peet, 1974; Taylor, 1978; Hawksworth, 1995; Gaston, 1996) and has been reviewed comprehensively by Magurran (1988, 2004).

The most commonly used measure and meaning of biodiversity still remains the number of species, or *species richness* (Gaston, 1996). Species richness is perhaps the most fundamental unit of biodiversity, and for non-scientists involved in conservation, the most common perception of biodiversity (Gaston, 1996). Despite the shortcomings of species richness in representing the various complex aspects of biodiversity (Harper and Hawksworth, 1995; Gaston, 1996), it remains the simplest and most universally recognized measure of biological diversity (Peet, 1974; Gaston, 1996).

Insects are well known for being the most diverse group of organisms on earth, but it is this rich diversity that provides one of the greatest obstacles to studying, measuring or even estimating insect biodiversity. Just under one million insect species have been described (Baillie *et al.*, 2004), with perhaps another order of magnitude of species yet to be discovered (May, 1990). Although the conservation status of all the described species of birds and amphibians – and most species (90%) of mammals has been evaluated by the International Union for Conservation of Nature (IUCN) – and the proportion of threatened species in each group is well-documented (Baillie *et al.*, 2004), only 0.06% of described insect species have been evaluated. This lack of knowledge and high insect diversity means that not only is it difficult to estimate changes in insect biodiversity, but it is also difficult to compare changes in insect biodiversity with that in other groups (McKinney, 1999; Baillie *et al.*, 2004; but see Thomas *et al.*, 2004).

2 Monitoring Biodiversity

In light of the large insect species richness and uncertainty surrounding total species richness, short cuts must be applied to measuring biodiversity, and priorities must be set for monitoring. Noss (1990) listed a number of categories of species that may warrant special interest in monitoring biodiversity, including: (i) *vulnerables*: species that are rare or prone to extinction; (ii) *indicators*: species that signal the effects of perturbations on a number of other species with similar habitat requirements; (iii) *flagships*: popular, charismatic species that serve as symbols and rallying points for conservation initiatives; and (iv) *umbrellas*: species with large area requirements, which, if given sufficient protected habitat, will bring many other species under protection. We also consider (v) *reference groups*: well-known groups that form the basis for extrapolation to other, less well-known groups (Hammond, 1995).

2.1 Vulnerable species

Monitoring abundance of vulnerable species is important for the conservation of individual species and maintaining overall species richness. From a practical standpoint, vulnerable insect species should at least have the advantage of having population sizes and distributions small enough to make abundance estimates feasible. Local populations, limited in size and space, can be assessed through quantitative field studies (e.g. Warren, 1991; Thomas *et al.*, 1992; Cowley *et al.*, 1999). However, even rare insect species with fairly specific habitat requirements can reach relatively high local densities while remaining cryptic in the environment, and therefore require a great deal of effort to census accurately (e.g. Purse *et al.*, 2003) and sometimes it may not even be possible to census them at all.

2.2 Indicator species

The use of indicator species is complicated by a lack of universal consensus about the concept (Simberloff, 1998; Caro and O'Doherty, 1999). Pearson (1994) distinguished between species that can be used to identify areas of high biodiversity (*biodiversity indicators*) and those that identify environmental changes (*environmental indicators*). Caro and O'Doherty (1999) further divided the latter group into species used to indicate changes in habitat (*ecological indicators*) and species used as indices of change in populations of other species, which they refer to as *population indicators*.

Ideally, an indicator should: be sensitive enough to provide an early warning of change; be distributed over a broad geographical area; be capable of providing continuous assessment over a wide range of conditions; be easy and cost-effective to measure, collect or estimate; display clearly the cycles or trends caused by the phenomena under study; and be ecologically relevant to the effects being studied (Pearson, 1995). Because no single indicator possesses all of these properties, a set of complementary indicators is often required (Noss, 1990).

2.3 Flagship species

Flagship species are used to attract conservation interest of the public (Western, 1987). As such, they are often large charismatic mammals, such as elephants or giant pandas (Western, 1987; Simberloff, 1998; Caro and O'Doherty, 1999), although butterflies are increasingly regarded as suitable flagships for insect conservation (New *et al.*, 1995). Insects such as butterflies, bumblebees and dragonflies may also serve as flagship groups, to encourage conservation of other invertebrates (New *et al.*, 1995: Samways, 2005).

2.4 Umbrella species

An umbrella species is one whose conservation is expected to confer protection to a large number of naturally co-occurring species (Roberge and Angelstam, 2004). Umbrella species provide a way to use species requirements as a basis for biodiversity conservation. Most umbrella species are large mammals and birds, but invertebrates are increasingly being considered (Roberge and Angelstam, 2004). Multispecies umbrella groups can provide greater representation of more diverse species (Lambeck, 1997; Roberge and Angelstam, 2004).

2.5 Reference groups

Hammond (1995) used reference groups as the basis for extrapolating ratios of species richness from a well-known group to a less well-known target group. Here, we refer to any well-known group from which biodiversity information can be extrapolated to another poorly known group as a reference group. In extrapolating potential extinction rates from UK butterflies to UK and world insect extinction rates, Thomas *et al.* (2004) and Thomas (2005) used butterflies as a reference group for extinction rates of insects in general. Providing this extrapolation is reasonable (see Thomas and Clarke, 2004; Thomas, 2005), monitoring changes in the reference group serves as a surrogate for monitoring wider insect biodiversity.

3 Data Sources for Long-term Monitoring

3.1 Expert opinion

One consequence of the species richness of insects is that there are many species about which very little is known. In the face of great uncertainty, the informed opinion of an expert can be a valuable tool for assessing population trends. Despite the quantitative rigour in current IUCN criteria for assigning species to categories of threat (Mace and Lande, 1991; IUCN World Conservation Union, 2001), subjective assessment by experts can provide similar results to a quantitative protocol (Keith *et al.*, 2004), and performance of objective criteria is improved when assessed by experienced researchers (Keith *et al.*, 2004; Regan *et al.*, 2005). The considered opinion of knowledgeable experts who have studied insect species or groups of species for a number of years is often the best and only information on their long-term population trends (e.g. van Swaay and Warren, 1999). Expert opinion therefore remains a valuable resource for judging insect population trends.

3.2 Inventories

Although the aim of monitoring is to uncover changes in ecosystem structure, composition or function over time, inventories document the spatial extent of populations, species or communities (Noss, 1990; Spellerberg, 1991; Kremen *et al.*, 1993). Species inventories are a commonly used method to catalogue the species richness of a given area (Goldstein, 2004; O'Connell *et al.*, 2004). Multiple inventories provide population 'snapshots' over time (Fleishman and MacNally, 2003). Multiple inventories at short intervals may be necessary when changes in the relative abundance of insect species in a community are large and rapid (Kremen *et al.*, 1993; Fleishman and MacNally, 2003; Oertli *et al.*, 2005). Changes in inventories between sampling periods can then be used to indicate abundance changes, providing standardized methods are used.

3.3 Long-term point samples

There are numerous naturalists, both professional and amateur, who record species abundance at a particular place for many years, and often very systematically. Such 'point samples', although spatially limited, provide good examples of how systematic monitoring over long time periods can be used to study changes in phenology (Forister and Shapiro, 2003; Ledneva *et al.*, 2004), community composition and species abundance (e.g. Moore, 1991, 2001; Shiffer and White, 1995).

When the quality of even casual observations from point samples can be assessed and combined with other more rigorous methods, the data can make a valuable contribution to species monitoring (Lepczyk, 2005). Data warehousing projects, such as the Global Population Dynamics Database (NERC Centre for Population Biology Imperial College, 1999), can organize such data-sets, ensure consistency of format and maintain minimum quality requirements. Although butterfly data have been included with vertebrate population data in a European species trend indicator (de Heer *et al.*, 2005), there are few insect point source data available for long-term trends monitoring. However, as long-term trends for regularly monitored insect populations accumulate, using insect-based indices from point samples to monitor wide-area trends is conceivable. Central coordination of point-sampled data remains the key to achieving the spatial coverage and data consistency suitable for long-term abundance monitoring.

3.4 Direct population monitoring

Biodiversity monitoring through long-term monitoring of species or population abundance requires statistically reliable estimates or indices of abundance. Estimates of abundance may be either *relative* or *absolute*, with absolute estimates usually requiring greater sampling effort from a clearly defined area (Southwood and Henderson, 2000). Simple count censuses fall somewhere between relative and absolute estimates of abundance. Intensive count censusing within a clearly defined area can produce estimates of absolute abundance and density. This, obviously, is most suitable for small populations with limited distributions, which may include vulnerable species. Less intensive censuses can be used to produce relative abundance estimates.

3.4.1 Relative abundance

Estimates of relative abundance permit the study of variation in abundance over time and space. Knowledge of total abundance is not necessary and relative changes can be estimated from indices of abundance. Standardized sampling protocols are essential. It must then be assumed that representative samples are taken and that the index chosen is representative of the true population size (Southwood and Henderson, 2000).

3.4.2 Mapping schemes

Mapping schemes are essentially continuous inventories and often include collation of long-term point samples. When carefully designed and broken down into appropriate intervals, these inventories can be used to estimate changes in relative abundance (O'Grady *et al.*, 2004b). The fundamental unit measured by mapping schemes is the area occupied by a species. Although more quantitative data may be collected, the essential information collected is the presence or absence of a species in a particular map unit.

In order to use mapped data to track changes in relative abundance, a positive abundance–occupancy relationship is assumed (Gaston, 1999; Conrad *et al.*, 2001). In general, the greater area a species occupies, the greater its abundance is likely to be. If the area occupied increases over time, the species is increasing in abundance, and if the area occupied shrinks, its abundance is declining (Gaston, 1999). The IUCN recognizes rapid declines in area of occupancy in its Red List criteria (IUCN World Conservation Union, 2001).

3.4.3 Butterfly atlas programmes

In the UK, insect mapping schemes originated using techniques established in successful plant mapping schemes (Harding, 1991). The most successful insect mapping scheme in the UK, and perhaps anywhere in the world, is the butterfly recording scheme, which has resulted in the publication of three butterfly atlases: *Atlas of Butterflies in Britain and Ireland* (using data for 1970–1982; Heath *et al.*, 1984); *The Millennium Atlas of Butterflies in Britain and Ireland* (using data for 1995–1999; Asher *et al.*, 2001); and *The State of Butterflies in Britain and Ireland* (using data for 2000–2004; Fox *et al.*, 2006a). These projects have mapped the distribution of resident and migrant butterflies by asking volunteers to record their presence in each of the 10 × 10 km grid squares of Britain and Ireland. All three survey periods achieved over 90% recording coverage of 10 km grid squares, and the two later surveys exceeded 95%. Comparison between these survey snapshots and pre-1970s historical data has enabled the changing distributions of butterflies to be assessed in Britain and Ireland, using both direct comparison and analytical methods to account for changing recording effort (Warren *et al.*, 2001; Thomas *et al.*, 2004; Fox *et al.*, 2006a).

The result of this survey work is a data-set of over 4.5 million butterfly distribution records for Britain and Ireland stretching from the present day back to the late 17th century. The aim of recent surveys coordinated by Butterfly Conservation (from 1995 onwards) has been to collate distribution data to support conservation efforts. These data can be used to assess species status and distribution trends through time, akin to monitoring species abundance. Indeed British butterfly distribution trends during the last three decades of the 20th century were closely correlated to trends derived from population monitoring using transects to measure relative abundance (Warren *et al.*, 2001 and see below). This finding supports the idea that a positive abundance–occupancy relationship (Gaston *et al.*, 2000) exists for British butterflies and that range changes, as measured by butterfly atlas schemes, may act as a surrogate for abundance monitoring (Thomas *et al.*, 2004; Thomas, 2005).

The most recent assessment of change in Britain revealed that the recorded distributions of 76% of resident butterflies ($n = 54$) decreased over the last three decades (comparison of survey data for 1970–1982 versus 1995–2004; Fox *et al.*, 2006a). This analysis made use of a subsampling technique to minimize the bias caused by a huge increase in recording effort (measured by number of records and number of recording visits) during the period. Six species (11%) had lost >50% of their past (1970–1982) distribution and one (*Glaucopsyche arion*, the large blue butterfly) had become extinct (although it was subsequently reintroduced to Britain during the same period). A further 15 butterflies (28%) suffered distribution decreases of >30%, including a number of formerly widespread species such as the dingy skipper (*Erynnis tages*), small pearl-bordered fritillary (*Boloria selene*), wall (*Lasiommata megera*) and grayling (*Hipparchia semele*).

Distribution data and the trends derived from them have been used in many ways to support conservation and in ecological research. Distribution trends have been used to define UK, national, regional and local priorities for conservation through the hierarchical Biodiversity Action Plan (BAP) process. Distribution records themselves are used at local scales to promote biodiversity conservation through site designation, targeting of habitat management or recreation, and in the planning (development control) system. The butterfly atlas data-sets have played a substantial role in the development of methods to understand and predict the impact of climate change on biodiversity both in the UK and across Europe (Hill *et al.*, 1999, 2002; Parmesan *et al.*, 1999; Thomas *et al.*, 2001; Warren *et al.*, 2001; Berry *et al.*, 2002; Davies *et al.*, 2005; Franco *et al.*, 2006; Menéndez *et al.*, 2006).

Similar mapping schemes have been attempted or are underway in a number of other European countries (Parmesan *et al.*, 1999; Maes and Van Dyck, 2001; van Swaay and Warren, 2003; Thomas, 2005).

3.4.4 Butterfly monitoring schemes

Long-term monitoring of the relative abundance of butterfly species is also achieved more directly using transect (fixed route) recording. The original UK Butterfly Monitoring Scheme (BMS) (Pollard *et al.*, 1975; Pollard and Yates, 1993) was launched in 1976 and made up of standardized transect counts of adult butterflies carried out at sites across the UK (initially ~30 sites, rising to over 130 by 2004). At each site, the transect is walked at least once a week from April to September, under conditions suitable for butterfly activity, and every sighting of each species made in an imaginary 5 m^3 is counted. Weekly counts from all transects are then used to provide national annual indices of abundance for each species, from which time series of changes in relative abundance can be generated (Moss and Pollard, 1993; Pollard and Yates, 1993; Pollard *et al.*, 1995; Thomas, 2005). A number of studies have tested the assumptions and statistical methodology of the transect scheme (Warren *et al.*, 2001; Thomas, 2005). The butterfly transect methodology has proved very popular and has been widely adopted by other individuals and organizations across the UK. Many use it solely for site-based monitoring of butterfly populations (e.g. in order to see whether habitat management benefits butterflies); others, as a tool in ecological research. The number of transects operating outside the BMS grew steadily at first, but increased rapidly after 1990. By 2003, over 500 transects were being recorded, with 80 new ones established in that year alone (Brereton *et al.*, 2006). Some county-based coordination of these additional transects had been established (e.g. in Hampshire, Hertfordshire and Middlesex, and Greater London), but national collation and analysis by Butterfly Conservation did not commence until 1998.

In 2006, the two transect monitoring schemes were joined to form the UKBMS (Greatorex-Davies and Roy, 2005; Brereton *et al.*, 2006). The UKBMS has collated data from over 1000 transects so far, representing nearly 150,000 weekly walks and records of over 10.5 million individual butterflies (Fox *et al.*, 2006a). Even so, there are additional transects that are not yet part of the UKBMS.

The BMS (and subsequently the Butterfly Conservation and UKBMS schemes) provided a standardized annual measure of the changing status of butterfly species, which could be used to generate short-term trends; something that cannot be derived from distribution recording (where long periods are required to achieve sufficiently comprehensive coverage). Furthermore, BMS data have played a key role in many of the advances in knowledge of butterfly ecology in the UK over the last three decades (Pollard and Yates, 1993). The data have unravelled the dependence of butterfly populations and ecology on weather and climate (Pollard, 1988; Pollard and Yates, 1993; Roy and Sparks, 2000; Roy *et al.*, 2001; Roy and Thomas, 2003; Brereton *et al.*, 2006). Not only has this paved the way for assessments of the impact of global climate change on biodiversity (including predictive modelling based on the BMS data: Roy *et al.*, 2001), but it has also greatly improved understanding of how landscape, land use and habitat changes affect butterflies. The analysis of BMS data can allow for the overriding effect of the weather, thus enabling other influences on particular butterfly populations to be detected (e.g. habitat management).

Despite these past achievements, the formation of the UKBMS has brought many advantages. For example, the BMS had been unable to calculate trends for many of the rarest butterflies, simply because these species occurred at very few monitored sites. The new data-set, on the other hand, has much better representation of these species and national population trends can be calculated for almost all species, including UKBAP Priority Species (see Fox *et al.*, 2006a). UKBMS trends have already contributed to the 2005/06 review of priorities in the UKBAP. The greater number and diversity of monitored sites in the new data-set has also allowed analysis of whether conservation initiatives such as Sites of Special Scientific Interest or agri-environment schemes have benefited butterflies (Brereton *et al.*, 2006). The UKBMS data have considerable potential for the development of policy-relevant biodiversity indicators suitable for governmental use at UK and county levels, to complement the 'Quality of Life' indicator based on populations of wild birds that is already in use. A headline indicator based on butterfly population trends would help to widen the representation of species (as a large part of UK biodiversity is made up of terrestrial insects) and biotopes (particularly open semi-natural habitats, such as grassland, heathland, woodland clearings and post-industrial 'brownfield' sites) within the government's package of sustainability indicators.

The annual monitoring of butterfly abundance has enabled the early detection of substantial trends, likely to be of relevance to policy development and implementation, before there is evidence of species' distribution change. Such a time lag between abundance trends assessed using population monitoring (transect) data and distribution data is expected (Conrad *et al.*, 2001), in part because distribution change is a product of population change and in part as an artefact of the different reporting periods (annual versus multi-year periods). For example, the UKBMS data indicate a significant long-term decrease in abundance of *Coenonympha pamphilus*, the small heath (52% decrease 1976–2004), which appears to have accelerated during the last 10 years (annual rate of abundance change 1976–2004 = –2.6%; 1995–2004 = –3.7%). This species remains one of the most widespread in Britain, and distribution data do not yet indicate a substantial decline. In contrast, *Polyommatus bellargus*, the Adonis blue, is a recovering species. UKBMS data show an abundance increase of 28% (1979–2004) and 63% (1995–2004). However, the butterfly's distribution remains less extensive than during the 1970s (and is assessed as a 19% decrease), even though the species has recolonized some of its historical range in recent years.

Transect monitoring schemes using the same methodology have also been established outside the UK, for example in Jersey, Catalonia (Spain), Flanders (Belgium), Finland, Germany, the Netherlands (van Swaay *et al.*, 2002) and Switzerland, whereas similar, but nationally uncoordinated recording schemes are also operated in North America and Japan (Thomas, 2005). In addition to national and sub-national monitoring of butterfly populations, the widespread adoption of similar methodologies in different countries has raised the possibility of international trends. Butterfly transect data are being used to develop pan-European biodiversity indicators to enable the European Union to assess performance in relation to international obligations and progress towards policy targets (de Heer *et al.*, 2005; van Swaay and van Strien, 2005).

3.4.5 The Rothamsted Insect Survey

The Rothamsted Insect Survey (RIS) has operated two sampling networks capable of providing estimates of relative abundance of insects since the 1960s (Woiwod and Harrington, 1994). Since 1965, the RIS suction trap network has used standard 12.2 m high suction traps, currently 16 in Britain and 56 traps of similar design, which are operated independently, in 19 other European counties.

Suction traps were developed as a quantitative method for studying aerial insect populations (Johnson and Taylor, 1955; Johnson, 1957), but when the RIS network was created in 1965, new 12.2 m traps were designed, with the primary purpose of monitoring migration and population dynamics of aphids, in order to provide an aphid pest-warning system (Taylor, 1989). Although the network is sparse, the suction traps appear to represent aphid dynamics over large areas (Macaulay *et al.*, 1988; Cocu *et al.*, 2005). Moreover, other than aphids, all of the insects captured are retained and analysis of these catches has proved useful as a long-term indicator of habitat-related changes in insect biodiversity and biomass (Benton *et al.*, 2002; Shortall *et al.*, 2006).

The second RIS network is a UK-wide network of standard Rothamsted light traps. Originally conceived as a means of studying spatial and temporal variation in insect abundance (e.g. Taylor and Woiwod, 1982), over 450 sites have been sampled since 1968 (Fig. 9.1), and traps have been operated at

Fig. 9.1. Distribution of sites sampled as part of the Rothamsted Insect Survey's (RIS) light trap network. Grey circles indicate sites sampled previously and black circles indicate traps running in 2005.

80–100 sites annually (Woiwod and Harrington, 1994; Conrad *et al.*, 2004; Woiwod *et al.*, 2005). The network uses standard Rothamsted traps, which are simple to operate and have been of a consistent design since the 1940s (Williams, 1948) to record over 630 species of larger (macro) moths. Although the traps are designed to catch only a relatively small sample, they are efficient in that they catch a high proportion of moths that fly near them (Bowden, 1982) and the sample is representative of the local moth community (Taylor and Brown, 1972; Taylor and French, 1974; Intachat and Woiwod, 1999). The small samples obtained are practical to handle without harming local moth populations (Williams, 1952).

Rothamsted light traps have been operated at various sites throughout the world as part of more general studies of moth diversity. In Europe this has involved sites in Denmark, Finland, France, Ireland and Gibraltar, but traps have also operated at more exotic sites in Malaysia (Barlow and Woiwod, 1989), Sulawesi (Barlow and Woiwod, 1990), Iraq, Seychelles and Tenerife.

Other coordinated light-trapping schemes exist in Hungary (Nowinszky, 2003) and Finland (Huldén *et al.*, 2000), and long-term moth-trapping data are available from Japan (Yamamura *et al.*, 2006) and Australia (White, 1991), but the Rothamsted network provides information on the greatest number of species, for the longest duration at the greatest geographic extent. Rothamsted light traps have proved ideal for long-term, quantitative and standardized national monitoring (Woiwod *et al.*, 2005).

The RIS moth data have been widely used in ecological research (Woiwod and Harrington, 1994; Woiwod *et al.*, 2005). A Rothamsted light trap had been operated at the edge of a field on the Rothamsted estate between 1933 and 1937 and again between 1946 and 1950. The data were used to study the effect of weather on insects and to develop and quantify the concept of diversity based on the observed species frequency distribution found in these samples (e.g. Fisher *et al.*, 1943; Williams, 1953).

A problem with many sampling programmes aimed at detecting change in diversity is that the most common measure, species richness, depends directly on sample size. One of the first attempts to derive a quantitative index that could be used to compare diversity from samples of different size was based on the light trap catches obtained at Rothamsted in the 1930s and 1940s (Fisher *et al.*, 1943; Williams, 1953). The value of α from the log-series distribution has proved to be a robust and reliable index of diversity which is relatively independent of sample size, has power to discriminate between sites, lacks sensitivity to unstable very abundant species or transient rare species, and responds well to environmental change (Taylor, 1978; Magurran, 2004). It has been possible to relate changes in the structure of moth diversity in Britain to urbanization (Taylor *et al.*, 1978) and relate the general pattern of diversity to the ecological land class stratification of Britain (Luff and Woiwod, 1995).

The data-set provides both spatial and abundance information, making it possible to study fundamental relationships between abundance and distribution. Early analyses established a species-specific power law between the large-scale spatial variance and mean density for 360 species of macro-moths (Taylor *et al.*, 1980). A similar analysis extended the result to temporal variability (Taylor *et al.*, 1980; Taylor and Woiwod, 1982). Recently, Conrad *et al.* (2001) demonstrated a time lag between abundance and occupancy changes for *Arctia caja*, the garden tiger, despite a significant positive relationship between the two variables – a result not possible to detect without a long time series. Further analyses of data for this moth species provided detailed information about changes in spatial structure of the population while its abundance declined sharply (Conrad *et al.*, 2006a).

Until recently, relatively little analysis has been directed at assessing species abundance trends from the RIS light trap data-set. The first detailed study of a declining species was of *A. caja* (Conrad *et al.*, 2002). Using the long time series and spatial information available from the data-set, Conrad *et al.* (2002, 2003, 2006a) showed that wet winters and warm springs were detrimental to this species and that changes in distribution and abundance are the consequence of recent climate change related to large-scale weather patterns over the North Atlantic.

The *A. caja* analysis suggested that other species formerly perceived as common might have undergone declines since the late 1960s. Conrad *et al.* (2004, 2006b) analysed records from the 35-year period from 1968 to 2002, using strict criteria to select adequately sampled sites and species, to estimate long-term population trends for 337 species of macro-moths.

The percentage of species with significant decreases (54%) was more than double those with significant increases (22%), and the total catch, summed across all species, has declined by nearly a third over the 35-year period (Fig. 9.2, and Conrad *et al.*, 2004, 2006b). The greatest proportion of decline was in the south, particularly the south-east, and the fewest declines in the north.

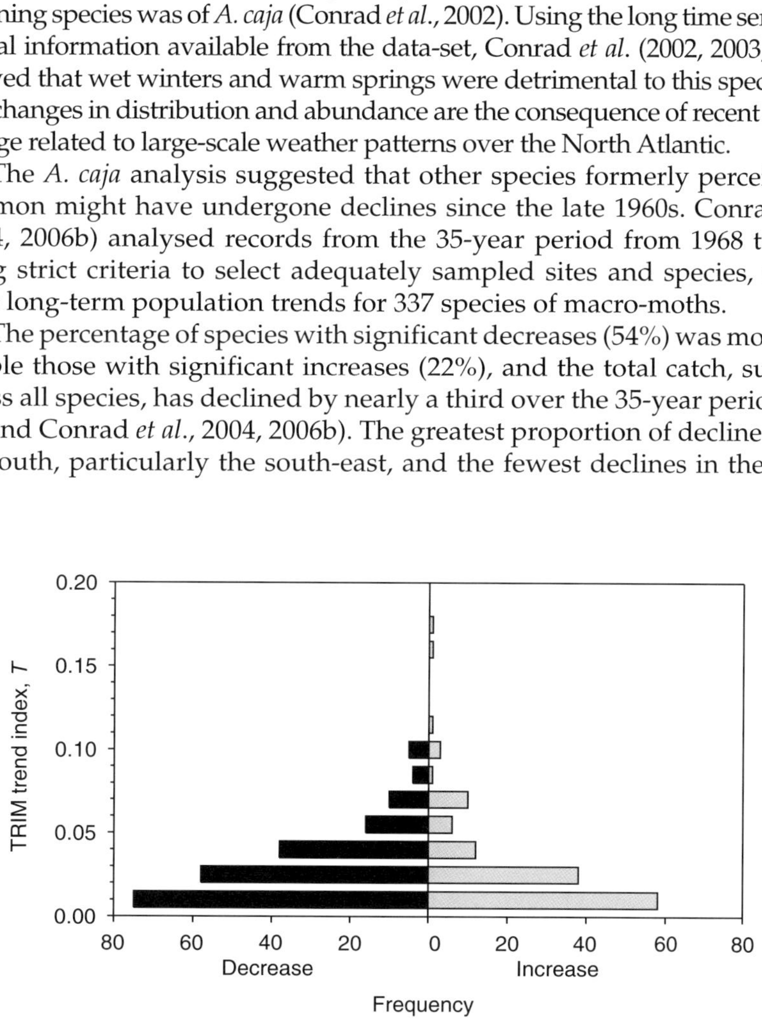

Fig. 9.2. Distribution of long-term trends of common British macro-moths. Grey bars show the number of species showing increasing long-term trends (positive *T*), whereas black bars indicate the number of species showing decreasing long-term trends (negative *T*). *T*, the 'TRIM trend index' is the overall slope of the regression of annual indices on a logarithmic scale (Pannekoek and Van Strien, 2001). For details see Conrad *et al.* (2004).

Total moth catches remained fairly stable in the north (Conrad *et al.*, 2004, 2006b). Perhaps most worrying is that if the rates of decline are compared against the decline criteria for IUCN threatened categories (IUCN World Conservation Union, 2001), 71 of the 337 species (21%) could be classified as 'threatened'. Prior to the study, all of these species were regarded as common and widespread and none was thought to warrant any conservation priority (Woiwod *et al.*, 2005). Although the increases found in some species can be linked to decreases in air pollution and increased planting of commercial and ornamental conifers, clear patterns among declining species have not emerged (Conrad *et al.*, 2004, 2006b).

Although it may initially seem counterintuitive to find that common species are declining rapidly, among the species tested there was no relationship between total catch and long-term trend (Fig. 9.3, open circles), indicating that trend is generally independent of total abundance. Species captured too infrequently for statistical analysis and species difficult to identify from external morphology, such as members of the genus *Epirrita* ('November moths'), were not included in the original analysis (Conrad *et al.*, 2004, 2006b). Data exist in the RIS database for an additional 313 such species. We ranked these species according to their number of captures and accumulated them in rank order to form eight groups of approximately 10,000 captures, with two additional groups of the remaining species. We then ana-

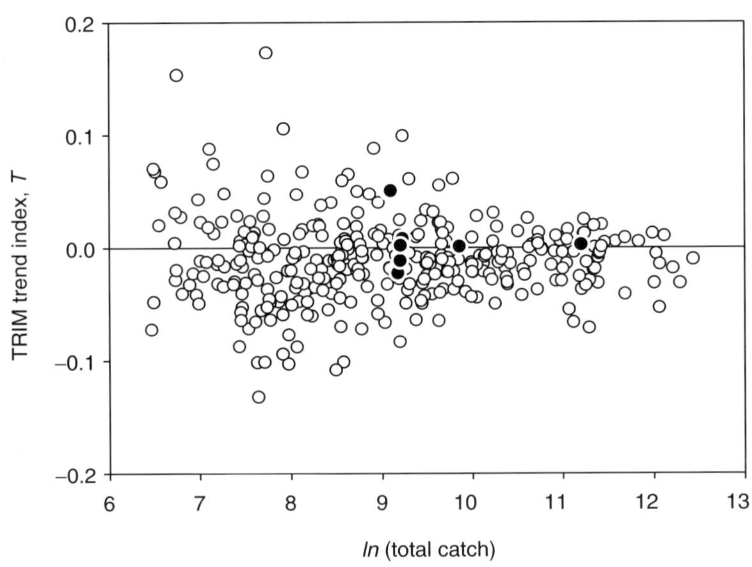

Fig. 9.3. The relationship between the long-term trend, *T*, and total number of a species caught by the Rothamsted Insect Survey (RIS) light trap network between 1968 and 2002. Open circles represent species caught relatively frequently and used to estimate long-term trends of common macro-moths by Conrad *et al.* (2004) and Fox *et al.* (2006b). Filled circles represent eight groups containing 313 rarely caught species or species not clearly identifiable from external morphology.

lysed these groups using the methodology of Conrad *et al.* (2004, 2006b). Although this analysis cannot indicate anything about decline rates in rare species, large values of *T* (+ve or −ve) for these species would indicate sharp changes in the probability of capturing rarer species in general. However, *T* values for these less common species are very similar to those of the more common species analysed earlier (Fig. 9.3, closed circles). It is reasonable to assume that the pattern observed for common species is representative of all UK macro-moths.

4 Discussion and Conclusions

Knowledge of threats to insect biodiversity lags behind those of vertebrates, but has been increasing rapidly in a few, well-studied groups (Conrad *et al.*, 2004; Thomas *et al.*, 2004; Thomas, 2005). Where insects are well studied, such as the in UK, they often have higher proportions of threatened species than many other taxa (McKinney, 1999; Conrad *et al.*, 2004, 2006b; Thomas *et al.*, 2004).

The assumption inherent in monitoring biodiversity through monitoring abundance of species is that changes in populations will reflect changes, or potential changes, in biodiversity itself (O'Grady *et al.*, 2004a). In the case of monitoring biodiversity through surrogates, such as indicator species, this assumption is explicit and requires experimental verification (McGeoch, 1998; McGeoch *et al.*, 2002). Where the species themselves are monitored for changes in biodiversity (e.g. Maes and Van Dyck, 2001; Gregory *et al.*, 2005), it must be assumed that species declines indicate a high probability of loss in biodiversity, or ultimately signal a high probability of extinction of the species in question (Mace and Lande, 1991; Keith *et al.*, 2000). Changes in population size, then, may serve as indices of change in biodiversity (cf. Buckland *et al.*, 2005; Thomas, 2005).

Abundance of British butterflies is perhaps better monitored than for any other insect taxon, with the atlas and transect monitoring schemes providing complementary results. New *et al.* (1995) proposed butterflies as insect conservation flagships, a view championed by Thomas *et al.* (2004) and Thomas (2005), who have suggested that butterflies should serve as a reference group for the extinction rate of other insects, although butterflies also serve as an umbrella group for conservation in many parts of Europe (van Swaay and Warren, 2003). Butterflies have been recognized as useful indicators, both for rapid and sensitive detection of subtle biotope or climatic change and as representatives for the diversity and responses of other taxa (Brown, 1991; Brown and Freitas, 2000; Kerr *et al.*, 2000; Parmesan, 2003; Thomas and Clarke, 2004; Thomas *et al.*, 2004; Fleishman *et al.*, 2005; Maes and Van Dyck, 2005; Maes *et al.*, 2005; Thomas, 2005; but see Ricketts *et al.*, 2002; Kremen *et al.*, 2003; Grill *et al.*, 2005).

Moths are often viewed as butterflies' 'poor cousins' (New, 2004b). Most macro-moths are not particularly charismatic and their nocturnal habits mean that even those that are brightly coloured and attractive tend not to be sufficiently well known to the public to serve as flagships (but see McQuillan, 2004; New, 2004a; Patrick, 2004). As an indicator group, common moth species possess most of the essential characteristics of successful indicators. They appear to be sensitive to environmental change, are distributed over a broad geographical area, have proved easy and cost-effective to collect, and can provide a continuous assessment of conditions (cf. Pearson, 1995). Common and widespread moth species serve as environmental indicators (see Pearson, 1994) or ecological indicators (see Caro and O'Doherty, 1999), showing a general decline in UK habitat for moths. Large declines in common moths indicate large losses in herbivore and prey biomass, which, in turn, must be a strong indication of the decline in the state 'ecosystem health'.

Rapidly declining common moth species could also serve as umbrella species, but their ubiquity would make conservation efforts difficult to focus on protecting particular biotopes. Further research is needed to isolate species or groups of species to serve as indicators of biodiversity change.

Although the figures are not directly compatible because of the different methods of estimation used, the proportion of declining British moth species (66%; Conrad *et al.*, 2004) is similar to the proportion of declining British butterfly species (71%; Thomas *et al.*, 2004), even though biodiversity of moths and butterflies may not correspond at local scales (Ricketts *et al.*, 2002). This supports the idea that British Lepidoptera can serve as a reference group for biodiversity declines in other insect species.

Although the declining trends found in moths were similar to those obtained for butterflies, an atlas mapping scheme to run in parallel with the RIS light trap network would provide still stronger confirmation and greater information on the spatial variation in patterns of decline (Conrad *et al.*, 2006a,b; Fox *et al.*, 2006b).

Discovering the forces that determine biodiversity has been suggested as one of the great questions of modern science (Pennisi, 2005). The study of biodiversity itself, however, is a relatively new concept. The idea of conserving biological diversity per se, as opposed to individual species or communities, was first published in 1980 (IUCN, 1980; Norse, 1980), and these earliest documents already outlined many of the concepts and concerns of present-day biodiversity conservation.

Insect monitoring programmes that have persisted are simple, inexpensive, centrally coordinated and have low intensity, both for recorders and the species recorded. The original purpose of the RIS was to study spatial change and variability in insect populations but now, nearly 40 years later, the long-term nature of the data has become as important. Using RIS data, changes in insect populations over spatial and temporal scales can now be studied in ways not previously possible. The challenge in long-term monitoring schemes is in maintaining them and justifying their continuation (Woiwod and Harrington, 1994). An important feature of most successful long-term studies is that it is difficult to predict the future uses of the data (Taylor, 1989; Woiwod, 1991). It is significant, therefore, that the schemes discussed here as the most successful for monitoring insect biodiversity were conceived before the idea of conserving biodiversity itself.

Acknowledgements

We would like to thank the thousands of people, many of whom were volunteers, who have contributed to butterfly and moth recording and monitoring schemes in the UK. We also thank Mark Parsons, Tom Brereton and Martin Warren (Butterfly Conservation), David Roy and Peter Rothery (Centre for Ecology and Hydrology), and Joe Perry and Suzanne Clark (Rothamsted Research) for their contributions to the butterfly and moth analyses reported. Rothamsted Research receives grant-aided support from the UK Biotechnology and Biological Sciences Research Council.

References

Asher, J.A., Warren, M.S., Fox, R., Harding, P., Jeffcoate, G. and Jeffcoate, S. (2001) *The Millennium Atlas of Butterflies in Britain and Ireland*. Oxford University Press, Oxford.

Baillie, J.E.M., Hilton-Taylor, C. and Stuart, S.N. (2004) *2004 IUCN Red List of Threatened Species. A Global Species Assessment*. IUCN, Gland, Switzerland and Cambridge.

Balmford, A., Crane, P., Dobson, A., Green, R.E. and Mace, G.M. (2005) The 2010 challenge: data availability, information needs and extraterrestrial insights. *Philosophical Transactions of the Royal Society of London Series B – Biological Sciences* 360, 221–228.

Barlow, H.S. and Woiwod, I.P. (1989) Moth diversity in lowland tropical forest in peninsular Malaysia. *Journal of Tropical Ecology* 5, 37–50.

Barlow, H.S. and Woiwod, I.P. (1990) Seasonality and diversity of Macrolepidoptera in two lowland sites in the Dumga-Bone national park, Sulawesi Utara. In: Knight, W.J. and Holloway, J.D. (eds) *Insects and the RainForests of South East Asia (Wallacea)*. The Royal Entomological Society of London, London, pp. 167–172.

Benton, T.G., Bryant, D.M., Cole, L. and Crick, H.Q.P. (2002) Linking agricultural practice to insect and bird populations: a historical study over three decades. *Journal of Applied Ecology* 39, 673–687.

Berry, P.M., Dawson, T.P., Harrison, P.A. and Pearson, R.G. (2002) Modelling potential impacts of climate change on the bioclimatic envelope of species in Britain and Ireland. *Global Ecology and Biogeography* 11, 453–462.

Bowden, J. (1982) An analysis of factors affecting catches of insects in light traps. *Bulletin of Entomological Research* 72, 535–556.

Brereton, T., Roy, D. and Greatorex-Davies, N. (2006) Thirty years and counting: the contribution to conservation and ecology of butterfly-monitoring in the UK. *British Wildlife* 17, 162–170.

Brown, K.S. (1991) Conservation of neotropical environments: insects as indicators. In: Collins, N.M. and Thomas, J.A. (eds) *The Conservation of Insects and Their Habitats*. Academic Press, London, pp. 350–404.

Brown, K.S. and Freitas, A.V.L. (2000) Atlantic forest butterflies: indicators for landscape conservation. *Biotropica* 32, 934–956.

Buckland, S.T., Magurran, A.E., Green, R.E. and Fewster, R.M. (2005) Monitoring change in biodiversity through composite indices. *Philosophical Transactions of the Royal Society of London Series B – Biological Sciences* 360, 243–254.

Callahan, J.T. (1984) Long-term ecological research. *Bioscience* 34, 363–367.

Caro, T.M. and O'Doherty, G. (1999) On the use of surrogate species in conservation biology. *Conservation Biology* 13, 805–814.

Cocu, N., Conrad, K., Harrington, R. and Rounsevell, M.D.A. (2005) Analysis of spatial patterns at a geographical scale over north-western Europe from point-referenced aphid count data. *Bulletin of Entomological Research* 95, 47–56.

Conrad, K.F., Perry, J.N. and Woiwod, I.P. (2001) An abundance-occupancy time-lag during the decline of an arctiid tiger moth. *Ecology Letters* 4, 300–303.

Conrad, K.F., Woiwod, I.P. and Perry, J.N. (2002) Long-term decline in abundance and distribution of the garden tiger moth (*Arctia caja*) in Great Britain. *Biological Conservation* 106, 329–337.

Conrad, K.F., Woiwod, I.P. and Perry, J.N. (2003) East Atlantic teleconnection pattern and the decline of a common arctiid moth. *Global Change Biology* 9, 125–130.

Conrad, K.F., Woiwod, I.P., Parsons, M., Fox, R. and Warren, M.S. (2004) Long-term population trends in widespread British moths. *Journal of Insect Conservation* 8, 119–136.

Conrad, K.F., Perry, J.N., Woiwod, I.P. and Alexander, C.J. (2006a) Large-scale temporal changes in spatial pattern during declines of abundance and occupancy in a common moth. *Journal of Insect Conservation* 10, 53–64.

Conrad, K.F., Warren, M.S., Fox, R., Parsons, M.S. and Woiwod, I.P. (2006b) Rapid declines of common, widespread British moths provide evidence of an insect biodiversity crisis. *Biological Conservation* 132, 279–291.

Cowley, M.J.R., Thomas, C.D., Thomas, J.A. and Warren, M.S. (1999) Flight areas of British butterflies: assessing species status and decline. *Proceedings of the Royal Society of London Series B – Biological Sciences* 266, 1587–1592.

Davies, Z.G., Wilson, R.J., Brereton, T.M. and Thomas, C.D. (2005) The re-expansion and improving status of the silver-spotted skipper butterfly (*Hesperia comma*) in Britain: a metapopulation success story. *Biological Conservation* 124, 189–198.

de Heer, M., Kapos, V. and ten Brink, B.J.E. (2005) Biodiversity trends in Europe: development and testing of a species trend indicator for evaluating progress towards the 2010 target. *Philosophical Transactions of the Royal Society of London Series B – Biological Sciences* 360, 297–308.

Dobson, A. (2005) Monitoring global rates of biodiversity change: challenges that arise in meeting the Convention on Biological Diversity (CBD) 2010 goals. *Philosophical Transactions of the Royal Society of London Series B – Biological Sciences* 360, 229–241.

Ehrlich, P.R., White, R.R., Singer, M.C., McKechnie, S.W. and Gilbert, L.E. (1975) Checkerspot butterflies: a historical perspective. *Science* 188, 221–228.

Fisher, R.A., Corbet, A.S. and Williams, C.B. (1943) The relation between the number of species and the number of individuals in a random sample of an animal. *Journal of Animal Ecology* 12, 42–58.

Fleishman, E. and MacNally, R. (2003) Distinguishing between signal and noise in faunal responses to environmental change. *Global Ecology and Biogeography* 12, 395–402.

Fleishman, E., Thomson, J.R., MacNally, R., Murphy, D.D. and Fay, J.P. (2005) Using indicator species to predict species richness of multiple taxonomic groups. *Conservation Biology* 19, 1125–1137.

Forister, M.L. and Shapiro, A.M. (2003) Climatic trends and advancing spring flight of butterflies in lowland California. *Global Change Biology* 9, 1130–1135.

Fox, R., Asher, J., Brereton, T., Roy, D. and Warren, M. (2006a) *The State of Butterflies in Britain and Ireland*. Pisces Publications, Newbury, UK.

Fox, R., Conrad, K.F., Parsons, M.S., Warren, M.S. and Woiwod, I.P. (2006b) *The State of Britain's Larger Moths*. Butterfly Conservation and Rothamsted Research, Wareham, UK.

Franco, A.M.A., Hill, J.K., Kitschke, C., Collingham, Y.C., Roy, D.B., Fox, R., Huntley, B. and Thomas, C.D. (2006) Impacts of climate warming and habitat loss on extinctions at species' low-latitude range boundaries. *Global Change Biology* 12, 1545–1553.

Gaston, K.J. (1996) Species richness: measure and measurement. In: Gaston, K.J. (ed.) *Biodiversity: A Biology of Numbers and Difference*. Blackwell, Oxford, pp. 77–113.

Gaston, K.J. (1999) Implications of interspecific and intraspecific abundance–occupancy relationships. *Oikos* 86, 195–207.

Gaston, K.J., Blackburn, T.M., Greenwood, J.J.D., Gregory, R.D., Quinn, R.M. and Lawton, J.H. (2000) Abundance–occupancy relationships. *Journal of Applied Ecology* 37, 39–59.

Goldstein, P.Z. (2004) Systematic collection data in North American invertebrate conservation and monitoring programmes. *Journal of Applied Ecology* 41, 175–180.

Greatorex-Davies, J.N. and Roy, D.B. (2005) *The Butterfly Monitoring Scheme: Report to Recorders 2004*. CEH Monks Wood and the Joint Nature Conservation Comittee, Huntingdon, UK.

Gregory, R.D., van Strien, A., Vorisek, P., Meyling, A.W.G., Noble, D.G., Foppen, R.P.B. and Gibbons, D.W. (2005) Developing indicators for European birds. *Philosophical Transactions of the Royal Society of London Series B – Biological Sciences* 360, 269–288.

Grill, A., Knoflach, B., Cleary, D.F.R. and Kati, V. (2005) Butterfly, spider, and plant communities in different land-use types in Sardinia, Italy. *Biodiversity and Conservation* 14, 1281–1300.

Hammond, P.M. (1995) Practical approached to the estimation of the extent of biodiversity in speciose groups. In: Hawksworth, D.L. (ed.) *Biodiversity: Measurement and Estimation*. Chapman & Hall and The Royal Society, London, pp. 119–136.

Harding, P.T. (1991) National species distribution surveys. In: Goldsmith, F.B. (ed.) *Monitoring for Conservation and Ecology*. Chapman & Hall, London, pp. 133–154.

Harper, J.L. and Hawksworth, D.L. (1995) Preface. In: Hawksworth, D.L. (ed.) *Biodiversity: Measurement and Estimation*. Chapman & Hall and The Royal Society, London, pp. 5–12.

Hawksworth, D.L. (1995) *Biodiversity: Measurement and Estimation*. Chapman & Hall and The Royal Society, London.

Heath, J., Pollard, E. and Thomas, J.A. (1984) *Atlas of Butterflies in Britain and Ireland*. Viking, Harmondsworth, UK.

Hellmann, J.J., Weiss, S.B., Mclaughlin, J.F., Boggs, C.L., Ehrlich, P.R., Launer, A.E. and Murphy, D.D. (2003) Do hypotheses from short-term studies hold in the long-term? An empirical test. *Ecological Entomology* 28, 74–84.

Hill, J.K., Thomas, C.D. and Huntley, B. (1999) Climate and habitat availability determine 20th century changes in a butterfly's range margin. *Proceedings of the Royal Society of London Series B – Biological Sciences* 266, 1197–1206.

Hill, J.K., Thomas, C.D., Fox, R., Telfer, M.G., Willis, S.G., Asher, J. and Huntley, B. (2002) Responses of butterflies to 20th century climate warming: implications for future ranges. *Proceedings of the Royal Society of London Series B – Biological Sciences* 269, 2163–2171.

Huldén, L., Albrecht, A., Itämies, J., Malinen, P. and Wettenhovi, J. (2000) *Atlas of Finnish Macrolepidoptera*. Lepidopterological Society of Finland and the Finnish Museum of Natural History, Helsinki, Finland.

Intachat, J. and Woiwod, I.P. (1999) Trap design for monitoring moth biodiversity in tropical rainforests. *Bulletin of Entomological Research* 89, 153–163.

IUCN (1980) *World Conservation Strategy: Living Resource Conservation for Sustainable Development*. International Union for the Conservation of Nature and Natural Resources, Gland, Switzerland.

IUCN World Conservation Union (2001) *IUCN Red List Categories and Criteria: Version 3.1*. IUCN Species Survival Commission, Gland, Switzerland and Cambridge.

Johnson, C.G. (1957) The distrubution of insects in the air and the empirical relation of density to height. *Journal of Animal Ecology* 26, 479–494.

Johnson, C.G. and Taylor, L.R. (1955) The development and use of large suction traps for airborne insects. *Annals of Applied Biolog* 43, 51–61.

Keith, D.A., Auld, T.D., Ooi, M.K.J. and Mackenzie, B.D.E. (2000) Sensitivity analyses of decision rules in World Conservation Union (IUCN) Red List criteria using Australian plants. *Biological Conservation* 94, 311–319.

Keith, D.A., McCarthy, M.A., Regan, H., Regan, T., Bowles, C., Drill, C., Craig, C., Pellow, B., Burgman, M.A., Master, L.L., Ruckelshaus, M., Mackenzie, B., Andelman, S.J. and Wade, P.R. (2004) Protocols for listing threatened species can forecast extinction. *Ecology Letters* 7, 1101–1108.

Kerr, J.T., Sugar, A. and Packer, L. (2000) Indicator taxa, rapid biodiversity assessment, and nestedness in an endangered ecosystem. *Conservation Biology* 14, 1726–1734.

Krebs, C.J. (1991) The experimental paradigm and long-term population studies. *Ibis* 133(Suppl. 1), 3–8.

Kremen, C., Colwell, R.K., Erwin, T.L., Murphy, D.D., Noss, R.F. and Sanjayan, M.A. (1993) Terrestrial arthropod assemblages: their use in conservation planning. *Conservation Biology* 7, 796–808.

Kremen, C., Lees, D.C. and Fay, J.P. (2003) Butterflies and conservation planning in Madagascar: from pattern to practice. In: Boggs, C.L., Watt, W.B. and Ehrlich, P.R. (eds) *Butterflies: Ecology and Evolution taking Flight*. University of Chicago Press, Chicago, Illinois, pp. 517–540.

Lambeck, R.J. (1997) Focal species: a multi-species umbrella for nature conservation. *Conservation Biology* 11, 849–856.

Ledneva, A., Miller-Rushing, A.J., Primack, R. and Imbres, C. (2004) Climate change as reflected in a naturalist's diary, Middleborough, Massachusetts. *The Wilson Bulletin* 116, 224–231.

Lepczyk, C.A. (2005) Integrating published data and citizen science to describe bird diversity across a landscape. *Journal of Applied Ecology* 42, 672–677.

Loh, J., Green, R.E., Ricketts, T., Lamoreux, J., Jenkins, M., Kapos, V. and Randers, J. (2005) The living planet index: using species population time series to track trends in biodiversity. *Philosophical Transactions of the Royal Society of London Series B – Biological Sciences* 360, 289–295.

Luff, M.L. and Woiwod, I.P. (1995) Insects as indicators of land use change: a European perspective, focusing on moths and ground beetles. In: Harrington, R. and Stork, N.E. (eds) *Insects in a Changing Environment*. Academic Press, London, pp. 399–422.

Macaulay, E.D.M., Tatchell, G.M. and Taylor, L.R. (1988) The Rothamsted insect survey '12-metre' suction trap. *Bulletin of Entomological Research* 78, 121–129.

Mace, G.M. and Lande, R. (1991) Assessing extinction threats: toward re-evaluation of IUCN threatened species categories. *Conservation Biology* 5, 148–157.

Maes, D. and Van Dyck, H. (2001) Butterfly diversity loss in Flanders (North Belgium): Europe's worst case scenario? *Biological Conservation* 99, 263–276.

Maes, D. and Van Dyck, H. (2005) Habitat quality and biodiversity indicator performances of a threatened butterfly versus a multispecies group for wet heathlands in Belgium. *Biological Conservation* 123, 177–187.

Maes, D., Bauwens, D., de Bruyn, L., Anselin, A., Vermeersch, G., van Landuyt, W., de Knijf, G. and Gilbert, M. (2005) Species richness coincidence: conservation strategies based on predictive modelling. *Biodiversity and Conservation* 14, 1345–1364.

Magurran, A.E. (1988) *Ecological Diversity and its Measurement*. Kluwer, Dordrecht, The Netherlands.

Magurran, A.E. (2004) *Measuring Biological Diversity*. Blackwell Science, Oxford.

May, R.M. (1990) How many species? *Philosophical Transactions of the Royal Society of London Series B – Biological Sciences* 330, 292–304.

McGeoch, M.A. (1998) The selection, testing and application of terrestrial insects as bioindicators. *Biological Reviews* 73, 181–201.

McGeoch, M.A., Van Rensburg, B.J. and Botes, A. (2002) The verification and application of bioindicators: a case study of dung beetles in a savanna ecosystem. *Journal of Applied Ecology* 39, 661–672.

McKinney, M.L. (1999) High rates of extinction and threat in poorly studied taxa. *Conservation Biology* 13, 1273–1281.

McQuillan, P.B. (2004) An overview of the Tasmanian geometrid moth fauna (Lepidoptera: Geometridae) and its conservation status. *Journal of Insect Conservation* 8, 209–220.

Menéndez, R., Megías, A., Hill, J., Braschler, B., Willis, S., Collingham, Y., Fox, R., Roy, D. and Thomas, C. (2006) Species richness changes lag behind climate change. *Proceedings of the Royal Society of London Series B – Biological Sciences* 273, 1465–1470.

Moore, N.W. (1991) The development of dragonfly communities and the consequences of territorial behaviour: a 27-year study on small ponds at Woodwalton Fen, Cambridgeshire, United Kingdom. *Odonatologica* 20, 203–231.

Moore, N.W. (2001) Changes in the dragonfly communities at the twenty ponds at Woodwalton Fen, Cambridgeshire, United Kingdom, since the study of 1962–1988. *Odonatologica* 30, 289–298.

Moss, D. and Pollard, E. (1993) Calculation of collated indices of abundance of butterflies based on monitored sites. *Ecological Entomology* 18, 77–83.

NERC Centre for Population Biology Imperial College (1999) The Global Population Dynamics Database. Available at: http://www.sw.ic.ac.uk/cpb/cpb/gpdd.html

New, T.R. (2004a) Concluding comment: looking to the future for moths. *Journal of Insect Conservation* 8, 275–276.

New, T.R. (2004b) A special issue on moths and conservation. *Journal of Insect Conservation* 8, 77.

New, T.R., Pyle, R.M., Thomas, J.A., Thomas, C.D. and Hammond, P.C. (1995) Butterfly conservation management. *Annual Review of Entomology* 40, 57–83.

Norse, E.A. (1980) Ecology and living resources: biological diversity. In: Norse, E.A. and McManus, R.E. (eds) *Environmental Quality 1980: The Eleventh Annual Report of the Council on Environmental Quality*. Council on Environmental Quality, Washington, DC, pp. 31–80.

Noss, R.F. (1990) Indicators for monitoring biodiversity: a hierarchical approach. *Conservation Biology* 4, 355–364.

Nowinszky, L. (2003) *The Handbook of Light Trapping*. Savaria University Press, Szombathely, Hungary.

O'Connell, A.F., Gilbert, A.T. and Hatfield, J.S. (2004) Contribution of natural history collection data to biodiversity assessment in national parks. *Conservation Biology* 18, 1254–1261.

O'Grady, J.J., Burgman, M.A., Keith, D.A., Master, L.L., Andelman, S.J., Brook, B.W., Hammerson, G.A., Regan, T. and Frankham, R. (2004a) Correlations among extinction risks assessed by different systems of threatened species categorization. *Conservation Biology* 18, 1624–1635.

O'Grady, J.J., Reed, D.H., Brook, B.W. and Frankham, R. (2004b) What are the best correlates of predicted extinction risk? *Biological Conservation* 118, 513–520.

Oertli, S., Müller, A. and Dorn, S. (2005) Ecological and seasonal patterns in the diversity of a species-rich bee assemblage (Hymenoptera: Apoidea: Apiformes). *European Journal of Entomology* 102, 53–63.

Pannekoek, J. and Van Strien, A.J. (2001) *TRIM 3 Manual (Trends & Indices for Monitoring Data)*. Statistics Netherlands, Voorburg, The Netherlands.

Parmesan, C. (2003) Butterflies as bioindicators for climate change effects. In: Boggs, C.L., Watt, W.B. and Ehrlich, P.R. (eds) *Butterflies – Ecology and Evolution Taking Flight*. University of Chicago Press, Chicago, Illinois, pp. 541–560.

Parmesan, C., Ryrholm, N., Stefanescu, C., Hill, J.K., Thomas, C.D., Descimon, H., Huntley, B., Kaila, L., Kullberg, J., Tammaru, T., Tennent, W.J., Thomas, J.A. and Warren, M. (1999) Poleward shifts in geographical ranges of butterfly species associated with regional warming. *Nature* 399, 579–583.

Patrick, B.H. (2004) Conservation of New Zealand's tussock grassland moth fauna. *Journal of Insect Conservation* 8, 199–208.

Pearson, D.L. (1994) Selecting indicator taxa for the quantitative assessment of biodiversity. *Philosophical Transactions of the Royal Society of London Series B – Biological Sciences* 345, 75–79.

Pearson, D.L. (1995) Selecting indicator taxa for the quantitative assessment of biodiversity. In: Hawksworth, D.L. (ed.) *Biodiversity: Measurement and Estimation*. Chapman & Hall and The Royal Society, London, pp. 75–79.

Peet, R.K. (1974) The measurement of species diversity. *Annual Review of Ecology and Systematics* 5, 285–307.

Pennisi, E. (2005) What determines species diversity? *Science* 309, 90.

Pollard, E. (1988) Temperature, rainfall and butterfly numbers. *Journal of Applied Ecology* 25, 819–828.

Pollard, E. and Yates, T. (1993) *Monitoring Butterflies for Ecology and Conservation*. Chapman & Hall, London.

Pollard, E., Elias, D.O., Skelton, M.J. and Thomas, J.A. (1975) A method of assessing the abundance of butterflies in Monks Wood National Nature reserve in 1973. *Entomologist's Gazette* 26, 79–88.

Pollard, E., Moss, D. and Yates, T.J. (1995) Population trends of common British butterflies at monitored sites. *Journal of Applied Ecology* 32, 9–16.

Purse, B.V., Hopkins, G.W., Day, K.J. and Thompson, D.J. (2003) Dispersal characteristics and management of a rare damselfly. *Journal of Applied Ecology* 40, 716–728.

Regan, T.J., Burgman, M.A., Mccarthy, M.A., Master, L.L., Keith, D.A., Mace, G.M. and Andelman, S.J. (2005) The consistency of extinction risk classification protocols. *Conservation Biology* 19, 1969–1977.

Ricketts, T.H., Daily, G.C. and Ehrlich, P.R. (2002) Does butterfly diversity predict moth diversity? Testing a popular indicator taxon at local scales. *Biological Conservation* 103, 361–370.

Roberge, J.-M. and Angelstam, P. (2004) Usefulness of the umbrella species concept as a conservation tool. *Conservation Biology* 18, 76–85.

Roy, D.B. and Sparks, T.H. (2000) Phenology of British butterflies and climate change. *Global Change Biology* 6, 407–416.

Roy, D.B. and Thomas, J.A. (2003) Seasonal variation in the niche, habitat availability and population fluctuations of a bivoltine thermophilous insect near its range margin. *Oecologia* 134, 439–444.

Roy, D.B., Rothery, P., Moss, D., Pollard, E. and Thomas, J.A. (2001) Butterfly numbers and weather: predicting historical trends in abundance and the future effects of climate change. *Journal of Animal Ecology* 70, 201–217.

Samways, M.J. (2005) *Insect Diversity Conservation*. Cambridge University Press, Cambridge.

Shiffer, C.N. and White, H.B. (1995) Four decades of stability and change in the Odonata populations at ten acre pond in central Pennsylvania. *Bulletin of American Odonatology* 3, 31–41.

Shortall, C., Harrington, R., Hall, M. and Clark, S.C. (2006) *Long-term Trends in Aerial Insect Populations*. Rothamsted Research and English Nature, Peterborough, UK.

Simberloff, D. (1998) Flagships, umbrellas, and keystones: is single-species management passé in the landscape era? *Biological Conservation* 83, 247–257.

Southwood, T.R.E. and Henderson, P.A. (2000) *Ecological Methods*. Blackwell Science, Oxford.

Spellerberg, I.E. (1991) *Monitoring Ecological Change*. Cambridge University Press, Cambridge.

Sykes, J.M. and Lane, A.M.J. (1997) *The United Kingdom Environmental Change Network: Protocols for Standard Measurements at Terrestrial Sites*. HMSO, London.

Taylor, L.R. (1978) Bates, Williams, Hutchinson – a variety of diversities. In: Mound, L.A. and Warloff, N. (eds) *Diversity of Insect Faunas*. Blackwell, Oxford, pp. 1–18.

Taylor, L.R. (1989) Objective and experiment in long-term research. In: Likens, G.E. (ed.) *Long-term Studies in Ecology: Approaches and Alternatives*. Springer-Verlag, New York, pp. 20–70.

Taylor, L.R. (1991) Proper studies and the art of the soluble. *Ibis* 133(Suppl. 1), 9–23.

Taylor, L.R. and Brown, E.S. (1972) Effects of light-trap design and illumination on samples of Moths in Kenya Highlands. *Bulletin of Entomological Research* 62, 91–112.

Taylor, L.R. and French, R.A. (1974) Effects of light-trap design and illumination on samples of moths in an English woodland. *Bulletin of Entomological Research* 63, 583–594.

Taylor, L.R. and Woiwod, I.P. (1982) Comparative synoptic dynamics. I. Relationships between inter- and intra-specific spatial and temporal variance/mean population parameters. *Journal of Animal Ecology* 51, 879–906.

Taylor, L.R., French, R.A. and Woiwod, I.P. (1978) The Rothamsted insect survey and the urbanization of land in Great Britain. In: Frankie, G.W. and Koehler, C.S. (eds) *Perspectives in Urban Entomology*. Academic Press, London, pp. 31–65.

Taylor, L.R., Woiwod, I.P. and Perry, J.N. (1980) Variance and the large-scale spatial stability of aphids, moths and birds. *Journal of Animal Ecology* 49, 831–854.

Thomas, C.D., Thomas, J.A. and Warren, M.S. (1992) Distributions of occupied and vacant butterfly habitats in fragmented landscapes. *Oecologia* 92, 563–567.

Thomas, C.D., Bodsworth, E.J., Wilson, R.J., Simmons, A.D., Davies, Z.G., Musche,

M. and Conradt, L. (2001) Ecological and evolutionary processes at expanding range margins. *Nature* 411, 577–581.

Thomas, J.A. (2005) Monitoring change in the abundance and distribution of insects using butterflies and other indicator groups. *Philosophical Transactions of the Royal Society of London Series B – Biological Sciences* 360, 339–357.

Thomas, J.A. and Clarke, R.T. (2004) Extinction rates and butterflies. *Science* 305, 1563–1564.

Thomas, J.A., Telfer, M.G., Roy, D.B., Preston, C.D., Greenwood, J.J.D., Asher, J., Fox, R., Clarke, R.T. and Lawton, J.H. (2004) Comparative losses of British butterflies, birds, and plants and the global extinction crisis. *Science* 303, 1879–1881.

van Swaay, C. and van Strien, A. (2005) Using butterfly monitoring data to develop a European grassland butterfly indicator. In: Keuhn, E., Thomas, J., Feldmann, R. and Settele, J. (eds) *Studies on the Ecology and Conservation of Butterflies in Europe*. Pensoft Publishers, Sofia, Bulgaria, pp. 106–108.

van Swaay, C.A.M. and Warren, M.S. (1999) *Red Data Book of European Butterflies (Rhopalocera)*. Council of Europe, Strasbourg, France.

van Swaay, C.A.M. and Warren, M.S. (2003) *Prime Butterfly Areas in Europe: Priority Sites for Conservation*. Nation Reference Centre of Agriculture, Nature and Fisheries, Ministry of Agriculture, Nature Management and Fisheries, Wageningen, The Netherlands.

van Swaay, C.A.M., Plate, C.L. and Van Strien, A.J. (2002) Monitoring butterflies in the Netherlands: how to get unbiased indices. *Proceedings of Experimental and Applied Entomology* 13, 21–27.

Warren, M.S. (1991) The successful conservation of an endangered species, the heath fritillary butterfly *Mellicta athalia*, in Britain. *Biological Conservation* 55, 37–56.

Warren, M.S., Hill, J.K., Thomas, J.A., Asher, J., Fox, R., Huntley, B., Roy, D.B., Telfer, M.G., Jeffcoate, S., Harding, P., Jeffcoate, G., Willis, S.G., Greatorex-Davies, J.N., Moss, D. and Thomas, C.D. (2001) Rapid responses of British butterflies to opposing forces of climate and habitat change. *Nature* 414, 65–69.

Western, D. (1987) Africa's elephants and rhinos: flagships in crisis. *Trends in Ecology and Evolution* 2, 343–346.

White, E.G. (1991) The changing abundances of moths in a tussock grassland, 1962–1989 and 50–70-year trends. *New Zealand Journal of Ecology* 15, 5–22.

Williams, C.B. (1948) The Rothamsted light trap. *Proceedings of the Royal Entomological Society of London Series A – General Entomology* 23, 80–85.

Williams, C.B. (1952) Some notes on killing insects for collections and for scientific research. *The Entomologist* 85, 271–279.

Williams, C.B. (1953) The relative abundance of different species in a wild animal population. *Journal of Animal Ecology* 22, 14–31.

Woiwod, I.P. (1991) The ecological importance of long-term synoptic monitoring. In: Firbank, L.G., Carter, N., Darbyshire, J.F. and Potts, G.R. (eds) *The Ecology of Temperate Cereal Fields*. Proceedings of the 32nd Symposium of the British Ecological Society with the Association of Applied Biologists. Blackwell Scientific, Oxford, pp. 275–304.

Woiwod, I.P. and Hanski, I. (1992) Patterns of density dependence in moths and aphids. *Journal of Animal Ecology* 61, 619–629.

Woiwod, I.P. and Harrington, R. (1994) Flying in the face of change: the Rothamsted Insect Survey. In: Leigh, R. and Johnston, A. (eds) *Long-term Experiments in Agricultural and Ecological Sciences*. CAB International, Wallingford, UK, pp. 321–342.

Woiwod, I.P., Gould, P. and Conrad, K.F. (2005) The Rothamsted light-trap network -shedding light on a common moth problem. *Atropos* 26, 5–18.

Yamamura, K., Yokozawa, M., Nishimori, M., Ueda, Y. and Yokosuka, T. (2006) How to analyze long-term insect population dynamics under climate change: 50-year data of three insect pests in paddy fields. *Population Ecology* 48, 31–48.

Yoccoz, N.G., Nichols, J.D. and Boulinier, T. (2001) Monitoring of biological diversity in space and time. *Trends in Ecology and Evolution* 16, 446–453.

10 The Conservation of Ecological Interactions

Jane Memmott, Rachel Gibson, Luisa Gigante Carvalheiro, Kate Henson, Rúben Hüttel Heleno, Martha Lopezaraiza Mikel and Sarina Pearce

School of Biological Sciences, University of Bristol, Woodland Road, Bristol BS8 1UG, UK

1 Introduction

The need to give the conservation of ecological processes an equal weighting to the conservation of patterns is repeatedly stressed but rarely implemented. Instead, conservation research tends to focus on the species as the unit of study, looking at the impact of habitat destruction on individual species, or assemblages of species from particular habitats. There is, however, increasing recognition that species and species lists are not the best units for study by conservation biologists, and that species interactions may be much more important. Although this issue was raised more than 30 years ago by Daniel Janzen (1974) stating that 'what escapes the eye, however, is a much more insidious kind of extinction: the extinction of ecological interactions', it is only recently that we have developed the empirical and analytical tools needed to study interactions at the appropriate scale: that of the whole community. All organisms are linked to at least one other species in a variety of critical ways, for example, as predators or prey, or as pollinators or seed dispersers with the result that each species is embedded in a complex network of interactions. Consequently, the extinction of one species can lead to a cascade of secondary extinctions in ecological networks in ways that we are only just beginning to understand (Sole and Montoya, 2001; Dunne *et al.*, 2002; Ives and Cardinale, 2004; Memmott *et al.*, 2004). Moreover, interactions between species can lead to 'community closure' after the loss of a species, with the result that a locally extinct species cannot re-establish itself if it is reintroduced (Lundberg *et al.*, 2000).

Ecosystem services are those ecological processes of use to mankind, with insects being key components of ecosystem services such as pollination, seed dispersal and pest control. The utilization and exploitation of ecosystem services by mankind is likely to be detrimentally affected by the loss of ecological interactions. This will impact, for example, upon biological control using indigenous natural enemies and upon crop yields via altered pollination rates. However, mammals and birds collectively still receive the majority of attention in terms of conservation efforts and remain flagship species for numerous conservation and agro-environmental programmes. While perhaps more charismatic than the average earwig, vertebrates contribute to far fewer ecosystem services than insects; indeed, Wilson (1987) succinctly described insects as 'the little things that run the world'.

The aim of this chapter is to review insect interactions and consider both their need for conservation and conservation's need for them. We begin by briefly reviewing three key categories of insect interactions and by describing a relatively new method of analysing the interactions between species. We then consider desirable interactions, i.e. those which can be exploited by man, undesirable interactions, such as those with alien species, and consider how best to restore lost interactions. As a measure of how seriously insect interactions are taken in conservation, we ask how many studies of rare plants consider the conservation of their insect pollinators. We end by outlining four areas we consider to merit particular attention for future work on ecological interactions.

2 Types of Ecological Interactions Involving Insects

Although food webs are the most commonly described ecological network, other types of interaction webs are investigated that include a variety of trophic and non-trophic interactions, such as pollination, seed dispersal, interference competition, habitat or shelter provisioning, recruitment facilitation or inhibition (Memmott *et al.*, 2006). For the purposes of this chapter we restrict ourselves to trophic and reproductive mutualisms, mentioning decomposition briefly to bemoan its status as the ugly sister of interaction networks, especially with respect to insects. We limit ourselves to trophic networks and reproductive mutualisms for ecological and methodological reasons. Ecologically, food webs are a fundamental component of any attempt to describe how complexes of species interact, and because reproductive mutualisms are essential for the survival of most plant species. Methodologically, these interactions have made the transition from the study of pairwise interactions to the study of networks of interactions.

2.1 Trophic interactions

Insects form an important component of many food webs, with food chains comprising green plants, insect herbivores and parasitoids including over half of all known species of metazoa (Strong *et al.*, 1984). Memmott and Godfray (1993) list eight insect food webs and since then there has been a veritable, albeit small, industry in producing these networks with the result that many of the problems with earlier food web studies (Cohen *et al.*, 1993) have been addressed. These new networks all focus on insect herbivores, their food plants and their natural enemies and they encompass a variety of habitats, from tropical rainforest to English meadow to Hawaiian swamp (Table 10.1). These webs have documented patterns and generated hypotheses concerning the underlying processes that structure communities, and in some cases, have led to tests of these hypotheses using manipulative field experiments. For example, Memmott *et al.* (1994) constructed a plant–leafminer–parasitoid

Table 10.1. Quantitative trophic networks, describing plant–herbivore–parasitoid communities.

Reference	Habitat	Location	Aims	Host herbivore
Clarke (2000)	Ancient woodlands	England	1. Effect of habitat Fragmentation on food web structure 2. Effect of fragment size	Leaf-mining insects
Henneman and Memmott (2001)	Swamp forest	Hawaii	1. Community-wide effects of introduced bicontrol agents	Lepidoptera
Lewis *et al.* (2002)	Moist tropical forest	Belize	1. Diversity of interactions 2. Compartmentalization 3. Seasonal variation 4. Indirect interactions (apparent competition)	Leaf-mining insects
Memmott *et al.* (1994)	Tropical dry forest	Costa Rica	1. Identify factors influencing diversity 2. Identify factors influencing diet breath at two trophic levels 3. Identify species with a strong function in the community	Leaf-mining insects
Muller *et al.* (1999)	Meadow	England	1. Identify changes in food web over time 2. Apparent competition 3. Compartmentalization	Aphids
Schönrogge and Crawley (2000)	Oak stands	England	1. Impact of invasion on structure and function of native communities 2. Apparent competition	Cynipid gall wasps
Rott and Godfray (2000)	Woodland	England	1. Variations between generations 2. Recruitment webs 3. Importance of host plant 4. Apparent competition	Phyllonorycter leaf miners
Valladares *et al.* (2001)	Native vegetation and zoological park	Argentina	1. Analyse the structure 2. Compartmentalization 3. Indirect interactions (apparent competition)	Dipteran leaf miners

web that described the trophic interactions in a Costa Rican tropical dry forest. On the basis of this web, a field experiment was used to investigate interactions between two groups of natural enemies (Memmott *et al.*, 1993). Similarly, Lewis *et al.* (2002) used a food web to characterize the structure of a leaf miner–parasitoid community in Belize and a field experiment was then used to test for the presence of indirect interactions (Morris *et al.*, 2004). New analytical techniques for dealing with the impact of extinction on networks (Sole and Montoya, 2001; Dunne *et al.*, 2002; Ives and Cardinale, 2004; Memmott *et al.*, 2004) bode well for giving food webs a predictive role in conservation biology, and these will be discussed in Section 3.

2.2 Pollination

Most higher plant species – up to 90% by some estimates (Nabhan and Buchmann, 1997) – rely on animals to pollinate their flowers. Although vertebrates such as birds, bats and marsupials can all act as pollinators, 'insects are undoubtedly the most important animal pollinators' (Proctor *et al.*, 1996). Insects visit flowers to obtain food, usually in the form of pollen or nectar. This is one side of a mutually beneficial relationship; the plants, in return, obtain the services of the pollinators in carrying pollen from one flower to another (Proctor *et al.*, 1996). Although plant–pollinator interactions have a long history of research, a community level network approach has only recently become popular. The study of pollination networks was, if not born, at least baptized by a seminal paper in 1996 by Nick Waser and colleagues and the first quantitative visitation web followed soon after (Memmott, 1999). Waser *et al.* (1996) assessed the evidence for generalization within plant–pollinator communities. Using a broad review drawing on studies of two American floras, and several surveys of pollinators of particular plants (e.g. of buttercups and orchids) and the plants used by particular pollinators (e.g. solitary bees), they deduced that moderate to substantial generalization is widespread. At the time of publication of this chapter, the paper by Waser *et al.* (1996) had been cited 296 times reflecting the huge surge of interest in this field. Since the publication of that paper, there has been a change in data quality, similar to that seen with trophic webs, as connectance webs have been replaced by webs that incorporate quantitative as well as qualitative information (Table 10.2).

2.3 Seed dispersal

Seed dispersal is the removal of seeds from a plant to another location, and plays a key role in regenerating natural communities (Christian, 2001). Ants are most commonly involved in seed dispersal, especially in drier habitats such as deserts, grassland and fynbos vegetation (Hölldobler and Wilson, 1990). They are especially attracted to seeds that offer food bodies known as elaiosomes, which are rich in amino acids, fatty acids and sugars (Hölldobler

Table 10.2. Quantitative pollination networks describing the frequency of interactions between plants and pollinators.

Reference	Habitat	Location	Aims
Dicks *et al.* (2002)	Meadow	England	1. Are plant–pollinator networks compartmentalized?
Forup and Memmott (2005)	Meadow	England	1. Does habitat restoration restore ecological interactions? 2. Pollen transport webs
Gibson *et al.* (2006)	Arable fields	England	1. What are the pollinator requirements of rare plants? 2. How to identify pollinators in visitation networks 3. Spatial and temporal variation in network structure
Memmott (1999)	Meadow	England	1. The structure of a plant–pollinator network
Olesen *et al.* (2002)	Oceanic islands	Azores and Mauritius	1. Testing for presence of invader complexes
Vazquez and Simberloff (2002)	Temperate forest	Argentina	1. Niche breadth in a disturbed habitat 2. Asymmetry in interactions

and Wilson, 1990). To date, published seed dispersal networks are dominated by bird dispersal (see references cited in Bascompte *et al.* 2003); indeed, we are not aware of a single published seed dispersal network dominated by ants, despite the popularity of research on granivorous ants.

3 Using Networks to Study the Conservation of Ecological Interactions

Using food webs such as those listed in Tables 10.1 and 10.2 as predictive tools in conservation biology has until recently been an unattainable goal. At first sight, webs appear labour-intensive to make, statistically intractable to analyse and of limited use to conservation ecologists. If, however, ecological webs are considered as networks, a suite of new analytical tools becomes available. A network is simply a set of nodes with connections between them, such as food webs, neural networks, social networks, the World Wide Web and co-authorship networks. Whilst ecologists have not yet used network techniques to study conservation, they have started to study its opposite: extinction. Species extinction is obviously what conservation ecologists are trying to avoid, but an understanding of the community-level impact of extinction is highly desirable. A complex systems approach simply involves asking what happens to the network when nodes (species in this context) are removed. This method allows ecologists to predict what happens to an ecological network when species go extinct. For example, does it have little effect on web structure or does it lead to a cascade of secondary extinctions?

Many complex systems, food webs included, display a surprising degree of tolerance against the loss of nodes. However, this tolerance comes at a high price, as these networks are then highly vulnerable to the removal of a few key nodes that play a vital role in maintaining the network's connectivity (Sole and Montoya, 2001; Dunne *et al.*, 2002). In particular, 'food webs show rivet-like thresholds past which they display extreme sensitivity to the removal of highly connected species' (Dunne *et al.*, 2002). This sensitivity is seen as a collapse of the entire network (i.e. 100% extinction). Thus, removing just a handful of well-connected species could, in theory at least, cause a cascade of secondary extinctions. This cascade cannot be predicted by independent data on the abundance and distribution of individual species. Memmott *et al.* (2004) developed this technique one step further by assigning different trophic groups differing risks of extinction. Combining computer models with a large plant–pollinator data set describing the interactions between 456 plants and 1428 pollinators they explored the vulnerability of plants to extinction of their pollinators. They found that many plants are pollinated by a diversity of insect species, whilst those that are specialized (i.e. visited by few pollinators) tend to use pollinators that are themselves generalists. By incorporating variation in extinction risk (i.e. increasing trophic rank leads to increasing risk of extinction) known from empirical work (e.g. Gilbert *et al.*, 1998), there was a profound effect on the extinction dynamics and the network became far more robust in terms of secondary extinctions. These features confer relative tolerance to extinction for plant species. However, as Memmott *et al.* (2004) point out, tolerance is not immunity and it is essential to conserve pollinators that are generalized, including bumblebees and some other types of bees.

4 Desirable Interactions and Their Utilization

Ecological interactions are effectively the currency of ecosystem services and these services provide gratis products, such as pest control, pollination and decomposition. The collection of data on interactions in applied systems usually involves data on simple food chains (e.g. crop–herbivore–predator and crop–herbivore–parasitoid) in which a pest herbivore is the focus of the study. This approach is used for the practical reason that it is relatively quick to use. However, by ignoring the community in which a pest herbivore operates, ecologists are likely to miss interactions that limit pest abundance or mitigate the impact of beneficial arthropods. Numerous reports list the major pest, predator and parasitoid species collected in different crop types, but only rarely is this information presented in the form of a food web illustrating the interactions between trophic levels (exceptions include Schoenly *et al.*, 1996 in rice and Mayse and Price, 1978 in soybean). Often a link between trophic levels is based only on previous studies in other systems and rarely quantified. This is disappointing given the potential of food webs not only in clarifying the links and importance of interactions between trophic levels, but also for evaluating the effects and sustainability of management strategies (Cohen *et al.*, 1994). It seems clear that the simple food chain (plant–herbivore–predator) on which most pest control theory is based is unrealistic (Rosenheim *et al.*, 1999). The good news is that food web science has developed to a stage where we are capable of sampling, visualizing and analysing complex interactions, although this technique is only just beginning to be applied to agroecosystems.

It is relatively straightforward, if time-consuming, to determine the parasitoids of a given host species. The host insect is reared in isolation until either an adult host or a parasitioid emerges (Memmott and Godfray, 1994; Memmott, 1999). However, in spite of the relative lack of obstacles (in comparison to predators, see below) only eight quantitative host–parasitoid webs have been published (Table 10.1), and none are from agricultural systems. Most are descriptive in nature (utilizing a single field site), and have general aims (e.g. to identify species with a strong function in the community) or seek to understand theoretical aspects of community ecology (e.g. compartmentalization). The use of trophic webs to investigate applied questions has so far been limited. One example is Henneman and Memmott (2001), who investigated the community-wide effects of introduced biocontrol agents in native Hawaiian habitats, these parasitoids having originally come from agricultural cane fields. They constructed quantitative food webs of interactions among plants, moths and their parasitoids in a native forest in order to examine the community-wide effects of introduced biocontrol agents on Kaua'i Island, Hawaii. About 83% of parasitoids reared from native moths were biological control agents, 14% were accidental immigrants and 3% were native species. On the positive side, all the biological agents reared were released before 1945 and there was no evidence that biocontrol agents released recently were attacking native moths. Their study highlights the importance of considering the potential damage caused by an introduced control agent, in addition to that being caused by the target alien species.

Even without food webs, though, parasitoids and their interactions with their host insects can be used as good indicators of ecological change at a variety of spatial scales (Tscharntke and Brandl, 2004). Thies and Tscharntke (1999) showed that in structurally complex agricultural landscapes (i.e. high percentage non-crop area), parasitism of the rape pollen beetle (*Meligethes aeneus*) was higher and crop damage lower than in simple landscapes. In a later study in wheat fields, Thies *et al.* (2005) found that complex landscapes were not only associated with increased aphid parasitism, but also higher rates of aphid colonization.

Whilst the action of parasitoids can be easily seen and studied in the field, the activity of predators is considerably harder to quantify. The 'hit-and-run style' of insect predation makes it very easy to miss predator–prey interactions when sampling. It is also uncommon for clear evidence to remain of who ate whom after predation occurs, further compounding the problem (Memmott *et al.*, 2000). However, molecular and immunological methods are increasingly used to detect prey in predator guts. For example, Symondson *et al.* (1996) used molecular probes to detect slugs in the diet of carabids; Harwood *et al.* (2005) used them to detect aphids in spider guts; and Sheppard *et al.* (2004) used them to detect non-target prey in the guts of predatory biocontrol agents. Techniques that can scan for multiple prey in a host in a single test have recently been developed (Harper *et al.*, 2005) and could revolutionize the way food webs are constructed.

In addition to biological control, insects provide a range of other ecosystem services and useful interactions. For example, without the services of dung beetles and dung flies the world would become a rather unsavoury place. The insects that act as decomposers, along with their predators and parasitoids, form a large component of soil food webs, and these webs drive ecosystem level processes such as energy flow and nutrient cycling (Wardle *et al.*, 1998). Pollination is obviously another key ecosystem service (Kremen *et al.*, 2002) and is discussed in more detail by Kremen and Chaplin-Kramer (Chapter 15, this volume).

5 Undesirable Interactions and Their Control

Not all ecological interactions are desirable. Many authors have stated that the introduction of alien species, and their interactions with native species, are one of the major threats to biodiversity (e.g. Schmitz and Simberloff, 1997; Mack and D'Antonio, 1998; Chittka and Schurkens, 2001). Aliens have been introduced at all trophic levels into both trophic and mutualistic networks. Their interactions with native species can occasionally be positive, for example, in Hawaii the pollinators of a native vine, *Freycinetia arborea*, are all extinct, but it is probably pollinated now by an alien bird, the Japanese white eye (Buchmann and Nabhan, 1996). Rather more frequently though, native–alien interactions are negative, for example, the disruption of native seed dispersal by Argentine ants in the South African fynbos (Christian, 2001). Most often ecologists simply do not know what most alien species are doing at the community level. Although extensive data exist on the distributions of alien species, their impact on native species as competitors, prey species, predators, pollinators and parasites, and even their impact upon ecosystem properties, there are rather little data on how aliens are accommodated into ecological networks. Five exceptions to this rule are Schönrogge and Crawley (2000) working on an alien cynipid wasp in the UK, Henneman and Memmott (2001) working on alien parasitoids in Hawaii, Munro and Henderson (2002) working on alien parasitoids in New Zealand, Memmott and Waser (2002) working on alien pollinators in the USA and Olesen *et al.* (2002) working on alien pollinators on two tropical islands.

Classical biological control involves the deliberate introduction of an alien species. For many years this was viewed as a sustainable, environmentally friendly form of pest control. However, over the last decade the environmental safety of biological control has become rather contentious with particular concerns about the potential interactions between biocontrol agents and 'non-target' species (Louda *et al.*, 1997; Thomas and Willis, 1998; Boettner *et al.*, 2000; Pemberton and Strong, 2000). A parasitoid maintained at high densities on a common pest insect can potentially drive a rare non-target species to extinction. Density dependence, which would ameliorate such an effect in a simple two-species interaction, is lacking in such cases (Simberloff and Stiling, 1996). There is a firm theoretical basis for this phenomenon, known as apparent competition (Holt, 1977). Non-target interactions can occur either directly, if an agent attacks a non-target host, or indirectly, when the agent affects non-target species via shared natural enemies. Food webs have been suggested as the appropriate model for research on non-target interactions in biological pest control (Strong, 1997; Pemberton and Strong, 2000; Henneman and Memmott, 2001). As described earlier, Henneman and Memmott (2001) successfully used this approach to quantify non-target effects in Hawaii.

The role of plant–pollinator interactions in promoting or constraining invasions is likely to vary considerably among invaded communities (Parker and Haubensak, 2002). Nevertheless, only a small proportion of possible invaders are known to have been restrained by lack of pollinators (Richardson *et al.*, 2000). Alien plants have the potential to influence native plants via shared pollinators. Alien and native plant species interact for pollinators in the same three ways that native plant species interact with other native plants: an alien plant species can compete for pollinators with a native plant species (Chittka and Schurkens, 2001; Ghazoul, 2002), facilitate attraction of pollinators (Moragues and Traveset, 2005) or even have no effect on visitation rates (Aigner, 2004). The extent to which the alien interaction affects the population of the native plant species will depend on whether seed set is pollen-limited and whether population size is limited by seed recruitment (Palmer *et al.*, 1997). Therefore, changes in pollination quantity (i.e. visitation rates), and even in pollination quality (i.e. reduction of conspecific pollen on stigmas or deposition of exotic pollen), will not necessarily affect seed set. In Thailand, Ghazoul (2002) reported that only one group of visitors, the butterflies, switched their diet from native *Dipterocarpus obtusifolius* to the alien *Chromolaena odorata*. This reduced visitation rates to *Dipterocarpus* along with the number of flowers receiving conspecific pollen. Moreover, it increased heterospecific pollen on stigmas. However, none of these factors influenced *Dipterocarpus* seed set. In comparison, Chittka and Schurkens (2001) found that the introduced *Impatiens grandulifera* (Himalayan Balsam) reduced visitation rates and seed set of the native *Stachys palustris* (Marsh Woundwort), this effect being mediated by shared bees. In summary, the effect of alien species on other plant species is species-specific and variable, depending on many ecological variables (Moragues and Traveset, 2005).

Several bee species (*Apis mellifera*, *Bombus* spp., *Megachile* spp. and *Osmia* spp.) have been deliberately introduced around the world, particularly for crop pollination, and in the case of the honeybee, also for its honey. *A. mellifera* and *Bombus terrestris* stand out for being widespread and very abundant. In general, these species tend to have a much wider diet than native bees, forage earlier and later in the day than native bees, outnumber any native species and use a large amount of the floral resources available (Goulson, 2003 and references therein; Potts *et al.*, 2001). Although there is no direct evidence at the population level, there are many indications that these bees compete with native species. For example, declines in native bee abundance are reported where exotic bees are present or most abundant (Aizen and Feinsinger, 1994; Dafni and Shmida, 1996; Kato *et al.*, 1999).

Although usually considered detrimental to local species, alien insect pollinators may successfully pollinate native plant species (e.g. Percival, 1974; Butz Huryn, 1997; Freitas and Paxton, 1998). However, there are also reports of reduction of pollination of native flora (e.g. Aizen and Feinsinger, 1994; Roubik, 1996). Nectar robbing by alien pollinators, e.g. *Bombus* spp., can potentially reduce visitation rates (McDade and Kinsman, 1980) and seed set (e.g. Irwin and Brody, 1999) of native plants. Moreover, *Apis* and *Bombus*, although having large flight ranges when compared with other visitor species, engage in few long flights while they are foraging, altering patterns of cross-pollination, which consequently may change the genetic structure of plant populations (Goulson, 2003 and references therein).

Field evidence for the community-level impacts of alien seed dispersers was anecdotal until recently. Christian (2001) reported that the invasion of South African shrublands by the Argentine ant, *Linepithema humile*, led to changes in plant community composition and the near extinction of two native ant species. The changes in plant community composition were due to a disproportionate reduction in large-seeded plant dispersal, as these rely on fewer more specialized dispersers for which services are not replaced. Whether parasitoids, pollinators or seed dispersers, all these community studies exemplify Simberloff's (2004) statement that 'most key issues in invasion biology…fall squarely at the community level', and that to answer how an invader affects a community 'often entails ingenious, detailed research on complicated systems because many impacts…are subtle even if devastating'.

6 The Restoration of Ecological Interactions

Restoration ecology can be viewed as the study of how to repair anthropogenic damage to the integrity of ecological systems (Cairns and Heckman, 1996) and one of the greatest challenges facing mankind will be maintaining ecological systems in working order as the human population rises towards 9 billion (UN, 2003). The São Paulo Declaration on Pollinators (Dias *et al.*, 1999) revealed serious declines in the number of native pollinator species in Central and North America and six European countries. It has been estimated that over the next 300 years, up to half a million insect species may become extinct (Mawdsley and Stork, 1995). Furthermore, the International Union for Conservation of Nature (IUCN) predicts a global loss of 20,000 flowering plant species over the next few decades, with inevitable consequences for the survival of their co-dependent pollinators. Such losses are likely to involve keystone plants or pollinators, affecting community structure and potentially exacerbating biodiversity degradation and losses of ecosystem services (Bronstein *et al.*, 1990; Cox *et al.*, 1991; Walker, 1992; Grime, 1997; Allen-Wardell *et al.*, 1998; Kearns *et al.*, 1998).

As succinctly described by Simberloff (1990), 'restoration is a game with a moving target whose trajectory cannot be accurately predicted, and the target in any event cannot quite be seen or characterised'. He refers to restoration as having a 'fuzzy target', as restoration practitioners often do not know exactly what they are hoping to restore, and indeed the resolution of the target itself has raised much debate. Furthermore, economic and social constraints have dictated the degree of partial restoration in order to minimize costs, resulting in managers focusing on restoring the 'superficial' vegetative structure of the system (Handel, 1997; Palmer *et al.*, 1997). Through reinstating the basic community structure, organisms associated with such habitats would be expected to arrive and establish themselves with time (Anderson, 1995; Handel, 1997; Palmer *et al.*, 1997; Hilderbrand *et al.*, 2005). Referred to as the 'field of dreams hypothesis' (Palmer *et al.*, 1997), this remains poorly tested, but relies on the redundancy of species present, particularly pollinator redundancy (the ability of other species to act as back-up pollinators in the absence of the main pollinator(s)), to restore ecosystem function (Zamora, 2000; Kearns, 2001). If this inherent functional redundancy is high, it is possible to set a minimum level of species diversity to be restored, which would ensure the gradual re-establishment of ecosystem function (Palmer *et al.*, 1997). Such thresholds have been tested through modelling (Lundberg and Ingvarsson, 1998), although in reality data to determine this baseline are either scarce or incomplete (Lambeck, 1997). Moreover, effective surrogate pollinators may also disperse pollen differently (Thomson and Thomson, 1992; Montalvo *et al.*, 1997; Aigner, 2001).

7 The Conservation of Interactions: Theory or Practice?

Insects form a large component of global biodiversity and as reviewed above form many links with other organisms. To determine the extent to which entomological interactions are considered in conservation ecology, we considered the case of pollination, asking how often this key link was investigated when conserving a rare plant species. To do this we searched the Web of Science for studies of named rare plant species between 2000 and 2005, and found 139 published studies. Although 113 of these had a conservation focus, we found that only 18 mentioned the importance of pollinators for the plant's survival, 14 went on to name the types of insects visiting the plant in question (e.g. bees or flies) and only 4 (Kang *et al.*, 2000; Peterson *et al.*, 2002; Evans *et al.*, 2003; Rodriguez-Perez, 2005) gave a full list of the insect species visiting the rare plant's flowers. Visitation does not equate to pollination, however, and none of the papers definitively identified the pollinator(s). From our survey of these 139 papers it is apparent that interactions are still not given the attention they deserve, at least by people working on plants. Without the conservation of interactions with pollinators, the conservation of a plant species is unlikely to be sustainable in the long term. In an effort to redress the balance, and using a community-level network approach we ran a project specifically designed to identify the pollinators of rare plants (Gibson *et al.*, 2006). One of the rare plants, *Galeopsis angustifolia* (Red Hemp Nettle) was visited by 22 insect species, but assessing the quantity and quality of the pollen on the flower visitors allowed us to uncover hidden links in the network, where the pollen carried on an insect's body acts as a record of its previous movements between flowers of different species. In addition, pollen transport data allowed us to eliminate 18 visitors which did not carry *G. angustifolia*'s pollen from the list of its potential pollinators (Fig. 10.1). Using a similar approach, Forup and Memmott (2005) asked whether a meadow restoration scheme had been successful. Working on plant–pollinator interactions on two old and two restored hay meadows, they found no difference in the proportion of plant species visited by potential pollinators, whilst all the

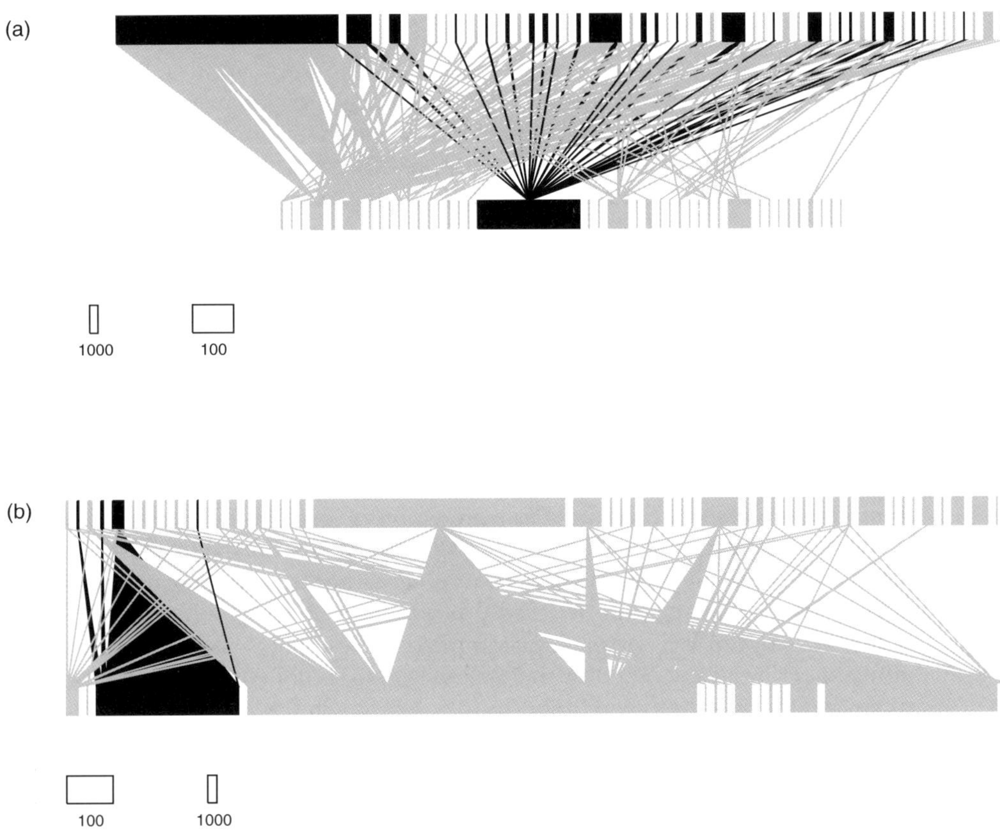

Fig. 10.1. (a) Visitation web for *Galeopsis angustifolia*. Each species of plant and insect is represented by a rectangle: the lower line represents flower abundance; the upper line represents insect abundance. The widths of the rectangles are proportional to their abundance at the field site, and the size of the lines connecting them represents the recorded frequency of the interaction. The target plant and the species it interacts with are shown in black. The scale bar represents number of floral units (1000) and number of insects (100). (b) Pollen transport web for *G. angustifolia* at GA1. The lower line represents pollen abundance; the upper line represents insect abundance. The scale bar represents the number of pollen grains (1000) and the number of insects (100). Note that in the visitation web, all interactions are shown, not just those involving insects identified to species. Consequently, (a) actually shows 25 insects visiting *G. angustifolia* (rather than 22 as stated in the text). (Reprinted from Gibson *et al.*, 2006, with permission from Blackwell Publishing.)

visited plant species were generalized, each having more than a single species of insect visitor. With regard to the restoration of interactions, as opposed to the restoration of species, is a field of research with very little data, but with considerable potential to make restoration more sustainable.

8 Conclusions

Insects form numerous key links with other species leading to complex networks of interactions. These links can take numerous forms: they may be critical to the survival of another species, e.g. pollination; they may form part of an ecosystem service and be of direct use to mankind, e.g. pest control; they may be undesirable links with alien species. However, as is apparent from our survey, considering the restoration of interactions is not a widespread practice. In this final section we will discuss potential future directions in the study of the conservation of ecological interactions, highlighting four fields in need of further attention and discussing some barriers impeding progress in these fields.

8.1 Interdisciplinary networks

Researcher comfort zones are a divisive force in ecology. Thus, while tropical pollinators are well studied at the levels of bees (by entomologists), hummingbirds (by ornithologists) and bats (by mammologists), there are no published data at the level of an entire tropical pollinator community. From the plant's point of view, insects, birds and bats are all simply pollen transporters and the taxonomic distinctions used by biologists are indiscernible. Similarly, ecosystem services are divided into different research fields, with pest control studied by agroecologists, pollination studied by pollination biologists and seed dispersal studied by ant ecologists and ornithologists. However, there are almost no data to say whether these services are actually independent of each other and the links between them could be of considerable importance (but see Larsen *et al.*, 2005 for a notable exception). A more interdisciplinary approach to the study of interactions would be very revealing. A handful of fascinating papers have been published: Bailey and Whitham (2003) link elk, aspen, sawflies and insectivorous birds; Knight *et al.* (2005) demonstrate that fish indirectly affect pollinators. These types of cascading interactions across very different groups of species (Bailey and Whitham, 2003) and across ecosystems (Knight *et al.*, 2005) are probably not uncommon, but just go unrecorded.

8.2 Clarity of measurements

With some interactions, ecologists do not measure what is assumed; for example, 'pollination networks' are without exception visitation networks. Nobody has yet unambiguously identified all the pollinators in a pollinator web (though see Gibson *et al.*, 2006 and Dicks, 2002 for identifying a few of the pollinators in their networks). Similarly, when working with birds, to be confident that seed dispersal is real dispersal, more than seed intake needs to be measured; intact seed also must be recovered from bird droppings.

8.3 Quantification

Quantitative data are needed for some types of networks, while others such as decomposition networks have received little study at all. Seed dispersal networks remain undeveloped in contrast to the publishing boom in high quality, quantitative pollination and trophic webs. Published bird networks are largely connectance webs and we are not aware of any published ant dispersal networks. Similarly, decomposition networks remain deeply unattractive as study systems, despite the decay of plant and animal remains being essential to the nutrient cycle, and indeed to life itself. Finally, while parasitoids are seen in food webs, parasites and diseases are not commonly represented, despite the fact that they are both diverse and abundant (e.g. Huxham and Raffaelli, 1995) and are ubiquitous in terrestrial, marine and aquatic systems.

8.4 Reality and validation of models

Considerable progress has been made using complex systems approaches to study the extinction of interactions at the community level. However, this approach needs to become more realistic and the models require validation in the field. Recent work by Ives and Cardinale (2004) that incorporates environmental stress into a network, making the species that survive more resilient to this stress, is an excellent example of how more realism can be introduced. Despite being computationally (relatively) straightforward, a complex systems approach has to date been largely used by theoretical ecologists. To validate these models in the field will require greater collaboration between field ecologists and theoretical biologists, or for field ecologists or theoreticians to master new techniques.

In summary, food webs, pollination webs and other ecological networks have not been widely applied to the field of conservation biology (but see Corbet, 2000 for a notable exception). Given the recent practical advances made in food web construction (e.g. ecoinformatics), the theoretical advances (e.g. models of extinction dynamics) and the ongoing loss of biodiversity, this is a very exciting time to begin to use the study of ecological interactions as a tool in conservation biology.

References

Aigner, P.A. (2001) Optimality modelling and fitness trade-offs: when should plants become pollinator specialists? *Oikos* 95, 177–184.

Aigner, P.A. (2004) Ecological and genetic effects on demographic processes: pollination, clonality and seed production in *Dithyrea maritima*. *Biological Conservation* 116, 27–34.

Aizen, M.A. and Feinsinger, P. (1994) Habitat fragmentation, native insect pollinators, and feral honey-bees in Argentine Chaco Serrano. *Ecological Applications* 4, 378–392.

Allen-Wardell, G., Bernhardt, P., Bitner, R., Burquez, A., Buchmann, S., Cane, J., Cox, P.A., Dalton, V., Feinsinger, P., Ingram, M., Inouye, D., Jones, C.E., Kennedy, K., Kevan, P., Koopowitz, H., Medellin, R., Medellin-Morales, S., Nabhan, G.P., Pavlik, B., Tepedino, V., Torchio, P. and Walker, S. (1998) The potential consequences of pollinator declines on the conservation of biodiversity and stability of food crop yields. *Conservation Biology* 12, 8–17.

Anderson, P. (1995) Ecological restoration and creation: a review. *Biological Journal of the Linnean Society* 56, 187–211.

Bailey, J.K. and Whitham, T.G. (2003) Interactions among elk, aspen, galling sawflies and insectivorous birds. *Oikos* 101, 127–134.

Bascompte, J., Jordano, P., Melian, C.J. and Olesen, J.M. (2003) The nested assembly of plant-animal mutualistic networks. *Proceedings of the National Academy of Sciences of the United States of America* 100, 9383–9387.

Boettner, G.H., Elkinton, J.S. and Boettner, C.J. (2000) Effects of a biological control introduction on three nontarget native species of saturniid moths. *Conservation Biology* 14, 1798–1806.

Bronstein, J.L., Gouyon, P.H., Gliddon, C., Kjellberg, F. and Michaloud, G. (1990) The ecological consequences of flowering asynchrony in monoecious figs: a simulation study. *Ecology* 71, 2145–2156.

Buchmann, S.L. and Nabhan, G.P. (1996) *The Forgotten Pollinators*. Island Press, Washington, DC.

Butz Huryn, V.M. (1997) Ecological impacts of introduced honey bees. *The Quarterly Review of Biology* 72, 275–297.

Cairns, J. Jr and Heckman, J.R. (1996) Restoration ecology: the state of an emerging field. *Annual Review of Energy and the Environment* 21, 167–189.

Chittka, L. and Schurkens, S. (2001) Successful invasion of a floral market: an exotic Asian plant has moved in on Europe's river-banks by bribing pollinators. *Nature* 411, 653.

Christian, C.E. (2001) Consequences of a biological invasion reveal the importance of mutualism for plant communities. *Nature* 413, 635–639.

Clarke, M. (2000) The impact of habitat fragmentation on community processes. PhD thesis, University of Bristol, Bristol, UK.

Cohen, J.E., Beaver, R.A., Cousins, S.H., Deangelis, D.L., Goldwasser, L., Heong, K.L., Holt, R.D., Kohn, A.J., Lawton, J.H., Martinez, N., Omalley, R., Page, L.M., Patten, B.C., Pimm, S.L., Polis, G.A., Rejmanek, M., Schoener, T.W., Schoenly, K., Sprules, W.G., Teal, J.M., Ulanowicz, R.E., Warren, P.H., Wilbur, H.M. and Yodzis, P. (1993) Improving food webs. *Ecology* 74, 252–258.

Cohen, J.E., Schoenly, K., Heong, K.L., Justo, H., Arida, G., Barrion, A.T. and Litsinger, J.A. (1994) A food-web approach to evaluating the effect of insecticide spraying on insect pest population-dynamics in a Philippine irrigated rice ecosystem. *Journal of Applied Ecology* 31, 747–763.

Corbet, S.A. (2000) Conserving compartments in pollination webs. *Conservation Biology* 14, 1229–1231.

Cox, P.A., Elmqvist, T., Pierson, E.D. and Rainey, W.E. (1991) Flying foxes as strong interactors in South-Pacific Island ecosystems: a conservation hypothesis. *Conservation Biology* 5, 448–454.

Dafni, A. and Shmida, A. (1996) The possible ecological implications of the invasion of *Bombus terrestris* (L.) (Apidae) at Mt. Carmel, Israel. In: Matheson, A., Buchmann, S., O'Toole, C., Westrich, P. and Williams, I.H. (eds) *The Conservation of Bees*. Academic Press, London, pp. 183–200.

Dias, B.S.F., Raw, A. and Imperatri-Fonseca, V.L. (1999) *Report on the Recommendations of the Workshop on the Conservation and Sustainable use of Pollinators in Agriculture, with Emphasis on Bees*. Brazilian Ministry of the Environment, São Paulo, Brazil.

Dicks, L.V. (2002) The structure and functioning of flower-visiting insect communities on hay meadows. PhD thesis, University of Cambridge, Cambridge.

Dicks, L.V., Corbet, S.A. and Pywell, R.F. (2002) Compartmentalization in plant–insect flower visitor webs. *Journal of Animal Ecology* 71, 32–43.

Dunne, J.A., Williams, R.J. and Martinez, N.D. (2002) Network structure and biodiversity loss in food webs: robustness increases with connectance. *Ecology Letters* 5, 558–567.

Evans, M.E.K., Menges, E.S. and Gordon, D.R. (2003) Reproductive biology of three sympatric endangered plants endemic to Florida scrub. *Biological Conservation* 111, 235–246.

Forup, M.L. and Memmott, J. (2005) The restoration of plant–pollinator interactions in hay meadows. *Restoration Ecology* 13, 265–274.

Freitas, B.M. and Paxton, R.J. (1998) A comparison of two pollinators: the introduced honey bee *Apis mellifera* and an indigenous bee *Centris tarsata* on cashew *Anacardium occidentale* in its native range of NE Brazil. *Journal of Applied Ecology* 35, 109–121.

Ghazoul, J. (2002) Flowers at the front line of invasion? *Ecological Entomology* 27, 638–640.

Gibson, R.H., Nelson, I.L., Hopkins, G.W. and Memmott, J. (2006) Pollinator webs, plant communities and the conservation of rare plants: arable weeds as a case study. *Journal of Applied Ecology* 43, 246–257.

Gilbert, F., Gonzalez, A. and Evans-Freke, I. (1998) Corridors maintain species richness in the fragmented landscapes of a microecosystem. *Proceedings of the Royal Society of London Series B – Biological Sciences* 265, 577–582.

Goulson, D. (2003) Effects of introduced bees on native ecosystems. *Annual Review of Ecology Evolution and Systematics* 34, 1–26.

Grime, J.P. (1997) Ecology – biodiversity and ecosystem function: the debate deepens. *Science* 277, 1260–1261.

Handel, S.N. (1997) *The Role of Plant–Animal Mutualisms in the Design and Restoration of Natural Communities*. Cambridge University Press, Cambridge.

Harper, G.L., King, R.A., Dodd, C.S., Harwood, J.D., Glen, D.M., Bruford, M.W. and Symondson, W.O.C. (2005) Rapid screening of invertebrate predators for multiple prey DNA targets. *Molecular Ecology* 14, 819–827.

Harwood, J.D., Sunderland, K.D. and Symondson, W.O.C. (2005) Monoclonal antibodies reveal the potential of the tetragnathid spider *Pachygnatha degeeri* (Araneae: Tetragnathidae) as an aphid predator. *Bulletin of Entomological Research* 95, 161–167.

Henneman, M.L. and Memmott, J. (2001) Infiltration of a Hawaiian community by introduced biological control agents. *Science* 293, 1314–1316.

Hilderbrand, R.H., Watts, A.C. and Randle, A.M. (2005) The myths of restoration ecology. *Ecology and Society* 10, U648–U658.

Hölldobler, B. and Wilson, E.O. (1990) *The Ants*. Harvard University Press, Cambridge, Massachusetts.

Holt, R.D. (1977) Predation, apparent competition and the structure of prey communities. *Journal of Theoretical Biology* 12, 197–229.

Huxham, M. and Raffaelli, D. (1995) Parasites and food-web patterns. *Journal of Animal Ecology* 64, 168–176.

Irwin, R.E. and Brody, A.K. (1999) Nectar-robbing bumble bees reduce the fitness of *Ipomopsis aggregata* (Polemoniaceae). *Ecology* 80, 1703–1712.

Ives, A.R. and Cardinale, B.J. (2004) Food-web interactions govern the resistance of communities after non-random extinctions. *Nature* 429, 174–177.

Janzen, D.H. (1974) The deflowering of Central America. *Natural History* 83, 48–53.

Kang, U., Chang, C.S. and Kim, Y.S. (2000) Genetic structure and conservation considerations of rare endemic *Abeliophyllum distichum* Nakai (Oleaceae) in Korea. *Journal of Plant Research* 113, 127–138.

Kato, M., Shibata, A., Yasui, T. and Nagamasu, H. (1999) Impact of introduced honeybees, *Apis mellifera*, upon native bee communities in the Bonin (Ogasawara) Islands. *Researches on Population Ecology* 41, 217–228.

Kearns, C.A. (2001) North American, dipteran pollinators: assessing their value and conservation status. *Conservation Ecology* 5(1), 5.

Kearns, C.A., Inouye, D.W. and Waser, N.M. (1998) Endangered mutualisms: the conservation of plant–pollinator interactions. *Annual Review of Ecology and Systematics* 29, 83–112.

Knight, T.M., McCoy, M.W., Chase, J.M., McCoy, K.A. and Holt, R.D. (2005) Trophic cascades across ecosystems. *Nature* 437, 880–883.

Kremen, C., Williams, N.M. and Thorp, R.W. (2002) Crop pollination from native bees at risk from agricultural intensification. *Proceedings of the National Academy of Sciences of the United States of America* 99, 16812–16816.

Lambeck, R.J. (1997) Focal species: a multi-species umbrella for nature conservation. *Conservation Biology* 11, 849–856.

Larsen, T.H., Williams, N.M. and Kremen, C. (2005) Extinction order and altered community structure rapidly disrupt ecosystem functioning. *Ecology Letters* 8, 538–547.

Lewis, O.T., Memmott, J., LaSalle, J., Lyal, C.H.C., Whitefoord, C. and Godfray, H.C.J. (2002) Structure of a diverse tropical forest insect-parasitoid community. *Journal of Animal Ecology* 71, 855–873.

Louda, S.M., Kendall, D., Connor, J. and Simberloff, D. (1997) Ecological effects of an insect introduced for the biological control of weeds. *Science* 277, 1088–1090.

Lundberg, P., Ranta, E. and Kaitala, V. (2000) Species loss leads to community closure. *Ecology Letters* 3, 465–468.

Lundberg, S. and Ingvarsson, P.K. (1998) Population dynamics of resource limited plants and their pollinators. *Theoretical Population Biology* 54, 44–49.

Mack, M.C. and D'Antonio, C.M. (1998) Impacts of biological invasions on disturbance regimes. *Trends in Ecology and Evolution* 13, 195–198.

Mawdsley, N.A. and Stork, N.E. (1995) *Species Extinctions in Insects: Ecological and Biogeographic Considerations*. Academic Press, London.

Mayse, M.A. and Price, P.W. (1978) Seasonal development of soybean arthropod communities in central Illinois. *Agro-Ecosystems* 4, 387–405.

McDade, L.A. and Kinsman, S. (1980) The impact of floral parasitism in two Neotropical hummingbird-pollinated plant species. *Evolution* 34, 944–958.

Memmott, J. (1999) The structure of a plant–pollinator food web. *Ecology Letters* 2, 276–280.

Memmott, J. and Godfray, H.C.J. (1993) Parasitoid webs. In: LaSalle, J. and Gauld, I.D. (eds) *Hymenoptera and Biodiversity*. CAB International, Wallingford, UK, pp. 217–234.

Memmott, J. and Godfray, H.C.J. (1994) The use and construction of parasitoid webs. In: Hawkins, B.A. and Sheehan, W. (eds) *Parasitoid Community Ecology*. Oxford University Press, Oxford, pp. 300–318.

Memmott, J. and Waser, N.M. (2002) Integration of alien plants into a native flower–pollinator visitation web. *Proceedings of the Royal Society of London Series B – Biological Sciences* 269, 2395–2399.

Memmott, J., Godfray, H.C.J. and Bolton, B. (1993) Predation and parasitism in a tropical herbivore community. *Ecological Entomology* 18, 348–352.

Memmott, J., Godfray, H.C.J. and Gauld, I.D. (1994) The structure of a tropical host parasitoid community. *Journal of Animal Ecology* 63, 521–540.

Memmott, J., Martinez, N.D. and Cohen, J.E. (2000) Predators, parasitoids and pathogens: species richness, trophic generality and body sizes in a natural food web. *Journal of Animal Ecology* 69, 1–15.

Memmott, J., Waser, N.M. and Price, M.V. (2004) Tolerance of pollination networks to species extinctions. *Proceedings of the Royal Society of London Series B – Biological Sciences* 271, 2605–2611.

Memmott, J., Alonso, D., Berlow, E., Dobson, A.P., Dunne, J.A., Sole, R. and Weitz, J.S. (2006) Biodiversity loss and ecological network structure. In: *Ecological Networks: Linking Structure to Dynamics in Food Webs*. A Volume in the Santa Fe Institute Studies in the Sciences of Complexity. Oxford University Press, Oxford.

Montalvo, A.M., Williams, S.L., Rice, K.J., Buchmann, S.L., Cory, C., Handel, S.N., Nabhan, G.P., Primack, R. and Robichaux, R.H. (1997) Restoration biology: a population biology perspective. *Restoration Ecology* 5, 277–290.

Moragues, E. and Traveset, A. (2005) Effect of *Carpobrotus* spp. on the pollination success of native plant species of the Balearic Islands. *Biological Conservation* 122, 611–619.

Morris, R.J., Lewis, O.T. and Godfray, H.C.J. (2004) Experimental evidence for apparent competition in a tropical forest food web. *Nature* 428, 310–313.

Muller, C.B., Adriaanse, I.C.T., Belshaw, R. and Godfray, H.C.J. (1999) The structure of an aphid-parasitoid community. *Journal of Animal Ecology* 68, 346–370.

Munro, V.M.W. and Henderson, I.M. (2002) Nontarget effect of entomophagous biocontrol: shared parasitism between native Lepidopteran parasitoids and the biocontrol agent *Trigonospila brevifacies* (Diptera: Tachinidae) in forest habitats. *Environmental Entomology* 31, 388–396.

Nabhan, G.P. and Buchmann, S.L. (1997) Services provided by pollinators. In: Daily, G.E. (ed.) *Nature's Services – Societal Dependence on Natural Ecosystems*. Island Press, Washington, DC, pp. 133–150.

Olesen, J.M., Eskildsen, L.I. and Venkatasamy, S. (2002) Invasion of pollination networks on oceanic islands: importance of invader complexes and endemic super generalists. *Diversity and Distributions* 8, 181–192.

Palmer, M.A., Ambrose, R.F. and Poff, N.L. (1997) Ecological theory and community restoration ecology. *Restoration Ecology* 5, 291–300.

Parker, I.M. and Haubensak, K.A. (2002) Comparative pollinator limitation of two non-native shrubs: do mutualisms influence invasions? *Oecologia* 130, 250–258.

Pemberton, R.W. and Strong, D.R. (2000) Safety data crucial for biological control insect agents. *Science* 290, 1896–1897.

Percival, M.S. (1974) Floral ecology of coastal scrub in southeast Jamaica. *Biotropica* 6, 104–129.

Peterson, A., Bartish, I.V. and Peterson, J. (2002) Genetic structure detected in a small population of the endangered plant *Anthericum liliago* (Anthericaceae) by RAPD analysis. *Ecography* 25, 677–684.

Potts, S.G., Dafni, A. and Ne'eman, G. (2001) Pollination of a core flowering shrub species in Mediterranean phrygana: variation in pollinator diversity, abundance and effectiveness in response to fire. *Oikos* 92, 71–80.

Proctor, M., Yeo, P. and Lack, A. (1996) *The Natural History of Pollination*. HarperCollins, London.

Richardson, D.M., Allsopp, N., D'Antonio, C.M., Milton, S.J. and Rejmanek, M. (2000) Plant invasions: the role of mutualisms. *Biological Reviews of the Cambridge Philosophical Society* 75, 65–93.

Rodriguez-Perez, J. (2005) Breeding system, flower visitors and seedling survival of two endangered species of *Helianthemum* (Cistaceae). *Annals of Botany* 95, 1229–1236.

Rosenheim, J.A., Limburg, D.D. and Colfer, R.G. (1999) Impact of generalist predators on a biological control agent, *Chrysoperla carnea*: direct observations. *Ecological Applications* 9, 409–417.

Rott, A.S. and Godfray, H.C.J. (2000) The structure of a leafminer–parasitoid community. *Journal of Animal Ecology* 69, 274–289.

Roubik, D.W. (1996) African honey bees as exotic pollinators in French Guiana. In: Matheson, A., Buchmann, S., O'Toole, C., Westrich, P. and Williams, I.H. (eds) *The Conservation of Bees*. Academic Press, London, pp. 173–182.

Schmitz, D.C. and Simberloff, D. (1997) Biological invasions: a growing threat. *Issues in Science and Technology* 13, 33–40.

Schoenly, K., Cohen, J.E., Heong, K.L., Litsinger, J.A., Aquino, G.B., Barrion, A.T. and Arida, G. (1996) Food web dynamics of irrigated rice fields at five elevations in Luzon, Philippines. *Bulletin of Entomological Research* 86, 451–466.

Schönrogge, K. and Crawley, M.J. (2000) Quantitative webs as a means of assessing the impact of alien insects. *Journal of Animal Ecology* 69, 841–868.

Sheppard, S., Henneman, M., Memmott, J. and Symondson, W.O.C. (2004) Infiltration by alien predators into invertebrate food-webs in Hawaii: a molecular approach. *Molecular Ecology* 13, 2077–2088.

Simberloff, D. (1990) Reconstructing the ambiguous: can island ecosystems be restored? In: Towns, D.R., Daugherty, C.H. and Atkinson, I.A.E. (eds) *Ecological Restoration of New Zealand Islands*. Conservation Sciences Publication No. 2, Department of Conservation, Wellington, New Zealand, pp. 37–51.

Simberloff, D. (2004) Community ecology: is it time to move on? *American Naturalist* 163, 787–799.

Simberloff, D. and Stiling, P. (1996) How risky is biological control? *Ecology* 77, 1965–1974.

Sole, R.V. and Montoya, J.M. (2001) Complexity and fragility in ecological networks. *Proceedings of the Royal Society of London Series B – Biological Sciences* 268, 2039–2045.

Strong, D.R. (1997) Ecology: fear no weevil? *Science* 277, 1058–1059.

Strong, D.R., Lawton, J.H. and Southwood, T.R.E. (1984) *Insects on Plants*. Blackwell Scientific Publications, Oxford.

Symondson, W.O.C., Glen, D.M., Wiltshire, C.W., Langdon, C.J. and Liddell, J.E. (1996) Effects of cultivation techniques and methods of straw disposal on predation by *Pterostichus melanarius* (Coleoptera: Carabidae) upon slugs (Gastropoda: Pulmonata) in an arable field. *Journal of Applied Ecology* 33, 741–753.

Thies, C. and Tscharntke, T. (1999) Landscape structure and biological control in agroecosystems. *Science* 285, 893–895.

Thies, C., Roschewitz, I. and Tscharntke, T. (2005) The landscape context of cereal aphid–parasitoid interactions. *Proceedings of the Royal Society of London Series B – Biological Sciences* 272, 203–210.

Thomas, M.B. and Willis, A.J. (1998) Biocontrol: risky but necessary? *Trends in Ecology and Evolution* 13, 325–329.

Thomson, J.D. and Thomson, B.A. (1992) Pollen presentation and viability schedules in animal-pollinated plants: consequences for reproductive success. In: Wyatt, R. (ed.) *Ecology and Evolution of Plant Reproduction*. Chapman & Hall, London, pp. 1–24.

Tscharntke, T. and Brandl, R. (2004) Plant–insect interactions in fragmented landscapes. *Annual Review of Entomology* 49, 405–430.

UN (2003) World Population Prospects: The 2004 Revision. Department of Economic and Social Affairs – Population Division. Available at: http://www.un.org/esa/population/unpop

Valladares, G.R., Salvo, A. and Godfray, H.C.J. (2001) Quantitative food webs of dipteran leafminers and their parasitoids in Argentina. *Ecological Research* 16, 925–939.

Vazquez, D.P. and Simberloff, D. (2002) Ecological specialization and susceptibility to disturbance: conjectures and refutations. *American Naturalist* 159, 606–623.

Walker, B.H. (1992) Biodiversity and ecological redundancy. *Conservation Biology* 6, 18–23.

Wardle, D.A., Verhoef, H.A. and Clarholm, M. (1998) Trophic relationships in the soil microfood-web: predicting the responses to a changing global environment. *Global Change Biology* 4, 713–727.

Waser, N.M., Chittka, L., Price, M.V., Williams, N.M. and Ollerton, J. (1996) Generalization in pollination systems, and why it matters. *Ecology* 77, 1043–1060.

Wilson, E.O. (1987) The little things that run the world (the importance and conservation of invertebrates). *Conservation Biology* 1, 344–346.

Zamora, R. (2000) Functional equivalence in plant–animal interactions: ecological and evolutionary consequences. *Oikos* 88, 442–447.

11 Insects and Climate Change: Processes, Patterns and Implications for Conservation

Robert J. Wilson,[1] Zoe G. Davies[2] and Chris D. Thomas[3]

[1]*Área de Biodiversidad y Conservación, Escuela Superior de Ciencias Experimentales y Tecnología, Universidad Rey Juan Carlos, Tulipán s/n, Móstoles, E-28933 Madrid, Spain;* [2]*Biodiversity and Macroecology Group (BIOME), Department of Animal and Plant Sciences, University of Sheffield, Sheffield S10 2TN, UK;* [3]*Department of Biology (Area 18), University of York, PO Box 373, York YO10 5YW, UK*

1 Introduction

The effects of climate change on biodiversity represent one of the most pressing challenges for conservationists in the 21st century. Although the great diversity of life has evolved and survived alongside continual climatic variation, the ability of biodiversity to respond to contemporary climate change is much more of an unknown, given the potentially unprecedented rate and magnitude of projected increases in the earth's surface temperature (IPCC, 2001; Root and Schneider, 2002; King, 2005; Lovejoy and Hannah, 2005). In the distant past, at least some of the comparable increases in temperature probably triggered mass extinction events (Hallam and Wignall, 1997; Benton and Twitchett, 2003). Coupled with the fact that many species are now restricted to very small areas of occupancy because of direct habitat loss and fragmentation caused by human activity (Vitousek *et al.*, 1997; Sanderson *et al.*, 2002; Gaston *et al.*, 2003), the stresses imposed by climate change on habitats, life histories and interactions between species may be such that widespread extinctions are inevitable unless climate change can be arrested or effective conservation measures can be adopted (C.D. Thomas *et al.*, 2004).

Recent reviews and meta-analyses show that a wide variety of ecological systems and taxa are already changing in ways consistent with climate change (Hughes, 2000; Walther *et al.*, 2002; Parmesan and Yohe, 2003; Root *et al.*, 2003), and many of the examples have been drawn from research conducted on insects. In this chapter we show how insect biodiversity is affected and potentially threatened, and the importance of insects as model systems for biological responses to climate change and associated conservation measures. We first present evidence of recent responses to climate change, concentrating on

 Insect Conservation Biology (eds A.J.A. Stewart, T.R. New and O.T. Lewis)

insect examples. We then examine evidence for the mechanisms behind those responses, before using an understanding of these mechanisms to address the likely future effects of climate change on insects, and the conservation actions that will be required to minimize negative effects on biodiversity. Global modelling of biogeographic responses to climate change suggests that there will be sweeping changes to local ecosystems and communities (Sala *et al.*, 2000; Peterson *et al.*, 2002), confirmed by palaeological and recent evidence that show the individualistic responses of species distributions to climate change (Thomas *et al.*, 2001; Coope, 2004). In this chapter, we particularly ask how the life histories of individual insect species influence their vulnerability, and propose adaptive strategies to identify susceptible species and manage for their well-being.

2 Recent Responses to Climate Change

Biological systems respond to a wide range of environmental drivers, of which climate change is only one. Current declines in the global distributions, population sizes and genetic diversity of species are associated with anthropogenic processes such as habitat loss and fragmentation, pollution, overexploitation of natural resources and the spread of invasive alien species (Sala *et al.*, 2000; J.A. Thomas *et al.*, 2004; Balmford and Bond, 2005). Given the effects of these alternative and interacting factors, the sensitivity of climate change as a political issue and positive bias towards the publication of significant results, an onus has been placed on scientists to identify an unequivocal role of climate change in driving biological changes. Meta-analyses of studies conducted for a wide variety of taxa and geographical regions have shown convincing evidence that biological systems are already changing in ways consistent with, and only satisfactorily explained by, climate change (Parmesan and Yohe, 2003; Root *et al.*, 2003). The two best-documented climate-related biological changes are shifts in species distributions and changes in phenology, with species shifting their ranges to higher latitudes and elevations, and life cycles beginning earlier in spring and continuing later in autumn associated with increasing temperatures (Hughes, 2000; Walther *et al.*, 2002). We now consider the evidence for these changes that has been provided by studies on insects.

2.1 Shifts in species distributions

The geographical ranges of most species have upper and lower latitudinal limits, and often have lower and upper elevational limits within particular regions (MacArthur, 1972; Gaston, 2003). These boundaries to geographic ranges are often set by regional climates that determine both the average availability of temperature, water and suitable conditions for growth and reproduction, and the most extreme conditions to which species and their essential biotic resources are exposed. As small, ectothermic organisms, insects are particularly sensitive to fluctuations in local temperature or moisture levels and, as a result, their distributions and habitat use are often closely related to climate.

For example, the northern range limits of British butterfly species are closely correlated with summer isotherms, reflecting the availability of warm conditions for development and adult activity at upper latitudinal range margins (Thomas, 1993). In addition, at increasing latitudes, butterflies become progressively more restricted to warm microhabitats characterized, for example, by south-facing slopes, short vegetation and bare ground, emphasizing the temperature limitation of species as they approach their 'cool', upper latitudinal margins (Thomas, 1993; Thomas *et al.*, 1998, 1999). There is also strong evidence that summer heat availability sets upper latitudinal limits to the distributions of many species of Hemiptera in the Arctic and northern Europe (Strathdee *et al.*, 1993; Hill and Hodkinson, 1995; Whittaker and Tribe, 1996; Miles *et al.*, 1997; Judd and Hodkinson, 1998; Hodkinson *et al.*, 1999). In contrast, insect distributions may be limited at their lower latitudinal margins by excessive temperatures or inadequate moisture availability, either directly through limits to their physiological tolerance or indirectly through climate effects on larval host plants in the case of herbivorous insects (Bale *et al.*, 2002; Hawkins *et al.*, 2003). Perhaps as a consequence of these two distinct patterns at 'cool' and 'warm' range margins, butterfly species richness in 220 × 220 km grid squares across Europe is closely correlated with actual evapotranspiration, a measure that takes into account both temperature and moisture availability. Species richness is greatest in warm, wet cells in central Europe, and declines both towards cool northern Europe and the hot dry Mediterranean, probably reflecting both declines in plant productivity and direct effects of temperature on insect physiology (Hawkins and Porter, 2003).

Given climatic limitation to species distributions, climate change is expected to shift the locations of suitable climates for species. Therefore, species distributions might expand into regions that become suitable and retract from regions that cease to be so. Recent climate warming is expected to cause range shifts to higher latitudes and elevations. The first documented example of such a range shift was provided by work on Edith's checkerspot butterfly *Euphydryas editha* (Parmesan, 1996, 2005), a non-migratory species that breeds in discrete localities in North America. By the 1990s, populations of *E. editha* had gone extinct from many locations, even though its larval host plants and apparently suitable habitat remained. Rates of local extinction were greatest at low latitudes and at low elevations, such that the average location of populations increased by 92 km northwards and 124 m upwards. In the same 100-year period mean annual isotherms moved 105 km northwards and 105 m upwards, suggesting a climatic link that is supported by the mechanisms involved in local extinctions in this species (Parmesan, 2005). Temperature and precipitation during spring determine: (i) whether *E. editha* adults emerge at a time when conditions are reliable for flight and reproduction (Singer and Thomas, 1996; Thomas *et al.*, 1996); and (ii) whether larvae reach diapause before summer host plant senescence (Weiss *et al.*, 1988). Drier, hotter and more extreme or unpredictable climatic conditions increase extinction risk at low latitudes and elevations (McLaughlin *et al.*, 2002a,b), leading to a northward and upward shift in the average latitudes and elevations of populations.

One of the first multispecies studies of range changes associated with climate change also showed a predominant pattern of poleward shifts in butterfly distributions. Species ranges shifted northwards during the 20th century for 22 (63%) of 35 non-migratory European butterflies that had data for both northern and southern margins (Parmesan *et al.*, 1999). Only two of the species showed southward shifts, and regional climate warming is the most likely explanation for the predominant pattern of colonization at upper latitudinal margins and/or extinction at lower latitudinal margins. For the species whose ranges shifted polewards, 21 (96%) showed northern range margin expansions and only 8 (36%) showed southern margin contractions. A larger sample of species that had data from at least one margin also showed a greater proportion of species with northern margin expansions (34 out of 52 species) than southern margin contractions (10 out of 40 species).

Following Parmesan *et al.* (1999), several studies have documented range expansions by butterflies beyond their former upper latitudinal margins (e.g. Hill *et al.*, 1999b, 2001; Crozier, 2003, 2004a,b). Butterflies have been valuable model systems because of a wealth of historical data about their distributions, and because they depend on thermal conditions throughout their life cycles. Insects vary greatly in their habitat use, thermal physiology and dispersal capacity, but recent research suggests that the upper latitudinal margins of many other insect taxa have also shifted northwards in response to recent climate change (Hickling *et al.*, 2005, 2006; Table 11.1). In this and other contexts, such as phenological change, large-scale and long-term monitoring schemes have provided invaluable evidence for the effects of climate change on a wide range of taxonomic groups (see Conrad *et al.*, Chapter 9, this volume). Experimental studies suggest that the mechanisms involved in range expansions have been similar across insect taxa (Crozier, 2003, 2004a,b; Musolin and Numata, 2003; Karban and Strauss, 2004; Battisti *et al.*, 2005, 2006; Table 11.1).

The relative paucity of evidence for contractions at warm, lower latitudinal margins is no cause for optimism about the fate of species where conditions are deteriorating. Range expansions are easier than contractions to detect because colonizations directly lead to species' presence in regions or large-scale grid cells, whereas local extinctions lead to the gradual decline of species to isolated populations within a region, which may be unlikely to persist in the long term (Wilson *et al.*, 2004). Many species may be suffering declines at their warm margins that go undetected because their regional populations persist but shift to higher elevations. Two studies have shown recent increases in the average elevations of atlas grid cells occupied by butterfly species. In Britain, four butterfly species at the southern margins of their distributions have gone extinct from low-elevation 10 km grid cells and colonized high-elevation cells, leading to a mean increase in elevation of 41 m between pre-1970 and 1999 (Hill *et al.*, 2002). In the Czech Republic, the average altitude of occupied atlas grid cells (~11 × 12 km) increased significantly for 15 butterfly species between 1950 and 2001, with 10 species retracting from low altitudes, 12 expanding at high altitudes and a mean upward shift of 60 m (Konvicka *et al.*, 2003). Actual recent changes in species' elevational ranges may be even greater than recorded in studies based on grid cells, since such cells may include wide altitudinal variation, particularly in mountainous regions. For example, in the study of Czech butterfly distributions, mean elevational range

Table 11.1. Examples of evidence for recent climate-related distributional shifts in insect species.

Evidence for climate-related range shift	Taxa (Location)	References
(a) Multispecies correlational studies in warming climates		
Poleward latitudinal shifts: expansions at upper margins; contractions at lower margins; increase in average latitude.	Butterflies (Europe) Odonata, Orthoptera, Hemiptera, Lepidoptera, Coleoptera (Britain)	Parmesan *et al.* (1999) Hickling *et al.* (2005, 2006)
Upward elevational shifts: colonizations at upper margins; extinctions at lower margins; increase in average altitude.	Butterflies (Britain; Czech Republic; Spain) Odonata, Orthoptera, Hemiptera, Coleoptera (Britain)	Hill *et al.* (2002); Konvicka *et al.* (2003); Wilson *et al.* (2005); Franco *et al.* (2006) Hickling *et al.* (2006)
(b) Mechanistic studies		
Extinctions at low elevations/ latitudes linked to rainfall decline and temperature increase; shift poleward (mean + 92 km) and upward (mean + 124 m).	Butterfly *Euphydryas editha* (Western North America)	Parmesan (1996, 2005) (See also McLaughlin *et al.*, 2002a,b; Thomas, 2005)
Extension of upper latitudinal margin linked to increased overwintering survival at warmer temperatures.	Bug *Nezara viridula* (Japan)	Musolin and Numata (2003)
Extension of upper latitudinal margin linked to warmer temperatures and higher humidity, increasing egg hatch and population size.	Bug *Philaenus spumarius* (California)	Karban and Strauss (2004)
Extension of upper latitudinal margin linked to increased overwintering survival at warmer temperatures; possible role of increased growth rate and voltinism in warmer summers.	Butterfly *Atalopedes campestris* (Pacific Northwest, USA)	Crozier (2003, 2004a,b)
Extension of upper latitudinal margin (+87 km) and upper elevational margin (+110–230 m) linked to: (i) increased winter larval survival; and (ii) increased summer adult dispersal at warmer temperatures.	Moth *Thaumetopoea pityocampa* (France, Italy)	Battisti *et al.* (2005, 2006)
Extension of habitat range linked to microclimate warming, resulting in increased habitat availability and habitat connectivity, permitting range expansion.	Butterfly *Hesperia comma* (Britain)	Thomas *et al.* (2001); Davies *et al.* (2005, 2006)

per cell was 250 m (Konvicka *et al.*, 2003). However, it is difficult to attribute these uphill range shifts solely to the effects of climate change, because habitat degradation is typically more severe at lower elevations.

Sampling discrete locations in different time periods has the potential to detect elevational shifts at a finer resolution and to control the effects of habitat degradation. Research on the elevational associations of butterflies in the Sierra de Guadarrama (a mountain range in central Spain) showed that the lower elevational limits of 16 species that were restricted to high altitudes (i.e. species at their warm range margins) rose on average by 212 m (± SE 60), accompanying a 1.3°C rise (equivalent to 225 m) in regional mean annual temperature between 1967–1973 and 2004 (e.g. Fig. 11.1a and b) (Wilson *et al.*,

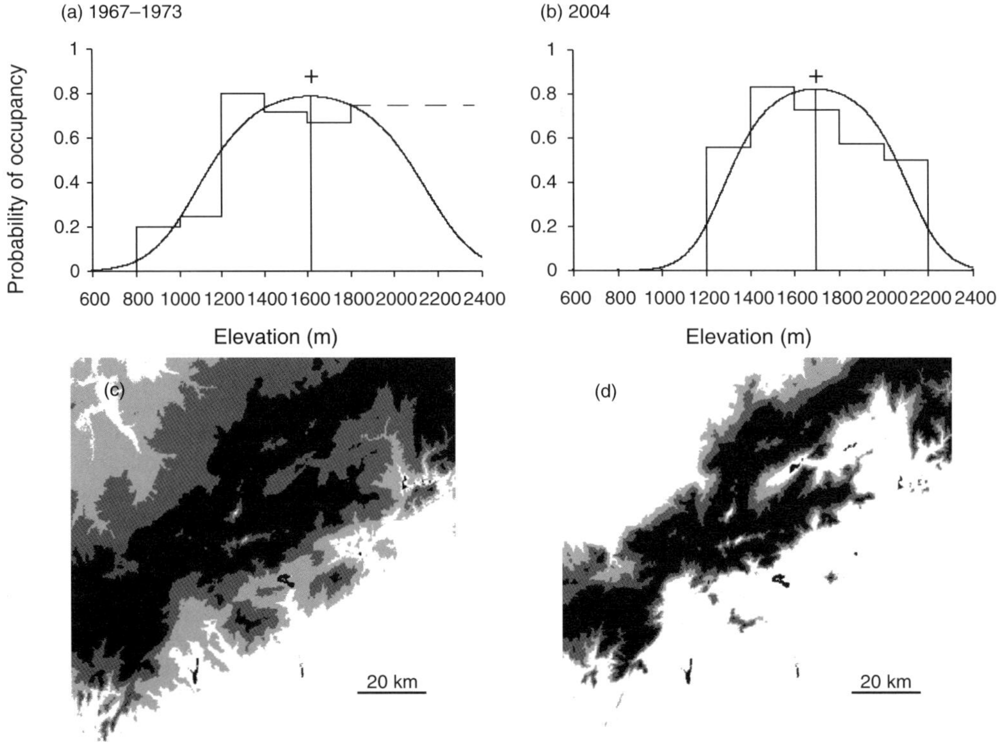

Fig. 11.1. Elevational associations of the butterfly *Lycaena alciphron* in the Sierra de Guadarrama (central Spain) in 1967–1973 and 2004. (a–b) Histograms of probability of occupancy in 200 m intervals (bars), and probability of occupancy modelled using logistic regression (curve) in (a) 1967–1973 and (b) 2004. Crosses show 'optimum' elevations with highest modelled probability of occupancy. In (a), dashed line denotes proportion of all four sites sampled above 1800 m. (c–d) Distributions of suitable elevations based on logistic regression models from (a) and (b), for (c) 1967–1973 and (d) 2004. Black: ≥50% probability of occupancy; dark grey: ≥20%; pale grey: ≥10%; white: <10%. For *L. alciphron*, optimum elevation increased from 1615 to 1694 m; lower elevational limit increased from 920 to 1320 m; and the area of suitable habitat (≥50% probability of occupancy) decreased by 38% between the surveys. (See Wilson *et al.*, 2005.)

2005). The close correlation between temperature increase and changes in lower elevational limits, coupled with the fact that the larval host plants of the study species were widespread in the region (and that widespread butterflies which used the same larval host plants showed no elevational range shifts), implied that climate rather than direct habitat change was the most important driver in the system. For these species, increases in upper elevational limits were non-significant between the two surveys, probably because many species already occupied high altitudes in the region during the first time period. As a result, there were overall reductions in the elevational ranges of the species and an average decline of 22% in the 'climatically suitable' area for each species over only 30 years (e.g. Fig. 11.1c and d). These rapid declines in distribution size show how elevational shifts at lower latitudinal range margins can mask range contractions, constraining species distributions to progressively smaller areas until they may face regional extinction.

Relatively fine resolution (1 × 1 km) surveys in 2004/05 of the four northern/montane butterfly species in Britain have also detected higher levels of retreat since 1970 (Franco *et al.*, 2006). Of the four species, *Erebia epiphron* retreated uphill by 130–150 m and showed no effects of habitat loss on its distribution; *E. aethiops* and *Aricia artaxerxes* retreated northwards by 70–100 km and showed combined impacts of climate change and habitat loss; and *Coenonympha tullia* declined through habitat loss, but showed no latitudinal or elevational shift. Averaged across the four species, it appears that climate change and habitat decline have been equally responsible for local extinctions near their range margins (Franco *et al.*, 2006).

2.2 Shifts in phenology

In addition to the shift in space of species distributions, recent climate change has led to an ecological shift in time, with changes to the seasonality of species' life cycles (phenology). Phenological studies have predominantly shown species becoming active, migrating or reproducing earlier in the year, associated with increases in temperatures that lead directly to increased growth rates or earlier emergence from winter inactivity (Menzel and Fabian, 1999; Roy and Sparks, 2000; Fitter and Fitter, 2002; Peñuelas *et al.*, 2002; Sparks and Menzel, 2002). Recent reviews of such studies show mean advances in the timing of spring events by 2.3–5.1 days per decade (Parmesan and Yohe, 2003; Root *et al.*, 2003), depending on the type of analysis and range of examples included. Increasing temperatures have also allowed a number of species to remain active for a longer period during the year (Sparks and Menzel, 2002) or to increase their annual number of generations (Roy and Sparks, 2000).

Long-term data from several insect-recording schemes in Europe and North America have provided evidence for advancement in appearance dates of adult insects as annual temperatures have increased (Table 11.2). In Britain, the annual first appearance dates from 1976 to 1998 for 28 out of 33 butterfly species were negatively related to temperature for at least 1 month of the year (i.e. earlier appearance at higher temperatures), and an increase in temperature of 1°C led to an average advance in first flight date of 4.5 days (Roy and Sparks, 2000). Conditions during early spring seem to be particularly important, with 22 species appearing significantly earlier associated with high February temperatures. The appearance dates of 11 species became significantly earlier in more recent years, even when taking account of monthly temperatures, suggesting either a progressive effect of some additional climatic or host plant effect or an evolutionary change. First appearance by butterflies has also advanced in California (North America) and Catalonia (Spain) associated with higher temperatures and lower rainfall in winter or spring (Forister and Shapiro, 2003; Stefanescu *et al.*, 2003). There is a similar negative relationship between temperature and insect appearance dates in Austria, with three butterfly species, the bee *Apis mellifera* and the cockchafer *Melolonthus melolonthus* showing 3- to 5-day advances associated with 1°C warmer February–April temperatures (Scheifinger *et al.*, 2005). However, in this case there was no temporal trend for earlier emergence, perhaps because population sizes of the species declined over time, leading to later first observations.

Mean flight dates (the estimated date of peak abundance during the adult flight period) for the first annual generations of species have advanced in conjunction with advances in first appearance date. For example, the peak of the

Table 11.2. Changes in annual appearance dates of insects associated with climate change.

Taxon	Location	Time period	Temperature increase	Change in appearance date	References
Butterflies	Britain	1976–1998	1.5°C (Feb–Apr mean); 1°C (May–July mean)	Advance, 26/35 spp. (13 significant, mean 8 days per decade)	Roy and Sparks (2000)
Butterfly (*Pieris rapae*)	NE Spain	1952–2000	1.4°C (annual mean)	Advance, 11.4 days	Peñuelas *et al.* (2002)
Butterflies	NE Spain	1988–2002	1–1.5°C (Feb, Mar, June mean)	Advance, 17/17 spp. (5 significant, mean 4.1 weeks)	Stefanescu *et al.* (2003)
Butterflies	California	1972–2002	1.2°C (annual daily max.)	Advance, 16/23 spp. (4 significant, mean 24 days)	Forister and Shapiro (2003)
Bee (*Apis mellifera*); Butterflies (*Aglais urticae*, *P. rapae*, *Gonepteryx rhamni*)	Austria	1951–1998	1.3°C (Feb–Apr mean)	Delay, 3–7 days	Scheifinger *et al.* (2005)

first generation of 104 common microlepidopteran species in the Netherlands advanced on average by 11.6 days between 1975 and 1994, accompanying a 0.9°C increase in annual mean temperature (Ellis *et al.*, 1997; Kuchlein and Ellis, 1997). In Catalonian butterflies, changes in mean flight date advanced between 1988 and 2002 for 16 out of 18 species, with an average advance of 2.5 weeks for the 8 species with significant relationships (Stefanescu *et al.*, 2003). In British butterflies, mean flight dates did not advance as much as first appearance dates, perhaps partly because mean flight date is affected by the number of generations that species have each year (Roy and Sparks, 2000). In univoltine species, mean flight date is closely correlated with first appearance date, but multivoltine species may increase their number of generations following early first emergence. For British butterflies, a trend in earlier first appearance was accompanied by a longer annual flight period in 24 species (overall average + 3 days per decade; $n = 35$ species), but this increase was particularly pronounced in several multivoltine species that were able to increase their number of generations in some parts of their range. For example, increases in average flight period of 8.9 days per decade for speckled wood *Pararge aegeria* and 13.1 days for comma *Polygonia c-album* reflect increased numbers of generations at higher latitudes.

Changes in insect phenology with year-to-year changes in temperature are mirrored by geographical relationships between phenology and regional temperature. For example, mean peak flight date for microlepidoptera is 5.1 days later in the north than in the south of the Netherlands, reflecting a 0.9°C difference in mean annual temperature (Ellis *et al.*, 1997). In Britain, 10 out of 29 butterfly species analysed had significantly earlier mean flight dates at warmer lower latitudes, with an average advance of 2.4 days/100 km moved south for these 10 species (equivalent to 6.0 days/1°C) (Roy and Asher, 2003). Insect emergence date also becomes delayed at higher elevations in mountainous regions (e.g. Hill and Hodkinson, 1995; Gutiérrez and Menéndez, 1998; Bird and Hodkinson, 1999; Fielding *et al.*, 1999), potentially restricting species to shorter periods of adult activity (Gutiérrez and Menéndez, 1998). In some insects these delays in activity can be avoided by local adaptations at cooler locations, for example in habitat selection for particularly warm microclimates (Thomas, 1993), or in faster growth rates and smaller adult sizes (Nylin and Svard, 1991; Ayres and Scriber, 1994).

3 Mechanisms behind Climate-related Shifts in Distributions and Phenology

3.1 Climate and population size

In addition to showing how increasing temperatures lead to advances in phenology, long-term butterfly monitoring data have shown the relationships between population sizes and weather conditions (e.g. Pollard, 1988; Roy *et al.*, 2001). These studies show that the annual population sizes of the vast majority of British butterflies are positively related to warm dry conditions during the spring and summer of flight, and warm wet conditions during the preceding year. However, the precise relationship depends on the life history of the species concerned. For example, the population sizes of several bivoltine species are most strongly associated with high temperatures in the current spring or summer, providing suitable conditions for larval and pupal development and adult activity. In contrast, hot or dry conditions in the previous year are associated with population declines in species such as ringlet *Aphantopus hyperantus* and speckled wood *P. aegeria*, whose larvae feed on plants growing in moist or partly shaded habitats and may be susceptible to increased drought stress. Similar negative relationships between population size and hot or dry conditions might be expected at the warm margins of species ranges. These year-to-year changes in population size represent the raw material for distributional shifts, with local extinctions occurring where population size declines, and range expansions where population size increases (weather conditions could also affect colonization rate through their effects on dispersal activity, e.g. White and Levin, 1981; Shreeve, 1992). Because of this link, Roy *et al.* (2001) were able to use models relating population size to weather conditions in 1976–1991 to predict historical changes in the abundance of three species in Britain over two centuries, based on historical meteorological and entomological records.

Detailed information on fluctuations in insect population distribution and abundance in Britain has also been provided by the Rothamsted Insect Survey (RIS) (see Conrad *et al.*, Chapter 9, this volume). Data from 406 light traps show the pronounced changes in populations of the garden tiger moth *Arctia caja* that have accompanied recent climate change (Conrad *et al.*, 2001, 2002, 2003). Population size of *A. caja* decreases in years with high rainfall or temperature in winter and early spring, and in spans of years with high index values for the East Atlantic (EA) teleconnection pattern, an atmospheric circulation system that affects winter weather in western Europe (Conrad *et al.*, 2003). Increasing winter temperature, rainfall and EA index values between 1968 and 1998 led to declines in *A. caja* local population density and distribution size, and a shift in its centres of distribution and abundance towards cooler, higher latitudes (Conrad *et al.*, 2002). A time lag in the response of species distribution to climate change was observed, with mean local population density falling abruptly between 1983 and 1984, and the proportion of occupied locations declining markedly between 1987 and 1988 (Conrad *et al.*, 2001). Increased *A. caja* mortality in warm, wet winters appears to be the cause of its distribution-level decline, but variation in weather systems like the EA teleconnection pattern, the El Niño Southern Oscillation and the North Atlantic Oscillation could lead to changes in insect population dynamics in a variety of ways (Holmgren *et al.*, 2001; Ottersen *et al.*, 2001). For example, increased rainfall on arid islands in the Gulf of California associated with the 1992–1993 El Niño event led to greatly increased plant productivity, a doubling of insect abundance relative to 1991 and a shift from an insect community composed largely of scavengers and detritivores to one dominated by herbivores (Polis *et al.*, 1997).

3.2 Direct effects of climate on growth, survival and fecundity

Temperature is the climatic variable for which there is most evidence of direct effects on insect life history (Bale *et al.*, 2002). The temperatures experienced by particular life stages of insects can have important effects on their growth, development, survival and fecundity. Whether climatic changes have a positive or negative effect on population sizes depends on whether the changes take insect life stages nearer to, or further from, the limits of their tolerance, and whether they increase or decrease the synchrony in space or time of insects with interacting species such as host organisms, competitors and natural enemies.

At temperate latitudes, where most insects grow or are active only during warm parts of the year, increasing temperatures often lead to an earlier breaking of winter diapause (Buse and Good, 1996; Miles *et al.*, 1997; Masters *et al.*, 1998; Fielding *et al.*, 1999), although in many species this process is at least partly under photoperiodic control (Hill and Hodkinson, 1996; Bradshaw and Holzapfel, 2001). Faster and earlier growth may allow multivoltine temperate insects with permanently available food supplies to increase population size by increasing their annual number of generations (Roy and Sparks, 2000). However, species that use only periodically available resources may not be able to increase activity periods or population sizes if there is no change in the temporal availability of food. Subtle differences between the cues involved in phenology at different trophic levels could lead to asynchrony between the emergence of larvae and the availability of their food. For herbivorous insects that feed on plant tissues whose palatability or nutrient richness changes over time, synchrony of larval emergence with plant growth can be critical (Feeny, 1970; Hill and Hodkinson, 1995; Hill *et al.*, 1998; Bale *et al.*, 2002; Hodkinson, 2005). Tree life cycles have only advanced by an average of 3 days accompanying recent climate change, compared with 5 days per decade for invertebrate life cycles (Root *et al.*, 2003), showing the potential for mismatches in the phenology of insects and arboreal host plants. For example, recent increases in mean winter temperature without an accompanying decline in the number of frost days have reduced synchrony between egg hatching by the winter moth *Operophtera brumata* and budburst by its host *Quercus robur* (Visser and Hollemann, 2001).

Changes in growth rate can also affect the level of synchrony between specialist insect parasitoids and their hosts. Synchrony between the Glanville fritillary butterfly *Melitaea cinxia* and its parasitoid *Cotesia melitaearum* decreases in cool years, because dark-coloured *M. cinxia* larvae increase their development rate by basking, whereas white, immobile *C. melitaearum* cocoons develop slowly in shaded microclimates. As a result, *M. cinxia* larvae pupate before adult parasitoid emergence and egg laying in cool years, reducing *C. melitaearum* population size, increasing its risk of local extinction and reducing its colonization rate (Van Nouhuys and Lei, 2004). Phenological change by insects can in turn affect levels of predation by other taxa at higher trophic levels: for example, great tits *Parus major* in the Netherlands have not advanced their egg-laying date to keep pace with changes to the temporal availability of caterpillars, their major food source (Visser *et al.*, 1998).

Relative synchrony in the growth of insects and their larval host plants can also determine the limits to species' geographic ranges, if at high temperatures host plants grow too quickly for insect exploitation of palatable tissues, or at low temperatures plants grow too slowly for insect development (Maclean, 1983; Bale *et al.*, 2002). At the local scale, host plants senesce before larvae of Edith's checkerspot butterfly *E. editha* reach summer diapause on south-facing slopes in hot years, whereas larval development is too slow on north-facing slopes during cool years (Weiss *et al.*, 1988). At the scale of species ranges, willow psyllids *Cacopsylla* spp. have a narrower range of larval host plant species and exploit a narrower range of plant tissues at the extremes of their distributional range because of the constraints of maintaining synchrony between larval and host development (Hodkinson, 1997; Hill *et al.*, 1998).

In addition to affecting the rate of growth, temperature directly influences mortality, with reduced survivorship towards both lower and upper thermal tolerances (Ratte, 1985). For example, the proportion of individuals developing to adulthood in the peacock *Inachis io* and comma *P. c-album* butterflies was >60% at temperatures of 15–30°C, but at temperatures of 9°C and 34°C, respectively 0% and 20–40% of individuals reached maturity (Bryant *et al.*, 1997). The upper latitudinal range margins of these species correspond to the 15°C July isotherm perhaps as a consequence of their requirements for sufficiently warm temperatures for summer larval survival and development.

Changes to ambient temperature, moisture availability and atmospheric CO_2 can have important effects on insect growth and larval host plant quality. Elevated CO_2 concentrations lead to reduced nitrogen levels and increased C/N ratios in leaves, and hence reduced insect performance (growth rate, weight gain and survival) (Coviella and Trumble, 1999; Zvereva and Kozlov, 2006). Most experimental studies show a positive effect of temperature on insect herbivore performance, such that there is no significant change in performance when CO_2 and temperature are increased together (for a review, see Zvereva and Kozlov, 2006). However, warming does not always mitigate the negative effects of elevated CO_2. For example, increased CO_2 levels do not affect survivorship in the leaf miner *Dialectica scalariella* at low ambient temperatures, because larvae feed for longer to compensate for reduced food quality. But at elevated temperatures development is accelerated and the short-time feeding on poor-quality food reduces survivorship and adult weight (Johns and Hughes, 2002). In a counter-example, increased temperatures at low CO_2 levels cause wilting and premature leaf loss in *Lantana camara*, reducing survivorship of the chrysomelid beetles *Octotoma championi* and *O. scabripennis*. At high temperatures, survival of the beetles is favoured by elevated CO_2 levels, because reduced water stress delays leaf loss (Johns *et al.*, 2003). Field-scale climate manipulation experiments show that interactions between the weather and plant biochemistry can exert marked effects on insect population dynamics. For example, the abundance of Auchenorrhyncha (Hemiptera) increased with summer rainfall and vegetation cover, but showed no decrease under drought conditions, even though vegetation cover became sparser, probably because drought-stressed foliage had a higher nutritional quality (Masters *et al.*, 1998). Food web models suggest that climate-induced changes to plant productivity and host plant quality could result in smaller and more variable herbivore population sizes, leading to weaker interactions between trophic levels (Emmerson *et al.*, 2004). For aphids, whose physiology has been studied in detail, the population dynamic effects of climate change have been modelled, taking account of climatic variables, CO_2 levels and interacting species (Hoover and Newman, 2004; Newman, 2004, 2005). These models predict that, under realistic CO_2 emission scenarios, changes to temperature and rainfall are the most important drivers of aphid population dynamics; but the prediction of the effects of higher emissions scenarios will require the modelling of sometimes complex interactions among variables (Newman, 2005).

For many temperate insects, mortality during the overwintering period may have important effects on population dynamics and the geographical limits to species distributions (Virtanen *et al.*, 1998; Bale *et al.*, 2002; Turnock and Fields, 2005). The minimum temperatures that can be experienced by overwintering stages may set the upper latitudinal limits to species ranges, and recent increases in winter temperatures have led to northward range expansions by increasing overwintering survival in insects such as the southern green stink bug *Nezara viridula* (Musolin and Numata, 2003) and the sachem skipper butterfly *Atalopedes campestris* (Crozier, 2003, 2004a,b). In contrast, low temperatures may be beneficial for species that spend winter in an inactive diapause, with reduced metabolic rate in cooler microhabitats associated with increased survival and fecundity in the goldenrod gall fly *Eurosta solidaginis* (Irwin and Lee, 2000, 2003). Survival by overwintering adults of the peacock butterfly *I. io* is affected both by temperature and moisture conditions, with greatly reduced survival at 10°C compared with 2°C, and in wet versus dry conditions (Pullin and Bale, 1989). The location of *I. io*'s southern geographic range margin near the 10°C January isotherm may result from its requirements for cold overwintering conditions (Bryant *et al.*, 1997).

The increasing severity or frequency of extreme climatic events such as droughts or unseasonal storms may be as important for long-term population survival as the effects of average changes to climatic conditions (Easterling *et al.*, 2000; Parmesan *et al.*, 2000). The potential for extreme events to impact on population dynamics is greatest in populations that are highly localized in space, or that breed in homogeneous habitats and are therefore uniformly exposed to the extreme conditions. Cold, wet weather at overwintering aggregations of the monarch butterfly, *Danaus plexippus*, can dramatically increase mortality (Oberhauser and Peterson, 2003). Two Californian populations of the butterfly *Euphydryas editha* went extinct in association with increased variability in precipitation, which reduced the temporal overlap between butterfly larvae and their host plants (McLaughlin *et al.*, 2002a). Of the two populations that were studied, population variation was greater and extinction was faster at the large, flat site than the smaller, more topographically variable site, where the topographic variation acted as a buffer against environmental extremes by increasing the annual period of host plant availability

(McLaughlin *et al.*, 2002b). Extreme events at individual locations are likely to affect species differently depending on their microclimatic associations, physiological tolerances and their position in the geographic range, which may potentially lead to changes in community composition. For example, drought conditions in Britain in 1995 and 1996 led to increases in the population sizes of southerly distributed butterfly species, but decreases in the abundance of carabid beetles that favoured low temperatures and wet soils (Morecroft *et al.*, 2002).

Temperature can also influence fecundity, through its effects on adult insect activity and the availability of suitable microhabitats for egg laying. The silver-spotted skipper butterfly *Hesperia comma* reaches its northern range margin in Britain, where it has been historically restricted to the hottest microclimates, laying its eggs on small tufts (<5 cm) of the larval host plant *Festuca ovina* in chalk grassland in southern England (Thomas *et al.*, 1986). This thermophilic habitat restriction was responsible for a pronounced decline in the British distribution of *H. comma* from the 1950s to the 1980s, when the abandonment of low-intensity livestock grazing and a rabbit decline caused by myxomatosis led to unsuitable tall vegetation across most sites in its former range. By 1982, the species was restricted to less than 70 refuge populations in England, nearly all of them on south-facing grassland with thin soils and a large amount of bare ground (Thomas *et al.*, 1986). However, following a recovery in rabbit populations and conservation grazing management in and around the refuge sites, *H. comma* spread its regional distribution and by 2000 had over 250 populations in England, many of them re-colonizations of formerly occupied localities (Thomas and Jones, 1993; Davies *et al.*, 2005). This range expansion was achieved partly as a result of improving habitat conditions in the refuge populations and surrounding sites, but is also related to warmer climates increasing fecundity. Field observation and experiments show that *H. comma* females lay a larger number of eggs in warmer conditions, and that the microhabitats used for egg laying change depending on ambient temperature: at low ambient temperatures, eggs are laid in particularly warm microhabitats, but at higher temperatures eggs are laid on plants growing in conditions that are no warmer, or even cooler, than ambient conditions (Fig. 11.2a) (Davies *et al.*, 2006). Between 1982 and 2001 (during which time local mean August temperature rose by 2°C), the typical microhabitat used for egg laying by *H. comma* changed (Fig. 11.2b), with the optimum proportion of bare ground declining from 41% to 21%, shown by logistic regression modelling of the probability of egg occurrence based on quadrats performed in the same locations 20 years apart. Most habitat patches in the networks of chalk grassland where *H. comma* occurs in England have a percentage cover of bare ground much closer to the new optimum for egg laying. As a result the species has been able to exploit larger areas of habitat in each grassland patch and colonize some habitat patches that would have been unsuitable under its earlier, more restrictive habitat requirements (including many sites on east, west and even north-facing slopes; Thomas *et al.*, 2001). Thus, climate warming has increased the availability of thermally suitable habitat for *H. comma* at the cool, northern edge of the species range, leading

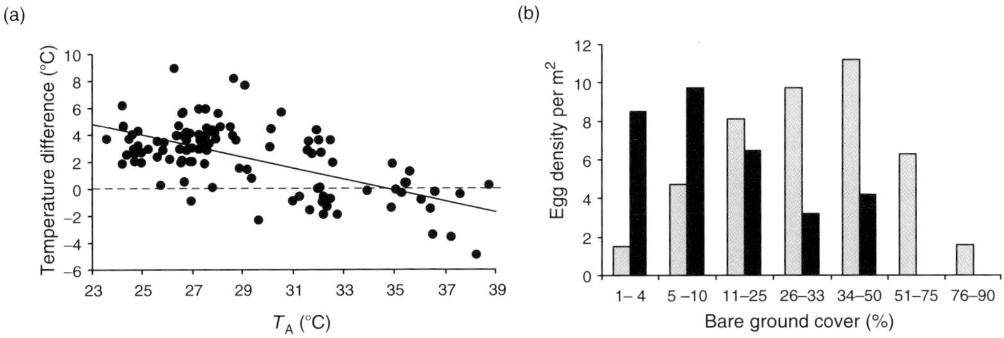

Fig. 11.2. Changing microhabitat choice in the butterfly *Hesperia comma*, associated with warming temperatures. (a) The temperature difference between sites selected for egg laying and ambient temperature (T_A) declined at increasing ambient temperature: Temperature difference = −0.41 (±0.05) × T_A + 14.16 (±1.51); $R^2 = 0.38$, $F_{1,103} = 63.12$, $P < 0.001$. (b) The density of eggs against percentage cover bare ground in 25 × 25 cm quadrats repeated at the same location in 1982 (grey bars) and 2001 (black bars). In 1982, eggs were associated with higher percentage cover bare ground (hotter microclimates) than in 2001. (Reproduced from Davies *et al.*, 2006, with permission from Blackwell Publishing.)

to increases in: (i) egg-laying rate; (ii) the effective area or population carrying capacity of habitat patches; and (iii) the number of habitat patches in the landscape that are available for colonization. Now that *H. comma* lays eggs in a wider variety of microhabitats, its population dynamics are also likely to be buffered against environmental variation: studies on butterflies (Sutcliffe *et al.*, 1997; McLaughlin *et al.*, 2002b) and the bush cricket *Metrioptera bicolor* (Kindvall, 1996) show that habitat heterogeneity can reduce the risk of local population extinction from fluctuating weather conditions (see Section 5).

3.3 Biotic interactions

In our discussion of the direct effects of climate on insects we have already considered some important interactions between climate, insects and their host organisms. Future distributions of insects will be constrained by the future distributions of their specific host species, or by climates in which they are phenologically synchronized with their food supplies (Hodkinson, 1999). Changes in host plant use across an insect species' range (e.g. Hodkinson, 1997) could interact with changing climates, with consequences for rates or patterns of range shifts. For example, recent northward range expansions in Britain by the brown argus *A. agestis* and comma *P. c-album* butterflies appear to have been facilitated by shifts in diet to incorporate increased use of widespread larval host plants (Thomas *et al.*, 2001; B. Braschler and J.K. Hill, 2004, unpublished data).

Interacting competitors, predators, parasitoids and pathogens could also affect the responses of species to climate change. A general effect of warmer temperatures could be increased growth rates, leading to reductions in mortality because of reduced exposure times of larvae to predation or parasitism (Bernays, 1997). The presence of particular interacting species can also influence the relationships between climatic conditions and species population size. For example, the fruit flies *Drosophila melanogaster*, *D. simulans* and *D. subobscura* coexist in the wild in southern Europe. Using laboratory microcosms, Davis *et al.* (1998) showed changes to the distribution and abundance of the three competing flies along a temperature cline of 10–25°C, depending on whether the species were alone or in the presence of their competitors. Population density of *D. subobscura* was reduced at temperatures of 15–20°C by the presence of *D. simulans*, whose population density was reduced at 10–15°C by *D. subobscura*; further addition of *D. melanogaster* caused the disappearance of *D. subobscura* at 25°C, and of *D. simulans* at 10°C. Addition of a parasitoid wasp *Leptopilina boulardi* led to further changes in abundance and distribution, with increases in population density of *D. melanogaster* at 20–25°C and of *D. subobscura* at 10–15°C, because of reductions in competitor density caused by the parasitoid. A 5°C increase in temperature, producing a cline of 15–30°C, led to the disappearance of *D. subobscura* from the 25°C treatment because of immigration by *D. melanogaster* and *D. simulans* from the 30°C cage, where the latter two species had high population density. In the same 15–30°C cline, there was an unexpected increase in *D. subobscura* abundance at 15°C.

The overall effect of species interactions may be to reduce the predictability of ecological responses to climate change, particularly when abundance and distributions are set by a few strong interactions (as opposed to a more diffuse pattern of interactions with many other species). Because species shift their distributions individualistically when climates change (e.g. Coope, 2004), shifting biotic interactions could alter the relationships of species population abundance and distribution with climate (Davis *et al.*, 1998). However, the importance of changing biotic interactions in predicting responses to climate change remains uncertain. Levels of predation or parasitism decline towards the upper latitudinal or elevational margins of species ranges, and if the ecophysiological limitations of species and their natural enemies are known, we might be able to predict the effects of climate change on the future ranges of both (Hodkinson, 1999). The problem with this approach is that species distributions and abundances change at different rates, depending on their dispersal ability and original abundance and distribution sizes, and this may make transient dynamics particularly difficult to predict. Time lags before natural enemies tracked the expanding ranges of herbivorous insects may have led to the increases in insect herbivory indicated by fossil plants following periods of climate warming (Wilf and Labandeira, 1999). The climatic or biotic limiting factors for particular interacting species may also be difficult to predict. For example, after increases in plant productivity associated with heavy rains in desert islands in the Gulf of California, spider densities doubled in 1992 in response to increased insect prey, but then were greatly reduced in 1993 because of parasitism by wasps whose populations increased because of increased nectar and pollen resources (Polis *et al.*, 1998).

3.4 Interactions of climate change with habitat loss and fragmentation

Evidence for 20th-century changes shows that many species have not been able to shift their distributions to track suitable climate space. For example, 46 non-migratory species of butterfly reach their upper latitudinal range margins in Britain, and recent increases in summer temperatures should have increased both local population densities and distribution sizes for these species (Roy *et al.*, 2001). However, between distribution surveys in 1970–1982 and 1995–1999, most butterfly species showed declines both in local population abundance and distribution size (Warren *et al.*, 2001). In particular, the distributions of sedentary, habitat specialists declined (24 out of 26 species), whereas half of the mobile, habitat generalist butterfly species expanded their range (9 out of 18 species). Even for relatively dispersive butterfly species, rates of range expansion into suitable climate space are constrained by the availability of suitable habitat (Hill *et al.*, 1999b, 2001).

One consequence of the differential abilities of species to track changing climates across anthropogenically altered landscapes could be a shift in the composition of ecological communities, away from habitat specialist and sedentary species towards wide-ranging, generalist species (Tilman *et al.*, 1994; Warren *et al.*, 2001; Menéndez *et al.*, 2006). The restructuring of ecological communities could have untold consequences for a wide range of ecological and evolutionary processes, particularly relating to ecosystem functioning and the effects of biotic interactions on species' responses to climate change (see Section 3.3.).

3.5 Interactions between phenological and temperature change

The consequences of phenological advancement have generally been considered in terms of a possible disruption of synchrony with host species, and a possible lengthening of the annual adult activity period (e.g. Roy and Sparks, 2000). A hitherto overlooked effect of phenological advancement is its effect on the temperatures that particular stages of insect life cycles experience. Whilst average annual temperatures have risen, certain stages of species' life cycles may encounter either cooler or warmer conditions as a result of the interaction between phenology shifts and changes in mean temperature. To illustrate this point, we investigated the net change in temperature that might have been experienced based on a combination of temperature changes and recent phenological advancement by adults of two univoltine butterfly species in the UK: the orange tip *Anthocharis cardamines*, which flies in spring, and the silver-spotted skipper *H. comma*, which flies in late summer. Peak flight date for each year between 1985 and 2004 (the longest period that had continuous records for both species from transects of the British Butterfly Monitoring Scheme) was calculated for each transect where at least four individuals were counted for either species. Change in phenology over time was calculated by the linear regression of peak date against year, giving advances of 5.5 days per decade for *A. cardamines* (peak date = −0.55 (± 0.09) *X* year + 1234.07 (± 182.73); $R^2 = 0.08$, $F_{1.432} = 36.66$, $P < 0.001$) and 4.7 days per decade for *H. comma* (peak date = −0.47 (± 0.11) *X* year + 1168.73 (± 223.12); $R^2 = 0.22$, $F_{1.62} = 17.82$, $P < 0.001$). At the nearest meteorological station (Mickleham) to the site where the *H. comma* egg-laying experiments were carried out (see Section 3.2), mean daily air temperature records had increased for given calendar dates between 1985 and 2004: by 1.1°C during the flight period of *A. cardamines* (mean first sighting date 18 April, mean last sighting 1 June) and by 2.0°C in *H. comma*'s flight period (1 August to 4 September). However, the concurrent advance in flight date of 11 days for *A. cardamines* led to its emergence earlier in spring, which would on average be 1.5°C cooler (the 'phenological shift', based on differences between dates during the flight period between 1985 and 2004), so that the temperature experienced by adult *A. cardamines* became cooler by 0.4°C (Fig. 11.3a). In contrast, the 9-day advance in *H. comma*'s flight period towards earlier August led to a 'phenological shift' in temperature of + 0.8°C because adults were now flying at a hotter time of the year. The phenological shift combined with the 2°C increase in August temperatures would lead to a net change of 2.8°C in the temperatures experienced by *H. comma* adults (Fig. 11.3b), potentially having a major effect on flight activity and habitat choice by the butterflies. Because *H. comma* egg-laying rate is positively correlated with temperature in Britain, this is likely to have resulted in a substantial increase in realized fecundity (Davies *et al.*, 2006).

We also estimated the net changes in temperature that would be experienced at different times of the year, based on changes to the Central England temperature between 1961 and 2000 (Manley, 1974; Parker *et al.*, 1992) and an average advance of 2.3 days per decade of phenological events (Parmesan and Yohe, 2003) (Fig. 11.3c). In agreement with the results presented based on local meteorological data and the flight periods of *A. cardamines* and *H. comma*, these results suggest that species or life stages that are active at dates from July until March will have experienced net increases in temperature, because of a synergy between year-to-year warming and phenological advance. In contrast, those parts of species life cycles that are active between April and June may have experienced a net reduction in temperature, even though spring temperatures are increasing. The reduction of net temperatures through late spring/early summer and exaggerated warming during the rest of the year may have repercussions for temperature-dependent activities of individual species. One consequence might be an increased effect of climate change on species whose most climate-sensitive stages are active at times of year experiencing large net changes in temperature. However, temperature will affect all stages of an insect life cycle, including diapause, so it is more satisfactory to consider the effects of net predicted changes on all individual stages. One concern is that there may be mismatches between the habitats selected, for instance, by egg-laying adults and those required by larvae, if microhabitat selection at different times of the year is based on the microclimates experienced (e.g. Roy and Thomas, 2003).

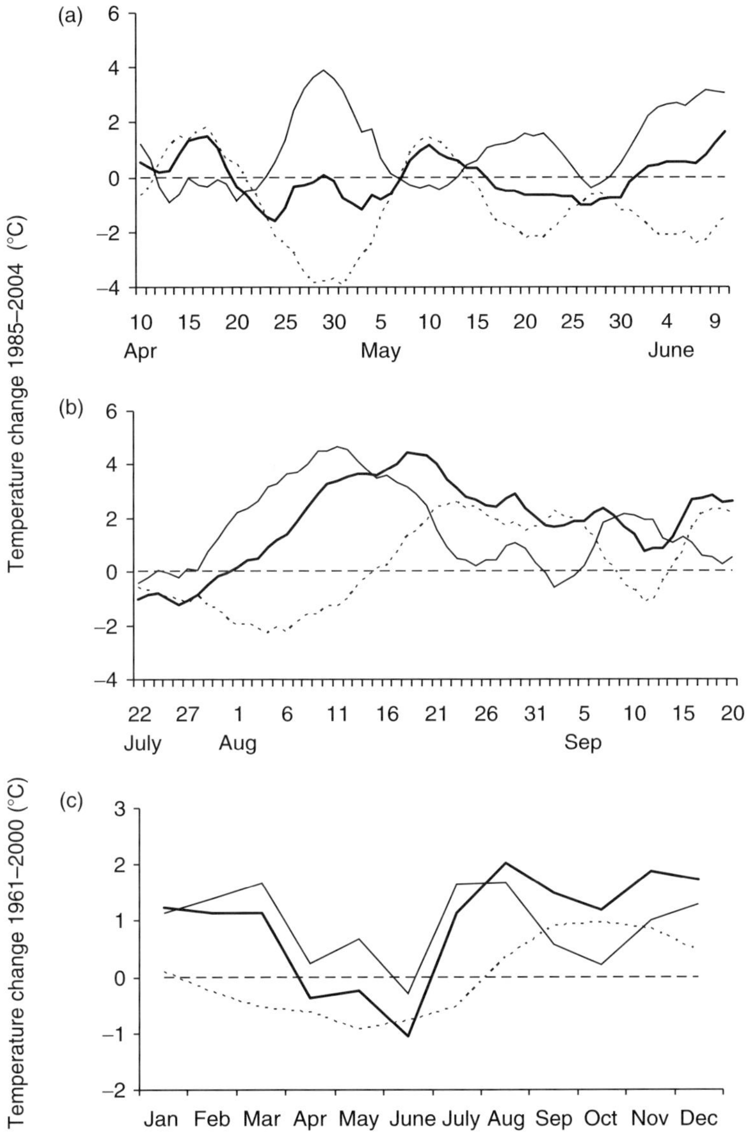

Fig. 11.3. Estimated net change in temperature experienced at different times of year as a result of climate warming and phenological advancement. Net change = thick continuous line; climate change = thin continuous line; direct effect of phenological advance = dotted line. (a) and (b) show week-long running mean changes based on temperature records from 1985 to 2004 at Mickleham in south-east England and phenological change in flight period over the same time for the butterflies: (a) *Anthocharis cardamines*, a spring-flying species; and (b) *Hesperia comma*, a late summer species. (c) shows estimated average change per activity month based on an 11-day advancement and changes in the Central England Temperature Series between 1961 and 2000. See Section 3.4 for details.

3.6 Adaptive responses

Most palaeological evidence suggests that insects have shifted their distributions to track suitable climates during periods of Quaternary climate change (the last 2 million years), rather than adapting *in situ* to changing conditions (Coope, 2004). Nevertheless, insects often have large population sizes and short generation times, and changes in selection may occur rapidly during periods of rapid climate change (Thomas, 2005). There may be selection for phenotypes that favour rapid expansion at range margins where climatic conditions improve, such as those associated with dispersal or the exploitation of novel or widespread resources. Contemporary evolutionary responses at expanding range margins include selection for dispersive forms of butterflies (Hill *et al.*, 1999a,c), ground beetles (Niemela and Spence, 1991) and bush crickets (Thomas *et al.*, 2001; Simmons and Thomas, 2004), and for increased egg laying on a widespread host plant relative to a more restricted former host by the brown argus butterfly *A. agestis* (Thomas *et al.*, 2001). These adaptations increase the rate at which species are able to track shifting suitable climate space, but once populations have been established, there may be a return to selection against dispersive forms, which may be associated with reduced fecundity (Hughes *et al.*, 2003; Simmons and Thomas, 2004). Therefore, forms adapted to range expansion may be favoured for a relatively short period and not readily detected by the fossil record.

The potential for adaptation during changing climates is dependent on the reservoir of genetic variation within populations of species. Many species show adaptations to the local climates experienced in different parts of their geographical range, for example in terms of size, growth rate, diapause induction or the range of plastic responses that can be elicited from individual genotypes (Ayres and Scriber, 1994; Nylin and Gotthard, 1998; Berner *et al.*, 2004). There are differences in preferred oviposition temperature, tolerance of drought and high temperatures, as well as longevity patterns for populations of the fruit fly *D. melanogaster* between hot, dry south-facing slopes and cooler, moister north-facing slopes in close proximity in Israel (Korol *et al.*, 2000). Species often show adaptive local variation in the day-length reduction that is required to induce winter diapause, with longer day lengths sufficient to induce diapause at locations, such as higher latitudes or elevations, where conditions deteriorate earlier in the year (e.g. Roff, 1980; Pullin, 1986; Gomi, 1997). The genetically controlled critical photoperiod for winter diapause induction in populations of the pitcher plant mosquito *Wyeomyia smithii* declined between 1972 and 1996, leading to later cessation of larval activity in conjunction with increasingly warm summers and later onset of autumn conditions (Bradshaw and Holzapfel, 2001).

Despite widespread genotypic and phenotypic variation across the geographical ranges of species, the ability of populations to adapt to new conditions will depend on their location in the current range. Populations at expanding range margins may be able to adapt relatively rapidly because of gene flow from the core of the species range. However, at the rear or trailing edge of a species distribution, the new prevailing conditions are less likely to have been experienced by populations of the species during its evolutionary past, such that the potential for pre-existing genetic variation to allow adaptation is much lower (Thomas, 2005). In addition, deteriorating conditions at the rear edge of species distributions are likely to reduce the extent of suitable habitat (e.g. Wilson *et al.*, 2005), leading to smaller, more isolated populations that contain reduced genetic variation and are prone to effects of inbreeding (Saccheri *et al.*, 1998). Species ranges have undergone successive shifts towards and away from the poles associated with Quaternary periods of warming and cooling: during these alternating shifts, isolated rear-edge populations may have developed local adaptations that were 'swallowed up' by gene flow from the core of the range when climatic conditions reversed (Coope, 2004). As a result, the greatest reservoir of genetic diversity occurs in parts of species ranges that have remained occupied during both glacial and interglacial periods (Hewitt, 2004; Schmitt and Hewitt, 2004). During current, interglacial conditions, this zone of greatest genetic diversity is located near the lower latitudinal margin of most species, where climate-related extinctions could represent a significant loss of future potentially adaptive variation (Hampe and Petit, 2005).

4 Modelling Future Effects of Climate Change

Geographic-scale correlations of species distributions with particular climatic conditions can be used to infer climatic constraints on species ranges, and thus to model 'bioclimate envelopes' for individual species (Pearson and Dawson, 2003). Climate envelope models have been constructed for a number of insects, allowing the prediction of the future locations of suitable climates based on their current climatic associations and realistic scenarios of climatic change (e.g. Hill *et al.*, 1999b, 2002; Beaumont and Hughes, 2002; Oberhauser and Peterson, 2003; Luoto *et al.*, 2005). Climate envelope models can be constructed using variables that have a priori associations with insect distributions, for example, annual cumulative temperature above a threshold level (that affects rates of growth and development), minimum winter temperatures (that affect overwintering survival) and moisture availability (that affects primary production) (e.g. Hill *et al.*, 2002; Luoto *et al.*, 2005).

Climate envelope models fit current species distributions well, both at upper and lower latitudinal range margins (Hill *et al.*, 2002), and appear to perform well for a variety of taxa (Huntley *et al.*, 2004). The models are relatively accurate for species whose distributions are contiguous, with the bounds likely to be set by climatic limitations either on the species itself or on some vital interacting species, such as a larval host plant. Models do not perform well for species that have widespread but scattered distributions, where habitat restrictions and/or local colonization–extinction dynamics may dominate distribution patterns within the climatically suitable range (Luoto *et al.*, 2005). Factors such as biotic interactions, local topographical variation and local evolutionary adaptation could also lead to discrepancies between observed distributions and those modelled based on coarse-scale climatic associations (e.g. Davis *et al.*, 1998; Hill *et al.*, 1999b, 2002).

Nevertheless, modelling future areas of suitable climate space for species, based on their current associations and future scenarios of climate change, allows very general conclusions to be drawn about the likely effects of climate change on species ranges, relative vulnerability of particular groups of species and relative effects of different scenarios of climate change or carbon emission levels (e.g. Beaumont and Hughes, 2002; Hill *et al.*, 2002; Peterson *et al.*, 2002; C.D. Thomas *et al.*, 2004). Modelling the current and future European distributions of 35 geographically widespread species of butterflies based on their current climate associations suggested that distribution sizes would not change significantly in the 21st century, as long as there were no geographical constraints to range shifts and species ranges were able to track suitable climate space perfectly (Hill *et al.*, 2002). However, if the 30 out of 35 species that have not shifted their distributions in conjunction with recent climate change were considered only to survive in areas of overlap between current and future favourable climates, the average predicted change in distribution size would be a 31% decline, and the five modelled species that were restricted to high latitudes in Britain and Europe would have an average predicted decline of 65% (Hill *et al.*, 2002). The climatically suitable ranges for 70 out of 77 Australian endemic butterfly species are predicted to decrease in size based on modelled climates in 2050, with areas of overlap of current and future distributions ranging from 63% to only 22% under conservative and more extreme scenarios of climate change (Beaumont and Hughes, 2002). It is evident that the distributions of species that are currently restricted to localized areas such as mountain ranges or islands may show little geographical overlap with locations that are predicted to be climatically suitable in the future. Species that have very narrow climatic tolerances and associated restricted geographical distributions will not be able to survive climate change, unless their populations can adapt to changing conditions.

The task of predicting future ranges is complicated by the dependence of most species on interacting host species, whose future distribution size or overlap with future modelled climate space for a species may also change. For example, overwintering sites for the monarch butterfly *D. plexippus* in Mexico are located in oyamel fir *Abies religiosa* forests that are characterized by cool, dry conditions between November and March. Survival at overwintering sites is a major determinant of annual abundance, and climate modelling suggests that unfavourable cold or wet conditions will prevail in 30 years time across the distribution of *A. religiosa* (Oberhauser and Peterson, 2003). Thus, although the migratory butterfly *D. plexippus* might itself have sufficient mobility to track changing climates, the geographical isolation of its overwintering habitat may prevent it from doing so. Models that include the effects of climate change both on the future distributions of focal species and their hosts may give increasingly realistic results. However, these models are likely to increase estimates of decline unless climate change allows the exploitation of novel hosts (e.g. Thomas *et al.*, 2001).

Most studies that have modelled the effects of climate on insects have taken advantage of the detailed information that is available on the distributions, habitat requirements and population dynamics of northern temperate taxa. The few lower latitude or southern hemisphere exceptions have usually modelled the responses of relatively well-known groups such as the Lepidoptera (Beaumont and Hughes, 2002; Erasmus *et al.*, 2002; Oberhauser and Peterson, 2003). Models for the effects of climate change on the huge diversity of insect taxa at tropical latitudes are hampered by a lack of information even of the basic biology of many species (see Lewis and Basset, Chapter 2, this volume). Latitudinal gradients in species richness represent one potential source of information about the potential effects of climate change on biodiversity at lower latitudes. Species richness in the tropics, subtropics and warm temperate zones is closely related to water availability, suggesting that any increase in temperature would need to be accompanied by increasing rainfall to avoid declines in species richness (Hawkins *et al.*, 2003).

One possible approach to estimate the vulnerability of taxa to climate change in poorly studied regions is to focus on the climate and habitat associations of species or morphospecies within an insect community. Andrew and Hughes (2004, 2005) sampled the Coleoptera and Hemiptera feeding on *Acacia falcata* at four latitudes from 26°7′S to 35°40′S on the east coast of Australia, and classified the species into four functional groups (named here in italics). *Cosmopolitan* species, which were found at more than one of the sample latitudes and on more than one host plant species, should be resilient to climate change. *Generalist feeders*, which were found only at one latitude but on more than one host plant, may be able to move their climate envelope by exploiting different hosts. The future distributions of *climate generalists*, which were only found on *A. falcata* but at more than one latitude, may be constrained more by their host plant than by climate change. *Specialists*, restricted to *A. falcata* at only one latitude, are expected to be most vulnerable to climate change, and constituted the most diverse group (50% of Coleoptera and 38% of Hemiptera). Many tropical insect species appear to be very rare, with high host plant specificity, localized distributions or low population density (Price *et al.*, 1995; Novotný and Basset, 2000) and may therefore struggle to respond to climate change. Well-designed taxonomic inventories such as those of Andrew and Hughes (2004, 2005) could be a valuable source of information about the effects of climate change on insect communities at tropical latitudes.

5 Climate Change and Insect Conservation

Evidence shows that insect species are shifting their ranges to accompany recent climate warming as they did in prehistoric periods of climate change (Wilf and Labandeira, 1999; Coope, 2004; Hewitt, 2004). A major challenge for conservation is to prevent species disappearing from climatically deteriorating parts of their range before they can colonize regions or habitats that become suitable. This challenge is compounded by additional drivers such as land use change and exotic species introductions that already threaten many species with extinction, and whose effects need to be borne in mind when designing conservation strategies (Sala *et al.*, 2000; Gabriel *et al.*, 2001;

J.A. Thomas *et al.*, 2004; Balmford and Bond, 2005). The foregoing discussion shows that species are likely to respond to climate change in individualistic ways, leading to sometimes unpredictable changes in distribution and abundance patterns, phenology and interactions between species. Conservation programmes may need to be similarly flexible and dynamic as a result, and may require modification to explicitly include the effects of climate change (Hannah *et al.*, 2002a,b; Hulme, 2005). We draw four general conclusions concerning insect conservation in a changing climate:

1. Climate change disproportionately threatens species with small or isolated populations or distribution sizes, narrow habitat requirements (or narrow distributions of resources in space or time) and poor dispersal abilities. These factors increase the likelihood that climate variation will result in declines in population size and local extinctions, and reduce the ability of species to exploit novel resources or colonize climatically favourable locations. It is evident that the same characteristics of species that make them particularly vulnerable to climate change also place them at risk from other anthropogenic effects such as habitat loss and fragmentation (e.g. Travis, 2003; Henle *et al.*, 2004; Kotiaho *et al.*, 2005). Therefore, climate change is likely to increase the vulnerability of most species that were already threatened.

2. Priority conservation management may be required in habitats or regions whose biodiversity is particularly sensitive to the effects of climate change. These regions or habitats can be identified by the modelling of species or biome responses to climate change (e.g. Hannah *et al.*, 2002a). At international scales, centres of endemism or biodiversity hotspots represent concentrations of species that are especially vulnerable to changes both in land use and climate (Myers *et al.*, 2000). High latitudes and elevations will experience the greatest changes in temperature, potentially shifting the suitable climate space for species to locations far outside their current ranges. Montane areas will be particularly vulnerable because they support a disproportionate number of rare or endemic species (e.g. Van Swaay and Warren, 1999; Williams *et al.*, 2003), and because they often represent the lower latitudinal margins of species ranges, which are especially vulnerable to climate warming (Wilson *et al.*, 2005) and which may be important reservoirs of genetic variability (Hampe and Petit, 2004). Conversely, mountainous areas may present opportunities for conservation, since: (i) they often retain comparatively intact habitats relative to lowland landscapes; (ii) steep elevational gradients may allow species ranges to track changing climates more quickly and over smaller distances than in the lowlands; and (iii) small-scale topographical variation may allow survival and adaptation in localized refugia. Minimizing the other threats to species in these regions may increase the likelihood that they will survive climate change.

3. At regional scales, landscape-scale habitat management of reserve networks and the wider environment will be important both to maintain current populations of species and to increase their likelihood of colonizing locations or habitats that become more favourable. Rates of range expansion by the butterfly *H. comma* in England were increased because grassland management in agri-environment schemes increased the area and connectivity of habitat at a landscape scale (Davies *et al.*, 2005). Climate-related changes in the habitat associations of *H. comma* meant that it was able to colonize many areas of grassland that would earlier not have been defined as ideal habitat for the species (Davies *et al.*, 2006). Thus, site protection or management may benefit species that are present not only at a site itself but in the surrounding landscape, and an appropriate large-scale approach may be required to identify and manage regions or habitat networks that support a large number of species of conservation priority (e.g. Moilanen *et al.*, 2005).

Management of the wider landscape to increase connectivity between populations will be least feasible for very sedentary species whose current distributions are very small or very isolated from locations that are expected to be suitable in the future. In this context, management of remnant networks of natural habitat combined with population translocations could be more cost-effective than the creation of wildlife corridors linking highly modified landscapes (Hulme, 2005). Using approaches such as those described in this chapter to model the locations of suitable climates and habitats could aid in the identification of priority species and regions for introductions. Introductions of insect species into suitable habitats beyond their current range have been successful on a number of occasions (e.g. Menéndez *et al.*, 2006). However, the scope of population translocations as a conservation tool may be limited to a relatively small number of flagship species by their cost and requirement for very detailed ecological data. It is essential that translocations do not cause more problems than they solve (e.g. bringing incompatible species into contact with one another).

4. The maintenance of habitat heterogeneity at local and landscape scales may favour species' persistence for two reasons. First, the habitat associations of species change with climate over time (Davies *et al.*, 2006) and over their geographic ranges (Thomas, 1993; Thomas *et al.*, 1998, 1999). As a result, the habitat conditions or management practices that benefit species may change between seasons (Roy and Thomas, 2003) or years (Kindvall, 1996; Sutcliffe *et al.*, 1997) and the provision of a variety of habitat or microhabitat types will allow species to exploit the conditions that are most favoured at a particular time. Careful monitoring may be increasingly necessary to detect the relationships of climate with the population sizes and habitat associations of species, as well as to ensure that habitat is not managed according to outdated prescriptions. Second, habitat heterogeneity could act as a buffer against extreme conditions, allowing populations to survive in some locations or habitats when others become temporarily unfavourable or uninhabitable. Habitats that have greater variation in topography or humidity support more persistent populations than more homogeneous habitats for the butterfly *E. editha* (McLaughlin *et al.*, 2002b) and the bush cricket *Metrioptera bicolor* (Kindvall, 1996). At a landscape scale, the use of two different types of habitat allowed a population network of *E. editha* to survive a succession of extreme climatic events (Singer and Thomas, 1996; Thomas *et al.*, 1996). All populations breeding in forest clearings went extinct in 1992, after early emergence subjected adults to unfavourable conditions in 1989 (asynchrony with nectar availability) and 1990 (mortality caused by a snowfall), and spring frosts killed host plants in 1992. However, populations survived in rocky outcrops where butterflies emerged later in the year and host plants were not killed by the 1992 frost. Rocky outcrops had supported lower population densities than forest clearings before the extreme climatic events, showing how locations and habitats that appear suboptimal based on current abundance patterns may be vital for long-term persistence in a changing climate.

The guidelines above can help to inform the adaptive management of biodiversity in the face of global change. However, in order to ensure that climate does not change so markedly that biological change is no longer manageable, conservationists also need to engage in political advocacy for reductions in greenhouse gas emissions (Hannah *et al.*, 2002b). Minimizing the amount of warming that takes place (climate change *mitigation*) is a prerequisite for the successful conservation management (*adaptation*) of the world's biodiversity in a changing climate.

6 Conclusions

Climate is an important determinant of the abundance and distribution of species. Species are associated with particular latitudes, elevations or habitats through the effects of climate both on the species themselves and on interacting taxa. For species to survive changing climates, they must either adapt *in situ* to new conditions or shift their distributions in pursuit of more favourable ones. Many insects have large population sizes and short generation times, and their phenology, fecundity, survival, selection and habitat use can respond rapidly to climate change. These changes to insect life history in turn produce rapid changes in abundance and distribution size, but some species fare much better than others, particularly in human-altered landscapes. In conjunction with recent climate change, widespread, generalist species at their cool range margins have expanded their distributions, whereas localized, habitat-specialist species and those at their warm margins have declined. In the face of these rapid changes to species, communities and ecosystems, the onus is placed on conservation to be equally dynamic. Landscape-scale conservation, with habitat heterogeneity providing a buffer against extreme conditions and changes in habitat use by threatened species, is an appropriate strategy to conserve species and to assist their colonization of areas that become more favourable as the climate changes.

Acknowledgements

We thank the British Atmospheric Data Centre (BADC) for access to the UK Meteorological Office Land Surface Observation Stations Data; and the UK Butterfly Monitoring Scheme (UKBMS), coordinated by Butterfly Conservation and the Centre for Ecology and Hydrology, for supplying butterfly transect data (particularly Tom Wigglesworth, Tom Brereton and David Roy). Víctor J. Monserrat provided information on butterfly sampling locations in the Sierra de Guadarrama during 1967–1973, and David Gutiérrez and Javier Gutiérrez assisted with analysis.

References

Andrew, N.R. and Hughes, L. (2004) Species diversity and structure of phytophagous beetle assemblages along a latitudinal gradient: predicting the potential impacts of climate change. *Ecological Entomology* 29, 527–542.

Andrew, N.R. and Hughes, L. (2005) Diversity and assemblage structure of phytophagous Hemiptera along a latitudinal gradient: predicting the potential impacts of climate change. *Global Ecology and Biogeography* 14, 249–262.

Ayres, M.P. and Scriber, J.M. (1994) Local adaptation to regional climates in *Papilio canadensis* (Lepidoptera: Papilionidae). *Ecological Monographs* 64, 465–482.

Bale, J.S., Masters, G.J., Hodkinson, I.D., Awmack, C., Bezemer, T.M., Brown, V.K., Butterfield, J., Buse, A., Coulson, J.C., Farrar, J., Good, J.E.G., Harrington, R., Hartley, S., Jones, T.H., Lindroth, R.L., Press, M.C., Symrnioudis, I., Watt, A.D. and Whittaker, J.B. (2002) Herbivory in global climate change research: direct effects of rising temperatures on insect herbivores. *Global Change Biology* 8, 1–16.

Balmford, A. and Bond, W. (2005) Trends in the state of nature and their implications for human well-being. *Ecology Letters* 8, 1218–1234.

Battisti, A., Stastny, M., Netherer, S., Robinet, C., Schopf, A., Roques, A. and Larsson, S. (2005) Expansion of geographic range in the pine processionary moth caused by increased winter temperatures. *Ecological Applications* 15, 2084–2096.

Battisti, A., Stastny, M., Buffo, E. and Larsson, S. (2006) A rapid altitudinal range expansion in the pine processionary moth produced by the 2003 climatic anomaly. *Global Change Biology* 12, 662–671.

Beaumont, L.J. and Hughes, L. (2002) Potential changes in the distributions of latitudinally restricted Australian butterfly species in response to climate change. *Global Change Biology* 8, 954–971.

Benton, M.J. and Twitchett, R.J. (2003) How to kill (almost) all life: the end-Permian extinction event. *Trends in Ecology and Evolution* 18, 358–365.

Bernays, E.A. (1997) Feeding by lepidopteran larvae is dangerous. *Ecological Entomology* 22, 121–123.

Berner, D., Körner, C. and Blanckenhorn, W.U. (2004) Grasshopper populations across 2000 m of altitude: is there life history adaptation? *Ecography* 27, 733–740.

Bird, J.M. and Hodkinson, I.D. (1999) Species at the edge of their range: the significance of the thermal environment for the distribution of congeneric *Craspedolepta* species (Sternorrhyncha: Psylloidea) living on *Epilobium angustifolium* (Onagraceae). *European Journal of Entomology* 96, 103–109.

Bradshaw, W.E. and Holzapfel, C.M. (2001) Genetic shift in photoperiodic response correlated with global warming. *Proceedings of the National Academy of Sciences of the United States of America* 98, 14509–14511.

Bryant, S.R., Thomas, C.D. and Bale, J.S. (1997) Nettle-feeding nymphalid butterflies: temperature, development and distribution. *Ecological Entomology* 22, 390–398.

Buse, A. and Good, J.E.G. (1996) Synchronization of larval emergence in winter moth (*Operophtera brumata* L.) and budburst in pedunculate oak (*Quercus robur* L.) under simulated climate change. *Ecological Entomology* 21, 335–343.

Conrad, K.F., Woiwod, I.P. and Perry, J.N. (2001) An abundance-occupancy time-lag during the decline of an arctiid tiger moth. *Ecology Letters* 4, 300–303.

Conrad, K.F., Woiwod, I.P. and Perry, J.N. (2002) Long-term decline in abundance and distribution of the garden tiger moth (*Arctia caja*) in Great Britain. *Biological Conservation* 106, 329–337.

Conrad, K.F., Woiwod, I.P. and Perry, J.N. (2003) East Atlantic teleconnection pattern and the decline of a common arctiid moth. *Global Change Biology* 9, 125–130.

Coope, G.R. (2004) Several million years of stability among insect species because of, or in spite of, ice age climatic instability. *Philosophical Transactions of the Royal Society of London Series B – Biological Sciences* 359, 209–214.

Coviella, C.E. and Trumble, J.T. (1999) Effects of elevated atmospheric carbon dioxide on insect–plant interactions. *Conservation Biology* 13, 700–712.

Crozier, L.G. (2003) Winter warming facilitates range expansion: cold tolerance of the butterfly *Atalopedes campestris*. *Oecologia* 135, 648–656.

Crozier, L.G. (2004a) Warmer winters drive butterfly range expansion by increasing survivorship. *Ecology* 85, 231–241.

Crozier, L.G. (2004b) Field transplants reveal summer constraints on a butterfly range expansion. *Oecologia* 141, 148–157.

Davies, Z.G., Wilson, R.J., Brereton, T.M. and Thomas, C.D. (2005) The re-expansion and improving status of the silver-spotted skipper butterfly (*Hesperia comma*) in Britain: a metapopulation success story. *Biological Conservation* 124, 189–198.

Davies, Z.G., Wilson, R.J., Coles, S. and Thomas, C.D. (2006) Changing habitat associations of a thermally constrained species, the silver-spotted skipper butterfly, in response to climate warming. *Journal of Animal Ecology* 75, 247

Davis, A.J., Jenkinson, L.S., Lawton, J.H., Shorrocks, B. and Wood, S. (1998) Making mistakes when predicting shifts in species range in response to global warming. *Nature* 391, 783–786.

Easterling, D.R., Meehl, G.A., Parmesan, C., Changnon, S.A., Karl, T.R. and Mearns, L.O. (2000) Climate extremes: observations, modeling, and impacts. *Science* 289, 2068–2074.

Ellis, W.N., Donner, J.H. and Kuchlein, J.H. (1997) Recent shifts in phenology of Microlepidoptera, related to climatic change. *Entomologische Berichten* 57, 66–72.

Emmerson, M., Bezemer, T.M., Hunter, M.D., Jones, T.H., Masters, G.J. and Van Dam, N.M. (2004) How does global change affect the strength of trophic interactions? *Basic and Applied Ecology* 5, 505–514.

Erasmus, B.F.N., Van Jaarsveld, A.S., Chown, S.L., Kshatriya, M. and Wessels, K.J. (2002) Vulnerability of South African animal taxa to climate change. *Global Change Biology* 8, 679–693.

Feeny, P. (1970) Seasonal changes in oak leaf tannins and nutrients as a cause of spring feeding by winter moth caterpillars. *Ecology* 51, 565–581.

Fielding, C.A., Whittaker, J.B., Butterfield, J.E.L. and Coulson, J.C. (1999) Predicting responses to climate change: the effect of altitude on the phenology of the spittlebug *Neophilaenus lineatus*. *Functional Ecology* 12(Suppl. 1), 65–73.

Fitter, A.H. and Fitter, R.S.R. (2002) Rapid changes in flowering time in British plants. *Science* 296, 1689–1691.

Forister, M.L. and Shapiro, A.M. (2003) Climatic trends and advancing spring flight of butterflies in lowland California. *Global Change Biology* 9, 1130–1135.

Franco, A.M.A., Hill, J.K., Kitschke, C., Collingham, Y.C., Roy, D.B., Fox, R., Huntley, B. and Thomas, C.D. (2006) Impacts of climate warming and habitat loss on extinctions at species' low-latitude range boundaries. *Global Change Biology* 12, 1545–1553.

Gabriel, A.G.A., Chown, S.L., Barendse, J., Marshall, D.J., Mercer, R.D., Pugh, P.J.A. and Smith, V.R. (2001) Biological invasions of Southern Ocean islands: the Collembola of Marion Island as a test of generalities. *Ecography* 24, 421–430.

Gaston, K.J. (2003) *The Structure and Dynamics of Geographic Ranges*. Oxford University Press, Oxford.

Gaston, K.J., Blackburn, T.M. and Goldewijk, K.K. (2003) Habitat conversion and global avian biodiversity loss. *Proceedings of the Royal Society of London Series B – Biological Sciences* 270, 1293–1300.

Gomi, T. (1997) Geographic variation in critical photoperiod for diapause induction and its temperature dependence in *Hyphantria cunea* Drury (Lepidoptera: Arctiidae). *Oecologia* 111, 160–165.

Gutiérrez, D. and Menéndez, R. (1998) Phenology of butterflies along an altitudinal gradient in northern Spain. *Journal of Zoology* 244, 249–264.

Hallam, A. and Wignall, P. (1997) *Mass Extinctions and Their Aftermath*. Oxford University Press, Oxford.

Hampe, A. and Petit, R.J. (2005) Conserving biodiversity under climate change: the rear edge matters. *Ecology Letters* 8, 461–467.

Hannah, L., Midgley, G.F. and Millar, D. (2002a) Climate change-integrated conservation strategies. *Global Ecology and Biogeography* 11, 485–495.

Hannah, L., Midgley, G.F., Lovejoy, T., Bond, W.J., Bush, M., Lovett, J.C., Scott, D. and Woodward, F.I. (2002b) Conservation of biodiversity in a changing climate. *Conservation Biology* 16, 264–268.

Hawkins, B.A. and Porter, E.E. (2003) Water–energy balance and the geographic pattern of species richness of western Palearctic butterflies. *Ecological Entomology* 28, 678–686.

Hawkins, B.A., Field, R., Cornell, H.V., Currie, D.J., Guégan, J.F., Kaufman, D.M., Kerr, J.T., Mittelbach, G.G., Oberdorff, T., O'Brien, E.M., Porter, E.E. and Turner, J.R.G. (2003) Energy, water, and broad-scale geographic patterns of species richness. *Ecology* 84, 3105–3117.

Henle, K., Davies, K.F., Kleyer, M., Margules, C. and Settele, J. (2004) Predictors of species sensitivity to fragmentation. *Biodiversity and Conservation* 13, 207–251.

Hewitt, G.M. (2004) Genetic consequences of climatic oscillations in the quaternary. *Philosophical Transactions of the Royal Society of London Series B – Biological Sciences* 359, 183–195.

Hickling, R., Roy, D.B., Hill, J.K. and Thomas, C.D. (2005) A northward shift of range margins in British Odonata. *Global Change Biology* 11, 502–506.

Hickling, R., Roy, D.B., Hill, J.K., Fox, R. and Thomas, C.D. (2006) The distributions of a wide range of taxonomic groups are expanding polewards. *Global Change Biology* 12, 450–455.

Hill, J.K. and Hodkinson, I.D. (1995) Effects of temperature on phenological synchrony and altitudinal distribution of jumping plant-lice (Hemiptera: Psylloidea) on dwarf willow (*Salix lapponum*) in Norway. *Ecological Entomology* 20, 237–244.

Hill, J.K. and Hodkinson, I.D. (1996) Effects of photoperiod and raised winter temperatures on egg development and timing of oviposition in the willow psyllid *Cacopsylla moscovita*. *Entomologia Experimentalis et Applicata* 78, 143–147.

Hill, J.K., Hamer, K.C. and Hodkinson, I.D. (1998) Variation in resource exploitation along an altitudinal gradient: the willow psyllids (*Cacopsylla* spp.) on *Salix lapponum*. *Ecography* 21, 289–296.

Hill, J.K., Thomas, C.D. and Blakeley, D.S. (1999a) Evolution of flight morphology in a butterfly that has recently expanded its geographic range. *Oecologia* 121, 165–170.

Hill, J.K., Thomas, C.D. and Huntley, B. (1999b) Climate and habitat availability determine 20th century changes in a butterfly's range margin. *Proceedings of the Royal Society of London Series B – Biological Sciences* 266, 1197–1206.

Hill, J.K., Thomas, C.D. and Lewis, O.T. (1999c) Flight morphology in fragmented populations of a rare British butterfly, *Hesperia comma*. *Biological Conservation* 87, 277–283.

Hill, J.K., Collingham, Y.C., Thomas, C.D., Blakeley, D.S., Fox, R., Moss, D. and Huntley, B. (2001) Impacts of landscape structure on butterfly range expansion. *Ecology Letters* 4, 313–321.

Hill, J.K., Thomas, C.D., Fox, R., Telfer, M.G., Willis, S.G., Asher, J. and Huntley, B. (2002) Responses of butterflies to twentieth century climate warming: implications for future ranges. *Proceedings of the Royal Society of London Series B – Biological Sciences* 269, 2163–2171.

Hodkinson, I.D. (1997) Progressive restriction of host plant exploitation along a climatic gradient: the willow psyllid *Cacopsylla groenlandica* in Greenland. *Ecological Entomology* 21, 47–54.

Hodkinson, I.D. (1999) Species response to global environmental change or why ecophysiological models *are* important: a reply to Davis *et al. Journal of Animal Ecology* 68, 1259–1262.

Hodkinson, I.D. (2005) Terrestrial insects along elevation gradients: species and community responses to altitude. *Biological Reviews of the Cambridge Philosophical Society* 80, 489–513.

Hodkinson, I.D., Bird, J., Miles, J.E., Bale, J.S. and Lennon, J.J. (1999) Climatic signals in the life histories of insects: the distribution and abundance of heather psyllids (*Strophingia* spp.) in the UK. *Functional Ecology* 13(Suppl.), 83–95.

Holmgren, M., Scheffer, M., Ezcurra, E., Gutiérrez, J.R. and Mohren, G.M.J. (2001) El Niño effects on the dynamics of terrestrial ecosystems. *Trends in Ecology and Evolution* 16, 89–94.

Hoover, J.K. and Newman, J.A. (2004) Tritrophic interactions in the context of climate change: a model of grasses, cereal aphids and their parasitoids. *Global Change Biology* 10, 1197–1208.

Hughes, C.L., Hill, J.K. and Dytham, C. (2003) Evolutionary trade-offs between reproduction and dispersal in populations at expanding range boundaries. *Proceedings of the Royal Society of London Series B – Biological Sciences* 270(Suppl. 2), S147–S150.

Hughes, L. (2000) Biological consequences of global warming: is the signal already apparent? *Trends in Ecology and Evolution* 15, 56–61.

Hulme, P.E. (2005) Adapting to climate change: is there scope for ecological management in the face of a global threat? *Journal of Applied Ecology* 42, 784–794.

Huntley, B., Green, R.E., Collingham, Y.C., Hill, J.K., Willis, S.G., Bartlein, P.J., Cramer, W., Hagemeijer, W.J.M. and Thomas, C.J. (2004) The performance of models relating species geographical distributions to climate is independent of trophic level. *Ecology Letters* 7, 417–426.

IPCC (2001) *The Third Assessment Report of the Intergovernmental Panel on Climate Change*. Cambridge University Press, Cambridge.

Irwin, J.T. and Lee, R.E. (2000) Mild winter temperatures reduce survival and potential fecundity in the goldenrod gall fly *Eurosta solidaginis* (Diptera: Tephritidae). *Journal of Insect Physiology* 46, 655–661.

Irwin, J.T. and Lee, R.E. (2003) Cold winter microenvironments conserve energy and improve overwintering survival and potential fecundity of the goldenrod gall fly *Eurosta solidaginis. Oikos* 100, 71–78.

Johns, C.V. and Hughes, L. (2002) Interactive effects of elevated CO_2 and temperature on the leaf-miner *Dialectica scalariella* Zeller (Lepidoptera: Gracillariidae) in Paterson's Curse, *Echium plantagineum* (Boraginaceae). *Global Change Biology* 8, 142–152.

Johns, C.V., Beaumont, L.J. and Hughes, L. (2003) Effects of elevated CO_2 and temperature on development and consumption rates of *Octotoma championi* and *O. scabripennis* feeding on *Lantana camara. Entomologia Experimentalis et Applicata* 108, 169–178.

Judd, S. and Hodkinson, I.D. (1998) The biogeography and regional biodiversity of the British seed bugs (Hemiptera: Lygaeidae). *Journal of Biogeography* 25, 227–249.

Karban, R. and Strauss, S.Y. (2004) Physiological tolerance, climate change, and a northward range shift in the spittlebug, *Philaenus spumarius. Ecological Entomology* 29, 251–254.

Kindvall, O. (1996) Habitat heterogeneity and survival in a bush cricket metapopulation. *Ecology* 77, 207–214.

King, D. (2005) Climate change: the science and the policy. *Journal of Applied Ecology* 42, 779–783.

Konvicka, M., Maradova, M., Benes, J., Fric, Z. and Kepka, P. (2003) Uphill shifts in distribution of butterflies in the Czech Republic: effects of changing climate detected on a regional scale. *Global Ecology and Biogeography* 12, 403–410.

Korol, A., Rashkovetsky, E., Iliadi, K., Michalak, P., Ronin, Y. and Nevo, E. (2000) Nonrandom mating in *Drosophila melanogaster* laboratory populations derived from closely adjacent ecologically contrasting slopes at 'Evolution Canyon'. *Proceedings of the National Academy of Sciences of the United States of America* 97, 12637–12642.

Kotiaho, J.S., Kaitala, V., Komonen, A. and Päivinen, J. (2005) Predicting the risk of extinction from shared ecological characteristics. *Proceedings of the National Academy of Sciences of the United States of America* 102, 1963–1967.

Kuchlein, J.H. and Ellis, W.N. (1997) Climate-induced changes in the microlepidoptera fauna of the Netherlands and the implications for nature conservation. *Journal of Insect Conservation* 1, 73–80.

Lovejoy, T.E. and Hannah, L. (2005) *Climate Change and Biodiversity*. Yale University Press, New Haven, Connecticut.

Luoto, M., Pöyry, J., Heikkinen, R.K. and Saarinen, K. (2005) Uncertainty of bioclimate envelope models based on the geographic distribution of species. *Global Ecology and Biogeography* 14, 575–584.

MacArthur, R.H. (1972) *Geographical Ecology: Patterns in the Distribution of Species*. Princeton University Press, Princeton, New Jersey.

Maclean, S.F. (1983) Life cycles and the distribution of psyllids (Homoptera) in arctic and subarctic Alaska. *Oikos* 40, 445–451.

Manley, G. (1974) Central England Temperatures: monthly means 1965–1973. *Quarterly Journal of the Royal Meteorological Society* 100, 389–405.

Masters, G.J., Brown, V.K., Clarke, I.P., Whittaker, J.B. and Hollier, J.A. (1998) Direct and indirect effects of climate change on insect herbivores: *Auchenorrhyncha* (Homoptera). *Ecological Entomology* 23, 45–52.

McLaughlin, J.F., Hellmann, J.J., Boggs, C.L. and Ehrlich, P.R. (2002a) Climate change hastens population extinctions. *Proceedings of the National Academy of Sciences of the United States of America* 99, 6070–6974.

McLaughlin, J.F., Hellmann, J.J., Boggs, C.L. and Ehrlich, P.R. (2002b) The route to extinction: population dynamics of a threatened butterfly. *Oecologia* 132, 538–548.

Menéndez, R., González Megías, A., Hill, J.K., Braschler, B., Willis, S.G., Collingham, Y., Fox, R., Roy, D.B. and Thomas, C.D. (2006) Species richness changes lag behind climate change. *Proceedings of the Royal Society of London Series B – Biological Sciences* 273, 1465–1470

Menzel, A. and Fabian, P. (1999) Growing season extended in Europe. *Nature* 397, 659.

Miles, J.E., Bale, J.S. and Hodkinson, I.D. (1997) Effects of temperature elevation on the population dynamics of the upland race of the heather psyllid *Strophingia ericae* (Curtis) (Homoptera: Psylloidea). *Global Change Biology* 3, 291–297.

Moilanen, A., Franco, A.M.A., Early, R.I., Fox, R., Wintle, B. and Thomas, C.D. (2005) Prioritizing multiple-use landscapes for conservation: methods for large multi-species planning problems. *Proceedings of the Royal Society of London Series B – Biological Sciences* 272, 1885–1891.

Morecroft, M.D., Bealey, C.E., Howells, O., Rennie, S. and Woiwod, I.P. (2002) Effects of drought on contrasting insect and plant species in the UK in the mid-1990s. *Global Ecology and Biogeography* 11, 7–22.

Musolin, D.L. and Numata, H. (2003) Timing of diapause induction and its life-history consequences in *Nezara viridula*: is it costly to expand the distribution range? *Ecological Entomology* 28, 694–703.

Myers, N., Mittermeier, R.A., Mittermeier, C.G., da Fonseca, G.A.B. and Kent, J. (2000) Biodiversity hotspots for conservation priorities. *Nature* 403, 853–858.

Newman, J.A. (2004) Climate change and cereal aphids: the relative effects of increasing CO_2 and temperature on aphid population dynamics. *Global Change Biology* 10, 5–15.

Newman, J.A. (2005) Climate change and the fate of cereal aphids in Southern Britain. *Global Change Biology* 11, 940–944.

Niemela, J. and Spence, J.R. (1991) Distribution and abundance of an exotic ground-beetle (Carabidae): a test of community impact. *Oikos* 62, 351–359.

Novotný, V. and Basset, Y. (2000) Rare species in communities of tropical insect herbivores: pondering the mystery of singletons. *Oikos* 89, 564–572.

Nylin, S. and Gotthard, K. (1998) Plasticity in life-history traits. *Annual Review of Entomology* 43, 63–83.

Nylin, S. and Svard, L. (1991) Latitudinal patterns in the size of European butterflies. *Holarctic Ecology* 14, 192–202.

Oberhauser, K. and Peterson, A.T. (2003) Modeling current and future potential wintering distributions of eastern North American monarch butterflies. *Proceedings of the National Academy of Sciences of the United States of America* 100, 14063–14068.

Ottersen, G., Planque, B., Belgrano, A., Post, E., Reid, P.C. and Stenseth, N.C. (2001) Ecological effects of the North Atlantic Oscillation. *Oecologia* 128, 1–14.

Parker, D.E., Legg, T.P. and Folland, C.K. (1992) A new daily Central England Temperature Series, 1772–1991. *International Journal of Climatology* 12, 317–342.

Parmesan, C. (1996) Climate and species range. *Nature* 382, 765–766.

Parmesan, C. (2005) Detection at multiple levels: *Euphydryas editha* and climate change. In: Lovejoy, T.E. and Hannah, L. (eds) *Climate Change and Biodiversity*. Yale University Press, New Haven, Connecticut, pp. 56–60.

Parmesan, C. and Yohe, G. (2003) A globally coherent fingerprint of climate change impacts across natural systems. *Nature* 421, 37–42.

Parmesan, C., Ryrholm, N., Stefanescu, C., Hill, J.K., Thomas, C.D., Descimon, H., Huntley, B., Kaila, L., Kullberg, J., Tammaru, T., Tennent, W.J., Thomas, J.A. and Warren, M. (1999) Poleward shifts in geographical ranges of butterfly species associated with regional warming. *Nature* 399, 579–583.

Parmesan, C., Root, T.L. and Willig, M.R. (2000) Impacts of extreme weather and climate on terrestrial biota. *Bulletin of the American Meteorological Society* 81, 443–450.

Pearson, R.G. and Dawson, T.P. (2003) Predicting the impacts of climate change on the distribution of species: are bioclimate envelope models useful? *Global Ecology and Biogeography* 12, 361–371.

Peñuelas, J., Filella, I. and Comas, P. (2002) Changed plant and animal life cycles from 1952 to 2000 in the Mediterranean region. *Global Change Biology* 8, 531–544.

Peterson, A.T., Ortega-Huerta, M.A., Bartley, J., Sánchez-Cordero, V., Soberón, J., Buddemeier, R.H. and Stockwell, D.R.B. (2002) Future projections for Mexican faunas under global climate change scenarios. *Nature* 416, 626–629.

Polis, G.A., Hurd, S.D., Jackson, C.T. and Sanchez-Piñero, F.S. (1997) El Niño effects on the dynamics and control of an island ecosystem in the Gulf of California. *Ecology* 78, 1884–1897.

Polis, G.A., Hurd, S.D., Jackson, C.T. and Sanchez-Piñero, F.S. (1998) Multifactor population limitation: variable spatial and temporal control of spiders on Gulf of California islands. *Ecology* 79, 490–502.

Pollard, E. (1988) Temperature, rainfall and butterfly numbers. *Journal of Applied Ecology* 25, 819–828.

Price, P.W., Diniz, I.R., Morais, H.C. and Marquis, E.S.A. (1995) The abundance of insect herbivore species in the tropics: the high local richness of rare species. *Biotropica* 27, 468–478.

Pullin, A.S. (1986) Effect of photoperiod and temperature on the life-cycle of different populations of the peacock butterfly *Inachis io*. *Entomologia Experimentalis et Applicata* 41, 237–242.

Pullin, A.S. and Bale, J.S. (1989) Effects of low temperature on diapausing *Aglais urticae* and *Inachis io* (Lepidoptera: Nymphalidae): cold hardiness and overwintering survival. *Journal of Insect Physiology* 35, 277–281.

Ratte, H.T. (1985) Temperature and insect development. In: Hoffmann, K.H. (ed.) *Environmental Physiology and Biochemistry of Insects*. Springer-Verlag, Berlin, pp. 33–66.

Roff, D. (1980) Optimizing development time in a seasonal environment: the 'ups and downs' of clinal variation. *Oecologia* 45, 202–208.

Root, T.L. and Schneider, S.H. (2002) Climate change: overview and implications for wildlife. In: Root, T.L. and Schneider, S.H. (eds) *Wildlife Responses to Climate Change: North American Case Studies*. Island Press, Washington, DC, pp. 1–56.

Root, T.L., Price, J.T., Hall, K.R., Schneider, S.H., Rozenzweig, C. and Pounds, J.A. (2003)

Fingerprints of global warming on wild animals and plants. *Nature* 421, 57–60.

Roy, D.B. and Asher, J. (2003) Spatial trends in the sighting dates of British butterflies. *International Journal of Biometeorology* 47, 188–192.

Roy, D.B. and Sparks, T.H. (2000) Phenology of British butterflies and climate change. *Global Change Biology* 6, 407–416.

Roy, D.B. and Thomas, J.A. (2003) Seasonal variation in the niche, habitat availability and population fluctuations of a bivoltine thermophilous insect near its range margin. *Oecologia* 134, 439–444.

Roy, D.B., Rothery, P., Moss, D., Pollard, E. and Thomas, J.A. (2001) Butterfly numbers and weather: predicting historical trends in abundance and the future effects of climate change. *Journal of Animal Ecology* 70, 201–217.

Saccheri, I., Kuussari, M., Kankare, M., Vikman, P., Fortelius, W. and Hanski, I. (1998) Inbreeding and extinction in a butterfly metapopulation. *Nature* 392, 491–494.

Sala, O.E., Chapin, F.S., Armesto, J.J., Berlow, E., Bloomfield, J., Dirzo, R., Huber-Sanwald, E., Huenneke, L.F., Jackson, R.B., Kinzig, A., Leemans, R., Lodge, D.M., Mooney, H.A., Oesterheld, M., Poff, N.L., Sykes, M.T., Walker, B.H., Walker, M. and Wall, D.H. (2000) Global biodiversity scenarios for the year 2100. *Science* 287, 1770–1774.

Sanderson, E.W., Jaiteh, M., Levy, M.A., Redford, K.H., Wannebo, A.V. and Woolmer, G. (2002) The human footprint and the last of the wild. *Bioscience* 52, 891–904.

Scheifinger, H., Koch, E. and Winkler, H. (2005) Results of a first look into the Austrian animal phenological records. *Meteorologische Zeitschrift* 14, 203–209.

Schmitt, T. and Hewitt, G.M. (2004) The genetic pattern of population threat and loss: a case study of butterflies. *Molecular Ecology* 13, 21–31.

Shreeve, T.G. (1992) Adult behaviour. In: Dennis, R.L.H. (ed.) *The Ecology of Butterflies in Britain*. Oxford University Press, Oxford, pp. 22–45.

Simmons, A.D. and Thomas, C.D. (2004) Changes in dispersal during species' range expansions. *American Naturalist* 164, 378–395.

Singer, M.C. and Thomas, C.D. (1996) Evolutionary responses of a butterfly metapopulation to human and climate-caused environmental variation. *American Naturalist* 148, S9–S39.

Sparks, T.H. and Menzel, A. (2002) Observed changes in seasons: an overview. *International Journal of Climatology* 22, 1715–1725.

Stefanescu, C., Peñuelas, J. and Filella, I. (2003) Effects of climatic change on the phenology of butterflies in the northwest Mediterranean Basin. *Global Change Biology* 9, 1494–1506.

Strathdee, A.T., Bale, J.S., Block, W., Coulson, S.J., Hodkinson, I.D. and Webb, N.R. (1993) Effects of temperature elevation on a field population of *Acyrthosiphon svalbardicum* (Hemiptera: Aphididae) on Spitsbergen. *Oecologia* 96, 457–465.

Sutcliffe, O.L., Thomas, C.D., Yates, T.J. and Greatorex-Davies, J.N. (1997) Correlated extinctions, colonisations and population fluctuations in a highly connected ringlet butterfly metapopulation. *Oecologia* 109, 235–241.

Thomas, C.D. (2005) Recent evolutionary effects of climate change. In: Lovejoy, T.E. and Hannah, L. (eds) *Climate Change and Biodiversity*. Yale University Press, New Haven, Connecticut, pp. 75–88.

Thomas, C.D. and Jones, T.M. (1993) Partial recovery of a skipper butterfly (*Hesperia comma*) from population refuges: lessons for conservation in a fragmented landscape. *Journal of Animal Ecology* 62, 471–481.

Thomas, C.D., Singer, M.C. and Boughton, D.A. (1996) Catastrophic extinction of population sources in a butterfly metapopulation. *American Naturalist* 148, 957–975.

Thomas, C.D., Bodsworth, E.J., Wilson, R.J., Simmons, A.D., Davies, Z.G., Musche, M. and Conradt, L. (2001) Ecological and evolutionary processes at expanding range margins. *Nature* 411, 577–581.

Thomas, C.D., Cameron, A., Green, R.E., Bakkenes, M., Beaumont, L.J., Collingham, Y.C., Erasmus, B.F.N., de Siqueira, M.F., Grainger, A., Hannah, L., Hughes, L., Huntley, B., van Jaarsveld, A.S., Midgley, G.F.,

Miles, L., Ortega-Huerta, M.A., Peterson, A.T., Phillips, O.L. and Williams, S.E. (2004) Extinction risk from climate change. *Nature* 427, 145–148.

Thomas, J.A. (1993) Holocene climate changes and warm man-made refugia may explain why a sixth of British butterflies possess unnatural early-successional habitats. *Ecography* 16, 278–284.

Thomas, J.A., Thomas, C.D., Simcox, D.J. and Clarke, R.T. (1986) Ecology and declining status of the silver-spotted skipper butterfly (*Hesperia comma*) in Britain. *Journal of Applied Ecology* 23, 365–380.

Thomas, J.A., Simcox, D.J., Wardlaw, J.C., Elmes, G.W., Hochberg, M.E. and Clarke, R.T. (1998) Effects of latitude, altitude and climate on the habitat and conservation of the endangered butterfly *Maculinea arion* and its *Myrmica* ant hosts. *Journal of Insect Conservation* 2, 39–46.

Thomas, J.A., Rose, R.J., Clarke, R.T., Thomas, C.D. and Webb, N.R. (1999) Intraspecific variation in habitat availability among ectothermic animals near their climatic limits and their centres of range. *Functional Ecology* 13, 55–64.

Thomas, J.A., Telfer, M.G., Roy, D.B., Preston, C.D., Greenwood, J.J.D., Asher, J. and Fox, R. (2004) Comparative losses of British butterflies, birds, and plants and the global extinction crisis. *Science* 303, 1879–1881.

Tilman, D., May, R.M., Lehman, C.L. and Nowak, M.A. (1994) Habitat destruction and the extinction debt. *Nature* 371, 65–66.

Travis, J.M.J. (2003) Climate change and habitat destruction: a deadly anthropogenic cocktail. *Proceedings of the Royal Society of London Series B – Biological Sciences* 270, 467–473.

Turnock, W.J. and Fields, P.G. (2005) Winter climates and cold hardiness in terrestrial insects. *European Journal of Entomology* 102, 561–576.

Van Nouhuys, S. and Lei, G. (2004) Parasitoid-host metapopulation dynamics: the causes and consequences of phenological asynchrony. *Journal of Animal Ecology* 73, 526–535.

Van Swaay, C.A.M. and Warren, M.S. (1999) *Red Data Book of European Butterflies (Rhopalocera)*. Nature and Environment, No. 99, Council of Europe Publishing, Strasbourg, France.

Virtanen, T., Neuvonen, S. and Nikula, A. (1998) Modelling topoclimatic patterns of egg mortality of *Epirrita autumnata* (Lepidoptera: Geometridae) with a Geographical Information System: predictions for current climate and warmer climate scenarios. *Journal of Applied Ecology* 35, 311–322.

Visser, M.E. and Hollemann, L.J.M. (2001) Warmer springs disrupt the synchrony of oak and winter moth phenology. *Proceedings of the Royal Society of London Series B – Biological Sciences* 268, 289–294.

Visser, M.E., van Noordwijk, A.J., Tinbergen, J.M. and Lessells, C.M. (1998) Warmer springs lead to mistimed reproduction in great tits (*Parus major*). *Proceedings of the Royal Society of London Series B – Biological Sciences* 265, 1867–1870.

Vitousek, P.M., Mooney, H.A., Lubchenco, J. and Melillo, J.M. (1997) Human domiation of earth's ecosystems. *Science* 277, 494–499.

Walther, G-R., Post, E., Convey, P., Menzel, A., Parmesan, C., Beebee, T.J.C., Fromentin, J.-M., Hoegh-Guldberg, O. and Bairlein, F. (2002) Ecological responses to recent climate change. *Nature* 416, 389–395.

Warren, M.S., Hill, J.K., Thomas, J.A., Asher, J., Fox, R., Huntley, B., Roy, D.B., Telfer, M.G., Jeffcoate, S., Harding, P., Jeffcoate, J., Willis, S.G., Greatorex-Davies, J.N., Moss, D. and Thomas, C.D. (2001) Rapid responses of British butterflies to opposing forces of climate and habitat change. *Nature* 414, 65–69.

Weiss, S.B., Murphy, D.D. and White, R.R. (1988) Sun, slope and butterflies: topographic determinants of habitat quality for *Euphydryas editha bayensis*. *Ecology* 69, 1486–1496.

White, R.R. and Levin, M.P. (1981) Temporal variation in vagility: implications for evolutionary studies. *American Midland Naturalist* 105, 348–357.

Whittaker, J.B. and Tribe, N.P. (1996) An altitudinal transect as an indicator of responses of a spittlebug (Auchenorrhyncha, Cercopidae) to climate change. *European Journal of Entomology* 93, 319–324.

Wilf, P. and Labandeira, C.C. (1999) Response of plant–insect associations to Paleocene-Eocene warming. *Science* 284, 2153–2156.

Williams, S.E., Bolitho, E.E. and Fox, S. (2003) Climate change in Australian tropical rainforests: an impending environmental catastrophe. *Proceedings of the Royal Society of London Series B – Biological Sciences* 270, 1887–1892.

Wilson, R.J., Thomas, C.D., Fox, R., Roy, D.B. and Kunin, W.E. (2004) Spatial patterns in species distributions reveal biodiversity change. *Nature* 432, 394–396.

Wilson, R.J., Gutiérrez, D., Gutiérrez, J., Martínez, D., Agudo, R. and Monserrat, V.J. (2005) Changes to the elevational limits and extent of species ranges associated with climate change. *Ecology Letters* 8, 1138–1146.

Zvereva, E.L. and Kozlov, M.V. (2006) Consequences of simultaneous elevation of carbon dioxide and temperature for plant–herbivore interactions: a meta-analysis. *Global Change Biology* 12, 27–41.

12 Conservation Genetics for Insects

David J. Thompson, Phillip C. Watts and Ilik J. Saccheri

Population and Evolutionary Biology Research Group, School of Biological Sciences, University of Liverpool, Crown Street, Liverpool L69 7ZB, UK

1 Introduction

Frankham *et al.* (2002) define conservation genetics as the application of genetics to preserve species as dynamic entities capable of coping with environmental change. Conservation genetics is both a basic and an applied science. Genetic studies supply conservation scientists and ecological managers with new insights relevant to two main aspects of population management: (i) the risks to population viability posed by low levels of genetic diversity, through inbreeding depression and reduced adaptive potential; and (ii) the structure of populations, including effective size, patterns and rates of gene flow and phylogeography at regional and global scales. At the operational level, this evolutionary dynamic perspective forces us to balance the benefits of local adaptation against those of genetic diversity, whilst at a strategic level it provides important information for prioritizing investment of scarce resources into populations with the highest conservation value.

Not everyone is convinced that conservation genetics has such an important role. In an influential paper, Lande (1988) argued that 'demography may usually be of more importance than population genetics in determining the minimum viable size of wild populations'. This statement has been extrapolated by others to mean that ecological and demographic factors drive populations to extinction before genetic factors have time to exert their influence. Following Lande's review the issue has been much debated in the literature but the number of critical empirical studies remains few. In recent reviews Spielman *et al.* (2004) and Frankham (2005) have argued that most species are not driven to extinction before genetic factors impact upon them.

Haig and Avise (1996) reported lower levels of genetic diversity in endangered bird species than in non-threatened species. This type of analysis was expanded significantly in the meta-analysis conducted by Spielman *et al.* which examined the hypothesis that threatened taxa showed less genetic

Table 12.1. Percentages of threatened taxa with lower heterozygosity than taxonomically related non-threatened taxa ($H_t < H_{nt}$) in a range of major taxa and the magnitudes of those differences. (From Spielman *et al.*, 2004.)

Taxon	$H_t < H_{nt}$ (%)	Median difference (%)	Mean difference (%)	*n*	*P*
All	77	40	35	170	<0.0005
Animals	78	38	35	134	<0.0005
Vertebrates	78	35	35	129	<0.0005
Homeotherms	81	43	40	94	<0.0005
Mammals	84	46	42	63	<0.0005
Birds	74	40	35	31	0.001
Poikilotherms	69	26	20	35	0.001
Invertebrates	80	67	37	5	0.140
Plants	75	57	38	36	<0.0005
Angiosperms	81	58	40	21	0.005
Gymnosperms	67	51	35	15	0.012

n = number of threatened taxa; *P* = probabilities based on Wilcoxon's signed rank tests.

diversity than related non-threatened taxa. They used the International Union for Conservation of Nature (IUCN) *Red List* criteria to select threatened species and compared genetic diversity between these and taxonomically related species. Their major finding is repeated in Table 12.1. The results are clear for most taxa: threatened species show less genetic diversity. The finding is consistent for most of the taxa studied by Spielman *et al.* (2004). However, it is not proven for invertebrates. Unfortunately, the sample size for this group is considerably lower than that for other groups and only two of the five invertebrate taxa are insects. The Uncompahgre fritillary butterfly, *Boloria acrocnema*, which is compared with two other *Boloria* spp., constitutes one taxon, and five different *Formica* ants, *F. aquilonia*, *F. lugubris*, *F. polyctena*, *F. rufa* and *F. uralensis*, which are compared with eight non-threatened *Formica* spp., constitute the other. The relationship observed in the Spielman *et al.* review is a correlation. Species that are endangered according to IUCN criteria show less heterozygosity. We need to explore whether decline in heterozygosity contributes directly to 'endangered' status or whether it is a side effect of little relevance to future persistence.

We were able to update this list slightly (Table 12.2). We observe no difference in levels of genetic diversity between threatened and non-threatened Hymenoptera (extrapolated from *Formica* ants), but the few studies available for Lepidoptera are symptomatic of the trend of Spielman *et al.* towards low diversity in threatened species. Similar correlations have been made for nationally 'threatened' (i.e. ones that are not on the current IUCN *Red List*) species (e.g. Cassel and Tammaru, 2003).

At present, there is no general 'insect signal', but there may be differences between families. It seems unlikely that insects should, on theoretical grounds, be any different from groups for which there are adequate data. The aim of

Table 12.2. Comparison of gene diversities (H_e) in IUCN-listed and non-listed insect taxa.

Listed taxa	IUCN	H_e	H_e	Non-listed taxa	References
Formica aquilonia	LR/nt	0.48	0.38	*Formica yessensis*	DeHeer and Herbers (2004)
Formica lugubris	LR/nt	0.47	0.64[a]	*Formica podzolica*	Goropashnaya *et al.* (2001)
Formica polyctena	LR/nt	0.72	0.39	*Formica truncorum*	Gyllenstrand *et al.* (2002, 2004)
Formica rufa	LR/nt	0.65	0.54	*Formica cinerea*	Hasegawa and Imai (2004)
			0.60	*Formica exsecta*	Mäki-Petäys *et al.* (2005); Seppa *et al.* (2004); Sundström *et al.* (2003)
Parnassius apollo	VU	0.45	0.76	*Parnassius smintheus*	Keyghobadi *et al.* (1999, 2002, 2005a,b); Petenian *et al.* (2005)
Maculinea alcon	LR/nt	0.52	0.60	*Maculinea nausithous*	Zeisset *et al.* (2005)
Maculinea teleius	LR/nt	0.41			
Maculinea alcon	LR/nt	0.09	0.29 (0.42)	*Maculinea nausithous*	Bereczki *et al.* (2005); Figurny-Puchalska *et al.* (2000)
Maculinea teleius	LR/nt	0.13			

[a]Only observed heterozygosity provided.
LR = lower risk; nt = near-threatened; VU = vulnerable.

this chapter is to explore whether there is a 'conservation genetics' for insects that might in some way be different from a conservation genetics for other organisms – at least those typically featuring in the conservation literature – by virtue of those features of insect life histories that have made them so successful. In order to illustrate the kind of data that we advocate should be collected and used by conservation managers, we introduce some information on population size and genetic variation in the UK of the damselfly *Coenagrion mercuriale*, one of Europe's highest profile odonate species from a conservation perspective.

2 Genetic Sources of Extinction Risk

There are two principal genetic threats associated with the small or declining population scenario with which conservationists are inevitably concerned. The first is reduced adaptive potential through random genetic drift, which generally acts slowly, but will be quicker in small populations; the second is a lowering of fitness due to the exposure of deleterious recessive alleles as homozygosity increases (inbreeding depression). Inbreeding depression is often most severe for major components of fitness (e.g. fertility, early development, physiological vigour). Adaptive variation may be most relevant to environmental pressures such as disease, climate change and chemical toxins. The relationships between population size, loss of genetic diversity and extent of inbreeding are described in Eq. 12.1 (for closed randomly mating populations):

$$H_t / H_0 = (1 - 1/[2\ N_e])^t = 1 - F_t \qquad (12.1)$$

where H_t is heterozygosity in generation t, H_0, the initial heterozygosity, N_e, the (genetically) effective population size and F_t, the inbreeding coefficient.

Thus, both drift and extent of inbreeding depend upon N_e, the effective population size, whereas the only statistic usually available to conservation practitioners is N_c, the census population size (referred to hereafter as N). In insects, even N is seldom very well known because data on insect population densities in the field over long periods are hard to find. Note that we draw a distinction between relative estimates of insect population sizes as determined from monitoring data and the densities of insect populations on which population dynamics operate. It is highly unlikely that there is a universal relationship between the relative estimates of population size obtained by, for example, Pollard walks and real density estimates in terms of numbers of insects per square metre.

The typical scenario faced by conservation practitioners is an estimate of N and knowledge of the literature of typical values of the ratio of N_e/N. Frankham (1995) in a meta-analysis of estimates of N_e/N found a mean value of 0.11, significantly lower than the first reviewed estimates of 0.25–0.5 (Mace and Lande, 1991). What this means is that a population in which $N = 5000$ would most likely have an effective population size of ~500, not the 1250–2500 estimated earlier. With this lowered estimate of N_e came the realization that many more populations than had been previously supposed were in danger of experiencing increased rates of drift and inbreeding.

3 How Does Reduced Genetic Variation and Increased Homozygosity Influence Population Size?

The combination of drift, inbreeding depression and other non-genetic demographic and environmental factors can send small populations into what Gilpin and Soulé (1986) termed an extinction vortex. A positive feedback loop is set in operation through impacts on the population growth rate, deepening inbreeding depression and causing further genetic homogenization of the population, compromising adaptive responses. If this 'genetic erosion' leads the population deeper into the vortex, there are two possible outcomes. The first is extinction. The second is what has become known as purging of deleterious genes such that the limited remaining genetic variation includes genes that enable the organism to cope with the present environmental conditions with which it finds itself. This may slow down the progression to extinction in the short term but in the longer term the reduced genetic diversity is likely to limit adaptive responses to environmental change. There are some well-publicized studies of organisms that have come back from the brink, most notably the Mauritius kestrel (*Falco punctatus*) that recovered from a single wild breeding pair in 1974 to number over 800 birds by 2000 (Groombridge *et al.*, 2000), but this is likely to be the exception and not the rule. Even if populations survive the extinction vortex in the short term and purge deleterious alleles from their genome, they may have insufficient genetic diversity to respond to environmental changes.

It is worth pointing out that much of the data on genetic diversity on threatened species has been provided by characterizing putative neutral molecular markers such as microsatellites. However, it is quantitative genetic variation that is the main determinant of the ability to evolve. Reed and Frankham (2001) explored the correlation between molecular and quantitative measures of genetic variation in a meta-analysis of 71 data-sets. They found that the mean correlation was weak ($r = 0.217$) and that there was no significant correlation between the two measures for life history traits ($r = -0.11$) or for heritability ($r = -0.08$), the quantitative measure generally considered the key indicator of adaptive potential.

Thus, it is unclear if the evolutionary potential is reduced in endangered species compared to comparable non-endangered species. Genome-wide estimates of genetic diversity based on a few molecular markers need to be interpreted with care and may not be an accurate predictor of the selection response to a specific environmental challenge. Increasingly, however, we will be in a position to measure adaptive diversity using quantitative trait loci and candidate genes (Fitzpatrick *et al.*, 2005).

4 Variation in Effective Population Size in Insects in the Field

The ratio of N_e/N is not a constant and may be influenced by a number of factors pertinent to insect populations:

1. Fluctuations in population size: if there are fluctuations, N_e will fall below the mean number of adults. N_e in a fluctuating population is not the average but the harmonic mean of the effective population sizes over t generations.

$$N_e \approx t \;/\; (\Sigma[1 \;/\; N_{ei}]) \tag{12.2}$$

where N_{ei} is the effective size in the ith generation.

2. Variation in family size: when variance in family size is greater than that predicted by the Poisson distribution, N_e will be less than the number of adults.

$$N_e \approx 2N \;/\; (1 + [\sigma^2 \;/\; k]) \tag{12.3}$$

where σ^2 is the variance in family size among individuals and k is the average number of offspring per individual.

3. Unequal sex ratio: N_e is biased towards the sex with the fewer individuals.

$$N_e = (4\, N_f\, N_m)/(N_f + N_m) \tag{12.4}$$

where N_f and N_m are numbers of reproductive females and males, respectively.

4.1 Variation in population size

Are insect population sizes more variable than those of other organisms? If so, fluctuations in population size are likely to render the effective population sizes of insect populations lower than might have been supposed. Hanski (1990) reviewed variability in population sizes in a number of taxa. His results are shown in Fig. 12.1. Even noting the caveat of Pimm and Redfearn (1988) that the variability of populations will increase over time, it is clear that variation in insect population sizes is considerably larger than that shown by vertebrates (and larger than the one order of magnitude speculated upon by Thomas, 1990). This characteristically high demographic variability may often lead to gross overestimation of N_e based on short runs of data, leading to potentially serious underestimates of minimum viable population (MVP) size.

4.2 Variation in family size

Variation in family size is known on theoretical grounds to reduce N_e. This has been carefully demonstrated with controlled experiments using *Drosophila* (see, e.g. Borlase *et al.*, 1993). Frankham *et al.* (2002) review examples from a range of species (all vertebrates). Other data on lifetime mating or reproductive success come from the behavioural ecology literature (Clutton-Brock, 1988), where the majority are once again vertebrates. However, there is a group of insects that has provided a large number

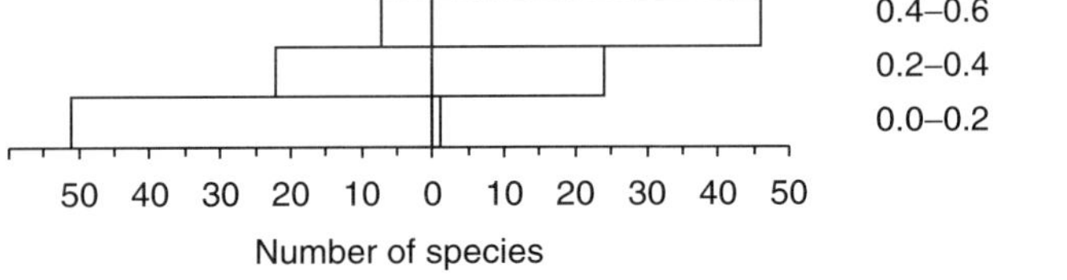

Fig. 12.1. Population variability in 91 species of terrestrial vertebrates (mammals, birds and lizards) and in 99 species of terrestrial arthropods (moths, aphids, hoverflies, grasshoppers, etc.). Variation has been measured for generations where possible (most studies). The distributions for vertebrates and arthropods are significantly different (two-tailed Kolmogorov-Smirnov statistic = 0.58, $P < 0.0001$). (Reproduced with permission from Hanski, 1990.)

of examples of lifetime mating success (LMS), and they are the Odonata. Odonates, by virtue of their (relatively) large size, confinement to discrete water bodies for reproduction and ability to be marked uniquely, with the marks being visible through close-focusing binoculars, are ideal organisms with which to study lifetime mating and reproductive success. Purse and Thompson (2005) demonstrated significant variation in family size in *C. mercuriale* at one of the UK's more threatened populations (Aylesbeare Common, Devon). Given variation in family sizes, $N_e \sim 8N/(V_{kf} + V_{km} + 4)$ where V_{kf} and V_{km} are the variances in reproductive success for females and males, respectively (Falconer and Mackay, 1996). Purse and Thompson's (2005) data on LMS were adjusted so that a stable population size is maintained (see Crow and Morton, 1955), and provide estimates of $V_{kf} = 7.4$ and $V_{km} = 13.5$ that generate an N_e/N ratio of 0.32; clearly this assumes that LMS is proportional to lifetime reproductive success but the validity of this assumption has not been tested.

4.3 Unequal sex ratio

Haplodiploid organisms have been considered immune to genetic load impacts because deleterious alleles are readily purged in haploid males, so the effect of genetic factors in contributing to their extinction has not been studied extensively. However Zayed (2004) has pointed out that complementary sex determination in the haplodiploid Hymenoptera leads to the production of inviable or effectively sterile diploid males when diploid progeny are homozygous at the sex-determining locus. This production of diploid males reduces the female population size and biases the breeding sex ratio in favour of haploid males, which in turn reduces N_e. This can lead, in small populations, to what Zayed and Packer (2005) term a novel extinction vortex (the diploid male vortex). This phenomenon has been demonstrated for the orchid bee *Euglossa imperialis* in lowland rainforests in Panama (Zayed *et al.*, 2004).

Sex-limited expression of deleterious alleles has an analogous effect on N_e. For example, in the satyrid butterfly *Bicyclus anynana* male fertility is acutely sensitive to inbreeding, with about 50% of sons from sib matings being completely sterile; female fertility on the other hand is insensitive to inbreeding (Saccheri *et al.*, 2005). Furthermore, sex-limited expression of deleterious alleles constrains the efficiency of purifying selection because the non-affected sex acts as carriers.

5 Evidence for Inbreeding Depression

The case for the occurrence of inbreeding depression under captive conditions either in zoos or in laboratories has frequently been made (see Lacy *et al.*, 1993). One of the most persuasive data-sets was gathered by Ralls *et al.* (1988), who examined pedigrees from 40 captive zoo populations belonging to 38 species and showed that the average increase in percentage mortality was 33% for inbred matings. Evidence for inbreeding depression in wild populations is less common, despite its importance in conservation biology and, for that matter, in evolutionary theory. Crnokrak and Roff (1999) examined inbreeding depression in wild species monitored in the field. They were able to obtain 169 estimates of inbreeding depression for 137 traits from seven birds, nine mammals, two fish, one snake, one snail and 15 plant species. They found significantly high levels of inbreeding depression, levels that could have biological importance. Notable by their absence from this review were data on insects. A later review (Keller and Waller, 2002) added one butterfly species. In the following sections we present some evidence for inbreeding depression in laboratory-maintained experimental insect populations, and then for insect populations in the field.

5.1 Evidence for inbreeding depression in insects in the laboratory

Saccheri *et al.* (1996) established inbred laboratory lines of the satyrid *B. anynana* with one, three and ten pairs of butterflies, which were subsequently allowed to increase to a maximum size of 300 butterflies. They measured fecundity, egg weight, egg hatching, adult emergence, adult size and the proportion of crippled adults in generations F2, F3, F5 and F7. Their most striking finding was an unexpectedly large decrease in egg hatching with increase in inbreeding (25% per 10% increase in inbreeding). This was a level of inbreeding not previously recorded in insect populations. Table 12.3 shows the regression coefficients of fitness component against expected inbreeding coefficient for *Bicyclus* and six other insect species, with a comparison between these six and *Bicyclus* underlining the severity of the inbreeding depression in the latter.

Bijlsma *et al.* (2000) reported that inbred populations of *Drosophila melanogaster* have a significantly higher short-term probability of extinction

Table 12.3. Regression coefficients (*b*) and their standard error for the regression of fitness component on expected inbreeding coefficient (*F*) for seven insects. (After Saccheri *et al.*, 1996.)

Species	Trait	Range of *F*	*b* (SE)	References
Bicyclus anynana	Egg hatching	0–0.27	−2.48 (0.13)	Saccheri *et al.* (1996)
Heleconius erato	Egg hatching	0–0.31	−1.07 (0.20)	Di Mare and Araújo (1986)
Dryas iulia	Egg hatching	0–0.38	−1.02 (0.16)	Haag and Araújo (1994)
Drosophila melanogaster	Egg–pupa viability	0–0.50	−0.30 (0.07)	García *et al.* (1994)
D. pseudoobscura	Egg–pupa viability	0–0.25	−0.52 (0.14)	Dobzhansky *et al.* (1963)
Musca domestica	Egg–pupa viability	0–0.25	−0.99 (0.51)	Bryant *et al.* (1986)
Tribolium castaneum	Egg–pupa viability	0–0.25	−0.45 (0.09)	Fernández *et al.* (1995)

than non-inbred populations, even for low levels of inbreeding. They also observed that extinction probability increases with greater levels of inbreeding. Moreover, the effect of inbreeding is enhanced in more stressful environments (high temperature, ethanol stress), demonstrating a synergistic impact of inbreeding and environmental stress. Subsequent work by Pedersen *et al.* (2005) has explored the role of heat shock proteins (Hsp) in coping with stressful conditions. Interestingly, these authors found that inbred *D. melanogaster* larvae upregulate the expression of Hsp70, possibly reflecting a cellular attempt to restore protein homeostasis.

5.2 Evidence for inbreeding depression in insects in the wild

Saccheri *et al.* (1998) looked at the effect of inbreeding on local extinction in a large metapopulation of the Glanville fritillary butterfly *Melitaea cinxia*. Adult butterflies were sampled from 42 populations across the Åland islands off south-western Finland. These populations ranged in character from small and isolated to large and non-isolated. Heterozygosity was investigated at seven polymorphic enzyme loci and one microsatellite locus. Seven of these 42 populations went extinct during the period under study (summer 1995 to summer 1996). When all the factors suspected of contributing to extinction risk were factored out, Saccheri *et al.* found that extinction risk increased significantly with decreasing heterozygosity and was responsible for 26% of the variation in extinction risk. Larval survival, adult longevity and egg-hatching rate were adversely affected by inbreeding and were the fitness components underlying the inbreeding–extinction relationship (see also Haikola *et al.*, 2001; Nieminen *et al.*, 2001).

Within their results were some factors that are likely to be operating within other insect populations. For example, there was a positive association ($P < 0.05$) between the date when females were sampled in the field and their heterozygosity, which suggested that short-lived females were more homozygous and inbred. In the field, females are able to produce up to seven clutches of 50–350 eggs. If their longevity is reduced due to inbreeding, there are likely to be significant effects on their population dynamics.

5.3 Augmentation: genetic rescue

In situations where a population's viability is clearly compromised by inbreeding or reduced genetic diversity, one way forward is to genetically augment the population with individuals from one or more populations with higher (or simply different) diversity. There has been some success with this approach.

Madsen *et al.* (1999, 2004) studied an isolated population of adders (*Vipera berus*) in southern Sweden. The adder population declined in the 1960s and showed all the indications of being inbred with low genetic variability and a high proportion of stillborn or deformed young. In 1992, 20 adult male adders were translocated from a large genetically viable population to Smygehuk study site. Genetic variability within the population increased as did the number of males censused. The proportion of stillborn offspring declined, indicating that the recruitment was due to higher juvenile survival. Thus, a dramatic recovery was witnessed in a seriously declining population by the introduction of new genetic material.

Not all augmentations have been quite so successful. For example, no genetic effect was detected in a Swedish population to which 47 Norwegian otters had been translocated (Arrendal *et al.*, 2004) and, despite population growth, genetic diversity was lower than before the release. At a second site, a release of seven otters may have altered the genetic composition of the resident population but the geographic spread of diversity appeared to be limited.

The success of insect introductions has been reviewed by Oates and Warren (1990). Understanding why some insect introductions are successful and some fail can be problematic. For example, only one of the two releases (each of 50 inseminated females) of the butterfly *Erebia epiphron* led to a successful colonization. Interestingly, these 50 individuals retained nearly as much (allozyme) diversity as the source and were able to establish a high density, viable population (Schmitt *et al.*, 2005).

6 Evidence for Ability to Cope with Environmental Change in Insects

One way in which insects may have a better chance of escaping the so-called extinction vortex should they ever be sucked into it is through their rapid generation times compared with the usual suspects in conservation. There is evidence for evolutionary responses in insects. Bradshaw and Holzapfel (2001) reported that over the last 30 years the genetically controlled photoperiod of the pitcher-plant mosquito, *Wyeomyla smithii*, has shifted towards shorter, more southern daylengths as growing seasons have become longer, and that this shift is detectable over a time period as short as 5 years.

Umina *et al.* (2005) investigated the latitudinal cline in the alcohol dehydrogenase polymorphism in *D. melanogaster*. This is one of the most thoroughly studied examples of a genetic latitudinal cline in any organism. The Adh^S allele increases in frequency with decreasing latitude in both hemispheres. Umina *et al.* found that the cline had shifted over 20 years in eastern coastal Australia, with southern high-latitude populations having the genetic constitution of more northerly populations, equivalent to a shift of around 4° in latitude (Fig. 12.2). In these two examples, the evolutionary potential to change was present, albeit only by a change in allele frequency at a single locus.

However, this is not always the case. Hoffmann *et al.* (2003) looked at desiccation resistance in *D. birchii*, which exhibits clinal variation in desiccation resistance over about 7° of latitude in north-eastern Australian rainforests, with resistance increasing with latitude. Hoffmann *et al.* estimated genetic variance in desiccation resistance in two different ways. First, in a selection

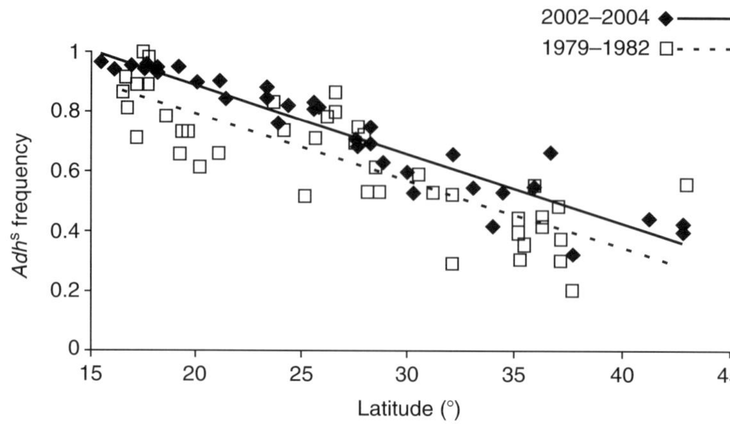

Fig. 12.2. The relationship between *Adh*^{s} frequency and latitude in *Drosophila melanogaster*. The dashed line and open symbols represent the years 1979–1982; the solid line and closed symbols represent the years 2002–2004. (Reproduced with permission from Umina *et al.*, 2005.)

experiment using the most resistant population, flies were unable to evolve further resistance despite intense selection for over 30 generations. Second, parent–offspring comparisons indicated low heritable variation for this trait but high levels of genetic variation in morphological traits such as wing size and wing aspect. *D. birchii* exhibited high levels of genetic variation at microsatellite loci, highlighting the importance of assessing evolutionary potential in targeted traits and re-emphasizing the care needed in interpreting results obtained from neutral markers when assessing evolutionary potential.

7 Conservation Genetics of *Coenagrion mercuriale*, the Southern Damselfly

We are interested in *C. mercuriale* (Charpentier) (Odonata: Zygoptera) because it is one of Europe's most threatened damselflies. It is listed on Annex II of the EC Habitats Directive and Appendix II of the Bern Convention and protected within Europe as a whole and by specific legislation in several countries. This species is one of two British resident odonates to be listed in the European Habitats directive that requires member states to designate Special Areas of Conservation for its protection, and therefore has a high conservation profile. The population centres are south-western Europe (Iberian Peninsula, France, Italy) and North Africa. *C. mercuriale* has either disappeared or is on the edge of extinction in Belgium, the Netherlands, Luxembourg, Slovenia, Romania, Poland and Austria (Grand, 1996) and is declining in other countries on the northern edge of its range, such as Germany and the UK. It is estimated that the species has suffered a 30% decline in the UK since 1960, largely due to anthropogenic changes in land use (Thompson *et al.*, 2003). In the UK *C. mercuriale* has a patchy distribution that is determined by the availability of specific breeding habitat, either small lowland heathland streams

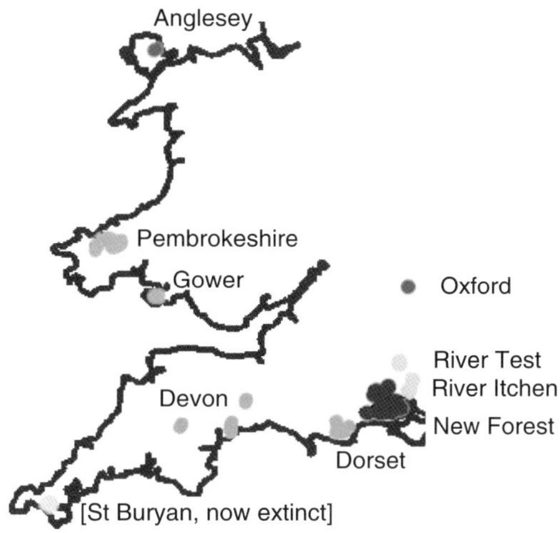

Fig. 12.3. Distribution of *Coenagrion mercuriale* populations in the UK.

emanating from base-rich substrate or in ditches on water–meadow systems on chalk streams. Within these biotopes *C. mercuriale* is confined to shallow, unshaded and permanently flowing small watercourses with perennial herbaceous aquatic vegetation.

We examined allelic variation at 14 unlinked microsatellite loci described by Watts *et al.* (2004a,b) to quantify the population structure of this species throughout almost all of its UK populations (Fig. 12.3). A consistent result, concomitant with its poor dispersal capability (Watts *et al.*, 2004c), is high levels of genetic differences between sites separated by relatively short distances (Watts *et al.*, 2004c, 2005, 2006). The spatial genetic variation among all UK sites may be summarized by principal component analysis (PCA) of allele frequencies and a plot of the sample scores (eigenvectors) of significant principal components. The first two principal components (Fig. 12.4) account for 24% and 17% of the variation within the data and are significant ($P < 0.001$ for each axis). The PCA plot is not amenable to explicit 'genetic interpretation' but reflects the distinctness and common ancestry of populations, such that geographically proximate sites tend to show the greatest similarity in their allele frequency profiles. It is notable that the New Forest samples form a large central cluster that corresponds with that observed in a continental European site and presumably the large amount of genetic variation that has been sustained by the larger populations in this region and thus its importance for biodiversity conservation. While a formal analysis is underway, it is evident that increased levels of genetic differentiation from the main New Forest sites is associated with some combination of: (i) geographic separation (e.g. Anglesey, in Figs 12.3 and 12.4) that presumably correlates with an increased temporal separation; (ii) inhospitable habitat matrix (e.g. Acres Down – ACD, Mariners Meadow – MAM); and (iii) small population size (e.g. ACD, Anglesey, Aylesbeare – AYB) where the effect of genetic drift in altering allele frequencies is more pronounced.

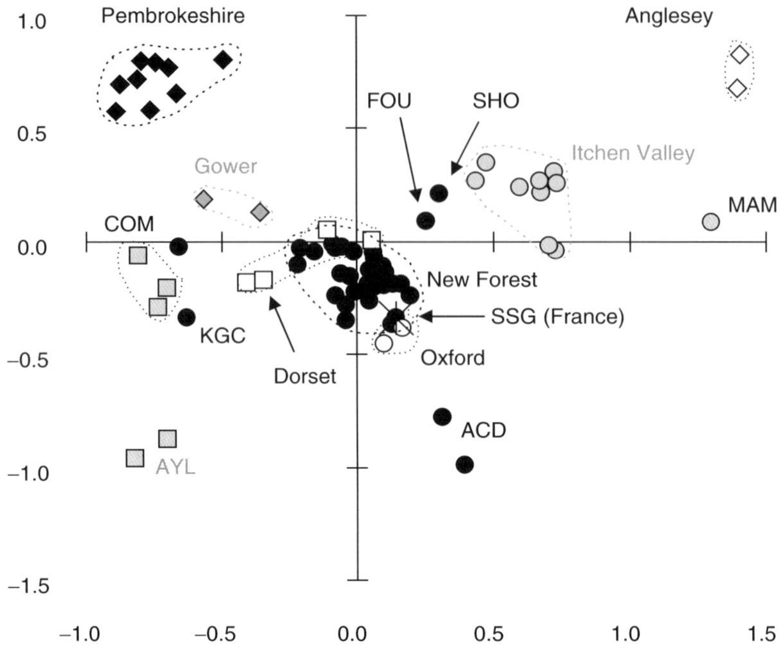

Fig. 12.4. Principal component analysis (PCA) plot showing spatial pattern of allele frequencies in the UK *Coenagrion mercuriale* populations (see Fig. 12.3 for distribution of sites). The symbols represent different centres of population. One French population (from Saint-Sulpice-de-Grimbouville, Normandy – SSG) is also plotted. Two Devon populations (Moortown Gidleigh Common and Aylesbeare Common, AYL), the Anglesey and Oxfordshire populations and Acres Down (ACD), New Forest, have two points representing sampling across 2 years. New Forest populations that separate from the main cluster are also indicated (FOU, SHO, COM and KGC). The most northerly of the Itchen Valley populations, Mariner's Meadow (MAM), is recognized separately.

Next, we measured genetic diversity in the UK populations in relation to their isolation (Fig. 12.5). The populations with highest expected heterozygosity and allelic richness are those in the centre of the *C. mercuriale* stronghold, in New Forest. Loss of diversity is greatest in peripheral sites than at the edges of population centres, particularly at the most isolated sites in the east Devon pebble beds, Dartmoor and in Anglesey. The extent to which this loss in diversity will 'drive' these populations to extinction or is merely a signal of isolation (at neutral markers) still needs to be explored. Partly to address this question, we examined gene diversity in museum specimens to look for evidence of diversity loss in extinct populations of *C. mercuriale*. One site from which *C. mercuriale* has been lost is the isolated site at St Buryan in Cornwall that is believed to have gone extinct in the late 1960s. Our historic specimens were collected in 1952 – less than ten generations before extinction. In view of the degree of isolation of this site, the genetic diversity displayed is substantially higher than expected (Fig. 12.5). Moreover, it is interesting that there is no signal of a substantial loss of diversity in a doomed population;

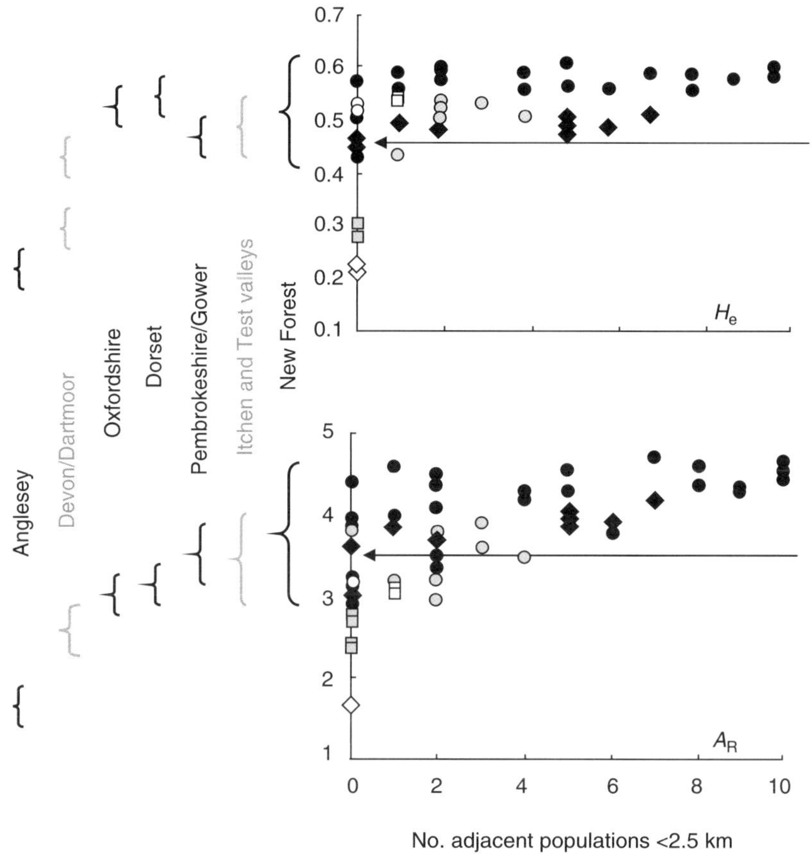

Fig. 12.5. The effect of isolation on expected heterozygosity (H_e) and allelic richness (A_R) in UK populations of *Coenagrion mercuriale*. The symbols and shadings correspond with the sites in Fig. 12.4. Solid circle = New Forest (●); solid diamond = Pembrokeshire (♦); grey circle = Itchen and Test valleys (◎); open square = Dorset (□); open circle = Oxfordshire (○); grey square = Devon (including Dartmoor) (▣); open diamond = Anglesey (◊). The arrow indicates the position of the extinct Cornish population for which DNA was extracted from museum specimens collected in 1952, less than 10 generations before extinction.

this perhaps reinforces the speed with which even large insect populations can be lost (e.g. through habitat loss) and highlights the dangers of focusing on a single factor (e.g. genetic erosion) rather than taking a holistic approach to conservation management.

8 Conclusions

The importance of inbreeding and genetic drift for population persistence is likely to vary considerably among insect species, depending on their genetic load (of deleterious mutations) and the need to adapt to environmental change over differing spatial and temporal scales. It would therefore be valuable to collect more data on inbreeding depression in insects and also to characterize the ecological context of selective environments, which determine the relative magnitude of hard versus soft selection (Wallace, 1975) and the demographic consequences of selection. While purely ecological management is aimed at maintaining a given census population size, genetic management is focused on the maintenance of effective population size. As we have discussed, these two measures of population size may differ by an order of magnitude or more, but in most insects both remain something of a mystery.

This said, we summarize the features that predispose many insects to such genetic effects as follows (we have considered the first two in detail in this chapter):

1. Small N_e/N ratio;
2. High genetic loads leading to large inbreeding depression;
3. Weak density dependence in ephemeral populations such that any reduction in fecundity, or increase in mortality, has an effect on population growth rate;
4. Fine-scale environmental heterogeneity (e.g. host plant variation) whose efficient use requires a reservoir of phenotypic variation;
5. Narrow habitat requirements and sensitivity to small environmental fluctuations (e.g. microclimate) that impose a limit on phenotypically plastic responses to environmental stress, placing greater importance on evolutionary adaptive responses.

Molecular markers are a valuable tool in conservation genetics for revealing information about population decline and isolation, neighbourhood size, barriers to dispersal and N_e. For example, studies on insects have correlated patch isolation with reduced genetic diversity (Cassel and Tammaru, 2003; Williams *et al.*, 2003; Krauss *et al.*, 2004; but cf. Schmitt *et al.*, 2005), although sometimes it can be difficult to attribute this effect to isolation per se rather than a consequence of small population size or patches occupying marginal habitats (e.g. Harper *et al.*, 2003; Krauss *et al.*, 2004; Schmitt *et al.*, 2005). Inferences about the causes of observed diversity patterns are further complicated by underlying (e.g. latitudinal) clines in diversity due to range expansion from southern refugia after the last ice age (see Hewitt, 2000). A detailed review of these effects is beyond the scope of this chapter but it is important to recognize the many difficulties of meta-analysis that may confound such effects.

Studies of population genetic structure typically represent a snapshot of a dynamic system which is unlikely to have reached genetic equilibrium. Moreover, heterozygosity is expected to approach equilibrium more slowly than the variance among populations (e.g. Crow and Aoki, 1984), which can lead to underestimates of future loss in diversity. For example, Keyghobadi *et al.* (2005a,b) observed that genetic differentiation among populations reflected the pattern of contemporary fragmentation among populations of the alpine butterfly *Parnassius smintheus*, while genetic diversity (H_e) was better correlated with forest landscape cover 40 years earlier. Our initial data exploration in *C. mercuriale* is suggestive of this, with allelic richness A_R showing a better correlation with patch isolation than H_e. Unfortunately, measures of allelic diversity are more sensitive to sample size than gene diversity. However, because rare alleles, which contribute little to heterozygosity, are lost more rapidly during a demographic reduction than more common alleles, allelic richness may be a better indicator of population diversity. More sophisticated statistical methods of analysing genetic data are currently under development (e.g. Goossens *et al.*, 2006).

A role for genetic management is most appropriate in situations where a population has gone locally extinct from part of its former range or where a population is isolated and shows signs of genetic erosion, either through low genetic diversity (relative to conspecific populations) or inbreeding depression. Attempts to rescue such populations, via reintroduction or augmentation with captive wild individuals from another source population, need to take a number of genetic considerations into account. The first is the evolutionary and ecological similarity of the target and source populations. In many cases the most closely related potential source will also share the most similar habitat, but in some cases a more phylogenetically distant source may be better adapted to the target habitat (Crandall *et al.*, 2000). The importance of local adaptation is illustrated by the changes in phenology observed in *Maculinea teleius* and *M. nausithous* reintroduced to the Netherlands from Polish stock (Wynhoff, 1998). In general, however, we should seek to preserve species-specific phylogeographies at local and regional scales (Avise, 2000). The problem is that these phylogeographies are only known in a handful of insects (e.g. Lunt *et al.*, 1998; Rubinoff and Sperling, 2004; Saccheri *et al.*, 2004; Vandewoestijne *et al.*, 2004), though in fact they are straightforward to obtain once the actual samples have been collected.

The second major genetic issue is the number of founders or immigrants that should be introduced. Inbreeding depression can be largely avoided with effective population sizes greater than 50 (1% inbreeding per generation), which may be equivalent to 100 or 1000 individuals. Maintaining genetic diversity, particularly the contribution of rare alleles, would require an effective population size closer to 1000 (Nunney and Campbell, 1993). In the longer-term Lande (1995) has argued that N_e of about 5000 is required to maintain potentially adaptive genetic variation in quantitative characters through mutation. Translating these largely theoretical guidelines for N_e into MVP sizes requires knowledge of the target population's demography, including variation in reproductive success among individuals, and environmental impacts (Frankham *et al.*, 2002).

We can illustrate (at least the first of) these issues in practical terms by returning to the endangered damselfly *C. mercuriale*. Thompson *et al.* (2003) and Rouquette and Thompson (2005) have described the habitat requirements of the species in its two most common biotopes in the UK: heathland streams and chalk stream flood plain ditches. Newly restored habitat close to existing sites can expect natural recolonization in ecological time (or could be augmented from existing strong populations in the Itchen Valley or Beaulieu

Heath, New Forest). Those populations in which genetic erosion has taken place, for example Nant Isaf in Anglesey and the Devon sites of Aylesbeare Common and Colaton Raleigh Common, should clearly be augmented from the UK stronghold sites. The issue is straightforward for the Devon sites where the habitat and phenology is similar to key sites within New Forest, so the source for the material to be reintroduced can be identified clearly. The issue is less clear-cut for the Nant Isaf site, which is one of only two fen sites for *C. mercuriale* in the UK. The other fen site, in Oxfordshire, is also genetically depauperate, whereas the UK stronghold sites are not fens. We can be less confident that augmentation would be successful, though there would appear to be no options other than waiting for the Nant Isaf population to become extinct.

References

Arrendal, J., Walker, C.W., Sundqvist, A.-K., Hellborg, L. and Vilà, C. (2004) Genetic evaluation of an otter translocation program. *Conservation Genetics* 5, 79–88.

Avise, J.C. (2000) *Phylogeography: The History and Formation of Species*. Harvard University Press, Cambridge, Massachusetts.

Bereczki, J., Pecsenye, K., Peregovits, L. and Varga, Z. (2005) Pattern of genetic differentiation in the *Maculinea alcon* species group (Lepidoptera, Lycaenidae) in Central Europe. *Journal of Zoological Systematics and Evolutionary Research* 43, 157–165.

Bijlsma, R., Bundgaard, J. and Boerema, A.C. (2000) Does inbreeding affect the extinction risk of small populations? Predictions from *Drosophila*. *Journal of Evolutionary Biology* 13, 502–514.

Borlase, S.C., Loebel, D.A., Frankham, R., Nurthen, R.K., Briscoe, D.A. and Daggard, G.E. (1993) Modelling problems in conservation genetics using captive *Drosophila* populations: consequences of equalisation of family sizes. *Conservation Biology* 7, 122–131.

Bradshaw, W.E. and Holzapfel, C.M. (2001) Genetic shift in photoperiodic response correlated with global warming. *Proceedings of the National Academy of Sciences of the United States of America* 98, 14509–14511.

Bryant, E.H., McCommas, S.A. and Combs, L.M. (1986) The effect of an experimental bottleneck upon quantitative genetic variation in the housefly. *Genetics* 114, 1191–1211.

Cassel, A. and Tammaru, T. (2003) Allozyme variability in central, peripheral and isolated populations of the scarce heath (*Coenonympha hero*: Lepidoptera, Nymphalidae); implications for conservation. *Conservation Genetics* 4, 83–93.

Clutton-Brock, T.H. (1988) *Reproductive Success: Studies of Individual Variation in Contrasting Breeding Systems*. University of Chicago Press, Chicago, Illinois.

Crandall, K.A., Bininda-Emonds, O.R.P., Mace, G.M. and Wayne, R.K. (2000) Considering evolutionary processes in conservation biology. *Trends in Ecology and Evolution* 15, 290–295.

Crnokrak, P. and Roff, D.A. (1999) Inbreeding depression in the wild. *Heredity* 83, 260–270.

Crow, J.F. and Aoki, K. (1984) Group selection for a polygenic behavioral trait: estimating the degree of population subdivision. *Proceedings of the National Academy of Sciences of the United States of America* 81, 6073–6077.

Crow, J.F. and Morton, N.E. (1955) Measurement of gene frequency drift in small populations. *Evolution* 9, 202–214.

DeHeer, C.J. and Herbers, J.M. (2004) Population genetics of the socially polymorphic ant *Formica podzolica*. *Insectes Sociaux* 51, 309–316.

Di Mare, R.A. and Araújo, A.M. (1986) A first survey of inbreeding effects in *Heliconius erato phyllis* (Lepidoptera: Nymphalidae). *Revista Brasileira de Genética* 9, 11–20.

Dobzhansky, T., Spassky, B. and Tidwell, T. (1963) Genetics of natural populations. XXXII. Inbreeding and the mutational and balanced loads in natural populations of *Drosophila pseudoobscura*. *Genetics* 48, 361–373.

Falconer, D.S. and Mackay, T.F.C. (1996) *Introduction to Quantitative Genetics*, 4th edn. Longman, Harlow, UK.

Fernández, A., Toro, M.A. and López-Fanjul, C. (1995) The effect of inbreeding on the redistribution of genetic variance of fecundity and viability in *Tribolium castaneum*. *Heredity* 75, 361–381.

Figurny-Puchalska, E., Gadeberg, R.M.E. and Boomsma, J.J. (2000) Comparison of genetic population structure of the large blue butterflies *Maculinea nausithous* and *M. teleius*. *Biodiversity and Conservation* 9, 419–432.

Fitzpatrick, M.J., Ben-Shahar, Y., Smid, H.M., Vet, L.E.M., Robinson, G.E. and Sokolowski, M.B. (2005) Candidate genes for behavioural ecology. *Trends in Ecology and Evolution* 20, 96–104.

Frankham, R. (1995) Effective population size/adult population size ratios in wildlife: a review. *Genetical Research* 66, 95–107.

Frankham, R. (2005) Genetics and extinction. *Biological Conservation* 126, 131–140.

Frankham, R., Ballou, J.D. and Briscoe, D.A. (2002) *Introduction to Conservation Genetics*. Cambridge University Press, Cambridge.

García, N., López-Fanjul, C. and García-Dorado, A. (1994) The genetics of viability in *Drosophila melanogaster*: effects of inbreeding and artificial selection. *Evolution* 48, 1277–1285.

Gilpin, M.E. and Soulé, M.E. (1986) Minimum viable populations: processes of species extinction. In: Soulé, M.E. (ed.) *Conservation Biology: the Science of Scarcity and Diversity*. Sinauer, Sunderland, Massachusetts, pp. 13–34.

Goossens, B., Chikhi, L., Ancrenaz, M., Lackman-Ancrenaz, I., Andau, P. and Bruford, M.W. (2006) Genetic signature of anthropogenic population collapse in orang-utans. *PLoS Biology* 4, 285–291.

Goropashnaya, A.V., Seppa, P. and Pamilo, P. (2001) Social and genetic characteristics of geographically isolated populations in the ant *Formica cinerea*. *Molecular Ecology* 10, 2807–2818.

Grand, D. (1996) Coenagrion mercuriale (Charpentier, 1840). In: van Helsdingen, P.J., Willemse, L. and Speight, M.C.D. (eds) *Background Information on Invertebrates of the Habitats Directive and the Bern Convention,* Council of Europe Publishing, Strasbourg, Germany 80, 245–253.

Groombridge, J.J., Jones, C.G., Bruford, M.W. and Nichols, R.A. (2000) 'Ghost' alleles of the Mauritius kestrel. *Nature* 403, 616.

Gyllenstrand, N., Gertsch, P.J. and Pamilo, P. (2002) Polymorphic microsatellite DNA markers in the ant *Formica exsecta*. *Molecular Ecology Notes* 2, 67–69.

Gyllenstrand, N., Seppa, P. and Pamilo, P. (2004) Genetic differentiation in sympatric wood ants, *Formica rufa* and *F. polyctena*. *Insectes Sociaux* 51, 139–145.

Haag, K.L. and Araújo, A.M. (1994) Inbreeding, genetic load and morphometric variation in natural populations of *Dryas iulia* (Lepidoptera: Nymphalidae). *Revista Brasileira de Genética* 17, 35–39.

Haig, S.M. and Avise, J.C. (1996) Avian conservation genetics. In: Avise, J.C. and Hamrick, J.L. (eds) *Conservation Genetics*. Chapman & Hall, New York, pp. 160–189.

Haikola, S., Fortelius, W., O'Hara, R., Kuussaari, M., Wahlberg, N., Saccheri, I., Singer, M. and Hanski, I. (2001) Inbreeding depression and the maintenance of genetic load in *Melitaea cinxia* metapopulations. *Conservation Genetics* 2, 325–335.

Hanski, I. (1990) Density dependence, regulation and variability in animal populations. *Philosophical Transactions of the Royal Society of London Series B – Biological Sciences* 330, 141–150.

Harper, G.L., Maclean, N. and Goulson, D. (2003) Microsatellite markers to assess the influence of population size, isolation and demographic change on the genetic structure of the UK butterfly *Polyommatus bellargus*. *Molecular Ecology* 12, 3349–3357.

Hewitt, G.M. (2000) The genetic legacy of the Quaternary ice ages. *Nature* 405, 907–913.

Hasegawa, E. and Imai, S. (2004) Characterization of microsatellite loci in red wood ants *Formica* (*s. str.*) spp. and the related genus *Polyergus*. *Molecular Ecology Notes* 4, 200–203.

Hoffmann, A.A., Hallas, R.J., Dean, J.A. and Schiffer, M. (2003) Low potential for climatic stress adaptation in a rainforest *Drosophila* species. *Science* 301, 100–103.

Keller, L.F. and Waller, D.M. (2002) Inbreeding effects in wild populations. *Trends in Ecology and Evolution* 17, 230–241.

Keyghobadi, N., Roland, J. and Strobeck, C. (1999) Influence of landscape on the population genetic structure of the alpine butterfly *Parnassius smintheus* (Papilionidae). *Molecular Ecology* 8, 1481–1495.

Keyghobadi, N., Roland, J. and Strobeck, C. (2002) Isolation of novel microsatellite loci in the Rocky Mountain apollo butterfly, *Parnassius smintheus*. *Hereditas* 136, 247–250.

Keyghobadi, N., Roland, J., Matter, S.F. and Strobeck, C. (2005a) Among- and within-patch components of genetic diversity respond at different rates to habitat fragmentation: an empirical demonstration. *Proceedings of the Royal Society of London Series B – Biological Sciences* 272, 553–560.

Keyghobadi, N., Roland, J. and Strobeck, C. (2005b) Genetic differentiation and gene flow among populations of the alpine butterfly, *Parnassius smintheus*, vary with landscape connectivity. *Molecular Ecology* 14, 1897–1909.

Krauss, J., Schmitt, T., Seitz, A., Steffan-Dewenter, I. and Tscharnke, T. (2004) Effects of habitat fragmentation on the genetic structure of the monophagous butterfly *Polyommatus coridon* along its northern range margin. *Molecular Ecology* 13, 311–320.

Lacy, R.C., Petric, A. and Warneke, M. (1993) Inbreeding and outbreeding in captive populations of wild animals. In: Thornhill, N.W. (ed.) *The Natural History of Inbreeding and Outbreeding: Theoretical and Empirical Perspectives*. University of Chicago Press, Chicago, Illinois, pp. 352–374.

Lande, R. (1988) Genetics and demography in biological conservation. *Science* 241, 1455–1460.

Lande, R. (1995) Mutation and conservation. *Conservation Biology* 9, 782–791.

Lunt, D.H., Ibrahim, K.M. and Hewitt, G.M. (1998) mtDNA phylogeography and post-glacial patterns of subdivision in the meadow grasshopper *Chorthippus parallelus*. *Heredity* 80, 633–641.

Mace, G.M. and Lande, R. (1991) Assessing extinction threats: towards a re-evaluation of IUCN threatened species categories. *Conservation Biology* 5, 148–157.

Madsen, T., Shine, R., Olsson, M. and Wittzell, H. (1999) Restoration of an inbred adder population. *Nature* 402, 34–35.

Madsen, T., Ujvari, B. and Olsson, M. (2004) Novel genes continue to enhance population growth in adders (*Vipera berus*). *Biological Conservation* 120, 145–147.

Mäki-Petäys, H., Zakharov, A., Viljakainen, L., Corander, J. and Pamilo, P. (2005) Genetic changes associated to declining populations of *Formica* ants in fragmented forest landscape. *Molecular Ecology* 14, 733–742.

Nieminen, M., Singer, M.C., Fortelius, W., Schöps, K. and Hanski, I. (2001) Experimental confirmation that inbreeding depression increases extinction risk in butterfly populations. *The American Naturalist* 157, 237–244.

Nunney, L. and Campbell, K.A. (1993) Assessing minimum viable population size: demography meets population genetics. *Trends in Ecology and Evolution* 8, 234–239.

Oates, M.R. and Warren, M.S. (1990) *A Review of Butterfly Introductions in Britain and Ireland*. World Wide Fund for Nature, Godalming, Surrey, UK.

Pedersen, K.S., Kristensen, T.N. and Loeschcke, V. (2005) Effects of inbreeding and rate of inbreeding in *Drosophila melanogaster* – Hsp70 expression and fitness. *Journal of Evolutionary Biology* 18, 756–762.

Petenian, F., Meglécz, E., Genson, G., Rasplus, J.Y. and Faure, E. (2005) Isolation and characterization of polymorphic microsatellites in *Parnassius apollo* and *Euphydryas aurinia* (Lepidoptera). *Molecular Ecology Notes* 5, 243–345.

Pimm, S.L. and Redfearn, A. (1988) The variability of population densities. *Nature* 334, 613–614.

Purse, B.V. and Thompson, D.J. (2005) Lifetime mating success in a marginal population of a damselfly, *Coenagrion mercuriale*. *Animal Behaviour* 69, 1303–1315.

Ralls, K., Ballou, J.D. and Templeton, A. (1988) Estimates of lethal equivalents and the cost of inbreeding in mammals. *Conservation Biology* 2, 185–193.

Reed, D.H. and Frankham, R. (2001) How closely correlated are molecular and quantitative measures of genetic variation? A meta-analysis. *Evolution* 55, 1095–1103.

Rouquette, J.R. and Thompson, D.J. (2005) Habitat associations of the endangered damselfly, *Coenagrion mercuriale*, in a water meadow ditch system in southern England. *Biological Conservation* 123, 225–235.

Rubinoff, D. and Sperling, F.A.H. (2004) Mitochondrial DNA sequence, morphology and ecology yield contrasting conservation implications for two threatened buckmoths (Hemileuca: Saturniidae). *Biological Conservation* 118, 341–351.

Saccheri, I., Kuussaari, M., Kankare, M., Vikman, P., Fortelius, W. and Hanski, I. (1998) Inbreeding and extinction in a butterfly metapopulation. *Nature* 392, 491–494.

Saccheri, I.J., Brakefield, P.M. and Nichols, R.A. (1996) Severe inbreeding depression and rapid fitness rebound in the butterfly *Bicyclus anynana* (Satyridae). *Evolution* 50, 2000–2013.

Saccheri, I.J., Boggs, C., Hanski, I. and Ehrlich, P.R. (2004) Genetics of checkerspot populations. In: Ehrlich, P.R. and Hanski, I. (eds) *On the Wings of Checkerspots: A Model System for Population Biology*. Oxford University Press, New York, pp. 199–218.

Saccheri, I.J., Lloyd, H., Helyar, S. and Brakefield, P.M. (2005) Inbreeding uncovers fundamental differences in the genetic load affecting male and female fertility in a butterfly. *Proceedings of the Royal Society of London Series B – Biological Sciences* 272, 39–46.

Schmitt, T., Cizek, O. and Konvicka, M. (2005) Genetics of a butterfly relocation: large, small and introduced populations of the mountain endemic *Erebia epiphron silesiana*. *Biological Conservation* 123, 11–18.

Seppa, P., Gyllenstrand, N., Corander, J. and Pamilo, P. (2004) Coexistence of the social types: genetic population structure in the ant *Formica exsecta*. *Evolution* 58, 2462–2471.

Spielman, D., Brook, B.W. and Frankham, R. (2004) Most species are not driven to extinction before genetic factors impact them. *Proceedings of the National Academy of Sciences of the United States of America* 101, 15261–15264.

Sundström, L., Keller, L. and Chapuisat, M. (2003) Inbreeding and sex-biased gene flow in the ant *Formica exsecta*. *Evolution* 57, 1552–1561.

Thomas, C.D. (1990) What do real population dynamics tell us about minimum viable population sizes? *Conservation Biology* 4, 324–327.

Thompson, D.J., Rouquette, J.R. and Purse, B.V. (2003) *Ecology of the Southern Damselfly,* Coenagrion mercuriale. Conserving Natura 2000 Rivers Ecology Series No. 8. English Nature, Peterborough, UK.

Umina, P.A., Weeks, A.R., Kearney, M.R., McKechnie, S.W. and Hoffmann, A.A. (2005) A rapid shift in a clinal pattern in *Drosophila* reflecting climate change. *Science* 308, 691–693.

Vandewoestijne, S., Baguette, M., Brakefield, P.M. and Saccheri, I.J. (2004) Phylogeography of *Aglais urticae* (Lepidoptera) based on DNA sequences of the mitochondrial COI gene and control region. *Molecular Phylogenetics and Evolution* 31, 630–646.

Wallace, B. (1975) Hard and soft selection revisited. *Evolution* 29, 465–473.

Watts, P.C., Thompson, D.J. and Kemp, S.J. (2004a) Cross-species amplification of microsatellite loci in some European zygopteran species (Odonata: Coenagrionidae). *International Journal of Odonatology* 7, 87–96.

Watts, P.C., Wu, J.H., Westgarth, C., Thompson, D.J. and Kemp, S.J. (2004b) A panel of microsatellite loci for the southern damselfly, *Coenagrion mercuriale* (Odonata: Coenagrionidae). *Conservation Genetics* 5, 117–119.

Watts, P.C., Rouquette, J.R., Saccheri, I.J., Kemp, S.J. and Thompson, D.J. (2004c) Molecular and ecological evidence for small-scale isolation by distance in an endangered damselfly, *Coenagrion mercuriale*. *Molecular Ecology* 13, 2931–2945.

Watts, P.C., Kemp, S.J., Saccheri, I.J. and Thompson, D.J. (2005) Conservation implications of genetic variation between spatially and temporally distinct colonies of the endangered damselfly *Coenagrion mercuriale*. *Ecological Entomology* 30, 541–547.

Watts, P.C., Saccheri, I.J., Kemp, S.J. and Thompson, D.J. (2006) Population structure and the impact of regional and local habitat isolation upon genetic diversity of the endangered damselfly *Coenagrion mercuriale* (Odonata: Zygoptera). *Freshwater Biology* 51, 193–205.

Williams, B.L., Brawn, J.D. and Paige, K.N. (2003) Landscape scale genetic effects of habitat fragmentation on a high gene flow species: *Speyeria idalia* (Nymphalidae). *Molecular Ecology* 12, 11–20.

Wynhoff, I. (1998) Lessons from the reintroduction of *Maculinea teleius* and *M. nausithous* in the Netherlands. *Journal of Insect Conservation* 2, 47–57.

Zayed, A. (2004) Effective population size in Hymenoptera with complementary sex determination. *Heredity* 93, 627–630.

Zayed, A. and Packer, L. (2005) Complementary sex determination substantially increases extinction proneness of haplodiploid populations. *Proceedings of the National Academy of Sciences of the United States of America* 102, 10742–10746.

Zayed, A., Roubik, D.W. and Packer, L. (2004) Use of diploid male frequency data as an indicator of pollinator decline. *Proceedings of the Royal Society of London Series B – Biological Sciences* 271, S9–S12.

Zeisset, I., Als, T.D., Settele, J. and Boomsma, J.J. (2005) Microsatellite markers for the large blue butterflies *Maculinea nausithous* and *Maculinea alcon* (Lepidoptera: Lycaenidae) and their amplification in other *Maculinea* species. *Molecular Ecology Notes* 5, 165–168.

13 Broadening Benefits to Insects from Wider Conservation Agendas

TIMOTHY R. NEW

Department of Zoology, La Trobe University, Melbourne, Victoria 3086, Australia

1 Introduction

The great diversity of contexts in which insect conservation is important has become abundantly clear in recent decades. Substantial advances have been made in both the theory and practice of conserving individual species and assemblages, and in appreciating the values of insects (and other invertebrates) in wider evaluation of biodiversity and environmental conditions. It is vital that this progress is accelerated and enhanced, towards a more satisfactory global agenda for insect conservation, and to overcome the predominance of small-scale or local decisions. Otherwise, conservation efforts for insects seem destined to remain as fragmented as many of the ecosystems with which we are concerned, and largely concentrated in the parts of the world already conserved most effectively and with the logistic capability to enhance those efforts further. Even with limited agreement over actual numbers of threatened species (against the wider backdrop of ignorance over insect species richness), a need for conservation is clear, and conservation must proceed as effectively as we can contrive in an environment of highly incomplete documentation of detail. The twin strands for advance appear: (i) to increase direct attention to insects themselves, to promote insect conservation per se; and (ii) to increase focus on the values of insects, as predominant constituents of biodiversity, as tools to enhance the effectiveness of wider conservation measures not directed primarily at insect conservation.

The full extent of need for insect conservation has not really been quantified, and can be discussed only in general terms. Their predominance 'constitutes much of the potential for use of biodiversity and offers a huge part of the complexity of managing this biodiversity' (Janzen, 1997, p. 424). Despite increasing numbers being 'listed' for conservation significance, using criteria, such as those advocated by the International Union for Conservation of Nature (IUCN), World Conservation Union and others relatively few insect species (certainly an extremely low proportion of the total and with many orders scarcely represented) have been signalled individually as threatened. Even for butterflies, the proportion of apparently threatened species is far lower than for birds or mammals (McKinney, 1999). None the less, many entomologists accept that numerous species are likely to become extinct and that far more than the few recorded have already done so as a consequence of human cupidity. It is possible, even likely, that insect extinction rates have been underestimated by up to three orders of magnitude, with the reality that up to a quarter of all insect species could be threatened with near-immediate extinction (McKinney, 1999). In this context, it is critical that we distinguish carefully the widespread calls for describing the world's biodiversity (in itself an entirely laudable aspect of human endeavour) from practical conservation of that biodiversity. The two topics are very different and, although describing and naming biodiversity in greater detail may provide firmer foundation for conservation activity, we cannot afford to await that as a prerequisite for action. As Moore (1995) says in a discussion of dragonfly conservation: '[T]ime is not on our side. Research and education are important, but they are long-term activities. They must never be used as excuses for not acting now.... In the absence of adequate data on distribution and requirements of (dragonfly) species, we can best protect species by ensuring that each and every country protects good examples of its main biotopes, with the conservation of healthy ecosystems given priority over the conservation of rare species as such.'

Even if the 20–30 years noted by some optimistic commentators is sufficient for description of the earth's insects, practical conservation must be enhanced in the interim period. Janzen (1994, 1997) asked, in relation to documenting tropical biodiversity, 'what do we not need to know?', and his provocative essay merits careful reading by insect conservationists. He noted, *inter alia*, that: (i) we do not need to enumerate all the taxa that exist, because we already know that there are thousands to millions of species and that most of these are wholly or almost wholly unknown; (ii) detailed distributional knowledge of all species is not needed, because the areas available for practical conservation are mostly already defined, and understanding the biota of those areas is paramount; (iii) detailed data on individual species are largely redundant in the wider biological and sociological contexts of conservation need; and (iv) realistic triage is needed to harmonize conservation with agroecology in the tropics with wider attention devoted to larger areas and, perhaps, less to small isolated fragments (because many taxa are probably beyond saving in the tropical landscape: Janzen wrote 'the living dead and the population fragments sprinkled across the tropical agroscape are slated for the dust bin'). Although full documentation of an insect assemblage or regional fauna at species (or similar) level remains utopian (and so cannot be a prerequisite for conservation), we might be in a position to define an initial portfolio of insect groups for effective taxonomic and ecological study and evaluation, and also to relate these to wider conservation agendas to alleviate the losses now apparent or suspected. In a climate of diminishing taxonomic support, formal documentation of those groups also needs focus. For many insect groups taxonomic knowledge is still very poor and resides largely within the realm of a few specialists. Concentrating on those groups for which knowledge is already greatest, with taxonomic attention then to be focused on the 'catch-up groups' (New, 1999a,b) with the aim of increasing the variety of 'well-known' groups of use in conservation assessment seems a pragmatic path to pursue, should taxonomy seek to become of greater relevance in insect conservation. It is thus not surprising that much of our understanding of insect species conservation has come from studies of representatives of 'popular' groups, particularly butterflies, some moths, beetles, ants, flies, orthopteroids and dragonflies – generally larger and attractive insects for which documentation of biology and conservation needs has accrued over (in some places) a century and more of interest from naturalists. This has led to development of increased capability to meet those needs, at least up to a point where our support systems become limiting, some general protocols for evaluating priority amongst numerous deserving species, and methods for increasing sound documentation and monitoring. Such species-level studies demonstrate amply the needs for subtle, and often highly individualistic management as a 'fine filter' level of conservation, and the spectrum of management across individual taxa helps to dictate the general principles involved. From this point of view (and not in any way diminishing their importance or interest in other contexts), it is distractive to attempt to incorporate 'black hole groups' in many practical conservation exercises. Attempts to incorporate these into practical conservation, other than to demonstrate enormous diversity, are commonly premature, however laudable the underlying motivation. Furthermore, capability to rely entirely on this fine filter approach is limited not only by the large number of deserving species and our lack of knowledge of most of these, but also the high costs of single species conservation programmes. For much of the world, the emphasis on single species focusing taken largely for granted in parts of the northern temperate zone simply cannot be the predominant strategy for conservation benefit, and broader ('coarse filter') approaches are more practicable.

The assemblage level of focus also reveals considerable advances in information and approach, some to facilitate forms of 'rapid biodiversity assessment', again as acknowledgement of the impracticability of interpreting complete inventory surveys. Increasingly sophisticated ways to enumerate insect diversity include careful use of 'morphospecies' (or similar category implying consistently recognizable entities) as a substitute for named species, progressively facilitated through digital imaging and interactive software to constitute a 'virtual voucher specimen bank' (see Oliver *et al.*, 2000). Data transfer by such means (incorporating measures such as bar coding labels on specimens) has potential to make all available information rapidly available to managers on the Internet with least possible delay. How it may be interpreted and applied is, of course, a more complex theme. Attention to 'functional groups' may also provide substantial information, and various forms of surrogacy for wider diversity have been sought, some focusing on particular taxa or levels of taxonomic analysis above the species level. The values of ants in Australia, for example, have been explored extensively for such wide applications and as indicators of environmental change and for monitoring restoration (Majer *et al.*, 2004). The desirable features of such groups have been debated extensively, but are encompassed broadly in Solis' (1997) advocacy for moths, for which positive attributes for biodiversity studies (in addition to considerable species richness) include: occurring in many habitats throughout the world; having many specialized habits and behaviours; being good indicators of areas of endemism; showing rapid responses to environmental disturbance and change; being easily sampled by quantitative methods; and including many taxa that are readily identifiable. In addition, applied values of many species – either as pests or beneficial taxa – may be employed to attract further study (New, 2001). Particularly in less-documented parts of the world, the major constraint on environmental evaluation using insect assemblages is the limited spectrum of 'better-known' insect groups, which are the major option for assemblage evaluation in conservation. However, the variety of contexts in which insects have been used to monitor or, broadly, indicate environmental conditions or management consequences in terrestrial and freshwater systems is itself testimony to their unique values as 'tools' in wider conservation. Increasing acknowledgement of these values leads to fuller recognition that insects can be politically persuasive in wider conservation agendas, both in strengthening advocacy and in facilitating more satisfactory management decisions.

If a truly global agenda for action on insect conservation is to emerge, increasingly holistic approaches are likely to feature strongly, including the integration of insects into wider conservation agendas (emphasizing the insects themselves, or the broader processes or contexts in which they are important). Four general trends or transitions towards a more holistic management are gradually being adopted: (i) species to assemblages; (ii) 'targets' to 'tools' with selection of species for wider flagship or umbrella values; (iii) incorporating insects with other organisms to complement and endorse conservation needs; and (iv) increasing awareness of insect importance to help transform them from being 'passengers' to 'drivers' in wider conservation policy.

What, then, might be reasonable aims towards integrating insects into wider conservation agendas; and how might these be approached? Priorities will clearly differ for different workers, but I suggest that the elements should include the following, however idealistic and generalized they may be:

1. Slowing rates of loss and decline of insect diversity, and attempting to reduce the extent of apparent and undocumented extinctions.
2. To a large extent, this devolves on preventing loss or alienation of natural habitats, perhaps particularly in the tropics.
3. In concert, improving the quality of degraded habitats through restoration or improved management, with increased focus on insect well-being. This entails measures from local to landscape scales, and incorporates private land, including agricultural land.
4. Defining better management principles, drawing both on expertise from single species studies and from wider assemblage assessment. The seven

management premises discussed by Samways (2005) provide a sound basis for discussion of optimal ways forward. Primary aims might be to reduce the effects of fragmentation, to enhance the effects of protected areas, to increase hospitality of the wider landscape for insects and to attempt to predict (and cater for) the impacts wrought by future climate changes.

5. Most of this must be underpinned by, and can only be effectively prosecuted through, wider community appreciation of the significance of insects in the natural world and to human welfare. Education is a critical component of conservation enhancement and must also include attempts to improve the 'image' of insects to the many people who do not appreciate their worth.

6. Expand from treating insects in isolation to integrating them more centrally in conservation planning and rendering them effective tools in such endeavour.

Four general points underpin much of what we need to consider:

1. The greatest values of insects in wider conservation and land management reflect their diversity, biomass and ecological variety. Paradoxically, the greatest barrier to incorporating them effectively in wider agendas is this same diversity and variety, together with the reality that heterogeneity in space and time is the rule, rather than the exception.

2. Equally difficult to communicate to non-entomologists is that every species is different, with even closely related taxa sometimes differing substantially in their biology and ecological optima. Although unity in effective general conservation protocols may be sought (and, to some extent, achieved), management for any one species is not likely to be optimal for others. Even those related taxa living in the same area or habitat may respond very differently to an equivalent environmental change. Most insects which have become conservation targets, through being threatened or listed in some way for attention, are relative ecological specialists and may be especially susceptible to imposed changes.

3. In seeking to enhance knowledge of insects for conservation, some form of closer focusing within the enormous array of taxa is necessary, with numerous groups of insects each having strong advocates for their priority values as indicators, surrogates or other signals of wider diversity or of environmental health.

4. The well-being of insect species and assemblages depends on the well-being of natural habitats, and on the management of these for compatibility with human uses of land and water. As the context for species well-being and evolution, the community level of conservation must become a prime focus for the future.

2 Wider Contexts and Enhanced Values for Species

2.1 Umbrella values

Many individual insect species gradually acquire wider values in conservation as their study proceeds, and those values can at times become important.

Thus, virtually any insect species associated with a specialized or vulnerable habitat can be promoted as an umbrella for that habitat, even if the practical values of the concept remain largely unproven or poorly defined (Caro and O'Doherty, 1999). Indeed, Haslett (1998) suggested that insect species to be nominated for priority listing under the Bern Convention in the future might profitably be selected to represent habitats that are currently underrepresented by the taxa already noted, hence increasing the overall ecological ambit of the species listed. Such a strategy may be important in focusing amongst the numerous deserving species and increasing conservation benefits – but ethically cannot replace citation of deserving taxa of other kinds. However, the principle of selecting priority species for such umbrella values, rather than simply allowing these to accrue more casually, merits careful thought.

2.1.1 An example: a birdwing butterfly

Broad ecological and sociological values of conservation may sometimes be combined effectively under an insect umbrella. One of the world's most charismatic insects and the largest butterfly is *Ornithoptera alexandrae*, Queen Alexandra's birdwing, native to a small area of Papua New Guinea (PNG) where its conservation is of considerable concern. In the early 1990s, the governments of Australia and PNG attempted to promote a far-sighted conservation plan for the Oro Province, based on this butterfly. The plan integrated human needs and welfare, and existing butterfly conservation measures coordinated through the Insect Farming and Trading Agency (see Parsons, 1992, for background). As part of Australia's programme of foreign aid (AusAID) to PNG, AUS$3.9 million was committed from 1995 to 2000 as probably the largest programme to date in which a single focal insect species has been used deliberately as an umbrella taxon. The complementary commitment from the PNG government was K815,000. With some changes in financial balance, the final project budget (June 1999) exceeded AUS$4.38 million. The overriding aim of the project was to encourage local landowners in this remote part of the country to conserve primary rainforest, by providing long-term economic incentives to reduce pressure to log forests for short-term gains. Ranching and marketing *O. alexandrae*, in conjunction with 'intensive research into the biology and ecology ... to determine its distribution and the foodplants and conditions it needs to survive' (AusAID, 1999), were the cornerstone of this ambitious programme, which had five main components:

1. Research, to enhance understanding of the distribution, biology and ecology of *O. alexandrae*;
2. Conservation of Primary Habitat Areas to maintain the existence of all important primary habitat areas;
3. Education and awareness: to promote knowledge of and concern for *O. alexandrae* throughout the country;
4. Economic and social issues: to provide economic and social incentives and measures for conserving *O. alexandrae* habitat;
5. Project management: to coordinate and manage inputs and implement activities.

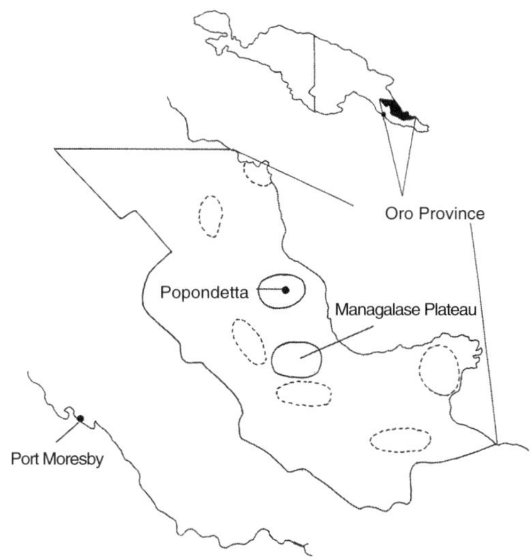

Fig. 13.1. Conservation of Queen Alexandra's birdwing butterfly, *Ornithoptera alexandrae*. The Oro Province of Papua New Guinea, indicating the major lowland (Popondetta) and highland (Managalese Plateau) areas in which the butterfly occurs. Dotted lines indicate other areas also surveyed during the recent Papua New Guinea Conservation Project (see text).

This project was thus highly innovative in closely integrating conservation and development aspects, and was the first to link economic and social opportunity directly with a butterfly in the region. Many uncertainties were evident in the initial project implementation document (Anon., 1996), and overall success depended on the cooperation of the traditional landowners of the restricted primary and secondary forest areas on which the butterfly depends. The history of conservation interest in *O. alexandrae*, summarized by Parsons (1992, 1996, 1999), led inexorably to the inference of increased endangerment through expansion of the oil palm industry in Oro Province (Fig. 13.1) (to which the butterfly is endemic and appears on the provincial flag) during the early 1990s. The PNG government's major purpose in seeking Australian cooperation was to establish a joint effort to ensure that this expansion would not harm *O. alexandrae*, and substantial foundation knowledge (such as the very comprehensive management plan prepared by Orsak, 1992) was available to formulate the project. Each of the above components was backed by detailed listing of objectives and actions, with the overall rationale being 'To ensure the survival of the remaining *O. alexandrae*, through a commitment to conservation which involves other improvements to the welfare of conservationist/landowners and their neighbours; which raises the possibility of ecotourism; and which at least postpones exploitation until resource extraction, resource management, returns to landowners and decision making by landowners are improved.' (Anon., 1996, p. 14).

Outcomes by 1999 benefited both the butterfly and local people (AusAID, 1999). It proved possible to breed *O. alexandrae* in captivity. The isolated clans of the Managalase Plateau agreed for their lands to be placed in a conservation zone, hence conserving the upland forests supporting a major population of the butterfly, and protecting these from industrial logging operations in the medium term. The conservation messages were spread through groups as diverse as Women Extension Officers dealing with health and nutrition and the newly formed Esse Rabbit Growers Association (based on development of smallholder rabbit rearing, primarily by local women, as a dietary improvement exercise) with a strong conservation clause in their Mission Statement, as well as direct components of the Oro Conservation Project. The lowland Popondetta plains population of the birdwing was under more immediate threat from logging and oil palm expansion, and accomplishing habitat security was more difficult there. The wider benefits of this project therefore emphasized social development, such as nutritional needs awareness, women's health issues, village birth attendant training courses, establishment of kitchen gardens, school science classes and generally increasing awareness of conservation. Rarely has a single insect species spearheaded such a diverse operation. The long-term outcomes are not fully assured (e.g. direct marketing of *O. alexandrae* will depend on revision of Convention on International Trade in Endangered Species (CITES) Appendix 1 listing) but the current management of the project by a local non-governmental organization (NGO), Conservation Melanesia, is enhancing 'pride of ownership' through the local community (Palangat, 2003).

The approach to insect conservation using flagship species for advocacy is inherently attractive (Lambeck, 1999) because, if it proves possible to manage entire ecosystems or assemblages (and even human intrusions) by focusing on the needs of one or few species, other needs may become redundant. Although it is intuitively unlikely that any such single (or few) species could serve to protect all critical functions of the wider system, they may have more practical relevance in conservation of their habitats, particularly of fragments (Launer and Murphy, 1994).

2.2 Wider contexts: processes

2.2.1 Importance of agroecosystems and their management

An agroecosystem focus for wider insect conservation benefit reflects that agriculture and associated activities involve modifications to ~36% of the earth's land area (Gerard, 1995), representing by far the largest single component of land use, often with massive changes to habitats and reliance on exotic species. Increasing compatibility between the historically polarized 'agricultural estate' and 'conservation estate' has received considerable attention in recent years, with increasing realization that many aspects of crop protection benefit from the presence and enhancement of native natural enemies and pollinators, which depend on natural habitats, and that landscape level considerations can indeed benefit agricultural productivity. A recent review by Tscharntke *et al.* (2005; see also Tscharntke *et al.*, Chapter 16, this volume; New, 2005) demonstrates the many ways in which agricultural practices can be managed for improved conservation impacts and the maintenance or enhancement of ecosystem services. Thus, area-wide pest management is advocated in an increasing variety of cases, with possible influences on non-target species acknowledged (Rothschild, 1998). Although the priorities of conservation biologists and agricultural producers differ, there is also much common ground (and some parallels) in the aims of sustaining productivity and reducing harmful side effects. Both parties seek to maintain insect populations at acceptable numbers. De-intensification of agriculture is occurring in many ways, with insects likely to be amongst the major beneficiaries, even though their well-being may not be amongst the stated primary aims of such reforms.

In some instances, this wider well-being of insects is the main aim of agroecosystem management; primary producers may have strong practical interests in the availability of native pollinators, as well as in predators and parasitoids as biological control agents for crop protection. The recent emphasis on conservation biological control (Barbosa, 1998), in which the major aims are to conserve, enhance and promote the influences of a wide range of native natural enemies, has led to some significant changes in agricultural land management practice. Such measures have helped to demonstrate the practical values of remnant natural vegetation within largely anthropogenic landscapes and its roles in providing critical resources (including supplementary foods and refuges). The discipline leads to wider appreciation of the values of native insect biodiversity and the need to sustain food webs for natural assemblages of insects (including the relatively rarer higher trophic level insects, such as specific parasitoids). It draws heavily on the principles of cultural control to achieve this. The discipline thereby necessitates some change in the perceived balance between 'planned biodiversity' and 'associated biodiversity' (see Altieri and Nicholls, 2004) in agroecosystems. The major relevance here is the broadening of interest to taxa and ecological associations not wholly based on cropping systems, and recognition of their roles in crop management. The scheme proposed by Poehling (1996) (Fig. 13.2a) summarized the context: essentially of integrated farm practices linking strongly with structural manipulation of habitats and serving jointly to increase diversity of natural enemies, enhancing their dispersal and access to cropping areas, and leading to reduced pest abundance. It is worth noting that natural enemies are largely of taxonomic groups that have failed to attract the focused attention of most insect conservation biologists (Shaw and Hochberg, 2001) and for which our level of ignorance renders species-focused conservation impracticable (Hochberg, 2000: 'anything but community or ecosystem conservation is unlikely to make much headway in the global conservation of insect parasitoids').

An underlying principle for this approach is to enable selected ecologically significant species (natural enemies) to remain functional, and their well-being enhanced, in environments sufficient to enable common species (of the focal taxa and their food species) to remain common. The components and linkages shown in Poehling's diagram (Fig. 13.2a) are thus relevant to wider conservation considerations and aims, possibly achievable by some change

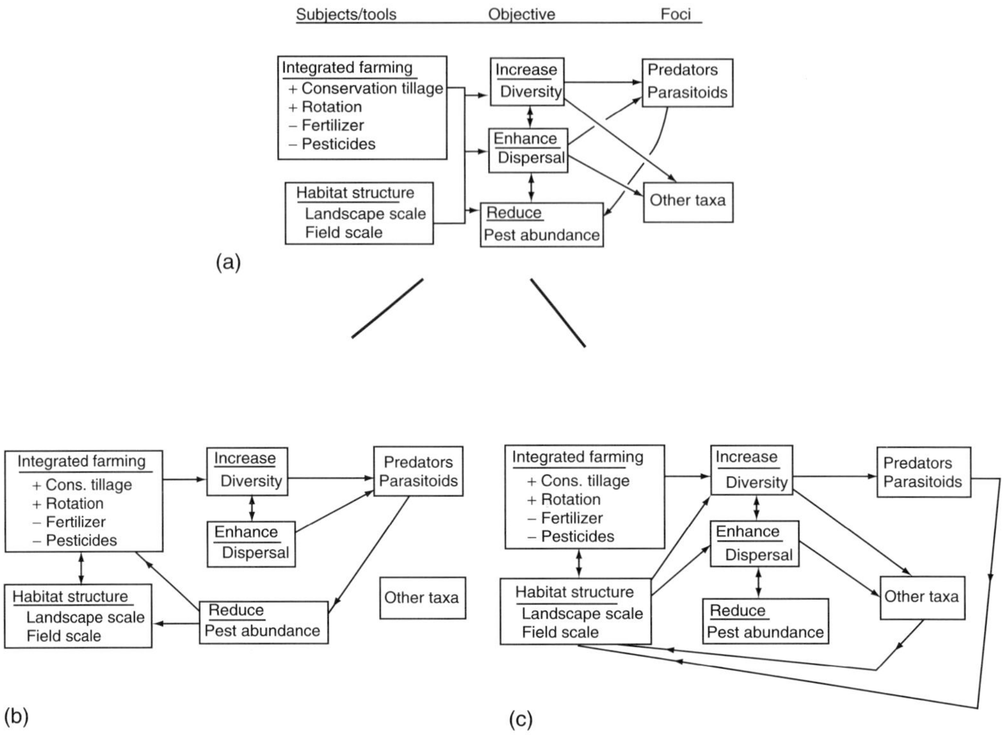

Fig. 13.2. Components and linkages for helping to integrate crop protection and conservation measures on farmlands: (a) a basic scheme, and how the balances may be shifted to benefit (after Poehling, 1996); (b) pest management through enhancing conservation biological control; and (c) wider insect conservation measures (b and c from New, 2005).

of emphasis (as in Fig. 13.2b and c) to harmonize more effectively the differing sectoral priorities. Whilst retaining the benefits of conservation biological control, conservation of the other, largely unheeded, insects may be enhanced by the umbrella effect of this practical priority. Recognition that relatively small changes in emphasis in the management of lands used for agriculture, forestry and industry can be important in conserving invertebrates has substantial ramifications for conservation. It is acknowledged widely that biodiversity in agricultural landscapes is affected by many factors, reflecting the presence, extent and composition of non-crop areas in the mosaic, in addition to farming practices themselves. However, and as Bengtsson *et al.* (2005) pointed out, practices such as 'organic farming' may have different effects in farms with different management intensities and 'geography', and the reactions of conservationists vary accordingly.

The suggestion of Lambeck (1999) of a dual approach to the practical management of agricultural landscapes has important ramifications for insect conservation, with most benefit likely to result from the first of his two categories, the less precise 'general enhancement' that utilizes general ecological principles as guidelines and thus obviates the need for detailed knowledge of any focal taxa. The major objectives are to maximize the number of species retained within the constraints of other land uses. Lambeck's second category, 'strategic enhancement', specifies more closely the targets of conservation and incorporates two broad objectives:

1. Retention of species, or of particular target biota, in the landscape.
2. Restoration of species that occurred in the landscape earlier but no longer do so. Aspects of this include restoration, reintroduction, translocation and other practices that necessitate knowledge of the target species.

The first of these encompasses the principle of 'keeping common species common' or, at least, present. The second also incorporates the more familiar consideration of species with 'rarity values', as the major foci in such programmes.

In essence, any such forms of ecological engineering (Gurr *et al.*, 2004) are designed, at least in part, to foster ecological sustainability. To a great extent, this is focused on reducing the polarization between the agricultural and conservation estates, and increasing variety in structure – for example, by providing more complex edges to agricultural areas (Haslett, 2001), hence enhancing the mosaic nature of the available habitats. Practices such as agroforestry could easily be tailored to have enhanced benefits for insects by more effective consideration of physical design and species composition. The wider task is to transfer insects from being largely the passive beneficiaries of these practices to harnessing the biological knowledge available on insects to increase their roles in active conservation management for even wider benefits. The transfer of the relevant knowledge from academia to land managers may be both the most difficult and the most urgent component of the exercise.

2.2.2 Landscape ecology

The major reforms to landscapes reflect the need to consider basic principles of landscape ecology in insect conservation to an ever-increasing extent, to counter the detrimental effects of fragmentation and habitat loss, and loss of connectivity, as effectively as possible. Many such reforms are necessarily generalized, but studies of insects have clarified many of the values involved, including the importance of small habitat patches in promoting diversity, and of effective connectivity, such as by establishing vegetation strips to facilitate interior or parallel dispersal. Many insects do not disperse extensively or easily, and even apparently low levels of habitat fragmentation may isolate populations effectively (references in New, 2005).

The likely ramifications of climate change emphasize the need for scenarios to conserve landscapes for future carrying capacity, as well as ensuring their current capability to sustain insects. Thus, future connectivity may necessitate strategic siting of any protected areas to be designated, with an underlying need to conserve 'ecological gradients', through which species may be displaced or moved. The principle of 'gradient analysis' (gradsect: Gillison and Brewer, 1985; see also Wessels *et al.*, 1998) also has implications for present-day surveys and documentation, as a basis for interpreting future compositional changes and conservation needs. Maintenance of altitudinal and latitudinal gradients, and of transition zones between major ecosystems or vegetation types is also important.

2.3 Conservation incentives on private lands

Linked with reformation of the agricultural estate, a variety of financial incentives for conservation have emerged. Practices such as set-aside (through which arable land is withdrawn from food production and farmers subsidized for doing this) have been widespread in Europe and the USA. Long-term set-aside in Britain can be an important option in providing and sustaining communities of ecological value and maintaining ecological function (Corbet, 1995). Although the agricultural estate is clearly a key arena for insect conservation, the more general contribution of wider 'private lands' in complementing the limited areas of reserves is also of critical importance.

Historically, there has been a substantial tendency to overvalue protected areas for insect conservation – commonly assuming that species living there will be conserved 'automatically' through the very act of habitat protection. Reservation of habitat indeed provides a firm basis from which to pursue insect conservation, but successional and other changes dictate that additional management is commonly necessary. However, Moore (1997) commented for Odonata: 'Not surprisingly the conservation of dragonflies has rarely been the primary purpose of establishing protected areas.' He also pointed out that Japan then had 24 such dedicated protected areas, reflecting strong traditional appreciation of dragonflies in Japanese culture. Indeed, specific reserves for butterflies and other, mainly 'charismatic', insects have been established in many countries, mostly on small or remnant sites (in Australia most such reserves are, at most, a few hectares in extent). Protected areas may be founded to support single notable species, typical or representative assemblages, or centres (hotspots) of richness or endemism, these sometimes fortuitously. Thus, for dragonflies, the Tambopata–Candamo Reserved Zone in Peru harbours what may be the highest 'spot richness' of Odonata on earth (Paulson, 1985). However, the reserves system of most countries is inadequate to conserve all native biota and is increasingly difficult to augment, even with formal national obligations to do so, as in Australia, so that integration of reserves with the wider landscape is integral to more holistic conservation effort. Commonly, the roles of protected areas for insect conservation are assumed rather than proven, and there have been rather few attempts even to prepare inventories of the species living there, or of selected focal groups.

Often, though, our knowledge of the insect fauna of major reserves is woefully inadequate, and no reasonably comprehensive inventory exists for most such areas. This situation is unlikely to change without more enlightened policy development. One practical problem of documentation, noted by Sands and New (2003) for Australia, is that regulations generally prohibit insect-collecting in national parks and some other reserves, or the considerable bureaucracy involved in obtaining such permits, often with severe restrictions on use, deters collectors from attempting to explore these important areas. In Australia, for example, records of even the best-documented insect groups (such as butterflies) from national parks are largely fortuitous rather than the result of systematic inventory surveys. The consequences include that: (i) knowledge of the richness of the areas is highly incomplete; (ii) the presence there of even notable species of individual conservation significance is largely unknown, as is any opportunity for their conservation based on presence in protected areas; leading to (iii) expensive land purchase, resumption or changes of tenure elsewhere to provide secure habitat on which to manage those species – all of which steps may be unnecessary if those species already have secure habitats in reserves. Although the limitations of protected areas as the conservation estate are understood and acknowledged widely, systematic surveys there, of selected insect groups, could contribute substantially in assessing their values for insects far more constructively than undertaken at present and also help to dictate conservation priorities elsewhere.

Speight and Castella (2001) enumerated some of the practical applications of 'inventory lists' in insect conservation, using European Syrphidae to demonstrate the considerable values of species lists integrated with relevant biological and habitat data to assess parameters such as: (i) the 'biological maintenance function' of a site, that is, a measure of the site quality, based on the relationship between the site species list, and a wider regional species pool that might be expected to occur in the array of habitats represented on the site; (ii) site biodiversity management, helping to specify the remedial measures needed for the 'missing' species to thrive, and what key habitats may be absent or under-represented; and (iii) regional biodiversity management through identifying habitats with high faunal diversity and those with high proportions of 'anthropophobic species, those which do not survive in habitats modified extensively by human activities'. They noted: 'Action taken to maintain biodiversity on protected sites, that does not take insects into account, is a contradiction in terms!'

However, most insects live outside protected areas, and even some British butterflies are reportedly inadequately represented in National Parks (Asher *et al.*, 2001), so that wider considerations are necessary for adequate conservation. The values of incorporating private land into the broader conservation estate are well-publicized. The concept of 'stewardship for biodiversity' implicitly incorporates ethical responsibility, and the broadening of conservation interests from protected areas alone to private lands has fostered a considerable array of 'incentives' for improved protection or management – most of them founded either on penalties for transgression (such as fines for clearing of native vegetation) or rewards for effective protection, or management by the landholders. The latter includes various forms of 'conservation credits' or 'biodiversity credits', although rarely acknowledging insects directly. The recognition of 'biodiversity assets' (in Victoria, Australia, these can include populations of known threatened species, representation of scarce ecological vegetation classes, or 'hotspots' of richness or endemism, as examples) is a tool of increasing value in according priority to areas for conservation action. However, in this form it is not always acceptable to landholders, because 'penalties' may still operate, and considerable thought is applied to the incentives side of the ledger. Recent moves in Australia implicitly recognize biodiversity values on private land with suggestions of incorporating 'butterfly credits' into this assessment following as part of the impetus from the recent national Action Plan for Australian Butterflies (Sands and New, 2002). The approach acknowledges the significance of butterfly conservation and draws unashamedly on their popularity, with additional sympathy gained from their general 'non-pest status', to draw attention to local hot spots of richness and the occurrence of significant species on private lands (New, 2006). In many cases, formal transfer of sites supporting such taxa to protected area status is impracticable. By offering such 'carrots' (rather than 'sticks') it is hoped that the goodwill of landholders towards insects may be increased effectively. Although still in its early stages, the possible options of rewards for: (i) managing sites or habitats; (ii) protecting sites or habitats; and (iii) not threatening sites or habitats are each under consideration.

3 Discussion

The above are simply examples of the many ways in which insects may participate in, and benefit from, wider conservation agendas leading to more effective ecosystem management. With the reality that insects will remain to a large extent passengers on wider conservation agendas, despite our efforts to incorporate them more centrally, we also need to consider how those passengers might be protected most effectively. With even limited defined values and knowledge that insects may have roles in endorsing or modifying approaches advocated more strongly for vertebrate animals and vascular plants. The twin approaches to wider conservation involve: (i) conservation of areas designated as hotspots or similar, with enhanced biodiversity values; and (ii) conservation of particular habitats, such as vegetation types, with an ecological rather then strictly geographical value.

Myers *et al.* (2002) emphasized the general need to 'support the most species at the least cost' in their definitions and advocacy for biodiversity hotspots where 'exceptional concentrations of endemic species are undergoing exceptional loss of habitat'. Using vascular plants and selected groups of vertebrates, 25 major global hotspots were recognized. Myers *et al.* suspected, in a view intuitively appealing to entomologists, that these rankings may be matched by similar concentrations of endemic insect species, extending the 'hotspots concept' effectively to include invertebrates, albeit by surrogacy. They used the specific example of fig-pollinators as a suite of obligate mutualisms involving ~900 host-specific wasp species to exemplify the richness of insect–plant interactions. There is little doubt that the strongest possible entomological endorsement of the biodiversity values of all 25 hotspots designated by Myers *et al.* (2002) could contribute significantly to insect conservation, should it aid in protecting those regions. It is pertinent to reiterate the fundamental premises underpinning this approach, with insects in mind. After Mittermeier *et al.* (1998):

1. The biodiversity of each and every nation is critically important to that nation's survival, and must be a fundamental component of any national or regional development strategy.
2. Some areas simply harbour far greater concentrations of biodiversity than others.
3. Many high-biodiversity areas exhibit very high levels of endemism.
4. Many high-biodiversity areas are under the most severe threat.
5. To achieve maximum impacts with limited resources, we must concentrate heavily (but not exclusively) on those areas highest in diversity and endemism and most severely threatened (Mittermeier *et al.*, 1998, p. 516).

Two somewhat analogous approaches addressing this question of 'where to conserve' based on 'what is there' have incorporated butterflies as the main focal group. The 'critical faunas approach' pioneered for milkweed butterflies (Nymphalidae: Danainae) (Ackery and Vane-Wright, 1984) and for swallowtails (Papilionidae) by Collins and Morris (1985) drew attention to the very limited distributions of many significant taxa, and use of this information to delineate regions with strong local endemism and richness, be this country, island or more limited areas. Collins and Morris (1985) cited a variety of earlier studies on butterflies (including those by Pyle, 1982, for Washington State, USA; Brown, 1982; Lamas, 1982, both for the Neotropics) that identified biogeographically and evolutionarily significant areas or centres of species richness. The above authors also developed the principles of complementarity to rank the importance for conservation of areas ranging from countries to smaller administrative units. For much of the tropical regions most important to both milkweed and swallowtail butterflies, land use conflicts are difficult to resolve, and losses of important natural habitats continue apace. Chances for augmenting effective reservation or protection of forests in tropical south-east Asia, for example, are low (e.g. MacKinnon, 1997) and, despite the importance of this approach to determining areas of considerable entomological importance, its impact in reality is likely to remain low.

More feasible, simply because of a well-defined fauna, geography and higher local interests, is the recent identification of 'Prime Butterfly Areas' (PBAs) in Europe (van Swaay and Warren, 2003), building on aspects of the European Habitats Directive to identify the areas of major conservation importance for butterflies in 37 countries and three archipelagos. In this context, butterflies are not a 'stand-alone' focus, but complement similar exercises for plants, birds and herpetofauna. One-third of Europe's 576 butterfly species are endemic to the continent. PBAs were selected on representation of 34 target species (those with global distribution limited to Europe, threatened in Europe, and/or listed on the Bern Convention Habitats Directive). These parameters led to the designation of 431 PBAs, covering ~1.8% of Europe's land area. Much stronger protection for all PBAs was recommended (56% of them were unprotected at that time and 47% of PBAs in the European Union

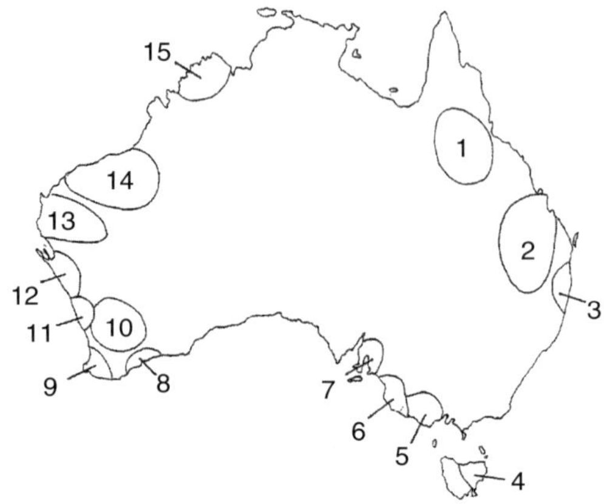

Fig. 13.3. Initial suite of areas delimited in Australia's national 'Biodiversity Hotspots Programme'. Key: (1) Einasleigh and Desert Uplands; (2) Brigalow North and South; (3) Border Ranges North and South; (4) Midlands of Tasmania; (5) Victorian Volcanic Plain; (6) South-east of South Australia and South-west Victoria; (7) Mt Lofty/Kangaroo Island; (8) Fitzgerald River Ravensthorpe; (9) Busselton Augusta; (10) Central and Eastern Avon Wheat Belt; (11) Mount Lesueur Eneabba; (12) Geraldton to Shark Bay sand plains; (13) Carnarvon Basin; (14) Hamersley/Pilbara; (15) North Kimberley. (After Department of the Environment and Heritage, 2005.)

were not protected under international law). Much of the formal conservation action needed must be undertaken at the national level.

One component of Australia's 'Biodiversity Hotspots Programme' involves establishment of a national biodiversity stewardship component, paying private landholders or lessees in hotspot regions to 'undertake above duty-of-care conservation activities' to deliver specific biodiversity outcomes and to secure conservation management of their properties in perpetuity. Initial focus is on the 15 national hotspots recognized by late 2003 (Fig. 13.3), collectively including the habitats of numerous listed threatened species, including insects, in many parts of the country. Actions that may be supported include feral animal and weed management in remnant native vegetation, exclusion or reduction of stock, ecological burning, habitat restoration and replacement of exotic vegetation by native species. The first major target of the programme has been the Mount Lofty Ranges, South Australia, in which a number of significant butterfly species occur (Sands and New, 2002), and for which conservation is likely to be enhanced by increasing protection for the 15% or so of native vegetation that has survived large-scale clearing.

With any such priority area designation, including the wider global hotspots, we are looking at the need for active conservation, with focus on retaining the biota and natural habitats present, and lessening future impacts of anthropogenic changes. Within much of the wider landscape, the mix of protection of existing habitat (including remnants) and restoration of degraded habitats must continue in tandem, and be enhanced strategically, with an aim of increasing effective representation of the full array of important habitats for insects in protected areas that can be managed effectively. Using southern Australia as an example to demonstrate need, the major reserves system (e.g. National Parks) has developed as much by accident as by design based on sound conservation or other ecological principles, and has thus not led to the ideal of an ecologically balanced and representative reserve system. Thus: (i) the conversion of vast areas of lowland Victoria for farming, including the large-scale cultivation of exotic grass species for 'improved pasture', has led to formerly extensive native grassland being reduced to <1% of its former area and regarded as the most endangered ecosystem (Kirkpatrick *et al.*, 1995), and grossly under-represented in reserves; and (ii) the numerous small fragments of relatively natural habitat remaining in Western Australia's wheat belt are largely areas topographically unsuitable for cultivation (background in Greenslade and New, 1991). In short, the ecological vegetation classes of such easily accessible lands have been largely cleared, creating an under-representation of these (and of associated insect and other fauna) in reserves. Such situations endorse, as a principle of much wider relevance, the need to improve the hospitality of such areas for native insects, and the wide importance of remnant refuges in peri-agricultural areas. Augmenting the current reserve system in this context, with insects helping to endorse the values of even very small areas of natural habitat, is gradually occurring. Thus, New Zealand has gradually developed a Protected Natural Areas Programme to address under-representation of natural habitats in the conservation estate (Kelly and Park, 1986) and Australia has a national policy to establish a Comprehensive, Adequate and Representative (CAR) reserves system based on a series of 85 bioregions (Brunkhorst *et al.*, 1998). Stages in this process involve identifying gaps (mainly in representation of vegetation associations) and setting priorities to redress these lacunae by identifying key sites, and establishing and managing them as reserves. Organizations such as Trust for Nature (TfN) (Victoria) have enormous importance in such endeavour – for example, this body has been instrumental in increasing, through direct purchase of private land, representation of native grasslands and other vulnerable associations in the permanent conservation estate. In some cases, presence of designated threatened species, including insects, has increased the priority for particular sites to be reserved. Thus, TfN took an important initiative in purchasing what is now the Nhill Sunmoth Reserve in western Victoria. This 4.5 ha remnant grassland site on the outskirts of the town had been subdivided into 21 building allotments, some to be developed imminently. This site is the only place where two species of Castniidae (*Synemon plana, S. selene*), important flagships for native grassland ecosystems, are known to co-occur. Both of these sunmoth species are regarded as threatened and the publicity gained for remnant grassland conservation from this case has been considerable; background information is given by Douglas (2004). The site is now protected fully from any form of development or other activity that could lead to habitat degradation. It is a clear focus for augmenting knowledge and management of habitat quality for sunmoths.

Identification of such broad-based critical habitats for insects in Australia is continuing. For butterflies, Sands and New (2002, following a parallel approach for birds by Garnett and Crowley, 2000) nominated a suite of particular vegetation types currently at risk and which are each the sole or predominant habitat for several endemic butterfly species, also under threat as these habitats succumb. The development of the 'bioregion' approach helps to transcend the limitations of political boundaries as divisions in favour of more solid ecological appraisal of ecosystem values, and in Australia represents a serious attempt to ensure protection of good examples of all significant major ecosystems. In formulating this approach, it was recognized that 'the likelihood of including functional assemblages of all species within a bioregion will be greatest when the full range of ecosystems present within an area is selected' and 'the most appropriate ecosystem classification for reserve design will include attributes of vegetation structure and flora/fauna composition in conjunction with environmental attributes' (Environment Australia, 1999). A stated aim of the programme was 'improvement of biodiversity conservation outside reserves' (Environment Australia, 1998). As elsewhere, such formal recognition and commitment may not lead to rapid adoption but, at least, an acceptable agenda has been recognized, in which it is likely that insects will gradually play an increasing part.

A point of persistent relevance in biodiversity studies is that biodiversity which underpins ecosystem services central to human health and livelihood should have high priority in conservation efforts. The benefits of conserving insect pollinators (Ricketts *et al.*, 2004, Kremen and Chaplin, Chapter 15, this volume) and some natural enemies in agricultural ecosystems can be quantified in dollars and set against other criteria (such as yield), and are a powerful demonstration of the values of insects to a productive economy. The numerous forms of subsidy noted earlier for improving conservation hospitality in agroecosystems are all relevant, but many of these have very local applications and seem unlikely to be extended to global scales. The critical but less economically quantified roles of insects in many of the more commonly overlooked ecosystem services suggest strongly that their places as 'ecosystem engineers' (or, in Coleman and Hendrix's (2000) parlance 'webmasters in ecosystems') afford a variety of ways in which advocacy for their well-being may be driven for greater conservation benefits. Efforts to incorporate insects effectively in wider conservation agendas, as exemplified in this chapter, assuredly have value both for the insects themselves and for the systems in which they play such wide and vital roles.

Acknowledgements

I thank Dr Meg Clarke and Dr Indra Thappa (Australian Agency for International Development), and Dr Ric Caven (URS Sustainable Development) for access to information on *O. alexandrae* conservation. Two reviewers provided very useful comments on a draft of this essay. My participation in the Insect Conservation Symposium was immensely facilitated by funding from the Royal Entomological Society.

References

Ackery, P.R. and Vane-Wright, R.I. (1984) *Milkweed Butterflies, their Cladistics and Biology*. Academic Press, London.

Altieri, M.A. and Nicholls, C.I. (2004) Effects of agroforestry systems on the ecology and management of insect pest populations. In: Gurr, G.M., Wratten, S.D. and Altieri, M.A. (eds) *Ecological Engineering for Pest Management*. CSIRO Publishing, Collingwood,Victoria, Australia, pp. 143–154.

Anon. (1996) Papua New Guinea Conservation Project. Oro Province. Project implementation document. Papua New Guinea Department of Environment and Conservation and Australian Agency for International Development, Canberra, Australia.

Asher, J., Warren, M., Fox, R., Harding, P., Jeffcoate, G. and Jeffcoate, S. (2001) *The Millenium Atlas of Butterflies in Britain and Ireland*. Oxford University Press, Oxford.

AusAID (1999) Papua New Guinea Conservation Project. Oro Province. Project Completion Report. Australian Agency for International Development, Canberra, Australia.

Barbosa, P. (ed.) (1998) *Conservation Biological Control*. Academic Press, San Diego, California.

Bengtsson, J., Ahnstrom, J. and Weibull, A.-C. (2005) The effects of organic agriculture on biodiversity and abundance: a meta-analysis. *Journal of Applied Ecology* 42, 261–269.

Brown, K.S. (1982) Historical and ecological factors in the biogeography of aposematic neotropical butterflies. *American Zoologist* 22, 453–471.

Brunkhorst, D.J., Thackway, R., Coyne, P. and Cresswell, I.D. (1998) Australian protected areas: toward a representative system. *Natural Areas Journal* 18, 255–260.

Caro, T.M. and O'Doherty, G. (1999) On the use of surrogate species in conservation biology. *Conservation Biology* 13, 805–814.

Coleman, D.C. and Hendrix, P.F. (eds) (2000) *Invertebrates as Webmasters in Ecosystems*. CAB International, Wallingford, UK.

Collins, N.M. and Morris, M.G. (1985) *Threatened Swallowtail Butterflies of the World*. IUCN, Gland, Switzerland and Cambridge.

Corbet, S.A. (1995) Insects, plants and succession: advantages of long-term set-aside. *Agriculture, Ecosystems and Environment* 53, 201–217.

Douglas, F. (2004) A dedicated reserve for conservation of two species of *Synemon* (Lepidoptera: Castniidae) in Australia. *Journal of Insect Conservation* 8, 221–228.

Environment Australia (1998) Australia's national report to the Fourth Conference of the parties to the Convention on Biological Diversity. Environment Australia, Canberra, Australia.

Environment Australia (1999) *Australian GuideLines for Establishing the National Reserve System*. Environment Australia, Canberra, Australia.

Garnett, S.R. and Crowley, G.M. (2000) *The Action Plan for Australian Birds 2000*. Environment Australia, Canberra, Australia.

Gerard, P.W. (1995) *Agricultural Practices, Farm Policy and the Conservation of Biological Diversity*. USDI Biological Sciences report, no. 4. Washington, DC.

Gillison, A.N. and Brewer, K.R.W. (1985) The use of gradient directed transects or gradsects in natural resource surveys. *Journal of Environmental Management* 20, 103–127.

Greenslade, P. and New, T.R. (1991) Australia: conservation of a continental insect fauna. In: Collins, N.M. and Thomas, J.A. (eds) *The Conservation of Insects and their Habitats*. Academic Press, London, pp. 33–70.

Gurr, G.M., Wratten, S.D. and Altieri, M.A. (eds) (2004) *Ecological Engineering for Pest Management*. CSIRO Publishing, Collingwood, Victoria, Australia.

Haslett, J.R. (1998) *Suggested Additions to the Invertebrate Species Listed in Appendix 2 of the Bern Convention*. Council of Europe, Strasbourg, France.

Haslett, J.R. (2001) Biodiversity and conservation of Diptera in heterogeneous land mosaics; a fly's eye view. *Journal of Insect Conservation* 5, 71–75.

Hochberg, M.E. (2000) What, conserve parasitoids? In: Hochberg, M.E. and Ives, A.R. (eds)

Parasitoid Population Biology. Princeton University Press, Princeton, New Jersey, pp. 266–277.

Janzen, D.H. (1994) Wildland diversity management in the tropics: where are we now, and where are we going? *Vida Silvestre Neotropical* 3, 3–15.

Janzen, D.H. (1997) Wildland biodiversity management in the tropics. In: Reaka-Kudla, M.L., Wilson, D.E. and Wilson, E.O. (eds) *Biodiversity II. Understanding and Protecting our Biological Resources*. Joseph Henry Press, Washington, DC, pp. 411–431.

Kelly, G.C. and Park, G.N. (1986) *The New Zealand Protected Natural Areas Programme: A Scientific Focus*. New Zealand Biological Resources Centre, publication No. 4. Department of Scientific and Industrial Research, Wellington, New Zealand.

Kirkpatrick, J.B., McDougall, K. and Hyde, M. (1995) *Australia's Most Threatened Ecosystems, the Southeastern Lowland Native Grasslands*. Surrey Beatty, Chipping Norton, New South Wales.

Lamas, G. (1982) A preliminary zoogeographical division of Peru, based on butterfly distributions (Lepidoptera: Papilionoidea). In: Prance, G.T. (ed.) *Biological Diversification in the Tropics*. Columbia University Press, New York, pp. 336–357.

Lambeck, R.J. (1999) *Landscape Planning for Biodiversity Conservation in Agricultural Regions*. Biodiversity Technical Paper no. 2. Department of the Environment and Heritage, Canberra, Australia.

Launer, A.E. and Murphy, D.D. (1994) Umbrella species and the conservation of habitat fragments: a case of a threatened butterfly and a vanishing grassland ecosystem. *Biological Conservation* 69, 145–153.

MacKinnon, J. (ed.) (1997) *Protected Areas Systems Review of the Indo-Malayan Realm*. World Conservation Monitoring Centre, Cambridge.

Majer, J.D., Shattuck, S.O., Andersen, A.N. and Beattie, A.J. (2004) Australian ant research: fabulous fauna, functional groups, pharmaceuticals, and the Fatherhood. *Australian Journal of Entomology* 43, 235–247.

McKinney, M.L. (1999) High rates of extinction and threat in poorly studied taxa. *Conservation Biology* 13, 1273–1281.

Mittermeier, R.A., Myers, N., Thomsen, J.B., da Fonseca, G.A.B. and Olivieri, S. (1998) Biodiversity hotspots and major tropical wilderness areas: approaches to setting conservation priorities. *Conservation Biology* 12, 516–520.

Moore, N.W. (1995) Message to the International Symposium on the conservation of dragonflies and their habitats. In: Corbet, P.S., Dunkle, S.W. and Ubukata, H. (eds) *Proceedings of the International Symposium on the Conservation of Dragonflies and their Habitats*. Japanese Society for the Protection of Birds, Kushiro, Japan, p. 69.

Moore, N.W. (compilor) (1997) *Dragonflies – Status Survey and Conservation Action Plan*. IUCN/SSC Odonata Specialist Group. IUCN, Gland, Switzerland.

Myers, N., Mittermeier, R.A., Mittermeier, C.G., da Fonseca, G.A.B. and Kent, J. (2002) Biodiversity hotspots for conservation priorities. *Nature* 403, 853–858.

New, T.R. (1999a) Limits to species focusing in insect conservation. *Annals of the Entomological Society of America* 92, 853–860.

New, T.R. (1999b) Entomology and nature conservation. *European Journal of Entomology* 96, 11–17.

New, T.R. (2001) The conservation of a discipline: traditional taxonomic skills in insect conservation. *Journal of Insect Conservation* 4(2000), 211–213.

New, T.R. (2005) *Invertebrate Conservation and Agricultural Ecosystems*. Cambridge University Pess, Cambridge.

New, T.R. (2006) Is there any future for butterfly credits in conservation planning? *Journal of Insect Conservation* 10, 1–3.

Oliver, I., Pik, A., Britton, D., Dangerfield, J.M., Colwell, R.K. and Beattie, A.J. (2000) Virtual biodiversity assessment systems: the application of bioinformatics technologies to the accelerated accumulation of biodiversity information. *BioScience* 50, 441–450.

Orsak, L. (1992) Saving the world's largest butterfly, Queen Alexandra's birdwing (*Ornithoptera alexandrae*). Five-year management action plan. Scientific Methods, Durham, California.

Palangat, R. (2003) The Queen Alexandra Birdwing Butterfly, largest butterfly on the planet. *Iko-Forestri Nius, Papua New Guinea* 5, 4–5.

Parsons, M.J. (1992) The world's largest butterfly endangered; the ecology, status and conservation of *Ornithoptera alexandrae* (Lepidoptera: Papilionidae). *Tropical Lepidoptera* 3(Suppl. 1), 35–62.

Parsons, M.J. (1996) Butterfly farming in the Indo-Australian region: an effective and sustainable means of combining conservation and commerce to protect tropical forests. In: Ae, S.A., Hirowatari, T., Ishii, M. and Brower, L.P. (eds) *Decline and Conservation of Butterflies in Japan III*. Yadoriga, Osaka, Japan, pp. 63–77.

Parsons, M.J. (1999) *The Butterflies of Papua New Guinea*. Academic Press, London.

Paulson, D.R. (1985) Odonata of the Tambopata Reserved Zone, Madre de Dios, Peru. *Review of Peruvian Entomology* 27, 9–14.

Poehling, H.M. (1996) Agricultural practices which enhance numbers of beneficial arthropods. In: Booij, C.J. and den Nijs, L.J.M.F. (eds) *Arthropod Natural Enemies in Arable Land. II. Survival, Reproduction and Enhancement*. Aarhus University Press, Denmark, pp. 269–275.

Pyle, R.M. (1982) Butterfly eco-geography and biological conservation in Washington. *Atala* 8, 1–26.

Ricketts, T.H., Daily, G.C., Ehrlich, P.R. and Michener, C.D. (2004) Economic value of tropical forest to coffee production. *Proceedings of the National Academy of Sciences of the United States of America* 101, 12579–12582.

Rothschild, G.H. (1998) Applied entomology: prospects and challenges for the next millennium. In: Zalucki, M.P., Drew, R.A.I. and White, G.G. (eds) *Pest Management: Future Challenges*. University of Queensland, Brisbane, Australia, pp. 1–10.

Samways, M.J. (2005) *Insect Diversity Conservation*. Cambridge University Press, Cambridge.

Sands, D.P.A. and New, T.R. (2002) *The Action Plan for Australian Butterflies*. Environment Australia, Canberra, Australia.

Sands, D.P.A. and New, T.R. (2003) Coordinated invertebrate surveys in Australia's National Parks: an important tool in refining invertebrate conservation management. *Records of the South Australian Museum*, Monograph series 7, 203–207.

Shaw, M.B. and Hochberg, M.E. (2001) The neglect of parasitic Hymenoptera in insect conservation strategies: the British fauna as a prime example. *Journal of Insect Conservation* 5, 253–263.

Solis, M.A. (1997) Snout moths: unraveling the taxonomic diversity of a speciose group in the Neotropics. In: Reaka-Kudla, M.L., Wilson, D.E. and Wilson, E.O. (eds) *Biodiversity II. Understanding and Protecting our Biological Resources*. Joseph Henry Press, Washington, DC, pp. 231–242.

Speight, M.C.D. and Castella, E. (2001) An approach to interpretation of lists of insects using digitized biological information about the species. *Journal of Insect Conservation* 5, 131–139.

Tscharntke, T., Klein, A.M., Kruess, A., Steffan-Dewenter, I. and Thies, C. (2005) Landscape perspectives on agricultural intensification and biodiversity in ecosystem service management. *Ecology Letters* 8, 857–874.

van Swaay, C. and Warren, M.S. (eds) (2003) *Prime Butterfly Areas in Europe: Priority Sites for Conservation*. National Reference Centre for Agriculture, Nature and Fisheries, Ministry of Agriculture, Nature Management and Fisheries, The Netherlands.

Wessels, K.J., van Jaarsveld, A.S., Grimbeek, J.D. and van der Linde, M.J. (1998) An evaluation of the gradsect biological survey method. *Biodiversity and Conservation* 7, 1093–1121.

14 The Extinction of Experience: A Threat to Insect Conservation?

Oliver D. Cheesman[1] and Roger S. Key[2]

[1]*108 Cholmeley Road, Reading, Berkshire RG1 3LY, UK;* [2]*Natural England, Northminster House, Peterborough PE1 1UA, UK*

1 Introduction

The need for, and constraints on, insect conservation have been widely discussed in recent years. Fundamental requirements include baseline taxonomic information and survey data, ongoing monitoring, and enhanced knowledge of insect life-histories. These needs can be met by entomologists acting in a professional capacity and/or through the voluntary efforts of professionals and amateurs. However, continued (and preferably enhanced) contributions of this kind require broad support. Professional posts must be made available, and suitably qualified and motivated individuals must exist to fill them. Those wishing to contribute in an amateur capacity need encouragement, support and appropriately designed tools (such as identification aids). Amongst those members of society not actively engaged in the study or conservation of insects, there must be some appreciation of their importance, as this underpins advocacy and influences policy making. Although approaches to insect conservation in temperate and tropical situations necessarily differ, the importance of engaging people at all levels of society is a common factor (Brown, 1991; Morris *et al.*, 1991).

Miller (2005) highlights an underappreciated aspect of the biodiversity crisis – the extinction of experience. This expression was originally used by Robert Pyle, a noted figure in insect conservation. Pyle (1993) applies the concept to North America, where it has been further explored by Louv (2005) as nature deficit disorder, but it has global relevance. Nearly half the world's people already live in urban areas, and the proportion that does is increasing. As a result, the majority of the human population will soon inhabit environments in which they are disconnected from nature. Predicted consequences include declining knowledge and appreciation of biodiversity, and declining support for its conservation amongst those who increasingly accept a biologically impoverished environment as the norm. This 'ratcheting down of expectations' is likely to be most pronounced between generations, as 'the environment encountered during childhood becomes the baseline against which environmental degradation is measured later in life' (Miller, 2005). It seems likely that a similar erosion, if not actually extinction, of experience (and expectations) occurs in rural communities, at least where agricultural intensification substantially alters the landscape, biodiversity and human responses to them. The increasing displacement of other cultural frameworks by a Western world view, including a tendency to assess the value of nature (and everything else) in monetary terms, may be a relevant underlying force. (For consideration of how the human experience of nature influences its meaningfulness in a cultural context, see, e.g. Maiteny, 2004.)

With notable exceptions, insects and other invertebrates (as well as those who study them!) already appear to attract relatively little public appreciation when compared to more charismatic groups. The extinction of experience may therefore represent a particular threat to the study and conservation of insects. In the UK, a tradition of natural history has long maintained some connection between people and nature, even through intense periods of urbanization and agricultural intensification (Barber, 1980; Allen, 1994). To varying extents, similar natural history traditions have developed in other parts of Europe and in North America, and it is no coincidence that the major principles of insect conservation have been derived from work in the UK and these other areas (Stewart and New, Chapter 1, this volume). It is often said that the UK has the best-studied flora and fauna in the world, largely thanks to the unpaid efforts of naturalists over many years, but there are now concerns over recruitment of the next generation. A joint Linnean Society/Botanical Society of the British Isles meeting in 2002 addressed this issue (Perring, 2002), as did the Yorkshire Naturalists' Union 2003 conference (Henderson, 2004). A further conference on the subject (Masters *et al.*, 2007) was held in 2006, under the auspices of Invertebrate Link (JCCBI), a forum for organizations concerned with invertebrate conservation in the UK (Cheesman and Phillips, 2004; Cheesman, 2006). The National Federation for Biological Recording (NFBR) has addressed related issues at meetings in 2004 (James, 2004) and 2006. Much can be learned from the British natural history tradition, both its historic vibrancy and current concerns over its decline. This chapter examines those concerns, their relationship to the extinction of experience concept and significance to insect conservation more widely.

2 The Importance of Naturalists

For the purposes of this chapter, the distinction between professionals and amateurs is not of primary importance; both need inspiration, encouragement, support and instruction. However, the focus is largely on those (loosely termed naturalists) who act in a voluntary capacity. They may be specialists (including taxonomists), or more general observers of nature. It has been recognized that conservation of insects and other taxa would benefit from greater participation by naturalists or 'citizen scientists', for example, in North

America (Opler, 1991; Stevenson *et al.*, 2003) and Australia (Greenslade and New, 1991; Stewart and New, Chapter 1, this volume). In Mexico, a developing biodiversity information network is drawing on experience in the UK and Sweden to stimulate public engagement and volunteer data recording (Way, 2004). Even in the neotropics, data collected by naturalists, albeit fragmentary in many cases, can be important in making initial assessments of insect biodiversity (Brown, 1991). There are also parallels between the recruitment, training and goals of naturalists and those of parataxonomists (Basset *et al.*, 2000, 2004; Janzen, 2004). Although usually associated with the tropics, parataxonomy may need to be embraced more formally in North America, if skills such as the collection, preparation, sorting, rudimentary identification and archiving of specimens are to serve their essential role in invertebrate conservation and monitoring programmes (Goldstein, 2004).

2.1 Taxonomy

Twice in the last 25 years, major reports have warned that professional taxonomy in the UK was in decline (House of Lords, 1991, 2002). Direct government funding to taxonomic institutions decreased between 1992 and 2002. Institutions outside the government grant-in-aid system, such as CAB International, also saw a major reduction in taxonomic expertise in this period, largely as a consequence of financial constraints, with entomology particularly badly affected (Table 14.1). With falling numbers of professionals, the UK increasingly relies on amateur experts (Anon., 2005; Lyal, 2005). However, there is a shortage of such expertise, the active community of amateur specialists is aging and the teaching of taxonomy is declining (House of Lords, 2002). Despite the warnings, taxonomic capacity in the UK for most animal and plant groups continues to deteriorate (Lyal, 2005), and the decline of taxonomy in schools and universities is likely to inhibit recruitment of the next generation of entomologists (Porter, 2001). The decline in taxonomy is not solely a UK phenomenon, but is part of a global trend (Lyal, 2005). Hence, measures to enhance taxonomic activity, whether through the creation of professional posts or stimulation of interest amongst amateurs, are of very broad relevance.

Table 14.1. Taxonomists employed by CAB International 1992–2002. (From House of Lords, 2002.)

	1992	2002
Bacteriology	1	0
Entomology/arachnology	12	0
Mycology	15	7
Nematology/parasitology	6	1

2.2 Biological recording

Knowledge of the distribution and status of species is fundamental to conservation. Where conservation is focused at the species level, typically in temperate regions, knowledge of changes in status over time is increasingly important as priority shifts from rarities to species demonstrably in decline (Porter, 2001; Stewart and New, Chapter 1, this volume). In tropical regions, knowledge of the distribution and status of insect species (or higher taxa) is a valuable component, for example, of site evaluation (Brown, 1991). Given their recognized value as sensitive indicators, information on many insect taxa is also important for monitoring wider environmental (including climate) change.

The UK is fortunate to have exceptional quantities of data on current as well as historical distribution and abundance of its flora and fauna. Burnett *et al.* (1995) estimated at least 80 million taxon-based records, with over 60,000 individuals (predominantly volunteers) actively involved in recording. This effort has allowed detailed national atlases to be produced for a range of taxa by the Biological Records Centre (BRC), including the most widely recorded insect groups (e.g. orthopteroids: Haes and Harding, 1997; butterflies: Asher *et al.*, 2001). A much wider range of 'provisional' atlases are available for other insect and invertebrate taxa. In addition to the now somewhat dated British *Red Data Book* (Shirt, 1987), the status of 'scarce and threatened' species of selected insect groups has been reviewed as part of a series published by the Joint Nature Conservation Committee (JNCC). Such resources have allowed the changing fortunes of British wildlife to be assessed (e.g. Hawksworth, 1974, 2001). Similarly, recording efforts in some other parts of Europe have enabled the production of invaluable resources. For Orthoptera alone, recent national atlases include those for France (Voisin, 2003), Germany (Maas *et al.*, 2002), Switzerland (Baur *et al.*, 2006) and the Netherlands (Kleukers *et al.*, 1997). Cooperative efforts across Europe are poorly developed for most insect taxa, but rather more advanced for butterflies. In North America, some data exist on the distribution and changing status of relatively well-studied insect taxa (e.g. dragonflies, butterflies, larger moths, scarab beetles and tiger beetles). Professional and amateur entomologists acting in a voluntary capacity have made important contributions to the collection of this information (Opler, 1991). Most knowledge of butterfly biology and distribution in Australia has come from a small number of amateurs, but other insect groups have attracted little attention; a similar situation exists in South Africa (Stewart and New, Chapter 1, this volume).

Even in countries with a strong natural history tradition, and a relatively small and taxonomically tractable fauna, some insects attract greater numbers of recorders than others. In the latter part of the 20th century, biological recording of certain 'popular' insect groups increased in the UK, notably butterflies and some moths (Fox, 2001) and dragonflies (Brooks, 2001). However, an increase in recording of easily recognized species can mask a simultaneous decline in recording of more 'difficult' species, even within the same higher taxon (e.g. hoverflies: Morris, 2005). In many cases, the recording of particular taxa remains the pursuit of only a small, or shrinking, number of individuals (Porter, 2001).

Biological recording depends on accurate species identification. Relevant skills can be developed at any age, but there are concerns that levels of ability and motivation are low amongst young people in the UK. Birchenough (2002) and Evans *et al.* (2005) found that, of 18 common birds, the majority of schoolchildren aged 7–16 could name less than 25%. Bebbington (2005) found that, of 10 common wild flowers, 86% of 16- to 18-year-old biology students could name only three or fewer, and 41% could name only one or none, noting that 'conversations with students suggest a general feeling that being able to name organisms is not important to them and that they have little interest in acquiring identification skills'. The problem persists, it appears, even amongst students who choose to study biology at university. Birchenough (2002) and Evans *et al.* (2005) found that first year undergraduate biologists could correctly name less than a third of common animal and plant species (a similar success rate was obtained by students studying English); final year biology undergraduates performed somewhat better, correctly identifying around half. None the less, lack of identification skills amongst graduates is now a significant barrier to employment by ecological consultancies and conservation agencies (Compton, 2004), who find it increasingly difficult to recruit new staff with field-based skills 'in which the UK used to be a world leader' (FSC/BES, 2004). Even 20 years ago, Speight (1986) suggested that 'an entire generation of European graduates has now appeared that is largely incapable of determining insect species or of even knowing how to begin to name an insect'. The depletion of such skills threatens to undermine the invaluable work of amateur naturalists in biological recording, and the ongoing accumulation of data on the distribution and abundance of species. Similar concerns have been expressed in North America, and linked to the extinction of experience concept. For example, Miller (2005) suggests that adolescents in Los Angeles were more likely to correctly identify automatic firearms by their report than birds by their call. However, as this latter example suggests, inability to identify common plants and animals does not seem to reflect a general inability to identify objects, it is simply that many young people's experience of the world around them is now focused on other stimuli. So, whilst they may be unable to identify more than a few common wild flowers, most Americans can identify hundreds of corporate logos (Miller, 2005), and British children as young as 8 years rapidly and enthusiastically learn to discriminate between the numerous fantasy 'creatures' represented on Pokémon trading cards (Balmford *et al.*, 2002; see Section 3.2).

3 Sources of Inspiration and Instruction

Early experience can lead to a lifetime interest or involvement in the study of animals and plants, or underpin their emergence later in life. Young children tend to approach insects with curiosity, rather than the revulsion that may be displayed by their parents, and an interest in entomology can develop at a very early age (Waring, 2004). If aspiring entomologists and other naturalists are to be encouraged and their potential realized, they need support. This must be made available at an early age for those who want it, but suitable facilities are also needed for those who (re)turn to natural history study later in life. This may be an increasingly important consideration in countries like the UK, with a growing proportion of older people in the population. Declining contact with nature is key to the extinction of experience concept. Birchenough (2002) and Evans *et al.* (2005) found that ability to identify animals and plants was greatest among social groups whose work (or, importantly, hobbies) brought them into closest contact with nature. With diminishing opportunities for (particularly young) people to explore nature freely, other points of contact, including the media and formal education system, may be increasingly significant sources of inspiration for young naturalists. These issues, as well as sources of early and ongoing support for entomologists in particular, are considered below.

3.1 Access to nature

The most obvious constraint on access to nature in urban situations is the paucity of even semi-wild spaces in the immediate environment (Key, 2004; Miller, 2005). However, more subtle societal trends may restrict access to (or activities in) natural settings, even when these are within reach. Miller (2005) notes how the accelerating 'pace of life' contributes to the estrangement of people from nature. The lives of American adults are 'overscheduled', and children's outdoor activities are increasingly structured in organized activities, leaving them less time to explore on their own. This problem is compounded by the fact that children are increasingly adopting sedentary lifestyles, and spending less time outdoors. In the UK, those who might previously have studied natural history during their working lives in an unrelated profession are increasingly constrained by a lack of leisure time in which to pursue such interests (Newbould, 2004; Jepson, 2005). Parents are less inclined to let children roam freely, citing fears over road safety or 'undesirable strangers', and even because modern concerns with cleanliness result in children being discouraged from activities where they might get dirty (Key, 2004; Bebbington, 2005). Many naturalists who developed their interest at an early age acknowledge the role of a particular, inspirational mentor. However, the mentoring of youngsters by older members of the community now seems to be regarded as a highly suspect practice (Howes, 2004), and increasingly restrictive rules about children's attendance at field meetings even create difficulties for organized clubs and societies (Key, 2004).

The encouragement of visitors to recognized wildlife and amenity sites to participate in biological recording can be a rewarding way to engage potential naturalists (Ely, 2004), although nature reserves may be less welcoming to children than in the past (Key, 2004). Many organizations stand to benefit from encouraging biological recording on their reserves, as the data obtained are important to inform conservation management. In the UK, even relatively large organizations like the Wildlife Trusts (WTs), with well-developed networks of nature reserves, do not have the in-house resources to monitor the wildlife on those sites, and are therefore reliant on volunteers (Hollings, 2004). However, partly (no doubt) because the value of their activities is not widely appreciated, fieldworkers can be treated with suspicion by members of the general public. Rather than being asked with interest what they are doing, naturalists may be more likely to be asked in an accusing tone if they have a licence to do it.

3.2 Collecting

With exceptions (such as butterflies and dragonflies), reliable identification of many insects and other invertebrates in the field presents particular challenges. Just as a botanist might collect a flowerhead or leaf in order to confirm an identification, the collection of invertebrate specimens is an important part of the process of learning to distinguish between taxa. In addition, the critical importance of retaining voucher specimens is widely acknowledged, for example, in relation to taxonomic research, ecological studies, biodiversity surveys, biological control and phytosanitary measures, as well as conservation programmes (e.g. Burnett *et al.*, 1995; Huber, 1998; Basset *et al.*, 2000, 2004; Salmon, 2000; Porter, 2001; Goldstein, 2004; James, 2004; Janzen, 2004).

Collecting insects and other invertebrates has very rarely resulted in species extinction, although (particularly when commercially driven) it may have contributed to the decline of some localized populations (Howarth and Ramsay, 1991; Salmon, 2000). Loss and degradation of habitats is the main driver of insect extinctions; where habitats remain in good condition, insect populations tend to be sufficiently robust that they can withstand the removal of even quite large numbers of individuals. Partly because of ongoing habitat degradation, many entomologists in the UK readily embraced the ethos of conservation during the 20th century, and recognized that restraint was required. A Code of Conduct for responsible collecting was published by representatives of the invertebrate study community in the 1960s, and has been widely respected (and occasionally revised) over subsequent years (Invertebrate Link, 2002).

Despite these considerations, there exists hostility towards collecting in some quarters (Porter, 2001), and some books aimed at a young audience of potential naturalists actively discourage it (Key, 2004). As well as largely misguided concerns over impacts on wild populations, accusations of cruelty to insects have changed little since the early 19th century (see Hollerbach, 1996). In addition to disregarding the practical importance of reference specimens, such perspectives seem to assume that collecting insects is analogous to killing birds, whilst overlooking, for example, the millions of insects that are squashed on car windscreens every year (Key, 2004). In Britain, the change in attitude towards collecting and collections has affected the recruitment of new recorders, particularly for those groups which require close examination (Porter, 2001; James, 2004). In North America, complacency over the need for voucher specimens, in combination with anti-collecting sentiments, threatens to decouple invertebrate species-based information from conservation management evaluation and monitoring (Goldstein, 2004). There is also concern (e.g. in Australia and continental Europe) that legal restrictions on collecting have been, or could be, counterproductive. Such legislation may draw the attention of commercial collectors to vulnerable populations, and/ or discourage much needed, responsible study of insects and other invertebrates (Balletto and Casale, 1991; Greenslade and New, 1991; Mikkola, 1991; Morris *et al.*, 1991).

No sensible entomologist would support collecting where this has a negative impact on (particularly small, remnant) wild populations. None the less, accumulation of natural history specimens has long provided an important source of inspiration for some of our most respected naturalists and conservationists. In combination with other changes, the decline in such harmless collecting has contributed to the general ignorance of even common species of animals and plants (Salmon, 2000). The finely honed ability of children to discriminate between numerous Pokémon characters was reported above. It is interesting to note that this world of fantasy 'species' was created by Satoshi Tajiri, who as a child in Japan had collected insects (especially beetles) and aspired to be an entomologist. By the late 1970s, the fields and ponds around the Tokyo suburbs that he had explored while growing up had disappeared under urban development. Tajiri saw Pokémon trading cards as a way for 'children of a new generation to have the chance to collect insects and other creatures the way he did' (Chua-Eaon and Larimer, 1999). Redirecting the collecting urge in this way may provide some opportunities for promoting natural history study amongst children. Balmford *et al.* (2002) suggest 'Ecomon', implying sets of trading cards featuring real species. In the UK at least, sets of Brooke Bond tea cards illustrating wild flowers or butterflies have played a useful role in the past (Key, 2004; Waring, 2004). One optimistic view is that advances in digital photography could result in a renaissance in 'collecting', with digital images replacing specimens (Jepson, 2005). However, this could further concentrate effort on relatively large, easily identified taxa, while the study of the majority that require closer examination continues to decline (Morris, 2005). Overall, accumulation of a small specimen collection remains an important (and largely harmless) way for children to engage with, and learn to identify, insects in many situations. However, appropriate codes of conduct, and restrictions (where they apply) should always be respected.

3.3 The media

For many people, television has become a primary source of entertainment and information. Wildlife documentaries have provided the public with unprecedented access to the natural world, and have done much to raise awareness of environmental and conservation issues. However, there are concerns that television has promoted passive interest in natural history at the expense of active participation in its study (Howes, 2004; Pickles, 2004). Bebbington (2005) suggests that programmes 'condense time and often highlight the spectacular, making the real world seem less interesting'. Entomology has long faced particular challenges in terms of its public image (e.g. Hollerbach, 1996). Insects and other invertebrates seem to attract relatively little public sympathy, often eliciting the 'monster in the bath' response (Speight, 1986), which is only exacerbated by unsympathetic media coverage. However, when allowed to bask in the warmth of a more positive light, invertebrates often attract a more favourable response. The recent BBC series *Life in the Undergrowth* attracted exceptionally high approval ratings from viewers in the UK (Masters and Smithers, 2006). Despite the dominance of television, the printed media also have a significant audience, whether for newspapers or magazines. Local media can do much to promote local societies (see below), specific local initiatives to promote invertebrates (Smithers, 2006), and local components of countrywide initiatives like the UK's National Insect Week (Haines and Rogers, 2006). None the less, naturalists, as well as invertebrates, are vulnerable to poor representation, with some sections of the media tending to portray them negatively as eccentrics (Jepson, 2005).

3.4 Museums and zoos

Museums and their collections have long played a major role in inspiring and instructing prospective entomologists and other naturalists in the UK (e.g. Waring, 2004). Local and provincial museums have often been particularly important (Barber, 1980; Palmer, 2004). Indeed, local museums are now increasingly required to perform a quantifiable educational function as a condition of their funding (Howes, 2004). However, this seems to have resulted in a proliferation of interactive displays, in place of more traditional exhibits. Natural history specimens, in particular, have tended to be used less (Palmer, 2004). Few European museums allocate much exhibition space to local insect faunas or their conservation (Speight, 1986). Although museums report large increases in numbers of young visitors, the majority are under 10 years, and come as part of school activities which are restricted to specific curriculum-based topics (Norris, 2004). Some local museums have also attempted to fulfil their educational role by organizing natural history talks and workshops. However, very few still employ curators and managers who are themselves active naturalists prepared to encourage protégés (Howes, 2004), and natural history collections tend to be the least well-supported area of museum holdings (Palmer, 2004). The changing demands placed on museums, particularly their new role as interactive learning centres, has also resulted in collections-based research becoming much less of a priority (Porter, 2001; Norris, 2004). Collections held in British museums are often of international importance, but are now at considerable risk unless increased government funding is found for basic curation (House of Lords, 2002). Around the world, developments in bioinformatics are gradually increasing the accessibility of data on museum holdings, geographical and ecological data associated with specimens, and illustrations to support accurate identifications (Pennisi, 2000; see below). However, digitization of this information is labour-intensive and resource-limited, and progress is likely to be particularly slow for speciose taxa like insects and other invertebrates (Pennisi, 2000).

As with museums, zoos (and botanic gardens) have the potential to inspire and educate, as well as containing increasingly important (in this case, living) collections. London Zoo lays claim to the first dedicated Insect House (from 1881), and the Invertebrate Conservation Unit there now occupies one of the most modern and sophisticated buildings on the Regent's Park site (Barrington-Johnson, 2005). Spencer (2006) describes the development of dedicated invertebrate facilities over recent years at Bristol Zoo, where surveys have demonstrated high levels of visitor appreciation and engagement. The potential for such centres to introduce the public to invertebrate diversity and its conservation is enhanced by the fact that many species are well suited to *ex situ* programmes (Pearce-Kelly *et al.*, Chapter 3, this volume). In some countries, butterfly houses (butterfly farms) provide another controlled environment in which the public can experience and learn about insect diversity (Morris *et al.*, 1991).

3.5 Formal education

With fewer opportunities for spontaneous exploration, outdoor learning in formal education would seem to be increasingly important. As well as providing early opportunities to develop, for example, identification skills, there are much broader benefits. A recent international review (Rickinson *et al.*, 2004) concluded that learning can be enhanced by the memorable setting of a field trip, and that fieldwork can promote positive attitudinal, interpersonal and social development. School fieldwork is often reported by students as being amongst the most enjoyable and rewarding aspects of their studies, can motivate students to pursue subjects at a higher level, and have a role in the subsequent adoption of environment-friendly behaviour (Fisher, 2001; Barker *et al.*, 2002). However, there is concern in a number of countries over diminishing opportunities for outdoor learning (Rickinson *et al.*, 2004). The UK has seen a decline in fieldwork as part of school and university teaching, despite recognition by teachers of its importance (e.g. Kinchin, 1993). Tilling (2004) found that fewer than 5% of schools provided 11- to 16-year-olds with residential field courses. Speight (1986) suggested that the majority of European children in this age group pass through school 'without even hearing mention of insects'. The quality and quantity of fieldwork undertaken by 16- to 18-year-olds has declined in recent years, and at least one in three such students does no fieldwork as part of his or her studies (Lock and Tilling, 2002; Tilling, 2005). Tilling (2004) found that numbers of residential biology groups had declined by one-third between 1970 and 2003, as a proportion of all school and university groups using Field Studies Council (FSC) centres. The FSC is the largest independent provider of field courses for UK schools (Tilling, 2005), but is part of a wider network of residential and day centres. In the early 1980s, over 2500 such centres were in operation, but the following decade saw the closure of many, particularly those owned by local education authorities (Rickinson *et al.*, 2004). These trends partly reflect a long-term shift in education in the UK, away from the teaching of whole-organism biology, towards (often more commercially orientated) molecular, genomic and cellular disciplines (Barker *et al.*, 2002; Compton, 2004; Tilling, 2005). It is interesting to note that decline in fieldwork appears to be much more severe in biology than in geography (Tilling, 2004, 2005). The decline in biology fieldwork could already be contributing to low numbers and quality of candidates for posts in the environmental sector, and could threaten environmental monitoring programmes that depend on volunteers.

The underlying causes of the decline of fieldwork in UK schools are complex (Tilling, 2005); principal factors identified by a number of studies are summarized in Table 14.2. Lack of confidence amongst teachers may reflect increasing numbers of non-specialists teaching biology (Kinchin, 1993; Fisher, 2001). Speight (1986) suggested that most European teachers had little knowledge of invertebrates, or training on how to incorporate them into teaching programmes. Furthermore, opportunities to gain confidence are diminishing, as fieldwork is declining as part of both teacher training and in-service experience (Fisher, 2001; Barker *et al.*, 2002; FSC/BES, 2004). Concerns over health and safety were identified as a principal factor by just one of the studies in Table 14.2. None the less, they appear to be a significant problem, not just in the UK (where the second-largest teaching union recently advised its members not to take children on school trips because of fears over pupil safety), but also in the USA and Australia; such concerns reflect a growing litigation culture which demands compensation for any act or omission that results in personal injury (Rickinson *et al.*, 2004). Despite the

Table 14.2. A comparison of principal factors reported to be influencing the decline of fieldwork in schools. (Adapted and updated from Barker *et al.*, 2002.)

Factor influencing decisions not to do fieldwork	Fido and Gayford (1982)	Kinchin (1993)	Fisher (2001)	Lock and Tilling (2002)	Tilling (2004)
Large class sizes	X	X			
Time/timetable constraints	X	X	X	X	X
Transport problems	X	X			
Lack of enjoyment, interest, motivation, confidence or ability of teacher		X	X	X	
Cost	X		X	X	X
Availability of suitable sites			X		
Curriculum/assessment constraints			X	X	X
Risk of accidents			X		
Students do not see the need			X		

concerns, 2004 data show that FSC centres have a safety record (one notifiable accident for every one million hours of fieldwork contact time) of which most schools would be proud (Tilling, 2005). Lack of time for fieldwork may result from inflexibility in the school timetable or curriculum, but may also reflect the need for teachers to invest increasing time in preparatory activities (Rickinson *et al.*, 2004). Barker *et al.* (2002) also note that the enthusiasm and dedication of teachers who organize out-of-school activities, often in their own time, is taken for granted. This can be demoralizing in itself, but is all the more so when combined with a blame culture which castigates such teachers when something goes wrong. Changes to the UK education system in recent years have resulted in fieldwork becoming an increasing financial burden on schools and parents (Lock and Tilling, 2002; Barker *et al.*, 2002). Also, teachers and students alike are increasingly unwilling to give up school holidays or weekends, partly because many students have part-time jobs and are reluctant to sacrifice income from these (Fisher, 2001; Barker *et al.*, 2002).

At university level, substantial reduction in the funding available for student tuition has put particular pressure on field courses (Compton, 2004). The need for students to fund their own fieldwork is likely to be particularly problematic at a time of increasing student indebtedness (Rickinson *et al.*, 2004). Fieldwork at universities has also suffered from increasing student/staff ratios, and constraints based on course design and assessment methods (Rickinson *et al.*, 2004). In relation to lifelong learning opportunities, Howes (2004) suggests that natural history courses have vanished from the prospectuses of many university extramural departments. Pilkington (2005) records a decline in Continuing Education provision for science in general, and links this (in part) to the increasing emphasis on accredited courses, driven by policy changes and funding rules, which may have alienated potential students and a number of tutors.

Outside the formal education sector, relevant training (e.g. in species identification) is available from a broad range of institutions in the UK, with the FSC playing a particularly important role. However, courses may be concentrated in particular localities, and those on insects tend to focus on butterflies and dragonflies, with very patchy coverage of other taxa (Howes, 2004). Further weaknesses in the diffuse training network include the lack of mechanisms for sharing best practice and for training of trainers, and a lack of overall coordination, including any central source of information on available courses and their contents, or measures to ensure that 'introductory' courses are matched to courses at a higher level (Pickles, 2004).

3.6 Societies and special interest groups

A key element of the support structure for British natural historians has long been a proliferation of clubs and societies, bringing together individuals with varying levels of experience (Allen, 1987, 1994; Salmon, 2000). Historically, many of these have been local (e.g. county level) and/or specialized (e.g. entomological) societies. In recent years, some national bodies (particularly those focused primarily on conservation) have attracted very large numbers of members (the Royal Society for the Protection of Birds has over a million; Butterfly Conservation has over 10,000), but this may have been achieved at the expense of smaller and/or local groups (Berry, 1987). Although some continue to flourish, there is much concern over aging and falling memberships amongst many local societies (Howes, 2004; Newbould, 2004), and a sense that such groups need to be more forthright in promoting their activities and achievements (Bowler, 2004; Newbould, 2004).

Local naturalists invariably have the greatest knowledge and understanding of local sites and of the species that inhabit them, and are best placed to monitor seasonal changes. In-depth knowledge of a given locality, based on year-round observations, and rooted in a strong social tie to the local community, is also a feature of the work of parataxonomists in the tropics (Janzen, 2004). Local entomological societies have been important in facilitating surveys of some taxa in parts of North America (Opler, 1991), as in the UK and elsewhere. However, it may require coordination of effort at a national level, for example, to produce a distributional atlas at country scale (e.g. Asher *et al.*, 2001). Such high-profile initiatives may also help to stimulate interest in particular taxa, and to promote the recruitment of new recorders. National societies are often able to instigate high-profile campaigns for which local societies lack the resources. However, a balance needs to be maintained between local and national bodies. Specialist and local societies can provide important channels for communication between individuals and large conservation organizations (Moore, 1991). The apparent decline of local natural history societies in the UK may be linked to societal changes. The once strong integrity of local communities is diminishing and the 'flexible job market' (i.e. reduced job security) increasingly requires individuals to move around the country in pursuit of short-term positions, preventing them from developing a detailed familiarity with the natural history of their surroundings.

Entomological societies exist all over the world, often as national bodies, but also at a more local (state, county) scale in some countries. The Australian Entomological Society has done much to promote insect conservation in a country where there are relatively few amateurs involved in the study of insects (Greenslade and New, 1991). None the less, the encouragement of amateurs and coordination of volunteers is an important role for specialist entomological societies in places like the UK. Here, a number of national societies and groups promote invertebrate studies, with varying degrees of specialization (see Table 14.3), and may provide support to their members through access to libraries and collections, and/or the organization of regular indoor and field meetings. Groups such as the Dipterists Forum and the Bees, Wasps and Ants Recording Society have done much in recent years to promote and coordinate studies within their particular taxonomic ambit. A number of substantial insect groups do not have a single society or special interest group promoting their study in Britain, notably the beetles, although *The Coleopterist* journal and its associated website does provide a focus for information exchange. Recently, Buglife (The Invertebrate Conservation Trust) was established, the first organization in Europe devoted to the conservation

Table 14.3. National societies, interest groups and recording schemes relevant to invertebrate studies in the UK.

National invertebrate special interest groups and societies	National recording schemes for invertebrates	
	Insects	*Other invertebrates*
Amateur Entomologists' Society Balfour-Browne Club Bees, Wasps and Ants Recording Society British Arachnological Society British Dragonfly Society British Entomological and Natural History Society British Myriapod and Isopod Group Buglife: The Invertebrate Conservation Trust Butterfly Conservation (British Butterfly Conservation Society) Conchological Society of Great Britain and Ireland Dipterists' Forum Malloch Society Royal Entomological Society **Other national bodies with an interest in invertebrate study/ conservation (excluding statutory bodies)** Ancient Tree Forum (Woodland Trust) Freshwater Biological Association Linnean Society National Trust for England, Wales and Northern Ireland Royal Horticultural Society Royal Society for the Protection of Birds (RSPB) The Wildlife Trusts	**Coleoptera**: aquatic species; Cantharoidea and Buprestoidea; Carabidae; Cerambycidae; Chrysomelidae and Bruchidae; Coccinellidae; Cryptophagidae (Atomariinae); Dermestoidea and Bostrichoidea; Elateroidea; Ptiliidae; Scarabaeoidea; Scirtidae; Scolytidae; Staphylinidae; Staphylinidae (Stenini) **Diptera**: Anthomyiidae; Brachycera (larger species); Chironomidae; Conopidae and Lonchopteridae; Culicidae; Dixidae; Drosophilidae; Empididae; Mycetophilidae; Nerioidea (Pseudopomyzidae, Micropezidae) and Diopsoidea (Tanypezidae, Strongylophthalmidae, Megamerinidae and Psilidae); Pipunculidae; Sciomyzidae; Simuliidae; Syrphidae; Tachinidae; Tephritidae; Tipuloidea and Ptychopteridae; Ulidiidae, Platystomatidae and Pallopteridae **Ephemeroptera**[*] **Heteroptera**: aquatic species; terrestrial species **Homoptera**: Auchenorrhyncha **Hymenoptera**: Aculeata; Symphyta **Lepidoptera**: Gelechiidae, Blastobasidae, Momphidae, Cosmopterigidae and Scythrididae; Incurvarioidea; Pyralidae and Pterophoridae; leaf-mining species; scarce macro-moths; various initiatives for butterfly monitoring **Neuroptera, Mecoptera and Megaloptera** **Odonata** **Orthoptera, Dermaptera, Dictyoptera and Phasmida** **Plecoptera**[*] **Siphonaptera** **Trichoptera**[*] [*]Increasingly working jointly as the Caddisfly, Mayfly & Stonefly Recording Scheme (CAMSTARS)	**Acarina**: Ixodoidea **Arachnida**: Araneae; Opiliones; Pseudoscorpiones **Collembola** **Crustacea**: hypogean species **Isopoda**: non-marine species **Mollusca**: non-marine species **Myriapoda**: Chilopoda; Diplopoda **Tricladida**: freshwater species; terrestrial species

of all invertebrates. Although not currently a membership organization in the traditional sense, Buglife provides access to a wide range of information via an excellent website. Some societies place particular emphasis on encouraging younger members. This is one focus of the Amateur Entomologists' Society (AES) in the UK, through their Bug Club, and is the purpose of the Young Entomologists' Society in the USA.

Taxon-specific national recording schemes (a range of which exist for invertebrates – see Table 14.3), supported by BRC, provide another valuable arm of the support structure for British natural historians. In recent years, the establishment of recording schemes, or perhaps more accurately the energy of the individuals who act as voluntary recording scheme coordinators, has greatly boosted the study of certain insect groups, including some dipteran families (Stubbs, 2001), orthopteroids (Marshall, 2001) and Auchenorrhyncha (Kirby *et al.*, 2001). The Butterfly Monitoring Scheme operates in a rather different way, but substantially enhanced recording of this group when it was established in 1976 (Fox, 2001). A national recording scheme for all macro-moths has recently been established, which will build upon historical data collected by the Rothamsted Insect Survey network (Conrad, Fox and Woiwod, Chapter 9, this volume). Local Records Centres (LRCs) provide another important element in the UK's biological recording network, and may have links with bodies such as local museums (James, 2004). The NFBR also plays a valuable facilitating role in promoting coordination between various players in the biological recording network.

3.7 Field guides and other publications

From the perspective of introducing children to insects at an early age (e.g. see Waring, 2004), it is interesting to note that Eric Carle's *The Very Hungry Caterpillar* is one of the most widely purchased books for youngest readers, estimated to have sold one copy every minute since its publication in 1969 (Casciani, 2005). Among books for older children, Dorling Kindersley publications are particularly well produced. For those children or adults who wish to study insects (or other taxa) in more detail, suitable introductory publications, especially aids to species identification, are an important resource. In Australia, 'the absence of a basic foundation of published natural history information … creates problems for beginners, amateur or student' (Greenslade and New, 1991). In addition to a wide range of books, British entomologists have long benefited from the regular publications of specialist and local societies, as well as small, independent journals. Other series of publications aim specifically to stimulate more detailed study as well as assisting in identification. These include AES publications (*Dipterists' Handbook, Hymenopterists' Handbook, Coleopterists' Handbook,* etc.), the *Naturalists' Handbooks* series, and the reinvigorated *New Naturalist* series. Barnard (1999) provides an annotated bibliography of key works for the identification of British insects and arachnids (see also Corke, 1999). Gilbert and Hamilton (1990) provide a much broader (but now somewhat outdated) guide to entomology information sources (cf. Harvey, 1999). However, key publications are often expensive (Newbould, 2004), out of date or out of print (Lyal, 2005), or not stocked by mainstream bookshops (Key, 2004).

Poorly produced identification guides (and dichotomous keys in particular) can be difficult for non-specialists to use (House of Lords, 2002); well-illustrated works that emphasize simple identification features are more accessible. Consequently, field guides, with numerous lifelike illustrations, often produced by naturalists rather than taxonomists, have become the primary identification tools for non-specialists (Stevenson *et al.*, 2003; Lyal, 2005). However, the economics of traditional publishing dictate that these must be commercially viable products. Where an assured market of enthusiastic amateurs is lacking, few guides are likely to be published (as in Australia: Greenslade and New, 1991). Where a market does exist, there is a tendency to produce numerous volumes covering the same, popular taxa. In the USA, for example, the number of field guides devoted to invertebrates is very small compared to those available nationally or regionally for birds or plants (Stevenson *et al.*, 2003). Amongst the insects, butterflies and dragonflies invariably receive disproportionate coverage, but this partly reflects their relative ease of identification in the field. An alternative commercial strategy is to cover large numbers of species and wide geographic areas (Stevenson *et al.*, 2003). For example, Collins' *Insects of Britain and Northern Europe* (Chinery, 1993), a key general work for insect identification by non-specialists, illustrates fewer than 800 of Britain's ~23,500 species (Barnard, 1999). Such guides can be of limited use to those working at a localized scale, and can result in false confidence in identifications made by comparing specimens with an inadequate range of illustrated species. If commercial viability is a constraint on general field guides, a much greater challenge confronts publishers of more detailed taxonomic material. In the UK, publication of such works largely relies on specialist publishers such as Harley Books, and dedicated bodies like the Ray Society. The Linnean Society's *Synopses of the British Fauna* series provides coverage of some non-insect invertebrate taxa, and the Freshwater Biological Association publishes a series of handbooks or keys to aquatic invertebrates including insects. For keys to the British insect fauna, an important resource is the Royal Entomological Society's (RES) *Handbooks for the Identification of British Insects* series. Unfortunately, this covers only about half of the total fauna, many volumes are out of print and the rate at which new volumes are published has declined in the last 20 years (Lyal, 2005). Recently, however, a partnership between the RES and FSC has started to reinvigorate the *Handbooks* series. FSC has a strong reputation for producing user-friendly keys, in their own *Aids for the Identification of Difficult Groups of Animals and Plants* (AIDGAP) series, which includes a number of valuable invertebrate identification works. For particular insect taxa, continental European publications have replaced UK-specific works as definitive aids to identification of the British fauna (Barnard, 1999). Important publications in this respect include, for example, selected volumes in the *Faune de France*, *Microlepidoptera of Europe*, and *Fauna Entomologica Scandinavica* series, amongst others. Continental European publications that include insect keys are summarized by Gaedike (2003), and earlier reviews in this series (Barnard,

1999). The absence of much-needed taxonomic works covering the European insect fauna, however, is partly a reflection of the shortage of appropriately trained entomologists (Speight, 1986).

3.8 Electronic and technological resources

Correspondence has long been an important means of exchanging information, and historically many eminent naturalists were happy to correspond with youngsters and beginners (Barber, 1980; Salmon, 2000). As a vehicle for correspondence, the postal system is increasingly being replaced by e-mail and thematic web-based electronic discussion groups. Rapid developments in electronic and technological resources have the potential to revolutionize many other aspects of information management relevant to naturalists. For example, Polaszek *et al.* (2005) call for the establishment of a web-based register of all new zoological names. Currently, the formal record of animal species names and descriptions is scattered across numerous, disparate publications; entomological taxonomic information alone is dispersed throughout at least 1100 specialized journals (Polaszek *et al.*, 2005). Lyal (2005) notes a number of recent global initiatives aimed at providing comprehensive, regularly updated checklists. Similarly, there are many potential benefits from the storage of biological records in electronic rather than paper-based systems (Burnett *et al.*, 1995). The UK's National Biodiversity Network (NBN) aims to provide a central, web-based repository for biological records and resources for biological recorders, and around 19 million records are currently available online via the NBN Gateway (Cooper *et al.*, 2005). Internationally, information held in museum collections is also increasingly available in electronic form, for example, via the Global Biodiversity Information Facility (GBIF) portal (Lyal, 2005).

Some out-of-print field guides and other key reference works are now available in electronic form. In the UK, Pisces Conservation Ltd has reproduced a number of such works on CD. Increasingly, new types of field guide and other aids to species identification are available, harnessing computer technology to provide larger volumes of information and illustrations than paper products, more flexible mechanisms for searching content, and links to an expanding range of rapidly updated Internet resources (Porter, 2001; House of Lords, 2002; Stevenson *et al.*, 2003; Ely, 2004; Gates, 2005). For example, some of Dorling Kindersley's recent books for children provide links to dedicated, web-based resources, including Burnie (2005) for young entomologists. Electronic products for an older audience include those available from ETI Bioinformatics in the Netherlands, including an *Interactive Guide to Butterflies of Europe*, and a range of more specialized entomological tools. Other examples of suppliers, products and web-based facilities are noted by Stevenson *et al.* (2003) and Gates (2005). Some of the problems associated with traditional field guides and handbooks (such as their tendency to cover only adult life stages: Lyal, 2005) can easily be overcome by electronic guides (Stevenson *et al.*, 2003), provided the underlying taxonomic research has been done. The use of computer databases, incorporating digital images, has proven to be a crucial factor in the rapid discrimination and identification of insect morphospecies by parataxonomists in the tropics (Basset *et al.*, 2000, 2004). Virtual resources can also provide valuable preparatory material for fieldwork in schools (FSC/BES, 2004), although they should not be seen as an effective substitute for activities in the field.

In the development of electronic field guides, available software is largely sufficient to manage content, and advances in digital photography have enhanced access to relevant images (e.g. see Jepson, 2005). A greater constraint on such tools as potential replacements for traditional paper products relates to hardware, and specifically the need to make electronic tools suited to use in the field (Burnett *et al.*, 1995). However, advances in the design of Personal Digital Assistants (PDAs) and Global Positioning System (GPS) equipment will increasingly enable the development of hand-held tools which provide access to high-quality images to assist identification, capture of electronic 'voucher' images and recording of associated information such as location data (Porter, 2001; Stevenson *et al.*, 2003; Gates, 2005). Tools of this kind are currently being developed, for example, at Oxford Brookes University, very much with recruitment of the next generation of naturalists in mind (e.g. Bailey and Thompson, 2005, unpublished data). Whilst most aspects of such developments are positive, there is some concern that the increased ability for individual recorders to manipulate and summarize (including by map production) their own data may be a disincentive for them to submit records to wider recording schemes and record centres (Porter, 2001).

Although significant challenges remain, substantial progress is being made in the development of automated identification systems (Gaston and O'Neill, 2004). Most such research has focused on recognition of morphological features, but some invertebrate taxa provide opportunities for alternative approaches. Some orthopterists have already embraced one new electronic tool, the ultrasonic (bat) detector, to amplify the songs of grasshoppers and their relatives (Baldock, 2000). They may be able to look forward to automated identifications, based on the acoustic characteristics of stridulation (Chesmore and Ohya, 2004). DNA barcoding provides the prospect of rapid identification of material based on genetic rather than morphological characters (Savolainen *et al.*, 2005). Such technologies should allow rapid processing of numerous specimens, removing a common constraint to many entomological studies. However, if made widely available, they may also inadvertently undermine the motivation of prospective naturalists to learn to identify species for themselves. Their likely impact on professional taxonomy has also been a source of considerable debate (Herbert and Gregory, 2005; Will *et al.*, 2005).

4 Conclusions

Naturalists (particularly amateurs) can make invaluable contributions to insect conservation, notably through their involvement in taxonomy and biological recording. In parts of the world with a limited history of such involvement, increased participation by entomological naturalists would be welcome.

Elsewhere, it is important to ensure continuation of a tradition of natural history study. Everywhere in the world, insect conservation would benefit from greater public support, in a general sense. Unfortunately, current global trends (particularly related to urbanization) result in an estrangement from nature, undermining appreciation for biodiversity and removing sources of inspiration for potential naturalists. This effect has been termed the extinction of experience (Pyle, 1993; Miller, 2005). Where specific support structures for aspiring entomologists are traditionally strong, as in the UK, many appear to be in decline, or at least changing rapidly. In order to preserve and enhance broad engagement with insect conservation, multiple solutions are required.

Opportunities (for children in particular) to explore and develop appreciation for nature are vital. Some urban centres retain impressive areas of natural habitats, including patches of Atlantic rainforest in Rio de Janeiro (Brazil), bushland remnants in Perth (Australia) and a network of habitats in Chicago (USA). Elsewhere, however, increasing measures should be taken to design wild spaces into urban environments. While formal parks and highly managed playgrounds may have some value, areas of undeveloped and unmanaged land may offer children more meaningful experiences of nature, provided they are allowed access to such sites (Miller, 2005). Even in relatively rural environments, opportunities (particularly for children) to gain first-hand experience of nature may be diminishing. The likely negative impacts of restrictions on collecting and access to the countryside have even been noted by Sir David Attenborough, probably the greatest advocate for natural history over the last five decades (Masters and Smithers, 2006). The media have a role to play, not only in promoting insects, natural history and naturalists, but in maintaining a sense of proportion over dangers to children playing outdoors.

Where opportunities are limited for young people to explore even semi-natural habitats spontaneously, better integration of outdoor learning into the formal education system may be particularly important. Unfortunately, in the UK, fieldwork has recently declined in schools and universities, and a range of factors will need to be addressed to reverse this trend. These include improved support for fieldwork in teacher training, better integration of fieldwork into curricula, updated resources, supportive (rather than obstructive) health and safety protocols, and better planned learning progression (within schools, but notably between school and university) (FSC/BES, 2004; Tilling, 2005). Some of the many benefits that students derive from outdoor learning appear to arise from their exposure to a new and unfamiliar environment. However, where (residential) trips to distant localities are not feasible, there are considerable benefits to be gained from activities within school grounds or elsewhere locally (Fisher, 2001; FSC/BES, 2004; Rickinson *et al.*, 2004). For example, use of local sites makes regular repeat visits more practical, allowing for monitoring of seasonal changes (Kinchin, 1993). Schools, perhaps in partnership with local clubs or societies, may also have a role in introducing young enthusiasts to appropriate mentors who can guide their further study of insects or other taxa. Mentoring may also be valuable to older aspiring naturalists, and could be incorporated into adult training programmes

(Hollings, 2004; Pickles, 2004). The broad applicability of such support is reflected in the fact that mentors can play an important part in the professional development of parataxonomists involved with tropical biodiversity inventory programmes (Janzen, 2004).

While early sources of inspiration are critical, ways of stimulating interest in insects among adults also need to be considered. In the UK, there appear to be increasing opportunities for amateurs with identification skills to obtain paid work for statutory agencies and the private sector (Newbould, 2004). However, financial reward is not a significant incentive for many naturalists, whose interests are more often driven by simple intellectual curiosity (Morris *et al.*, 1991; Anon., 2005). Indeed, the increasing commercialization of biological recording in the UK has had a negative impact on the submission of data to record centres by some volunteers (Porter, 2001). Although knowledge of (and interest in) local biodiversity may be under threat from the extinction of experience, general awareness of broad environmental issues appears to be increasing. Public involvement in the monitoring of climate change impacts could become an important way of reconnecting people with nature (Gates, 2005; Jepson, 2005). For example, a wide range of non-specialists submit observations to the UK Phenology Network (Collinson and Sparks, 2005; Sparks and Collinson, 2006), and such participation has been promoted on television by the BBC's *Springwatch* programmes. Given their recognized value as bioindicators, insects could benefit in particular from a 'recruitment drive' for naturalists based on such environmental monitoring. The availability of skilled amateurs to monitor the progress of invasive species, such as the Harlequin Ladybird *Harmonia axyridis* in the UK (Roy *et al.*, 2005), may also become increasingly important in many countries.

Perhaps most, if not all, people are born potential entomologists (Waring, 2004). Those whose hobbies and pastimes bring them into relatively close contact with nature should probably be seen as particularly likely recruits. Gardens are increasingly recognized for their value to biodiversity, and for their role in introducing the public to wildlife (Gaston *et al.*, 2004). Brooks (2001) notes that enthusiasm for wildlife gardening has increased interest in dragonflies, and boosted their numbers through the digging of appropriately managed garden ponds. Similarly, there is much interest amongst some gardeners in selecting flowering plants to attract butterflies and bees, and in providing overwintering places for predatory insects such as ladybirds and lacewings. Anglers have also been identified as a group well placed to study insects in the UK, and a number have now enthusiastically received training in the identification and monitoring of 'riverflies' (caddisflies, mayflies and stoneflies) (Waterton, 2003; Lapsley, 2005; FSC, 2006). Some birdwatchers are developing interests in butterflies, moths and dragonflies (Newbould, 2004). Dragonflies are particularly amenable to study by birdwatchers, as they are associated with the same wetland sites as a number of interesting bird species, and can be relatively easily identified through binoculars (Brooks, 2001). Once engaged, it may be possible to encourage those with an interest in butterflies and dragonflies to embrace less-studied insect groups (Key, 2004), going some way to redressing the imbalance in recording that currently exists between taxa.

A key element in overcoming barriers created by the extinction of experience is greater coordination and cooperation between relevant stakeholders. This may be between conservationists and planners, in relation to urban design (Miller, 2005), or between educational establishments and others (such as societies, museums and conservation bodies) who can provide teachers and researchers with support (Fisher, 2001; Bowler, 2004; Gates, 2005). There also needs to be cooperation and coordination between providers of training outside the formal education system. Such partnerships already form the basis of many courses in the UK, where FSC provides administrative and other facilities (including accommodation for longer courses), but relies on a network of external trainers to deliver course content. Such cooperation could be facilitated by establishing a register of specialists willing to act as tutors in training events (Key, 2004). Coordination is also important between the disparate elements of any biological recording network where this already exists (e.g. in the UK: Burnett *et al.*, 1995; and in parts of North America: Opler, 1991). Worryingly, it appears that some elements of the biological recording network in the UK are actually drifting apart (James, 2004).

Future generations of naturalists will require a network of bodies to provide support to beginners and experts alike; both local and national societies have a role to play. Perhaps particular lessons can be drawn from organizations like the Wildlife Trusts and Butterfly Conservation in the UK, which both operate at a national level but are structured into a network of local (county) branches. Other important sources of support include identification aids. There has been relatively little formal study of the ways in which field guides serve as learning and identification tools. Future incarnations might seek design input from cognitive psychologists, anthropologists and educators (Stevenson *et al.*, 2003) to ensure that such products are 'intellectually ergonomic'. This may be particularly important as increasing numbers of traditional, paper identification aids are displaced by highly information-rich electronic alternatives. Although the final demise of the book is still a long way off, the time may soon come when some electronic identification and recording tools become standard pieces of field equipment.

Research suggests that contact with nature is good for human quality of life, based on physical, emotional and intellectual indicators, as well as promoting appreciation of biodiversity (Miller, 2005). A range of studies show that outdoor learning teaches children much more than identification skills (Rickinson *et al.*, 2004). Some of the fundamental activities which are needed to encourage a new generation of entomologists, naturalists and citizen scientists, therefore, have much wider benefits. Logically, the converse is also true: if we fail to address the threats inherent in the extinction of experience, it will not be insect conservation alone that suffers.

Acknowledgements

For their insightful comments on an earlier version of this manuscript, we are very grateful to Sally Corbet, Rebecca Farley, Ian Kinchin and two anonymous referees. Albert Henderson, John Newbould, Rebecca Farley, Stewart Evans and Ian Kinchin kindly provided reference material.

References

Allen, D.E. (1987) The natural history society in Britain through the ages. *Archives of Natural History* 14, 243–259.

Allen, D.E. (1994) *The Naturalist in Britain: A Social History*, 2nd edn. Princeton University Press, Princeton, New Jersey.

Anon. (2005) *Nature: Who Knows?* English Nature, Peterborough; Lancaster University, Lancaster; Natural History Museum, London.

Asher, J., Warren, M., Fox, R., Harding, P., Jeffcoate, G. and Jeffcoate, S. (2001) *The Millennium Atlas of Butterflies in Britain and Ireland*. Oxford University Press, Oxford.

Baldock, D. (2000) Songs of bush-crickets and grasshoppers and the use of ultrasound detectors. *British Wildlife* 11, 319–323.

Balletto, E. and Casale, A. (1991) Mediterranean Insect Conservation. In: Collins, N.M. and Thomas, J.A. (eds) *The Conservation of Insects and their Habitats*. 15th Symposium of the Royal Entomological Society of London. Academic Press, London, pp. 121–142.

Balmford, A., Clegg, L., Coulson, T. and Taylor, J. (2002) Why conservationists should heed Pokémon. *Science* 295, 2367.

Barber, L. (1980) *The Heyday of Natural History 1820–1870*. Jonathan Cape, London.

Barker, S., Slingsby, D. and Tilling, S. (2002) *Teaching Biology in the Classroom: Is it Heading for Extinction?* A Report on Biology Fieldwork in the 14–19 Curriculum. FSC Occasional Publication 72, Field Studies Council/British Ecological Society, Shrewsbury, UK.

Barnard, P.C. (ed.) (1999) *Identifying British Insects and Arachnids: An Annotated Bibliography of Key Works*. Cambridge University Press, Cambridge.

Barrington-Johnson, J. (2005) *The Zoo: The Story of London Zoo*. Robert Hale, London.

Basset, Y., Novotny, V., Miller, S.E. and Pyle, R. (2000) Quantifying biodiversity: experience with parataxonomists and digital photography in Papua New Guinea and Guyana. *Bioscience* 50, 899–908.

Basset, Y., Novotny, V., Miller, S.E., Weiblen, G.D., Missa, O. and Stewart, A.J.A. (2004) Conservation and biological monitoring of tropical forests: the role of parataxonomists. *Journal of Applied Ecology* 41, 163–174.

Baur, B., Baur, H., Roesti, C., Roesti, D. and Thorens, P. (2006) *Sauterelles, Grillons et Criquets de Suisse*. Haupt Verlag, Bern, Switzerland.

Bebbington, A. (2005) The ability of A-level students to name plants. *Journal of Biological Education* 39, 62–67.

Berry, R.J. (1987) Scientific natural history: a key base to ecology. In: Berry, R.J. and Crothers, J.H. (eds) *Nature, Natural History and Ecology*. Proceedings of a symposium organized jointly by the Linnean Society of London and the Field Studies Council. Linnean Society of London and Academic Press, London, pp. 17–29.

Birchenough, A. (2002) Improving environmental knowledge. In: *Engaging the Public in Science*. Extracted papers from ENSUS 2002, 2nd International Conference in Marine Science and Technology for Environmental Sustainability. School of Marine Science and Technology, University of Newcastle, Newcastle upon Tyne, UK.

Bowler, P. (2004) The wider scene: promoting training and the interest in natural history. In: Henderson, A. (ed.) *Recruiting and Training the Next Generation of Field Naturalists*. Yorkshire Naturalists' Union Conference 2003 Proceedings. *Yorkshire Naturalists' Union Bulletin* 41(Suppl.), 59–62.

Brooks, S.J. (2001) Dragonflies and damselflies. In: Hawksworth, D.L. (ed.) *The Changing Wildlife of Great Britain and Ireland*. Systematics Association Special Volume Series 62. Taylor & Francis, London, pp. 340–354.

Brown Jr, K.S. (1991) Conservation of neotropical environments: insects as indicators. In: Collins, N.M. and Thomas, J.A. (eds) *The Conservation of Insects and Their Habitats*. 15th Symposium of the Royal Entomological Society of London. Academic Press, London, pp. 349–404.

Burnett, J., Copp, C. and Harding, P. (1995) *Biological Recording in the United Kingdom: Present Practice and Future Development*. Summary Report. Department of the Environment, London.

Burnie, D. (2005) *e.explore: Insect*. Dorling Kindersley, London.

Casciani, D. (2005) Counting on the caterpillar. Available at: http://news.bbc.co.uk/1/hi/magazine/4717503.stm

Cheesman, O.D. (2006) Invertebrate Link (JCCBI) and the Royal Entomological Society. *Antenna* 30, 99–100.

Cheesman, O.D. and Phillips, J.W. (2004) Invertebrate conservation in the UK: the role of Invertebrate Link (JCCBI) and the British Entomological and Natural History Society. *British Journal of Entomology and Natural History* 17, 10–14.

Chesmore, E.D. and Ohya, E. (2004) Automated identification of field-recorded songs of four British grasshoppers using bioacoustic signal recognition. *Bulletin of Entomological Research* 94, 319–330.

Chinery, M. (1993) *Field Guide to the Insects of Britain and Northern Europe*, 3rd edn. Collins, London.

Chua-Eaon, H. and Larimer, T. (1999) Beware of the Poké Mania. *Time* 154, 80–84, 86, 93.

Collinson, N. and Sparks, T. (2005) Nature's calendar: 2004 results from the UK Phenology Network. *British Wildlife* 16, 251–256.

Compton, S.G. (2004) The position of taxonomy and systematics in formal education today: a personal perspective. In: Henderson, A. (ed.) *Recruiting and Training the Next Generation of Field Naturalists.* Yorkshire Naturalists' Union Conference 2003 Proceedings. *Yorkshire Naturalists' Union Bulletin* 41(Suppl.), 28–31.

Cooper, J., Brewer, A., Ostler, R., Rowland, F. and Wilkinson, S. (2005) Britain's wildlife on the Internet: the NBN gateway. *British Wildlife* 17, 1–5.

Corke, D. (1999) Identifying insects. *British Wildlife* 10, 153–163.

Ely, W.A. (2004) The next generation. In: Henderson, A. (ed.) *Recruiting and Training the Next Generation of Field Naturalists.* Yorkshire Naturalists' Union Conference 2003 Proceedings. *Yorkshire Naturalists' Union Bulletin* 41(Suppl.), 34–35.

Evans, S.M., Dixon, S. and Heslop, J. (2005) Environmental knowledge: how good is it and where does it come from? In: Evans, S.M. and Foster-Smith, J. (eds) *Promoting Knowledge and Understanding of the Environment.* Extracted papers from ENSUS 2005, 3rd International Conference in Marine Science and Technology for Environmental Sustainability. School of Marine Science and Technology, University of Newcastle, Newcastle upon Tyne, UK, pp. 1–15.

Fido, H.S.A. and Gayford, C.G. (1982) Fieldwork and the biology teacher: a survey in secondary schools in England and Wales. *Journal of Biological Education* 16, 27–34.

Fisher, J.A. (2001) The demise of fieldwork as an integral part of science education in United Kingdom schools: a victim of cultural change and political pressure? *Pedagogy, Culture and Society* 9, 75–96.

Fox, R. (2001) Butterflies and moths. In: Hawksworth, D.L. (ed.) *The Changing Wildlife of Great Britain and Ireland.* Systematics Association Special Volume Series 62. Taylor & Francis, London, pp. 300–327.

FSC/BES (2004) *Creating the Right Balance: Delivering Fieldwork for Effective 16–19 Ecology Teaching.* FSC Occasional Publication 86, Field Studies Council/British Ecological Society, Shrewsbury, UK.

FSC (2006) *River Invertebrate Monitoring for Anglers.* AIDGAP Guide – 2006 Test Version. Field Studies Council, Shrewsbury, UK.

Gaedike, R. (2003) Bibliography of keys to European insects (1996–2000). *Nova Supplementa Entomologica* 16, 1–121.

Gaston, K.J. and O'Neill, M.A. (2004) Automated species identification: why not? *Philosophical Transactions of the Royal Society of London Series B – Biological Sciences* 359, 655–667.

Gaston, K.J., Smith, R.M., Thompson, K. and Warren, P.H. (2004) Gardens and wildlife: the BUGS project. *British Wildlife* 16, 1–9.

Gates, P. (2005) The amateur naturalist: a species recovery programme. In: Evans, S.M. and Foster-Smith, J. (eds) *Promoting Knowledge and Understanding of the Environment.* Extracted papers from ENSUS 2005, 3rd International Conference in Marine Science and Technology for Environmental Sustainability. School of Marine Science and Technology, University of Newcastle, Newcastle upon Tyne, UK, pp. 85–91.

Gilbert, P. and Hamilton, C.J. (1990) *Entomology: A Guide to Information Sources*, 2nd edn. Mansell Publishing, London.

Goldstein, P.Z. (2004) Systematic collection data in North American invertebrate conservation and monitoring programmes. *Journal of Applied Ecology* 41, 175–180.

Greenslade, P. and New, T.R. (1991) Australia: conservation of a continental insect fauna. In: Collins, N.M. and Thomas, J.A. (eds) *The Conservation of Insects and Their Habitats.* 15th Symposium of the Royal Entomological Society of London. Academic Press, London, pp. 33–70.

Haes, E.C.M. and Harding, P.T. (1997) *Atlas of Grasshoppers, Crickets and Allied Insects in Britain and Ireland.* ITE Research Publication No.11. The Stationery Office, London.

Haines, C. and Rogers, L. (2006) Will you be creating a 'buzz' about insects? *Antenna* 30, 64–66.

Harvey, J.M.V. (1999) Sources of information. In: P.C. Barnard (ed.) *Identifying British Insects and Arachnids: An Annotated Bibliography of Key Works.* Cambridge University Press, Cambridge, pp. 1–20.

Hawksworth, D.L. (ed.) (1974) *The Changing Flora and Fauna of Britain.* Systematics Association Special Volume No. 6. Academic Press, London.

Hawksworth, D.L. (ed.) (2001) *The Changing Wildlife of Great Britain and Ireland.* Systematics Association Special Volume Series 62. Taylor & Francis, London.

Henderson, A. (ed.) (2004) Recruiting and training the next generation of field naturalists. Yorkshire Naturalists' Union Conference 2003 Proceedings. *Yorkshire Naturalists' Union Bulletin* 41(Suppl.), 52–57.

Herbert, P.D.N. and Gregory, T.R. (2005) The promise of DNA barcoding for taxonomy. *Systematic Biology* 54, 852–859.

Hollerbach, A.L. (1996) Of Sangfroid and Sphinx Moths: cruelty, public relations, and the growth of entomology in England, 1800–1840. *Osiris* (2nd series) 11, 201–220.

Hollings, Y. (2004) Training requirements of the Yorkshire Wildlife Trust. In: Henderson, A. (ed.) *Recruiting and Training the Next Generation of Field Naturalists.* Yorkshire Naturalists' Union Conference 2003 Proceedings. *Yorkshire Naturalists' Union Bulletin* 41(Suppl.), 40–43.

House of Lords (1991) *Systematic Biology Research.* Select Committee on Science and Technology, 1st Report 1991–1992. HL Paper 22-I.

House of Lords (2002) *What on Earth? The Threat to the Science Underpinning Conservation.* Select Committee on Science and Technology, 3rd Report 2001–2002. HL Paper 118(i).

Howarth, F.G. and Ramsay, G.W. (1991) The conservation of island insects and their habitats. In: Collins, N.M. and Thomas, J.A. (eds) *The Conservation of Insects and Their Habitats.* 15th Symposium of the Royal Entomological Society of London. Academic Press, London, pp. 71–107.

Howes, C.A. (2004) From Lit & Phil to Bionet. In: Henderson, A. (ed.) *Recruiting and Training the Next Generation of Field Naturalists.* Yorkshire Naturalists' Union Conference 2003 Proceedings. *Yorkshire Naturalists' Union Bulletin* 41(Suppl.), 8–27.

Huber, J.T. (1998) The importance of voucher specimens, with practical guidelines for preserving specimens of the major invertebrate phyla for identification. *Journal of Natural History* 32, 367–385.

Invertebrate Link (JCCBI) (2002) A code of conduct for collecting insects and other invertebrates. *British Journal of Entomology and Natural History* 15, 1–6.

James, T. (ed.) (2004) *Natural Partners: Biodiversity Observations and Collections.* Report of a conference held at the National Museum & Galleries of Wales, Cardiff, 2–3 July 2004. Available at: http://www.nfbr.org.uk/assets/multimedia/NFBR-NBN_Conference_2004_report.pdf

Janzen, D. (2004) Setting up tropical biodiversity for conservation through non-damaging use: participation by parataxonomists. *Journal of Applied Ecology* 41, 181–187.

Jepson, P. (2005) Natural history remastered. *British Wildlife* 17, 27–31.

Key, R. (2004) Report on seminar session: Group 2 Invertebrate Workshop. In: Henderson, A. (ed.) *Recruiting and Training the Next Generation of Field Naturalists.* Yorkshire Naturalists' Union Conference 2003 Proceedings. *Yorkshire Naturalists' Union Bulletin* 41(Suppl.), 52–57.

Kinchin, I.M. (1993) Teaching ecology in England and Wales: a survey of current

practice. *Journal of Biological Education* 27, 29–33.
Kirby, P., Stewart, A.J.A. and Wilson, M.R. (2001) True bugs, leaf- and plant-hoppers, and their allies. In: Hawksworth, D.L. (ed.) *The Changing Wildlife of Great Britain and Ireland*. Systematics Association Special Volume Series 62. Taylor & Francis, London, pp. 262–299.
Kleukers, R.M.J.C., van Nieukerken, E.J., Odé, B., Willemse, L.P.M. and van Wingerden, W.K.R.E. (1997) *De Sprinkhanen en Krekels van Nederland (Orthoptera)*. Nederlandse Fauna 1. Nationaal Natuurhistorisch Museum, KNNV Uitgeverij and EIS-Nederland, Leiden. [Partially revised edition, 2004.]
Lapsley, P. (2005) River flies: beacons for environmental quality. *Fly-Fishing and Fly-Tying* February, 42–45.
Lock, R. and Tilling, S. (2002) Ecology fieldwork in 16 to 19 biology. *School Science Review* 84, 79–87.
Louv, R. (2005) *Last Child in the Woods: Saving Our Children from Nature Deficit Disorder*. Algonquin Books, Chapel Hill, North Carolina.
Lyal, C.H.C. (2005) Crumbling foundations: the erosion of the taxonomic base of biology. In: Evans, S.M. and Foster-Smith, J. (eds) *Promoting Knowledge and Understanding of the Environment*. Extracted papers from ENSUS 2005, 3rd International Conference in Marine Science and Technology for Environmental Sustainability. School of Marine Science and Technology, University of Newcastle, Newcastle upon Tyne, UK, pp. 23–36.
Maas, S., Detzel, P. and Staudt, A. (2002) *Gefährdungsanalyse der Heuschrecken Deutschlands: Verbreitungsatlas, Gefährdungseinstufung und Schutzkonzepte*. Bundesamt für Naturschultz, Bad Godesberg, Bonn, Germany.
Maiteny, P. (2004) Perceptions of nature by indigenous communities. In: Burley, J., Evans, J. and Youngquist, J.A. (eds) *Encyclopedia of Forest Sciences*. Elsevier, Oxford, pp. 462–471.
Marshall, J.A. (2001) Grasshoppers, crickets and allied insects. In: Hawksworth, D.L. (ed.) *The Changing Wildlife of Great Britain and Ireland*. Systematics Association Special Volume Series 62. Taylor & Francis, London, pp. 328–339.
Masters, G. and Smithers, P. (2006) Profile: Sir David Attenborough. *Antenna* 30, 60–63.
Masters, G.J., Smithers, P. and Cheesman, O.D. (2007) Taking entomology to a wider audience: who will watch the small things that run the world? *Antenna* 31(2).
Mikkola, K. (1991) The conservation of insects and their habitats in Northern and Eastern Europe. In: Collins, N.M. and Thomas, J.A. (eds) *The Conservation of Insects and Their Habitats*. 15th Symposium of the Royal Entomological Society of London. Academic Press, London, pp. 109–119.
Miller, J.R. (2005) Biodiversity conservation and the extinction of experience. *Trends in Ecology and Evolution* 20, 430–434.
Moore, N.W. (1991) Observe extinction or conserve diversity. In: Collins, N.M. and Thomas, J.A. (eds) *The Conservation of Insects and Their Habitats*. 15th Symposium of the Royal Entomological Society of London. Academic Press, London, pp. 1–8.
Morris, M.G., Collins, N.M., Vane-Wright, R.I. and Waage, J. (1991) The utilisation and value of non-domesticated insects. In: Collins, N.M. and Thomas, J.A. (eds) *The Conservation of Insects and Their Habitats*. 15th Symposium of the Royal Entomological Society of London. Academic Press, London, pp. 319–347.
Morris, R. (2005) Natural history remastered. *British Wildlife* 17, 150–151.
Newbould, J.A. (2004) Recruiting and training the next generation of field naturalists. In: Henderson, A. (ed.) *Recruiting and Training the Next Generation of Field Naturalists*. Yorkshire Naturalists' Union Conference 2003 Proceedings. *Yorkshire Naturalists' Union Bulletin* 41(Suppl.), 3–7.
Norris, A. (2004) The role of natural history museums and natural history collections. In: Henderson, A. (ed.) *Recruiting and Training the Next Generation of Field Naturalists*. Yorkshire Naturalists' Union Conference 2003 Proceedings. *Yorkshire Naturalists' Union Bulletin* 41(Suppl.), 32–33.
Opler, P.A. (1991) North American problems and perspectives in insect conservation. In: Collins, N.M. and Thomas, J.A. (eds) *The*

Conservation of Insects and Their Habitats. 15th Symposium of the Royal Entomological Society of London. Academic Press, London, pp. 9–32.

Palmer, C. (2004) The function of museums and records centres and how they have changed. In: James, T. (ed.) *Natural Partners: Biodiversity Observations and Collections*. Report of a conference held at the National Museum & Galleries of Wales, Cardiff, 2–3 July 2004, pp. 4–5. Available at: http://www.nfbr.org.uk/assets/multimedia/NFBR-NBN_Conference_2004_report.pdf

Pennisi, E. (2000) Taxonomic revival. *Science* 289, 2306–2308.

Perring, F.H. (ed.) (2002) *How Do We Find and Train the Next Generation of Field Botanists?* Joint Linnean Society/BSBI meeting, 4th April 2002 – abstracts. Available at: http://www.linnean.org/index.php?id = 57

Pickles, S. (2004) Future prospects: a potential training partnership. In: Henderson, A. (ed.) *Recruiting and Training the Next Generation of Field Naturalists*. Yorkshire Naturalists' Union Conference 2003 Proceedings. *Yorkshire Naturalists' Union Bulletin* 41(Suppl.), 45–50.

Pilkington, M. (2005) *Science in the Countryside: Lifelong Learning for Ecological Citizenship*. National Institute of Adult Continuing Education, Leicester, UK.

Polaszek, A., Agosti, D., Alonso-Zarazaga, M., Beccaloni, G., de Place Bjorn, P., Bouchet, P., Brothers, D.J., the Earl of Cranbrook, Evenhuis, N., Godfray, H.C.J., Johnson, N.F., Krell, F.-T., Lipscomb, D., Lyal, C.H.C., Mace, G.M., Mawatari, S., Miller, S.E., Minelli, A., Morris, S., Ng, P.K.L., Patterson, D.J., Pyle, R.L., Robinson, N., Rogo, L., Taverne, J., Thompson, F.C., van Tol, J., Wheeler, Q.D. and Wilson, E.O. (2005) A universal register for animal names. *Nature* 437, 477.

Porter, K. (2001) Tracking future trends: the biodiversity information network. In: Hawksworth, D.L. (ed.) *The Changing Wildlife of Great Britain and Ireland*. Systematics Association Special Volume Series 62. Taylor & Francis, London, pp. 422–434.

Pyle, R. (1993) *The Thunder Tree: Lessons from an Urban Wildland*. Houghton Mifflin, Boston, Massachusetts.

Rickinson, M., Dillon, J., Teamey, K., Morris, M., Choi, M.Y., Sanders, D. and Benefield, P. (2004) *A Review on Outdoor Learning*. Field Studies Council, Shrewsbury, UK.

Roy, H., Rowland, F., Brown, P., Ware, R. and Majerus, M. (2005) Ecology of the Harlequin Ladybird: a new invasive species. *British Wildlife* 16, 403–407.

Salmon, M.A. (2000) *The Aurelian Legacy: British Butterflies and Their Collectors*. University of California Press, Berkeley, California.

Savolainen, V., Cowan, R.S., Vogler, A.P., Roferick, G.K. and Lane, R. (2005) Towards writing the encyclopaedia of life: an introduction to DNA barcoding. *Philosophical Transactions of the Royal Society of London Series B – Biological Sciences* 360, 1805–1811.

Shirt, D.B. (ed.) (1987) *British Red Data Books, Vol. 2: Insects*. Nature Conservancy Council, Peterborough, UK.

Smithers, P. (2006) Wild about insects: taking entomology to the public in the south-west. *Antenna* 30, 96–97.

Sparks, T. and Collinson, N. (2006) The UK Phenology Network: some highlights from 2005. *British Wildlife* 17, 237–241.

Speight, M.C.D. (1986) Attitudes to insects and insect conservation. *Proceedings of the 3rd European Congress of Entomology, Amsterdam* (24–29 August 1986) 3, 369–380.

Spencer, W. (2006) Bugworld: Bristol Zoo's invertebrate display. *Antenna* 30, 80–84.

Stevenson, R.D., Haber, W.A. and Morris, R.A. (2003) Electronic field guides and user communities in the eco-informatics revolution. *Conservation Ecology* 7 online article 3. Available at: http://www.consecol.org/vol7/iss1/art3

Stubbs, A.E. (2001) Flies. In: Hawksworth, D.L. (ed.) *The Changing Wildlife of Great Britain and Ireland*. Systematics Association Special Volume Series 62. Taylor & Francis, London, pp. 239–261.

Tilling, S. (2004) Fieldwork in UK secondary schools: influences and provision. *Journal of Biological Education* 38, 54–58.

Tilling, S.M. (2005) The decline in biology fieldwork: fact or fiction? In: Evans, S.M. and Foster-Smith, J. (eds) *Promoting Knowledge and Understanding of the Environment*. Extracted papers from ENSUS 2005,

3rd International Conference in Marine Science and Technology for Environmental Sustainability. School of Marine Science and Technology, University of Newcastle, Newcastle upon Tyne, UK, pp.17–21.

Voisin, J.-F. (ed.) (2003) *Atlas des Orthoptères et des Mantides de France*. Patrimoines Naturels Vol. 60. Muséum National d'Histoire Naturelle, Paris.

Waring, P. (2004) Entomologists: born or made? *Entomologist's Record and Journal of Variation* 116, 122–124.

Waterton, C. (2003) Messing about on the river. *Salmo Trutta* 6, 56–58.

Way, L. (2004) Your data going global: what is in store now that we can connect biodiversity data locally, nationally and internationally. In: James, T. (ed.) *Natural Partners: Biodiversity Observations and Collections*. Report of a conference held at the National Museum & Galleries of Wales, Cardiff, 2–3 July 2004, pp. 14–16. Available at: http://www.nfbr.org.uk/assets/multimedia/NFBR-NBN_Conference_2004_report.pdf

Will, K.W., Mishler, B.D. and Wheeler, Q.D. (2005) The perils of DNA barcoding and the need for integrative taxonomy. *Systematic Biology* 54, 844–851.

15 Insects as Providers of Ecosystem Services: Crop Pollination and Pest Control

Claire Kremen and Rebecca Chaplin-Kramer

Department of Environmental Sciences, Policy and Management, University of California, Berkeley, CA 94720, USA

1 Introduction

Ecosystem services are the set of ecological functions that are critical for human survival, such as climate regulation, crop pollination and pest control, and that enhance human well-being, by providing aesthetic or recreational pleasure (Daily, 1997; Daily *et al.*, 2000). Recently, ecosystem services have been grouped into four classes. Provisioning services produce goods such as timber, food and water; regulating services modulate essential processes such as climate control and flood protection; cultural services provide aesthetic, inspirational or recreational opportunities; and supporting services provide the basis upon which all other services depend such as soil production (MEA, 2003).

In this broad definition of ecosystem services, insects, as the most diverse multicellular organisms on the planet, both at species and higher taxonomic levels (Hawksworth and Kalin-Arroyo, 1995, available at: http://tolweb.org/), are evidently of critical functional importance. Their many ecological roles include pollination, population regulation and pest control, decomposition, seed dispersal and protection, and provision of food to other organisms, including humans. Insect pollinators are required for more than 65% of the world's angiosperm species (Axelrod, 1960). Many wild plant populations are more limited by their lack of pollinators than by other resources (Burd, 1994; Ashman *et al.*, 2004). Insects are important regulators of other organisms, primarily other insects and plants (Leigh, 1996); as such they can provide both direct benefits to human welfare through regulation of crop pests and losses through crop damage (Pimentel, 1998). Insects are major contributors to decomposition of vegetable and animal materials (Daily *et al.*, 1997; Losey and Vaughan, 2006), from dung beetles that bury dung, carrion beetles and flies that feed on dead animals, termites and leafcutter ants that process large amounts of wood and leaves, to myriad insects in the soil and

 Insect Conservation Biology (eds A.J.A. Stewart, T.R. New and O.T. Lewis)

water that process decomposing materials. Insects such as ants and dung beetles also disperse seeds or provide secondary dispersal for seeds, while protecting them from seed predation by burial (MacMahon *et al.*, 2000). Some insects are ecosystem engineers – building large mounds above the ground (e.g. termites) or excavating galleries in the earth (e.g. leafcutter ants), and thus creating habitat for other animals and determining the flow of materials within and between ecosystems. Insects provide food for other insects and most vertebrate groups, either directly, or through products that they produce (e.g. honey). They produce many other products useful to humans (Eisner, 1991), and are harnessed by humans to provide forensic information (Catts and Goff, 1992) or deliver chemicals to precise locations. Insects dominate most terrestrial ecosystems, not simply in terms of their diversity, but also their sheer abundance and biomass (Fittkau and Klinge, 1973; Wilson, 1987; Nee, 2004).

In this chapter, however, we will limit our discussion to two relatively well-known regulating services (see MEA, 2003) that affect human survival and well-being directly via agricultural production: i.e. pest control and pollination services in agroecosystems. In addition, we focus primarily on the component of these services that flow from unmanaged ecosystems, as opposed to services provided by specific organisms that are cultivated expressly for that purpose (e.g. biocontrol agents). Thus we study wild bees that pollinate crops and natural enemies that control crop pests, and how these organisms and the services they provide may be altered by agricultural intensification.

Why examine services in agroecosystems and agricultural intensification? Agriculture occupies approximately 38% of the terrestrial land surface (DeFries *et al.*, 2004) and, as the world's largest industry, is a major cause of habitat transformation (Vitousek *et al.*, 1997; Robertson and Swinton, 2005). Agriculture produces ecosystem goods and services such as food, forage and carbon sequestration, as well as disservices including pollution, biodiversity loss and water wastage (Tilman *et al.*, 2001 and references therein). Successful crop production also depends on ecosystem services such as crop pollination, pest control, soil production and nutrient cycling (Tscharntke *et al.*, 2005). By virtue of the enormous area occupied by agriculture, better management of these landscapes could produce huge environmental benefits in human health, food security and biodiversity conservation (Drinkwater *et al.*, 1995; Daily *et al.*, 2001; Kremen *et al.*, 2002b; McNeely and Scherr, 2002; Kleijn *et al.*, 2004). Similarly, because agriculture is a dominant habitat type, understanding its impacts on the population dynamics and persistence of wild species of plants and animals is of fundamental value in ecology.

2 Global Importance of Pest Control and Pollination Services

The amount spent globally on pesticides is perhaps the best indicator of how much we value pest control: world pesticide expenditure reaches more than US$30 billion annually, and insecticides alone account for nearly one-third of that (Kiely *et al.*, 2004). Naylor and Ehrlich (1997) added the cost of plant resistance breeding programmes to the amount spent on pesticides in their estimate, arguing that these costs serve as a proxy for the value of services formerly provided by natural enemies, which would set a lower bound for the value of natural pest control at US$54 billion. Furthermore, pesticide usage is expected to triple by 2050 (Tilman *et al.*, 2001). Arthropod pests constitute a serious threat to food security and the economy, destroying an estimated 37% of potential crops in the USA annually even with pesticide use (Pimentel *et al.*, 1992). While these figures give a ballpark estimate for the value of natural pest control, it is difficult to quantify directly for two reasons: first, the use of pesticides is so widespread that a baseline cannot be established for the level of pest control by natural enemies in the absence of pesticides; conversely, some natural enemies are bound to be present in the system despite pesticide application, so it is unclear how well pesticides could contain pests in complete absence of their natural enemies. Essentially, we cannot determine how much of the pest control that does occur is due to pesticides and how much is due to natural enemies.

Almost 15–30% of food consumed by humans in developed countries requires an animal pollinator (Townsend, 1974; McGregor, 1976; Crane, 1990) either directly or indirectly (e.g. production of lucerne seed requires an insect vector; lucerne seed is grown to make lucerne hay to feed cattle). Proportions are likely to be similar for undeveloped countries. Two-thirds of the 103–108 crops that make up 90% of the national per capita food supplies for 146 countries (based on determination of Prescott-Allen and Prescott-Allen, 1990) require animal pollinators (Nabhan and Buchmann, 1997) and 70% of the world's crops requires a pollinator for at least one cultivar (Roubik, 1995). Although many of the staple crops are wind-pollinated (Prescott-Allen and Prescott-Allen, 1990), human health and well-being depends not only on the quantity of food that requires animal pollinators, but also on the diversity of such food. Loss of pollination services might have relatively minor affects on food quantity (Southwick and Southwick Jr, 1992; Ghazoul, 2005), but would certainly have negative effects on food diversity, quality, prices and security (Steffan-Dewenter *et al.*, 2005). As one example, fruits and vegetables provide the majority of vitamin C in the human diet, and 60 out of 65 globally important fruit and vegetable crops either require, or benefit from, animal pollination (Klein *et al.* in press). Thus, loss of pollinators might have a much stronger effect on food quality than on caloric production.

3 Crop Pollination Services

Historically, crop pollination needs were met by wild pollinators living within the farming landscape (Kevan and Phillips, 2001), and this is still true in less intensive agricultural systems (e.g. Ricketts *et al.*, 2004; Morandin and Winston, 2005). However, in many modern crop production systems requiring an animal pollinator, pollination is now managed as intensively as other aspects of the farming system, by bringing large numbers of commercial pollinators directly to the field where pollination is needed. The honeybee, *Apis mellifera* (a native of Africa, Europe and the Middle East), is utilized for at least 90% of managed pollination services (Williams, 1996; N. Calderone, Cornell University, 2005, email communication) around the globe for pollination of hundreds of crops for food or seed production (Crane, 1990). Estimates of commodity-specific, regional or global values of pollination services are difficult to determine for a variety of reasons (Kevan and Phillips, 2001; Losey and Vaughan, 2006) and have focused on managed *A. mellifera*. Nabhan and Buchmann (1997) have suggested that contributions from wild bees would be similar to those of managed bees.

The extent of our reliance upon this single species for such an important service is risky. In the USA, managed stocks of the honeybee have declined by 50% over the last 50 years (USDA National Agriculture Statistics Service, 1997; Losey and Vaughan, 2006) due primarily to the *Varroa destructor* mite (Morse and Goncalves, 1979; Beetsma, 1994), which both weakens individuals and transmits disease. Recently, *Varroa* mites have developed resistance to the miticides (reviewed in Elzen and Hardee, 2003), leading in 2005 to rates of colony mortality of up to 50% in many areas of the USA (E. Mussen, Davis, California, 2005, personal communication). *Varroa* has affected not only the USA, but also Europe and the Middle East (Griffiths, 1986; Komeili, 1988).

These losses of honeybees can lead to shortages of pollinators at critical moments, as in the almond orchards of California in 2004 and 2005 (E. Mussen, Davis, California, 2005, personal communication). This US$1 billion/year industry depends entirely on managed honeybees for its success, and nearly 1 million honeybee colonies are imported each year into California from across the USA to provide the necessary services. Pollination shortages in this crop are exacerbated by four conditions: (i) varieties in cultivation in the USA are self-incompatible and entirely dependent on insect vectors for fruit set; (ii) it blooms in February when few native pollinators are available; (iii) it is typically planted in large monocultures; and (iv) farmers are continuing to plant almond orchards despite the volatility in the supply of honeybee pollinators, which are a critical input for the success of their operation.

Reliance on honeybees for pollination in North and South America has been complicated by hybridization of the European subspecies with a more aggressive African form, *A. mellifera scutellata*, which was introduced to Brazil in 1956, and has expanded northwards as far as the southern USA. *A. mellifera scutellata* appears to be more resistant to tracheal and *Varroa* mites than feral European *Apis* (Loper, 2001; Martin and Medina, 2004) but makes bee-keeping more difficult. Although bee-keepers in South and Central America and Southern Africa ultimately adapted their bee-keeping style to the Africanized bees (DeJong, 1996), it is expected that widespread Africanization of US honeybee colonies would hasten the collapse of the honeybee industry in the USA, due to fear of liability issues arising from aggressive encounters between Africanized bees and humans or livestock (Parker *et al.*, 1987; Kevan and Phillips, 2001). In addition, recent arrival of Africanized bees in Florida, the primary queen and package (worker) producing state, suggests that bee-keepers replacing *Varroa*-infested colonies may soon be receiving replacement queens and workers that are Africanized (N. Calderone, Washington, DC, 2005, personal communication).

There are two non-exclusive alternatives to our overreliance on the honeybee – domestication and commercialization of additional species (reviewed in Parker *et al.*, 1987 and in Kevan *et al.*, 1990), and conservation and enhancement of populations of wild pollinators on or near farms (Batra, 2001). We focus on the latter aspect in this chapter.

3.1 Services provided by wild bee communities

3.1.1 Mechanisms

We do not know how many unmanaged species contribute to crop pollination, or what percentage of crop pollination results from pollinating visits by unmanaged species. Bees are the most important pollinators of many crops, and are recorded visitors to 73% of the crop species requiring pollinators (Nabhan and Buchmann, 1997). Thousands of bee species visit crop plants globally (Free, 1993), but few exhaustive surveys have been conducted. As many as 190 species of bees are associated with lowbush blueberry in north-eastern North America alone (Kevan *et al.*, 1990). In a single location in California, workers recorded 66 bee species visiting selected spring and summer crops (Kremen *et al.*, 2002a). Other wild visitors to crops are flies, wasps, butterflies, moths, midges, thrips, beetles, birds and bats (banana), including 37 invertebrate and 7 vertebrate genera (Roubik, 1995; Nabhan and Buchmann, 1997).

Wild pollinators can contribute to crop pollination in four ways. They can substitute for the services normally provided by commercially managed pollinators, replacing them either fully or partially. They can enhance the services provided by managed pollinators, through behavioural interactions that increase the pollination efficiency of the managed pollinator. They can provide services to plants that are not efficiently pollinated by a managed pollinator. Finally, they can enhance productivity in plants that self-pollinate, and for which pollination is consequently rarely managed. In contrast, wild pollinators can also detract from crop pollination in several ways, either through nectar robbery, by competing for pollen with other, potentially superior pollinators, or by transferring non-conspecific pollen that clogs stigmas.

When wild bees provide an equivalent (redundant) service to that of the managed pollinator, they can partially or fully substitute for that pollinator. In watermelon production in northern California, honeybees are often imported to fields to provide pollination services. Although their pollination efficiency is low relative to other bee visitors (Fig. 15.1a.), their contribution to overall pollination is high due to their high abundance under these circumstances (Fig. 15.1b). About 30 native bee species also visit watermelon flowers in this area, and contribute to pollination (Fig. 15.1b, see also Kremen *et al.*, 2002b). Although none of these species are abundant compared to the artificially high abundance of the honeybee, these species collectively provide an average of 28–100% of pollination needs for watermelon (range = 6–100%), depending on the farm environment (Fig. 15.1b). Organic farms

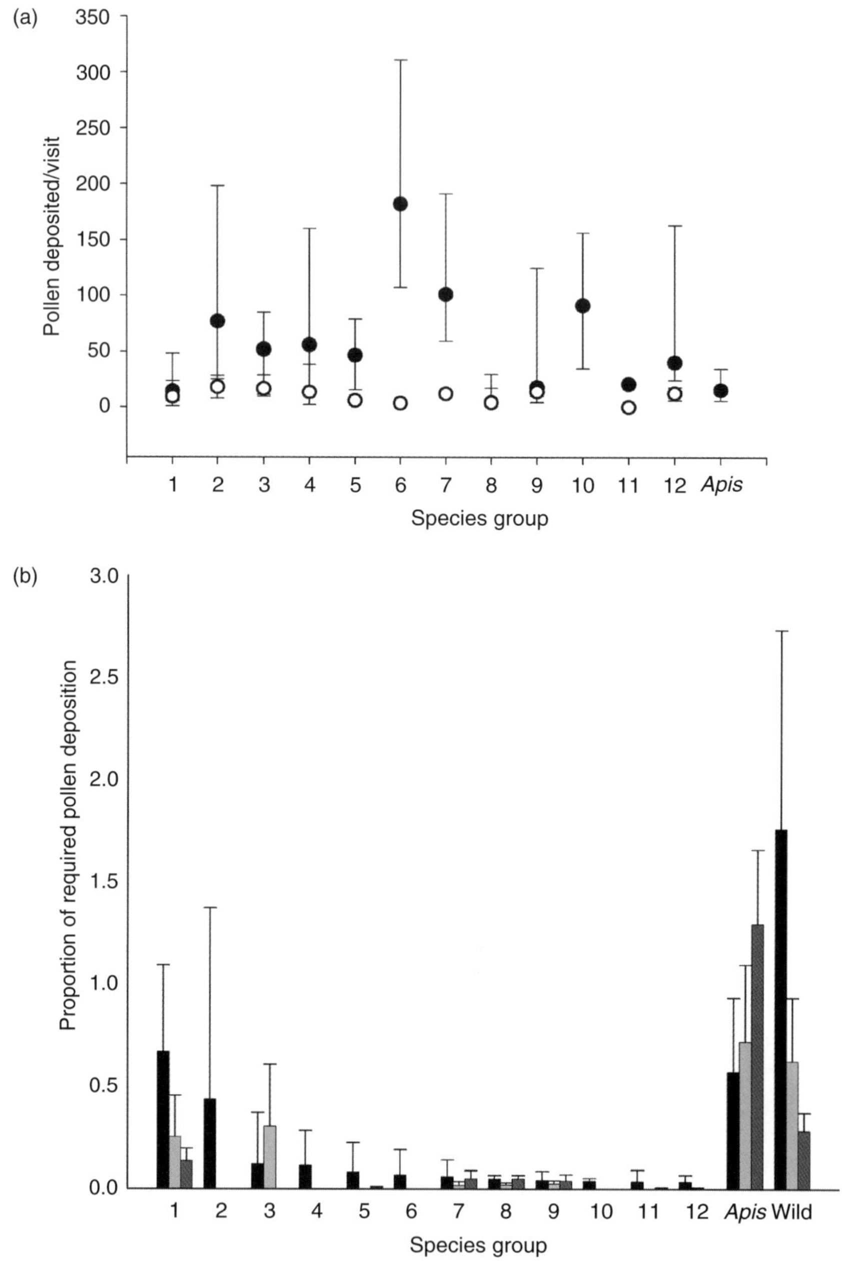

Fig. 15.1. (a) Pollination efficiencies of wild and honeybee visitors to watermelon (pollen grains deposited/visit). A total of 30 wild bee species are grouped into 12 ID groups. Female = black dot; male = white dot; medians with quartiles are shown. (b) Proportional contribution of each species required to set a marketable watermelon on different farm types (Organic Near = black; Organic Far = light grey; Conventional Far = medium grey), and of all wild species combined, based on the assumption that 1000 pollen grains are needed to set marketable fruit. Note that no farms meeting criteria for Conventional Near existed in the study area. (Modified from Kremen *et al.*, 2002b.)

close to natural habitat (lowest agricultural intensity) reliably receive a large proportion of their pollination requirements from the wild bee community; these farmers never import honeybees to their farms, and the honeybee contribution on its own (Fig. 15.1b) is not sufficient to provide them with the services they need. Therefore, such farmers clearly rely on wild pollinators to some extent. At the other end of the agricultural intensification gradient, conventional farmers far from natural habitat never receive sufficient pollination from wild bees; such farmers always import honeybees to provide pollination services. Nevertheless, they do receive some benefits from the wild bee visitors to their crops, although they may be unaware of these benefits (Kremen *et al.*, 2002a,b, 2004).

Wild bees can enhance the services provided by managed honeybees via behavioural interactions that increase honeybee pollination efficiency. There is a single documented example of this phenomenon (Greenleaf and Kremen, 2006b), but it is likely to be widespread in cropping systems that require movements between cultivars for successful fruit or seed production (e.g. hybrid seed production systems, many orchard crops). In hybrid sunflower seed production, farmers plant four rows of male-sterile, nectar-producing (female) cultivars for every 6–10 rows of male-fertile, pollen- and nectar-producing (male) cultivars in a repeating pattern. Honeybees are stocked at 2–2.5 colonies/ha, but lack of pollination is a major factor cited by farmers for underproduction. Individual honeybees tend to forage either for pollen or for nectar (Free, 1963). Honeybees had low pollination efficiency on sunflower relative to the most efficient wild bee visitors (mean of 3 seeds/visit compared to 19). There was a strong linear relationship, however, between honeybee pollination efficiency and the richness and abundance of wild bees present, increasing honeybee efficiency up to fivefold. Various observations suggested that honeybee foragers rarely crossed between sunflower rows, unless they had just interacted with a wild bee. So on average, while wild bees contributed only a small proportion of total sunflower pollination directly, they doubled the effectiveness of honeybees by enhancing their per-visit pollination efficiency, and thus the value of the pollination services honeybees provide (Greenleaf and Kremen, 2006b).

Wild bees are more effective pollinators than *A. mellifera* for some crops that depend on animal pollinators for fruit set, including lucerne, blueberry and cranberry (Parker *et al.*, 1987; Delaplane and Mayer, 2000). In these crops, *Apis* cannot reliably work the floral mechanism allowing pollination (Proctor *et al.*, 1996). Growers often import large numbers of honeybees hoping that increasing the frequency of encounters will increase the number of successful pollination events. Alternative pollinators have been domesticated in some cases, including *Megachile rotundata* and *Nomia melanderi* for lucerne, and *Osmia* spp. for blueberry, but the use of these alternative pollinators is not widespread (see Table 8.5 in Crane, 1990). Alternatively, growers of these crops rely on wild pollinators that are much more effective than the honeybee on their crops. In the 1970s in Canada, blueberry growers became acutely aware of their reliance on native pollinators when applications of fenitrothion to nearby forests for spruce budworm control greatly reduced many pollinator populations, which was then correlated with significant crop losses (Kevan and Plowright, 1989).

The majority of economically important fruit and vegetable crops that self-pollinate may also benefit from pollination provided by insect vectors by enhanced fruit set and/or size (e.g. coffee, Klein *et al.*, in press; cherry tomato, Greenleaf and Kremen, 2006a). The mechanism may be due to increased deposition of self-pollen, cross-pollen or both, reflecting both genetic and physiological factors' contribution to fertilization, fruit set and fruit growth (Proctor *et al.*, 1996; Delaplane and Mayer, 2000). Growers of self-pollinating plants generally do not import pollinators (except in cultivation of greenhouse tomatoes, whose flowers need vibration, either by wind or an insect, to release their pollen); thus, enhanced fruit production due to animal-mediated pollination in self-pollinating field crops is generally due to visitation by wild bees (e.g. Klein *et al.*, 2003a; Ricketts *et al.*, 2004; Chacoff and Aizen, 2006; Greenleaf and Kremen, 2006a).

Visitation by some insects may actually be detrimental for crop pollination. Insects that cut holes at the base of the flower's corolla in order to obtain nectar resources may reduce a flower's attractiveness and deter other insects from visiting and pollinating the plant (Irwin *et al.*, 2001). Insects that visit multiple flowering species may transfer non-conspecific pollen during visits to crop flowers, which could then clog stigmas, reducing both the effectiveness of that visit and of subsequent visits by the same or other pollinators. In general, non-*Apis* individuals are thought to exhibit lower flower constancy than *Apis* individuals (reviewed in Slaa and Biesmeijer, 2005). Non-*Apis* wild pollinators could therefore reduce pollination services provided by *A. mellifera*, although we are not aware of documented examples in crops.

Insects (usually bees, but also pollen-eating beetles) that remove large amounts of pollen while depositing only tiny amounts can be negative, rather than positive, for pollination function in crops. The extent to which a given species (whether wild or managed) is detrimental versus beneficial for crop pollination services depends on three things: (i) its species-specific behaviour, leading to a given ratio of pollen removal to deposition; (ii) the composition of the pollinator community; and (iii) whether the amount of available pollen is a limiting factor. Under limiting conditions (i.e. all pollen produced is removed), if one visiting species has a high ratio of pollen removal/deposition relative to other community members, its contribution to pollination will be negative, because it removes pollen from the system that other, better pollinators could otherwise deposit. If it has a low removal/deposition ratio relative to other species, or if there are no other pollinating species, it becomes positive for pollination services (Thomson and Thomson, 1992; Thomson and Goodell, 2001). If the amount of pollen is not limiting, however, more visits from any visitor that deposits any amount of pollen adds to the total pollen deposited on the crop. Pollen supply will depend greatly on the cultivar, crop-breeding system and other details of cultivation (e.g. proportion of pollenizers planted).

Although wild bee pollinators may augment or, in some cases, substitute for the services provided by commercially managed pollinators, it is important to recognize some inherent limitations to services provided by wild, unmanaged bees. Wild pollinator populations are notably variable in space and time (Roubik, 2001; Williams *et al.*, 2001); thus, services they provide may not be consistent enough to meet the needs of large-scale intensive agriculture. Unlike *A. mellifera*, which forms permanent colonies of 30,000–50,000 individuals, non-*Apis* bees often have relatively small population sizes, particularly at the beginning of the flight season for multi-voltine species.

Under what circumstances might wild pollinator communities partially or fully substitute for managed pollinators? First, if local and/or landscape conditions favour diverse and abundant wild pollinator communities (see below), such communities may partially or fully provide the necessary services (e.g. Kremen *et al.*, 2002b; Morandin and Winston, 2005). Second, social bees with their larger abundances can provide important services to crops. Social bees are the most abundant native visitors to coffee flowers in Indonesia and Costa Rica (Klein *et al.*, 2003b; Ricketts, 2004) and macadamia (Heard and Exley, 1994) and longan (Blanche *et al.*, in press) flowers in Australia, and are thus probably more important contributors to enhanced fruit set (Vazquez *et al.*, 2005). Feral, non-native *A. mellifera* also provide significant pollination services to various crops in the Neotropics and Australia (e.g. Cunningham *et al.*, 2002; Ricketts, 2004; Chacoff and Aizen, 2006; Blanche *et al.*, in press), and formerly provided them in the USA, but have reportedly been virtually extirpated in the wild in some areas due to *Varroa*. Third, crops which require an animal pollinator but benefit little from *A. mellifera* (e.g. blueberry, cranberry) often get most of their pollination from wild bees.

Commercially managed pollinators are clearly critical to the success of modern agriculture, but wild, unmanaged pollinators, despite the caveats noted above, could spread the risk of depending overly on just one or a few species. Risks of relying on only a few species come from: (i) the challenges of maintaining a stable supply of commercial pollinators, in the face of problems managing the genetics, pathogens and parasites of *Apis* and other domesticated species (National Research Council of the National Academies, 2006); and (ii) limitations in the qualities of pollination services provided by only a few pollinator species (see also Section 3.1.2). For example, *Apis* workers communicate with each other about the spatial location and quality of foraging resources. On the plus side, this can lead to massive recruitment of workers to a crop that is rewarding in pollen and nectar. On the minus side, *Apis* workers may concentrate in focal areas of the crop, ignoring other areas needing pollination, or, in the worst case, leaving the crop to forage on non-crop resources, providing no benefit to the grower. Although less numerous and certainly patchy in their distributions, foraging by communities of wild bees may complement the services provided by *Apis* and spread them over a larger area of the crop (Proctor *et al.*, 1996). Given their small, patchy populations, however, the goal of managing for wild pollinators should be to augment services provided by commercial pollinators by maintaining diverse communities that collectively provide more stable services than any individual wild pollinator species could (Tilman *et al.*, 1998; Kremen *et al.*, 2002b, 2004; Klein *et al.*, 2003b).

3.1.2 Role of diversity

A more diverse community of wild pollinators can provide a greater amount of pollination services to a greater number of crops with greater stability. More diverse communities of pollinators in agricultural systems had greater aggregate abundances and visit frequencies (Steffan-Dewenter and Tscharntke, 1999, also see Klein *et al.*, 2003b; Ricketts, 2004; Greenleaf, 2005; Larsen *et al.*, 2005; Pritchard, 2005; Chacoff and Aizen, 2006). The strikingly consistent positive relationship between abundance and richness across these studies suggests that the loss of richness will generally reduce the number of visits and hence the level of pollination services provided to crops by the wild bee community. This could be particularly important because it has been shown, through meta-analysis, that visit frequency is a more important predictor of total level of pollination services than per-visit effectiveness (Vazquez *et al.*, 2005).

While many pollinator species that visit crops are generalists, different crop species nevertheless attract different, albeit partially overlapping, sets of pollinator species from the local species pool. Therefore, maintaining diverse pollinator communities locally is important for providing pollination services to a more diverse set of crops. Within a crop, diversity of the pollinator community is also important for ensuring the stability of pollination services across time and space. Several lines of reasoning support this assertion. From theoretical principles, we know that more diverse communities whose populations fluctuate in a random, uncorrelated fashion will provide more consistent services than less diverse communities (the portfolio effect; Tilman *et al.*, 1998). Empirical work supports the claims of theory, although few studies have yet been conducted. Richer communities provided more stable pollination services to watermelon crops from day to day within a season (C. Kremen, 2006, unpublished data) and from bush to bush within a coffee field (Klein *et al.*, 2003b; Steffan-Dewenter *et al.*, 2006). High-diversity communities may include an array of species with broader physiological and behavioural ranges that are able to fly and pollinate flowers under varied environmental circumstances, and thus provide greater consistency than lower-diversity communities (Herrera, 1995; Bishop and Armbruster, 1999; Klein *et al.*, 2003b). Insect populations, especially bees, fluctuate greatly in the wild from year to year, as well as within seasons and across space (Herrera, 1988; Wolda, 1988; Roubik, 2001; Williams *et al.*, 2001). Such transient losses are unlikely to affect pollination services to a given plant species as long as the system is relatively diverse (Williams *et al.*, 2001; Morris, 2003; Memmott *et al.*, 2004). In the watermelon system, entirely different bee species predominated in their visit frequencies (abundances), and hence their contributions to watermelon pollination, in 2 successive years, but in both years the community collectively provided sufficient services (Kremen *et al.*, 2002b). In Costa Rica, decline in 1 year in the abundance of feral non-native *A. mellifera scutellata* was partially balanced by increases in abundances of native species (Ricketts, 2004). In these systems, managing for wild bee richness, rather than for the abundance of a particular species, is an important factor in maintaining a consistent level of service.

3.1.3 Economic value

We know of only five estimates calculating the economic value of wild bee pollination services. In three of these cases, all services came from wild bees, facilitating economic evaluation, while in the others, economic values were partitioned between managed honeybee and wild bee contributions. In Costa Rica, the enhanced value of coffee production attributed to wild bee pollination was estimated as US$393/ha of forest, and this value was both an order of magnitude above existing national schemes to pay for environmental services, and equivalent to other competing land uses (Ricketts *et al.*, 2004). In contrast, in Indonesia, the enhanced value of coffee production attributed to wild bee pollination was estimated at only US$55/ha of forest. This order of magnitude difference is due to the differing landscape configurations in which these studies were done, particularly a higher proportion of forest area in Indonesia, and our lack of knowledge of the dependence of wild bee abundances on patch area (Priess *et al.*, in press). In canola rapeseed production in northern Canada, Morandin and Winston (2006) found that profits lost by retiring up to 30% of the field area from production would be more than compensated by the increased yields due to better pollination found in fields near uncultivated areas, which supports a more diverse and abundant pollinator community. For the entire hybrid sunflower seed production industry in the USA, Greenleaf and Kremen (2006b) estimated that wild bees contributed on average US$2 million through direct pollination, and US$12 million indirectly by enhancing the pollination services provided by wild bees, as described above, while honeybees contributed another US$12 million in unassisted pollination (Greenleaf and Kremen, 2006b). For US production of fruits and vegetables, Losey and Vaughan (2006) estimated the contribution of wild bees at US$3.07 billion, and the contribution of honeybees at US$17.01 billion.

3.2 Effects of agricultural land use on wild bee communities and pollination services

Agricultural land use may have either positive or negative effects on pollinator communities and the services they provide, depending both on the intensity of agricultural land use, spatial scale (Tscharntke *et al.*, 2005) and the biome type in question, although too few studies have been conducted to predict these effects with certainty. Both local and landscape-scale factors may be important, although the relative importance of each varies with study system and pollinator guild. In a Mediterranean biome in California, agricultural intensification, which included both the reduction of nearby natural habitat and the predominance of large-scale industrialized agriculture (for a definition, see Tscharntke *et al.*, 2005), led to dramatic reductions in the species richness and abundance of wild bee pollinators on watermelon (Kremen *et al.*, 2002b, 2004), tomato (Greenleaf and Kremen, 2006a) and sunflower (Greenleaf and Kremen, 2006b), with concomitant estimated reductions in the services wild bees provide to these crops. In all of these studies, the chief factor influencing wild bee distributions appeared to be the area of nearby natural habitats (chaparral and oak woodlands) within several kilometres of the farm site; the proportional area or proximity of natural habitat was positively correlated with bee species richness, abundance, number of nesting bees found on farms, and magnitude and stability of pollination services provided by wild bees (Kremen *et al.*, 2004; Greenleaf, 2005; Kim *et al.*, in press). Local farm management type (organic versus conventional) only weakly affected these community response variables once the landscape level effects were factored out. Individual bee species were differentially sensitive to the gradient of agricultural intensification, but none increased in response to it (Kremen, 2004). The species that were more effective pollinators were also the most sensitive to agricultural intensification; thus, their loss exacerbated the effects on pollination services (Larsen *et al.*, 2005).

Similarly, in the neotropics, distance to wild patches significantly influenced the richness and abundance of wild bees visiting and pollinating coffee in Costa Rica (Ricketts, 2004) and grapefruit in Argentina (Chacoff and Aizen, 2006). These wild bees included indigenous solitary and social bees, and feral colonies of introduced *A. mellifera scutellata.* Over a span of 100 m from the forest edge, visitation dropped precipitously by 75% (Ricketts *et al.*, 2004), although a decline in pollination services was not observed till 1600 m (Ricketts, 2004). Similarly, richness and visit frequency of native wild bees visiting grapefruit (*Citrus paradisi*) declined precipitously with distance from the forest edge in Argentina (by eightfold within 1000 m) while the visit frequency of feral *A. mellifera scutellata,* which accounted for 95% of visits, dropped by twofold over the same distance (Chacoff and Aizen, 2006). In coffee fields in Sulawesi, Indonesia, distance to wild habitat affected the richness of native social but not solitary bees, while light levels within the fields were strongly positively correlated with solitary bee richness and with the abundance of both solitary and social bees. Fruit set was significantly correlated to both factors (Klein *et al.*, 2003b). In macadamia (*Macadamia integrifolia*) orchards in southern Queensland and New South Wales in Australia, the abundance of its most common native pollinator, *Trigona carbonaria*, but not of managed *A. mellifera* correlated with the proportional area of *Eucalyptus* forests within 1 km of orchards (Heard and Exley, 1994); however, on the Atherton Tablelands in northern Queensland, where *Apis* was not managed for pollination, distance from rainforest corresponded with a decline in both feral *A. mellifera* visits, and in fruit set of macadamia, although there was no correlation between fruit set per raceme and *Apis* abundance per site (Blanche *et al.*, in press). In the same area, beetle visitors to custard apple (*Annona squamosa* × *A. cherimola* cultivar) declined in diversity and abundance with distance from rainforest habitat, with a corresponding decline in fruit production (Pritchard, 2005; see also Blanche and Cunningham, 2005); wild stingless bees (*Trigona* spp.), but not feral *A. mellifera*, declined in abundance with distance from rainforest in longan (*Dimocarpus longan*) orchards, with a corresponding decrease in fruit set (Blanche *et al.*, in press). In these mosaic environments of tropical forest and agriculture, forest patches again appear to play an important role in providing essential habitat for native and non-native bee pollinators of crops, and thus for pollination services.

In temperate agricultural landscapes in Europe with calcareous grasslands, woods, meadows and other semi-natural areas, distance to grasslands influenced diversity, abundance and pollination services provided by social and solitary bees to two self-incompatible plants, mustard (*Sinapsis arvensis*) and radish (*Raphanus sativus*), effectively halving the reproductive output of each plant at 250 m and then again at 1000 m (Steffan-Dewenter and Tscharntke, 1999). In contrast, however, the abundance of common *Bombus* spp. in this same environment did not correlate with the proportional area of semi-natural habitat but did correlate with the proportional area of mass-flowering crops like oilseed rape, clover and sunflower, suggesting that the enormous pulse of pollen and nectar resources provided by large crop fields can promote abundance of selected bee species (Westphal *et al.*, 2003), particularly if these resources are staggered across the bumblebee flight seasons.

Although reported studies of wild bees on crops generally show a decline of diversity, abundance and services with agricultural intensification (see Tscharntke *et al.*, 2005), not all studies of bee communities (including bees visiting non-crop resources) show the same diversity and abundance trends. For example, in the Atlantic Coastal Pine Barren's ecoregion of the north-eastern USA, bee species richness and abundance in fragments of Pine Barren's habitat increased significantly when surrounded by a predominantly agricultural matrix, compared to a predominantly forested matrix (Winfree *et al.*, in press). Agricultural habitats also had significantly greater richness and abundance than naturally forested habitats, and more species were found to be unique to the agricultural areas compared to the forested areas. Forests in the Pine Barren forests comprised a pine overstory with a low-diversity, ericaceous understory. Both floral richness and abundance were higher in agricultural areas than within Pine Barren forests. In this system, agriculture apparently enhances rather than detracts from bee richness and abundance, although it must be noted that the intensity of agricultural land use is relatively low (at most, ~30% of land uses within 1.6 km of sample sites) compared with other study systems (Winfree *et al.*, in press).

In summary, based on reported studies, pollination services provided by wild bees are most likely being reduced in many of the areas where they could be contributing to crop pollination. At the same time, numbers of commercially managed colonies of *Apis* have also declined. Yet, there are comparatively few documented instances of shortages in pollination services. This indicates that we are not yet in crisis, but precaution suggests that we should be worrying about this before a crisis does unfold. In particular, our heavy reliance on *A. mellifera*, especially given its management issues and volatility, makes us vulnerable to sudden, unforeseen changes in its abundance (NAS-NRC report).

4 Pest Control Services

The intensification and expansion of agriculture in the latter part of the 20th century has amplified the age-old competition between humans and arthropod herbivores for food produced from crops. Despite soaring prices and additional costs from risks to human and ecosystem health, synthetic pesticides have been used as the main line of defence against arthropod pests since the 1940s. The effect of pesticides on arthropod communities has created an unintentional experiment, illuminating many of the subtleties in the relationships between pests and their predator and parasitoid natural enemies. The irony of pesticides is that they can exacerbate the very pest problems they were designed to solve, often doing more damage to natural-enemy populations than to pests. The migratory abilities of pests are often better than enemies, allowing them to recolonize fields more quickly than their natural enemies after spraying (Pontius *et al.*, 2002). Additionally, natural enemies have not evolved to cope with toxic compounds to the extent that many herbivorous pest insects have (Ehrlich and Raven, 1964), and are therefore more susceptible to pesticides (Naylor and Ehrlich, 1997). Predators and parasitoids also occupy higher trophic levels and are thus generally characterized by smaller populations than herbivores, resulting in a slower recovery from pesticide applications (Naylor and Ehrlich, 1997). By impacting natural enemies more severely than pests, pesticides can create conditions where pest populations temporarily grow unchecked by enemy control. As a result, formerly unimportant herbivore species can begin to cause serious losses and/or established pest species may rebound later in the season to levels greater than those prior to pesticide application.

The unintended consequences of pesticides have been well documented around the world. Krishna *et al.* (2003) reported several instances of the emergence of novel pest species following aggressive pesticide regimes in India. For example, the bollworm (*Helicoverpa armigera*) in Andhra Pradesh, India, which originally only posed an economic threat to chickpea, expanded following the 'injudicious and indiscriminate' use of pesticides on cotton, causing a 66% reduction in yield of cotton. Target pests can also become more of a problem with the use of pesticides. The cottony-cushion scale (*Icerya purchasi*) in California citrus and the bollworm in cotton in Peru are two famous examples of resurging pest populations whose natural enemies were annihilated by pesticides (Ruttan, 1999). Another classic case of post-pesticide pest rebounds can be found in the brown plant-hopper (*Nilaparvata lugens*) in rice (Kenmore, 1980; Naylor and Ehrlich, 1997; Pontius *et al.*, 2002). Brown plant-hopper eggs are unaffected by pesticide sprays that decimate their natural enemies (Pontius *et al.*, 2002); thus plant-hopper outbreaks ravaged the rice fields in Indonesia during the 1970s and 1980s after pesticides were introduced. Beyond the initial loss of natural-enemy populations, Settle *et al.* (1996) demonstrated a mechanism for these pesticide-induced outbreaks that requires an understanding of the ecology of the entire rice ecosystem. In the absence of pesticides, the natural enemies in this system were able to keep plant-hopper populations at consistently low levels because the enemy populations themselves were supported by other non-pest invertebrate populations, such as detritivores and filter-feeders (Pontius *et al.*, 2002). The use of broad-spectrum pesticides also decimated these alternative food sources. Recent advances in food web theory have suggested that top-down control would not exist in terrestrial systems without alternative prey 'subsidies' that keep enemy densities high and stable (Polis and Strong, 1996). Pesticides can destabilize the interactions between natural enemies and their multispecies prey base, creating a window of opportunity for pest populations to explode.

The recognition of the risks of disrupting agricultural arthropod communities led to the development of integrated pest management (IPM) in the 1950s and 1960s. IPM was intended to be an interdisciplinary approach to controlling pests through the management of all aspects of the ecosystem. Unfortunately, this original intent has not been truly realized, as many farmers continue to apply pesticides preemptively rather than as an informed decision based on careful monitoring of pest populations (Ehler and Bottrell, 2000). Furthermore, IPM could potentially be improved if it more effectively mimicked nature; for example, while most successful cases of biological control in agriculture have involved specialist parasitoids, generalist predators are thought to be the agents primarily responsible for top-down control in nature (Hawkins *et al.*, 1999). While IPM has made progress in identifying biocontrol agents and augmenting their populations through the rearing and release of individuals, more comprehensive and successful pest control programmes may require considering the natural-enemy populations in landscapes surrounding the farm. Studying arthropod relationships at the interface of agricultural and natural systems may help elucidate how best to regain aspects of natural pest control.

Habitat modifications to agricultural systems such as beetle banks, weed strips, intercropping and hedgerows have been successfully implemented to improve pest control in Europe and Asia (Landis *et al.*, 2000). Habitat modifications may be successful because they enhance natural-enemy population sizes, or perceived enhancements may simply be due to local concentration of natural enemies that already existed within the broader landscape near target crop sites (Gurr *et al.*, 1998). On-farm habitat modifications will probably be insufficient for natural enemies that also require off-farm resources. Natural areas near agricultural areas could provide these resources and could further enhance existing enemy populations by replenishing populations decimated by pesticides. If natural habitat is necessary to serve as a source for natural enemies, habitat modifications at within-crop and within-farm scales may function poorly in providing pest control, unless a sufficient area of natural habitat is also present (Kruess and Tscharntke, 1994). Research in pest management has begun to focus on this interplay between landscape and local scale: in other words, how landscape structure across scales affects natural-enemy communities and the pest control services they provide on farm sites.

Studies investigating landscape-level effects on natural enemies have used many different definitions of the term landscape. 'Landscape complexity' can mean perimeter to area ratio of the field, distance to natural habitat, or proportional area of natural or non-crop habitat or habitat heterogeneity within foraging or dispersal range of the natural enemy. These landscape variables are of course interrelated. Both percentage of non-crop area and, more specifically, of natural habitat are thought to be positively correlated with habitat diversity (Sunderland and Samu, 2000). There also appears to be a general correlation between the percentage of non-crop habitat and perimeter/area ratios of fields (Menalled *et al.*, 2003). The term 'landscape complexity' in this chapter will therefore include each of these variables.

4.1 Services provided by natural-enemy communities

In assessing the pest control services provided by natural enemies, it is important to recognize that an increase in enemy abundance or even in predation rates does not always translate to a concomitant increase in pest control. A reduction in pest densities resulting from an increase in predator abundance could trigger a density-dependent compensatory response, such as increased pest reproduction or dispersal, leading to no net change in pest population. Additionally, both natural enemies and pests may respond in a similar manner to the same landscape features; for example, Thies *et al.* (2005) found that landscape complexity increased both parasitism of cereal aphids and the rate of aphid colonization, resulting in no net change in aphid population size. Landscape complexity affected the relationship between predator and prey in the bird cherry-oat aphid system, but the effect changed over time (Ostman *et al.*, 2001a). Early in the season, aphid establishment was lower in complex landscapes due to increased predator abundance relative to simple landscapes. Later in the season, however, predators of aphids also fed on numerous alternate prey in complex landscapes, resulting in a larger impact on pests in simpler landscapes. Predation in the establishment phase was more important to total aphid population and ultimately to crop yields than predation later in the season, leading to greater control of aphids in complex landscapes. These studies illustrate the complexity of predator–prey relationships, underscoring the importance of measuring changes in enemy populations, pest populations and pest control function concomitantly.

Many studies that have examined predator–prey relationships have shown pest populations to respond to changes in natural-enemy abundance. A review of 181 manipulative field experiments revealed that generalist predators reduce pest numbers significantly in 75% of cases (Symondson *et al.*, 2002). Sunderland and Samu (2000) reviewed several studies documenting greater pest control with increased spider densities. Other studies have directly demonstrated increases in predation rate with greater predator abundance. Ostman (2004) found aphid removal rate to increase with generalist predator abundance in barley fields. Ostman *et al.* (2001a) showed that the presence of predators reduce the number of establishing aphids, and ultimately the number of total aphid-days in spring barley. Chang and Snyder (2004) found egg predation rates of Colorado potato beetle to be significantly correlated with total predator density, regardless of which predators were present. Manipulating parasitoid abundance is more difficult than manipulating predator abundance, because parasitoids are so mobile that releases tend to diffuse rapidly. Most of the evidence of the impact of increased abundance of parasitoids on pest populations comes from the many successful cases of pest control achieved through the introduction of exotic parasitoids as biological control agents (120 successful cases out of 223 tabulated by DeBach, 1974). Measuring the effect of naturally occurring parasitoids as a function of their abundance also suggests a link between increased parasitoid density and pest control. Murdoch and Briggs (1996) reviewed recent work showing that aggregation in parasitoids improves the ability of parasitoids to reduce pest densities. Other work investigating variable parasitoid density over landscape gradients has found correlations between greater parasitoid abundance and increased parasitism rates, which will be further discussed in Section 4.1.1 (Thies and Tscharntke, 1999; Thies *et al.*, 2003).

For both predators and parasitoids, behavioural considerations such as search efficiency and capture success rate can potentially outweigh abundance in determining pest control. A rare natural enemy is not necessarily an ineffective one if it is more efficient than other more abundant natural enemies (DeBach, 1974). This leads to another important factor in natural-pest control: the diversity of the natural-enemy community.

4.1.1 The role of diversity

A more diverse natural-enemy community can enhance or protect pest control function through a variety of mechanisms. Tscharntke *et al.* (2005) reviewed four ways in which biodiversity is tied to function: species complementarity, when more than one type of predator or parasitoid adds to the control of a pest species; sampling effect, a particularly effective natural enemy is more likely by chance alone to occur when more species are present; redundancy, having more species will buffer against disturbance or ecosystem change; and idiosyncrasy, when the whole is greater than the sum of the parts due to interaction between species. These expected benefits of diversity to pest control may be countered by antagonistic effects between different natural-enemy species, such as competition and intraguild predation. Many species may compete for the same prey at different life stages, and theory suggests that the natural enemy that wins in competition may not necessarily be the most effective at suppressing the pest at the stage most crucial to control (Murdoch and Briggs, 1996). If the addition of a new natural enemy to a community causes significant mortality to one or more of the natural enemies controlling pest populations, more natural enemies could actually result in reduced pest control (Rosenheim *et al.*, 1993, 1995). On the other hand, intraguild predation could help supplement natural-enemy diets when pest densities are low, potentially maintaining a higher natural-enemy population size more capable of controlling sudden pest outbreaks. Furthermore, intraguild predation can be mitigated by vegetative complexity, which may provide different natural-enemy species with refuges from each other, enhancing co-occurrence (Finke and Denno, 2002).

Despite the complexities of the effects of natural-enemy diversity on pest control, there are many examples of enhanced pest control in diverse systems. Perfecto *et al.* (2004) demonstrated the potential for increased pest control function with a more complete predator guild. While this study investigated birds, not insects, it is still instructive: higher vegetative diversity on coffee farms was correlated with higher bird species richness and better pest control. Entomological studies have found similar correlations in arthropod communities. Snyder and Ives (2003) recorded additive effects of pest control by parasitoids and generalist predators despite some evidence for intraguild predation between the two groups. Furthermore, they noted that each guild may have different roles in pest control: the generalist predators caused an immediate decline in prey growth rates though they did not exert density-dependent control, while the parasitoids were capable of more long-term density-dependent control after a delay corresponding to the generation time of the parasitoid. This study illustrates that different natural-enemy groups may complement one another temporally. Other studies have demonstrated interactions (idiosyncrasies) with greater than additive effects. Schmidt *et al.* (2003) reported aphid densities to be 18% higher without ground predators, 70% higher without flying predators and parasitoids, and 172% higher without both guilds of enemies. Losey and Denno (1998) noted a potential for facilitation between flying and ground predators, as aphid dropping behaviour in response to lady beetle foraging made aphids more vulnerable to predation by ground-dwelling carabid beetles. In cage experiments, Cardinale *et al.* (2003) showed that three enemy species together were able to suppress pea aphid more than the summed effects of each alone, translating to a non-additive increase in crop production. Additionally, a diverse enemy community could potentially control a greater richness of pests on diverse crops, though this has not yet been tested.

4.1.2 Economics of pest control services by natural enemies

Does improved pest control by natural enemies mean economic savings in crop yield due to reduced herbivore damage? Studies aimed at answering this question suggest that natural-enemy communities could indeed be valuable to crop production. Through a predator removal experiment, Ostman *et al.* (2003) showed that the presence of natural enemies increased barley yields by 303 kg/ha, preventing 52% of yield loss due to aphids. Riechert and Bishop (1990) found broccoli plants without spiders to have more than three times the damage of plants with spiders. It is difficult to disentangle the effects of pesticides and natural enemies to calculate the value of natural pest control across all crops (see above). However, estimates can be made based on current research on pest demographics. Losey and Vaughan (2006) calculate that insect pests cost the USA $18.77 billion every year, from damages to crops combined with the cost of insecticides. They estimate that 39% of pest species in the USA are native, and thus calculate that losses due to native pests are 39% of $18.77 billion, or $7.32 billion. They further estimate that only 35% of native potential pest species reach damaging levels, which means that $7.32 billion represents only 35% of the losses that would happen in absence of all natural pest control. They therefore conclude that having no natural control of native insect pests would cost the USA $20.92 billion annually, though pathogens and host-plant resistance account for some of that natural pest control. However, their estimate of $20.92 billion encompasses only the potential costs of native pests; the bulk of pest species in the USA are non-native and many of them may be under natural pest control, e.g. by generalist predators, to some degree (Symondson *et al.*, 2002).

4.2 Effect of landscape on natural enemies

The strongest evidence that landscape complexity can enhance natural-enemy populations on farms comes from the many studies linking increased complexity to increased enemy abundance and richness. In 26 studies investigating abundance of natural enemies with some metric of landscape complexity, all found increases in some species; only three of those studies found the results to be inconsistent across years, regions or species (Table 15.1). These studies encompass a wide range of cropping systems and arthropod groups. Specialists and generalists, both predators and parasitoids, in annual and perennial systems seem to be enhanced by natural habitat or landscape complexity. While this may be due in part to a bias towards publishing studies that show some effect, it is striking that no studies have documented a decline in natural-enemy abundance with increasing landscape complexity. The effect of landscape complexity on natural-enemy species richness is not as clear-cut, but most studies show a positive correlation. Of 16 studies 12 found increased richness in more complex landscapes, 2 recorded inconclusive or no effects and 2 found decreased richness (Table 15.2). It is interesting to note that all four studies that showed no increase of enemy richness with landscape complexity involved specialist parasitoids; these may rely less on natural habitat than generalists whose life histories demand a variety of food sources.

In addition to enhancing abundance and species richness (quantity effects), year-round access to natural habitat outside the fields could also improve the quality of natural enemies. Ostman *et al.* (2001b) showed that predatory beetle ‘condition’, defined here as fat reserves, was higher in fields with greater perimeter/area ratios, encouraging dispersal into fields. Costamagna and Landis (2004) showed that parasitism and parasitoid longevity is limited by carbohydrate resources, suggesting that enemy condition can influence pest control function. Access to alternative habitat can also improve enemy reproduction; Bommarco (1998) found that fields in more complex landscapes with more perennial habitat had larger beetles whose fecundity was tripled compared with individuals in simpler landscapes. These few studies suggest that future research should assess quality as well as quantity response variables when tracking landscape effects on natural-enemy populations. If the additional resources provided by natural habitat enhance not only the number of natural enemies but also their function, their impact on the pest community could be greater than expected (Hagen *et al.*, 1976).

Studies cited in Section 4.1 showed a general connection between enemy abundance or richness and pest control, suggesting that these landscape-mediated increases in enemies could therefore improve pest control. Further support for this hypothesis is lent by studies that have directly quantified the

Table 15.1. Effects of landscape complexity on predator and parasitoid abundances.

Location	Enemy	Crop	Abundance	Study
European grassland				
England	Linyphiid spiders	Grasses	+	Bell *et al.* (2002)
Germany	Spiders	Wheat	+	Lemke and Poehling (2002)
Germany	Linyphiid spiders	Wheat	+	Schmidt and Tschartnke (2005)
Norway	Carabid, staphylinid	Cereals	+	Andersen (1997)
England, Norway	Aphid predators	Cereals	+	Dennis and Fry (1992)
Sweden	Many	Wheat, barley, rape	+	Lagerlof and Wallin (1993)
Switzerland	Carabid, staphylinid	Wheat	+	Frank and Reichhart (2004)
North American grassland				
Iowa	Carabids	Maize	+	Varchola and Dunn (2001)
Michigan	Coccinellids	Lucerne, maize, wheat	+	ColungaGarcia *et al.* (1997)
Michigan	*C. septempunctata*	Lucerne, maize, wheat	n.s.	ColungaGarcia *et al.* (1997)
Michigan	*Eriborus**	Maize	+	Dyer and Landis (1997)
Michigan	Carabids	Maize	+	Lee *et al.* (2001)
Michigan	*M. communis**	Maize	+	Menalled *et al.* (2003)
Michigan	*G. militaris**	Maize	n.s.	Menalled *et al.* (2003)
Oklahoma	Carabids	Wheat	+	French *et al.* (2001)
South Dakota	Coccinellids	Maize	+	Elliot *et al.* (2002a)
South Dakota	Aphid predators	Lucerne	+	Elliot *et al.* (2002b)
Washington	*C. septempunctata*	Broccoli	+	Banks (1999)
Washington	*P. melanarius*	Broccoli	n.s.	Banks (1999)
Mediterranean/chaparral				
France	Mites	Vineyard	+	Tixier *et al.* (1998)
California	Many	Tomato	+	LeTourneau and Goldstein (2001)
Texas	Generalist predators	Cotton	+	Prasifka *et al.* (2004)
	Spiders	*Review*	+	Sunderland and Samu (2000)
	Linyphiid spiders	*Model*	+	Halley *et al.* (1996)
	Carabids	*Model*	+	Bilde and Topping (2004)
	Coccinellids	*Model*	+	Bianchi and Van Der Werf (2003)

*Denotes parasitoid species.

Table 15.2. Effects of landscape complexity on predator and parasitoid species richness.

Location	Enemy	Crop	Diversity	Study
European grassland				
England, Norway	Aphid predators	Cereals	+	Dennis and Fry (1992)
Germany	Parasitoid	Red clover	+	Kruess and Tscharntke (1994)
Germany	Rape-pollen beetle parasitoid	Rape	–	Tscharntke *et al*. (2002)
Germany	Natural enemies of bees	Bioindicator system	+	Tscharntke *et al*. (1998)
Sweden	Many	Wheat, barley, rape	+	Lagerlof and Wallin (1993)
Sweden	Many	Leys, cereals	+	Weibull *et al*. (2003)
Switzerland	Carabids, staphylinids	Wheat	+	Frank and Reichhart (2004)
Midwestern grassland				
Iowa	Carabids	Maize	+	Varchola and Dunn (2001)
Michigan	Coccinellids	Lucerne, maize, wheat	+	ColungaGarcia *et al*. (1997)
Michigan	Army worm parasitoid	Maize	–	Costamagna *et al*. (2004)
Michigan	Army worm parasitoid	Maize	n.s.	Marino and Landis (1996)
Michigan	Army worm parasitoid	Maize	+	Menalled *et al*. (1999)
Michigan	Army worm parasitoid	Maize	+/–	Menalled *et al*. (1999)
Ohio	Spiders	Soybean, maize	+	Buddle *et al*. (2004)
Oklahoma	Carabids	Wheat	+	French and Elliot (2001)
Mediterranean/chaparral				
California	Many	Tomato	+	LeTourneau and Goldstein (2001)

relationship between landscape complexity and natural-enemy function (predation and parasitism rates) in addition to, or instead of, enemy abundance. Of 14 studies 11 found increased pest mortality due to parasitism in more complex landscapes (Table 15.3). However, 2 of those 11 reported a lack of effect in certain regions or in certain years (Menalled *et al*., 1999, 2003). The remaining 3 studies found no effect at all, although the authors suggested that the effect of landscape complexity on parasitism, while often positive, is not always captured at the appropriate temporal and spatial scales. Far fewer studies have quantified predation rates, due to difficulties in the methodology of tracking

Table 15.3. Effects of landscape complexity on rates of parasitism or predation.

Location	Prey	Crop	Parasitism	Study
European grassland				
Belgium	Aphid	Wheat	+	Langer and Hance (2004)
Germany	Weevil	Red clover	+	Kruess and Tscharntke (1994)
Germany	*M. aeneoventris*	Wheat	+	Kruess (2003)
Germany	Rape-pollen beetle	Rape	+	Thies and Tscharntke (1999)
Germany	Rape-pollen beetle	Rape	+	Thies *et al.* (2003)
Germany	Aphid	Wheat	+	Thies *et al.* (2005)
Germany	Rape-pollen beetle	Rape	+	Tscharntke *et al.* (2002)
Germany	Bees (as bioindicator)	Grasslands	+	Tscharntke *et al.* (1998)
Midwestern grassland				
Michigan	Army worm	Maize	n.s.	Costamagna *et al.* (2004)
Michigan	Army worm	Maize	+	Marino and Landis (1996)
Michigan	Army worm	Maize	+/ n.s.	Menalled *et al.* (1999)
Michigan	Army worm	Maize	+/–/n.s.	Menalled *et al.* (2003)
Ohio	Green cloverworm	Soybean	n.s.	Pavuk and Barrett (1993)
Tropical forest				
Indonesia	Bees (as bioindicator)	Coffee	n.s.	Klein *et al.* (2003a)

or directly observing predators. Tscharntke *et al.* (1998) used trap-nesting bees and wasps and their natural enemies as indicators to study the effects of isolation from natural habitat on predator–prey relationships. They reported predation and parasitism to be enhanced by proximity to natural habitat fragments, and discovered that the effect of the habitat increased with its age. Ostman *et al.* (2001a) quantified predation rates on an important pest species. Through exclosure experiments and counting the removal rates of aphids glued to paper in the field, they documented a positive correlation between landscape complexity and predation rate. Thies and Tscharntke (1999) and Thies *et al.* (2003) documented landscape-mediated augmentations of natural-enemy function on crop yields. Using two different definitions of landscape complexity (adjacent fallows in 1999, percentage of non-crop area in 2003) in an oilseed rape system, they concluded that crop damage is lower in complex landscapes with increased abundances of natural enemies and parasitism of pests.

While there seems to be substantial evidence that landscape complexity enhances natural-enemy populations that perform valuable pest control services, the potential for negative effects (ecosystem disservices) must also be considered. Natural habitat may disrupt the dispersal of enemies between and into fields. Mauremooto *et al.* (1995) showed that movement by predatory carabid beetles is markedly slowed by hedgerows, and Wratten *et al.* (2003) found the same effect in hoverflies. This disruption could create a temporal asynchrony between pest and natural-enemy populations, providing pests with a brief but potentially important refuge from predation (Kareiva, 1987). Prey may therefore build up in small pockets of natural habitat within cropland, resulting in spillover of prey from natural habitat to cropland (With *et al.*, 2002). Indeed, the worry that natural habitat could serve as sources of pests as well as enemies is an important consideration in assessing services and disservices provided by natural ecosystems. In a survey of several taxonomic groups of insects, Zabel and Tscharntke (1998) showed that herbivore richness increased with the patch size of natural habitat in an agricultural landscape. However, they also found that predator richness decreased with the isolation of natural habitat patches, and that the effect was of greater magnitude than the effect on herbivores. Likewise, Kruess and Tscharntke (2000) found that parasitoids suffer more from habitat loss than herbivores. In both cases, the greater habitat requirements discovered for natural enemies led the authors to suggest that increasing habitat connectivity should primarily promote natural-enemy populations. Thus, in some cases, the disservice of supplying pests may be mitigated by the service of supplying enemies.

Although gaps in the research remain, particularly in perennial systems (the vast majority of studies reviewed here were conducted in annual cereals), the available data suggest that farms could greatly benefit from proximity to natural habitat. The complexities of multi-trophic arthropod food webs make it difficult to draw any universal conclusions. Still, many natural-enemy populations respond positively to more complex landscapes, and the enhanced abundance or diversity of natural enemies has been shown in several cases to improve pest control and reduce yield loss. Further study is needed to determine the economic value of this ecosystem service in different regions and for different crop systems, as well as to quantify the relationship between natural habitat area and services provided. Such research has important implications for land use decisions, agricultural policy and the sustainability of food production for the future.

5 Managing for Services from Beneficial Insects

Under what circumstances would managing for beneficial insects make sense economically? The bottom line is that farmers will not take valuable land out of production to provide habitat for wild beneficial insects unless it is worth their while. The benefits to farmers include the added fruit set due to the presence of wild bees, the reduction in crop damages due to the presence of natural enemies, and the insurance against potential shortages in the supply of commercially managed pollinators or the emergence of novel or pesticide-resistant pest species (Naylor and Ehrlich, 1997; Armsworth and Roughgarden, 2003; Kremen, 2005). The costs include lost revenues from any lands taken out of production, and labour and material costs for developing habitat for beneficial insects, although programmes now exist in some countries, notably the European Union and the USA, that provide incentives for more sustainable forms of agriculture that both support and rely on biodiversity (available at: www.europa.eu.int/comm/agriculture/envir/index_en.htm). Farmers are most likely to be motivated to manage for beneficial insects if they are already experiencing pollination shortages or pest problems, and/or if their farms are situated in favourable environments, far from sources of pesticides that they cannot control, and close to set-aside lands. Alternatively, managing for beneficial insects could become cost-effective, and thus attractive to a wider range of farmers, through development of small-scale, low-cost management actions that take place primarily in the non-cropped portions of a farm (e.g. along irrigation ditches or roads, around barns and tail-water ponds), and thus incur little or no opportunity costs. Altering certain practices within cultivated fields, like introducing multi-cropping, allowing cover crops to flower or permitting weedy borders, could promote floral resources for beneficial insects at little or no cost to farmers (Pickett and Bugg, 1998; Vaughan *et al.*, 2004; Pywell *et al.*, 2006). Hedgerows and weed strips can provide habitat and alternate prey sources for both pollinators and natural enemies, although farmers may be concerned that these also serve as sources of pests and thus more research may be necessary to determine when such habitat modifications are most beneficial (Gurr *et al.*, 1998). Some small-scale changes that might initially incur small annual costs may accumulate and ultimately more than pay for themselves, by allowing farmers to reduce rental fees for honeybees or to weather periods of scarcity of commercially managed pollinators (Kremen *et al.*, 2002b), and by reducing costs of pesticides and/or the release of purchased biocontrol agents (Gurr *et al.*, 1998).

Which crops are particularly vulnerable to shortages of pollinators or pest control agents? Crops or cultivars that are entirely dependent on animal pollinators include plants with separate male and female flowers (e.g. watermelon) and self-incompatible crops (e.g. many cultivars of orchard crops including almond and apple). The production of hybrid seeds in some crops requires producing separate male and male-sterile cultivars (Agrawal, 1998), followed by cross-breeding them via an animal pollinator; such hybrid seed production is vulnerable to pollinator shortages (e.g. sunflower, onions, carrot, cabbage). The situation in which a crop is grown can also increase its vulnerability to pollinator shortages. Growing such crops in large monocultures requires having a large supply of commercially managed bees in a given place at a given time, and can cause farmers to compete with one another for available pollinators, particularly if the flowering period is brief and synchronous (e.g. almond in California); in addition, wild bee pollinators are less likely to be present in the interiors of large-sized fields, unless they are actually nesting within the field. Crops particularly vulnerable to shortages of natural enemies can be identified by the intensity of pesticide applications. The US National Center for Food and Agricultural Policy maintains a national pesticide use database that quantifies the amount of active ingredient of each type of pesticide applied and the total number of acres treated for all major crops in the USA (available at: http://www.ncfap.org/database/national/default.asp). Although not all pesticides are equal in toxicity, most crops use enough different types of pesticides that toxicity differences between individual pesticides should even out over all crops. Ranking crops in order of the total amount of pesticide used (by active ingredient), amount used per acre and number of different types of pesticides used summarizes the pesticide intensity for each crop, which indicates at least in part the severity of pest problems faced

Table 15.4. Top pesticide-intensive crops in the USA, 1997. (Compiled from The National Center for Food and Agricultural Policy's Pesticide Use Database, available at: http://www.ncfap.org/database/national/default.asp.)

Crop	Total lbs a.i. applied	Rank	Lbs a.i. applied per acre	Rank	Number of pesticides	Rank
Strawberries	9,165,742	10	564	1	25	8
Citrus	74,846,118	1	254	13	27	6
Grapes	49,078,231	2	432	2	23	14
Tomatoes	17,656,433	5	235	15	31	2
Sweet peppers	4,330,932	15	297	8	30	3
Peaches	9,512,728	9	370	6	23	14
Cotton	23,317,538	3	62	34	44	1
Potatoes	13,900,051	8	182	21	25	9
Apples	16,503,137	6	68	33	28	4

a.i. = Active ingredient.

by crops. Omitting fungicides and herbicides, strawberries, citrus, grapes, tomatoes, sweet peppers, peaches, cotton, potatoes and apples are the most pesticide-intensive crops in the USA, ranking in the top ten in at least two of the three above criteria (Table 15.4). These crops would therefore make good experimental subjects to test the effects of differing degrees of natural pest control in the absence of pesticides.

6 Conclusions

In general, it seems clear from a review of the literature that complex landscapes that include natural habitat components improve the magnitude and/or stability of these two ecosystem services in agroecosystems. Nevertheless, it is also important to note that complex landscapes may also favour pest insects, reduce the ability of natural enemies to detect prey and favour one service such as pollination while not favouring the other (e.g. Steffan-Dewenter *et al.*, 2001). System-wide studies that provide detailed information on the ecology of multiple ecosystem services will be the best way to improve our understanding of the likely trade-offs that may occur, and management actions that should be taken to promote the desired mix of ecosystem services in agricultural landscapes (Kremen, 2005).

References

Agrawal, R.L. (1998) *Fundamentals of Plant Breeding and Hybrid Seed Production*, Science Publishers, Enfield, New Hampshire.

Andersen, A. (1997) Densities of overwintering carabids and staphylinids (Coleoptera: Carabidae and Staphylinidae) in cereal and grass fields and their boundaries. *Journal of Applied Entomology (Zeitschirist fur Angewandte Entomologie)* 121, 77–80.

Armsworth, P.R. and Roughgarden, J.E. (2003) The economic value of ecological stability. *Proceedings of the National Academy of*

Sciences of the United States of America 100, 7147–7151.

Ashman, T.-L., Knight, T.M., Steets, J.A., Amarasekare, P., Burd, M., Campbell, D.R., Dudash, M.R., Johnston, M.O., Mazer, S.J., Mitchell, R.J., Morgan, M.T. and Wilson, W.G. (2004) Pollen limitation of plant reproduction: ecological and evolutionary causes and consequences. *Ecology* 85, 2408–2421.

Axelrod, D.I. (1960) *The Evolution of Flowering Plants*. University of Chicago Press, Chicago, Illinois.

Banks, J.E. (1999) Differential response of two agroecosystem predators, *Pterostichus melanarius* (Coleoptera: Carabidae) and *Coccinella septempunctata* (Coleoptera: Coccinellidae), to habitat-composition and fragmentation-scale manipulations. *Canadian Entomologist* 131, 645–657.

Batra, S. (2001) Coaxing pollen bees to work for us. In: Stubbs, C. and Drummond, F. (eds) *Bees and Crop Pollination-Crisis, Crossroads, Conservation*. Entomological Society of America, Lanham, Maryland.

Beetsma, J. (1994) The *Varroa* mite, a devastating parasite of western honeybees and an economic threat to beekeeping. *Outlook on Agriculture* 23, 169–175.

Bell, J.R., Johnson, P.J., Hambler, C., Haughton, A.J., Smith, H., Feber, R.E., Tattersall, F.H., Hart, B.H., Manley, W. and Macdonald, D.W. (2002) Manipulating the abundance of *Lepthyphantes tenuis* (Araneae: Linyphiidae) by field margin management. *Agriculture Ecosystems and Environment* 93, 295–304.

Bianchi, F. and Van Der Werf, W. (2003) The effect of the area and configuration of hibernation sites on the control of aphids by *Coccinella septempunctata* (Coleoptera: Coccinellidae) in agricultural landscapes: a simulation study. *Environmental Entomology* 32, 1290–1304.

Bilde, T. and Topping, C. (2004) Life history traits interact with landscape composition to influence population dynamics of a terrestrial arthropod: a simulation study. *Ecoscience* 11, 64–73.

Bishop, J.A. and Armbruster, W.S. (1999) Thermoregulatory abilities of Alaskan bees: effects of size, phylogeny and ecology. *Functional Ecology* 13, 711–724.

Blanche, R. and Cunningham, S.A. (2005) Rainforest provides pollinating beetles for atemoya crops. *Journal of Economic Entomology* 98, 1193–1201.

Blanche, R., Ludwig, J.A. and Cunningham, S.A. (2006) Proximity of rainforest enhances pollination and fruit set in macadamia and longan orchards in north Queensland, Australia. *Journal of Applied Ecology* 43, 1182–1187.

Bommarco, R. (1998) Reproduction and energy reserves of a predatory carabid beetle relative to agroecosystem complexity. *Ecological Applications* 8, 846–853.

Buddle, C.M., Higgins, S. and Rypstra, A.L. (2004) Ground-dwelling spider assemblages inhabiting riparian forests and hedgerows in an agricultural landscape. *American Midland Naturalist* 151, 15–26.

Burd, M. (1994) Bateman's principle and plant reproduction: the role of pollen limitation in fruit and seed set. *Botanical Review* 60, 81–109.

Cardinale, B.J., Harvey, C.T., Gross, K. and Ives, A.R. (2003) Biodiversity and biocontrol: emergent impacts of a multi-enemy assemblage on pest suppression and crop yield in an agroecosystem. *Ecology Letters* 6, 857–865.

Catts, E.P. and Goff, M.L. (1992) Forensic entomology in criminal investigations. *Annual Review of Entomology* 37, 253–272.

Chacoff, N. and Aizen, M. (2006) Edge effects on flower-visiting insects in grapefruit plantations bordering premontane subtropical forest. *Journal of Applied Ecology* 43, 18–27.

Chang, G.C. and Snyder, W.E. (2004) The relationship between predator density, community composition, and field predation of Colorado potato beetle eggs. *Biological Control* 31, 453–461.

ColungaGarcia, M., Gage, S.H. and Landis, D.A. (1997) Response of all assemblage of Coccinellidae (Coleoptera) to a diverse agricultural landscape. *Environmental Entomology* 26, 797–804.

Costamagna, A.C. and Landis, D.A. (2004) Effect of food resources on adult *Glyptapanteles militaris* and *Meteorus communis* (Hymenoptera: Braconidae), parasitoids of *Pseudaletia unipuncta* (Lepidoptera: Noctuidae). *Environmental Entomology* 33, 128–137.

Costamagna, A.C., Menalled, F.D. and Landis, D.A. (2004) Host density influences parasitism of the armyworm *Pseudaletia unipuncta* in agricultural landscapes. *Basic and Applied Ecology* 5, 347–355.

Crane, E. (1990) *Bees and Beekeeping: Science, Practice and World Resources*. Comstock, Ithaca, New York.

Cunningham, S.A., FitzGibbon, F. and Heard, T.A. (2002) The future of pollinators for Australian agriculture. *Australian Journal of Agricultural Research* 53, 893–900.

Daily, G., Matson, P.A. and Vitousek, P.M. (1997) Ecosystem services supplied by soil. In: Daily, G. (ed.) *Nature's Services: Societal Dependence on Natural Ecosystems*. Island Press, Washington, DC, pp. 113–132.

Daily, G.C. (1997) *Nature's Services: Societal Dependence on Natural Ecosystems*. Island Press, Washington, DC.

Daily, G.C., Soderqvist, T., Aniyar, S., Arrow, K., Dasgupta, P., Ehrlich, P.R., Folke, C., Jansson, A., Jansson, B.O., Kautsky, N., Levin, S., Lubchenco, J., Maler, K.G., Simpson, D., Starrett, D., Tilman, D. and Walker, B. (2000) Ecology: the value of nature and the nature of value. *Science* 289, 395–396.

Daily, G.C., Ehrlich, P.R. and Sanchez-Azofeifa, G.A. (2001) Countryside biogeography: use of human-dominated habitats by the avifauna of southern Costa Rica. *Ecological Applications* 11, 1–13.

DeBach, P. (1974) *Biological Control by Natural Enemies*. Cambridge University Press, Cambridge.

DeFries, R., Foley, J.A. and Asner, G.P. (2004) Land-use choices: balancing human needs and ecosystem function. *Frontiers in Ecology and the Environment* 2, 249–257.

DeJong, D. (1996) Africanized honey bees in Brazil, forty years of adaptation and success. *Bee World* 77, 67–70.

Delaplane, K.S. and Mayer, D.F. (2000) *Crop Pollination by Bees*. CAB International, Wallingford, UK.

Dennis, P. and Fry, G.L.A. (1992) Field margins: can they enhance natural enemy population-densities and general arthropod diversity on farmland. *Agriculture Ecosystems and Environment* 40, 95–115.

Drinkwater, L.E., Letourneau, D.K., Workneh, F., van Bruggen, A.H.C. and Shennan, C. (1995) Fundamental differences between conventional and organic tomato agroecosystems in California. *Ecological Applications* 5, 1098–1112.

Dyer, L.E. and Landis, D.A. (1997) Influence of noncrop habitats on the distribution of *Eriborus terebrans* (Hymenoptera: Ichneumonidae) in cornfields. *Environmental Entomology* 26, 924–932.

Ehler, L.E. and Bottrell, D.G. (2000) The illusion of integrated pest management. *Issues in Science and Technology* 16, 61–64.

Ehrlich, P.R. and Raven, P.H. (1964) Butterflies and plants: a study in coevolution. *Evolution* 18, 586–608.

Eisner, T. (1991) Chemical prospecting: a proposal for action. In: Bormann, F.H. and Kellert, S.R. (eds) *Ecology, Economics, Ethics: the Broken Circle*. Yale University Press, New Haven, Connecticut, pp. 196–202.

Elliott, N.C., Kieckhefer, R.W. and Beck, D.A. (2002a) Effect of aphids and the surrounding landscape on the abundance of Coccinellidae in cornfields. *Biological Control* 24, 214–220.

Elliott, N.C., Kieckhefer, R.W., Michels, G.J., Jr and Giles, K.L. (2002b) Predator abundance in alfalfa fields in relation to aphids, within-field vegetation, and landscape matrix. *Environmental Entomology* 31, 253–260.

Elzen, G.W. and Hardee, D.D. (2003) United States Department of Agriculture – Agricultural Research Service Research on managing insect resistance to insecticides. *Pest Management Science* 59, 770–776.

Finke, D.L. and Denno, R.F. (2002) Intraguild predation diminished in complex-structured vegetation: implications for prey suppression. *Ecology* 83, 643–652.

Fittkau, E.J. and Klinge, H. (1973) On biomass and trophic structure of the central Amazonian rain forest ecosystem. *Biotropica* 5, 2–14.

Frank, T. and Reichhart, B. (2004) Staphylinidae and Carabidae overwintering in wheat and sown wildflower areas of different age. *Bulletin of Entomological Research* 94, 209–217.

Free, J.B. (1963) The flower constancy of honeybees. *Journal of Animal Ecology*, 119–131.

Free, J.B. (1993) *Insect Pollination of Crops*. Academic Press, San Diego, California.

French, B.W. and Elliot, N.C. (2001) Species diversity, richness, and evenness of ground beetles (Coleoptera: Carabidae) in wheat fields and adjacent grasslands and riparian zones. *Southwestern Entomologist* 26, 315–324.

French, B.W., Elliott, N.C., Berberet, R.C. and Burd, J.D. (2001) Effects of riparian and grassland habitats on ground beetle (Coleoptera: Carabidae) assemblages in adjacent wheat fields. *Environmental Entomology* 30, 225–234.

Ghazoul, J. (2005) Buzziness as usual? Questioning the global pollination crisis. *Trends in Ecology & Evolution* 20, 367–373.

Greenleaf, S.A.S. (2005) *Local-scale and Foraging-scale Affect Bee Community Abundances, Species Richness, and Pollination Services in Northern California*. Princeton University Press, Princeton, New Jersey.

Greenleaf, S.S. and Kremen, C. (2006a) Wild bee species increase tomato production but respond differently to surrounding land use in Northern California. *Biological Conservation* 133, 81–87.

Greenleaf, S.S. and Kremen, C. (2006b) Wild bees enhance honey bees' pollination of hybrid sunflower. *Proceedings of the National Academy of Sciences of the United States of America* 103, 13890–13895.

Griffiths, D. (1986) Summary of the present status of *varroatosis* in Europe. In: Cavalloro, R. (ed.) *European Research on Varroatosis Control: Proceedings of a Meeting of the EC Experts' Group, Bad Homburg,* 15–17 Octorber 1986. Balkema, Brookfield, Vermont, pp. 11–13.

Gurr, G.M., van Emden, H.F. and Wratten, S.D. (1998) Habitat manipulation and natural enemy efficiency: implications for the control of pests. In: Barbosa, P. (ed.) *Conservation Biological Control*. Academic Press, San Diego, California, pp. 155–183.

Hagen, K.S., Bombosch, S. and McMurtry, J.A. (1976) The biology and impact of predators. In: Huffaker, C.B. and Messenger, P.S. (eds) *Theory and Practice of Biological Control*. Academic Press, New York, pp. 93–142.

Halley, J.M., Thomas, C.F.G. and Jepson, P.C. (1996) A model for the spatial dynamics of linyphiid spiders in farmland. *Journal of Applied Ecology* 33, 471–492.

Hawkins, B.A., Mills, N.J., Jervis, M.A. and Price, P.W. (1999) Is the biological control of insects a natural phenomenon? *Oikos* 86, 493–506.

Hawksworth, D.L. and Kalin-Arroyo, M.T. (1995) Magnitude and distribution of biodiversity. In: Heywood, V.H. and Watson, R.T. (eds) *Global Biodiversity Assessment*. Cambridge University Press, Cambridge, pp. 107–193.

Heard, T.A. and Exley, E.M. (1994) Diversity, abundance, and distribution of insect visitors to Macadamia flowers. *Environmental Entomology* 23, 91–100.

Herrera, C.M. (1988) Variation in mutualisms: the spatio-temporal mosaic of a pollinator assemblage. *Biological Journal of the Linnean Society* 35, 95–125.

Herrera, C.M. (1995) Microclimate and individual variation in pollinators: flowering plants are more than their flowers. *Ecology* 76, 1516–1524.

Irwin, R.E., Brody, A.K. and Waser, N.W. (2001) The impact of flower larceny on individuals, populations, and communities. *Oecologia* 129, 161–168.

Kareiva, P. (1987) Habitat fragmentation and the stability of predator–prey interactions. *Nature* 326, 338–390.

Kenmore, P.E. (1980) *Ecology and Outbreaks of a Tropical Insect Pest of the Green Revolution, the Rice Brown Planthopper,* Nilaparvata lugens *Stal.* University of California, Berkeley, California.

Kevan, P.G. and Phillips, T.P. (2001) The economic impacts of pollinator declines: an approach to assessing the consequences. *Conservation Ecology* 5, 8.

Kevan, P.G. and Plowright, R.C. (1989) Fenitrothion and insect pollinators. In: Environment, Canada, Conservation and Protection (ed.) *Environmental Effects of*

Fenitrothion Use in Forestry: Impacts on Insect Pollinators, Songbirds, and Aquatic Organisms. Environment, Canada, Conservation and Protection, Dartmouth, pp. 13–42.

Kevan, P.G., Clark, E.A. and Thomas, V.G. (1990) Insect pollinators and sustainable agriculture. *American Journal of Alternative Agriculture* 5, 12–22.

Kiely, T., Donaldson, D. and Grube, A. (2004) Pesticide Industry Sales and Usage: 2000 and 2001 Market Estimates. Biological and Economic Analysis Division, Office of Pesticide Programs, EPA., Washington, DC.

Kim, J., Williams, N. and Kremen, C. (2006) Effects of cultivation and proximity to natural habitat on ground-nesting native bees in California sunflower fields. *Journal of the Kansas Entomological Society* 79, 309–320.

Kleijn, D., Berendse, F., Smit, R., Gilissen, N., Smit, J., Brak, B. and Groeneveld, R. (2004) Ecological effectiveness of agri-environment schemes in different agricultural landscapes in the Netherlands. *Conservation Biology* 18, 775–786.

Klein, A.M., Steffan-Dewenter, I. and Tscharntke, T. (2003a) Bee pollination and fruit set of *Coffea arabica* and *C-canephora* (Rubiaceae). *American Journal of Botany* 90, 153–157.

Klein, A.M., Steffan-Dewenter, I. and Tscharntke, T. (2003b) Fruit set of highland coffee increases with the diversity of pollinating bees. *Proceedings of the Royal Society of London Series B – Biological Sciences* 270, 955–961.

Klein, A.M., Vaissière, B., Cane, J.H., Steffan-Dewenter, I., Cunningham, S.A., Kremen, C. and Tscharntke, T. (2007) Importance of crop pollinators in changing landscapes for world crops. *Proceedings of the Royal Society of London Series B – Biological Sciences* 274, 303–313.

Komeili, A.B. (1988) The impact of the *Varroa* mite on Iranian commercial beekeeping. *American Bee Journal* 128, 423–424.

Kremen, C. (2004) Pollination services and community composition: does it depend on diversity, abundance, biomass, or species traits? In: Freitas, B.M. and Pereira, J.O.P. (eds) *Solitary Bees: Conservation, Rearing and Management for Pollintion*. Universidade Federal do Ceará, Ceará, Brazil, pp. 115–124.

Kremen, C. (2005) Managing ecosystem services: what do we need to know about their ecology? *Ecology Letters* 8, 468–479.

Kremen, C., Bugg, R.L., Nicola, N., Smith, S.A., Thorp, R.W. and Williams, N.M. (2002a) Native bees, native plants and crop pollination in California. *Fremontia* 30, 41–49.

Kremen, C., Williams, N.M. and Thorp, R.W. (2002b) Crop pollination from native bees at risk from agricultural intensification. *Proceedings of the National Academy of Sciences of the United States of America* 99, 16812–16816.

Kremen, C., Williams, N.M., Bugg, R.L., Fay, J.P. and Thorp, R.W. (2004) The area requirements of an ecosystem service: crop pollination by native bee communities in California. *Ecology Letters* 7, 1109–1119.

Krishna, V.V., Byju, N.G. and Tamizheniyan, S. (2003) Integrated pest management in Indian agriculture: a developing economic perspective. In: Radcliff, E.B. and Hutchson, W.D. (eds) *IPM World Textbook*. St Paul, Minnesota.

Kruess, A. (2003) Effects of landscape structure and habitat type on a plant–herbivore–parasitoid community. *Ecography* 26, 283–290.

Kruess, A. and Tscharntke, T. (1994) Habitat fragmentation, species loss, and biological control. *Science* 264, 1581–1584.

Kruess, A. and Tscharntke, T. (2000) Species richness and parasitism in a fragmented landscape: experiments and field studies with insects on *Vicia sepium*. *Oecologia* 122, 129–137.

Lagerlof, J. and Wallin, H. (1993) The abundance of arthropods along 2 field margins with different types of vegetation composition: an experimental study. *Agriculture Ecosystems and Environment* 43, 141–154.

Landis, D.A., Wratten, D.A. and Gurr, G.M. (2000) Habitat management to conserve natural enemies of arthropod pests in agriculture. *Annual Review of Entomology* 45, 175–201.

Langer, A. and Hance, T. (2004) Enhancing parasitism of wheat aphids through apparent competition: a tool for biological control.

Agriculture Ecosystems and Environment 102, 205–212.

Larsen, T.H., Williams, N. and Kremen, C. (2005) Extinction order and altered community structure rapidly disrupt ecosystem functioning. *Ecology Letters* 8, 538–547.

Lee, J.C., Menalled, F.B. and Landis, D.A. (2001) Refuge habitats modify impact of insecticide disturbance on carabid beetle communities. *Journal of Applied Ecology* 38, 472–483.

Leigh, E., Jr (ed.) (1996) *The Ecology of a Tropical Forest: Seasonal Rhythms and Long-Term Changes*, 2nd edn. Smithsonian Institution Press, Washington, DC.

Lemke, A. and Poehling, H.M. (2002) Sown weed strips in cereal fields: overwintering site and "source" habitat for *Oedothorax apicatus* (Blackwall) and *Erigone atra* (Blackwall) (Araneae: Erigonidae). *Agriculture Ecosystems and Environment* 90, 67–80.

LeTourneau, D. and Goldstein, B. (2001) Pest damage and arthropod community structure in organic vs. conventional tomato production in California. *Journal of Applied Ecology* 38, 557–570.

Loper, G.M. (2001) Nesting sites, characterized and longevity of feral honey bee colonies in the Sonoran desert of Arizona: 1991–2000. In: Erickson, E.H. Jr, Page, R.E. Jr and Hanna, A.A. (eds) *Proceedings of the 2nd International Conference On Africanized Honey Bees and Bee Mites*. A.I. Root Company, Medina, Ohio.

Losey, J.E. and Denno, R.F. (1998) Positive predator–predator interactions: enhanced predation rates and synergistic suppression of aphid populations. *Ecology* (Washington, DC) 79, 2143–2152.

Losey, J.E. and Vaughan, M. (2006) The economic value of ecological services provided by insects. *BioScience*, 311–323.

MacMahon, J.A., Mull, J.F., Crist, T.O., Fautin, D.G., Futuyma, D.J. and James, F.C. (2000) Harvester ants (*Pogonomyrmex* spp.): their community and ecosystem influences. *Annual Review of Ecology and Systematics* 31, 265–291.

McGregor, S.E. (1976) *Insect Pollination of Cultivated Crop Plants*. USDA-ARS, Washington, DC.

McNeely, J. and Scherr, S.J. (2002) *Ecoagriculture: Strategies to Feed the World and Save Wild Biodiversity*. Island Press, Washington, DC.

Marino, P.C. and Landis, D.A. (1996) Effect of landscape structure on parasitoid diversity and parasitism in agroecosystems. *Ecological Applications* 6, 276–284.

Martin, S. and Medina, L. (2004) Africanized honeybees have unique tolerance to *Varroa* mites. *Trends in Parasitology* 20, 112–114.

Mauremooto, J.R., Wratten, S.D., Worner, S.P. and Fry, G.L.A. (1995) Permeability of hedgerows to predatory carabid beetles. *Agriculture Ecosystems and Environment* 52, 141–148.

MEA (Millennium Ecosystem Assessment) (2003) *Ecosystems and Human Well-being: A Framework For Assessment*. Island Press, Washington, DC.

Memmott, J., Waser, N.M. and Price, M.V. (2004) Tolerance of pollination networks to species extinctions. *Proceedings of the Royal Society of London Series B – Biological Sciences* 271, 2605–2611.

Menalled, F.D., Marino, P.C., Gage, S.H. and Landis, D.A. (1999) Does agricultural landscape structure affect parasitism and parasitoid diversity? *Ecological Applications* 9, 634–641.

Menalled, F.D., Costamagna, A.C., Marino, P.C. and Landis, D.A. (2003) Temporal variation in the response of parasitoids to agricultural landscape structure. *Agriculture Ecosystems and Environment* 96, 29–35.

Morandin, L. and Winston, M. (2006) Pollinators provide economic incentive to preserve natural land in agroecosystems. *Agriculture Ecosystems and Environment* 116, 289–292.

Morandin, L.A. and Winston, M.L. (2005) Wild bee abundance and seed production in conventional, organic, and genetically modified canola. *Ecological Applications* 15, 871–881.

Morris, W. (2003) Which mutualists are most essential? Buffering of plant reproduction against the extinction of pollinators. In: Kareiva, P. and Levin, S.A. (eds) *The Importance of Species: Perspectives on Expendability and Triage*. Princeton

University Press, Princeton, New Jersey, pp. 260–280.

Morse, R.A. and Goncalves, L.S. (1979) *Varroa* disease, a threat to world beekeeping. *Bee Culture* 107, 179–181.

Murdoch, W.W. and Briggs, C.J. (1996) Theory for biological control: recent developments. *Ecology* (Washington, DC) 77, 2001–2013.

Nabhan, G.P. and Buchmann, S. (1997) Services provided by pollinators. In: Daily, G.C. (ed.) *Nature's Services: Societal Dependence on Natural Ecosystems*. Island Press, Washington, DC, pp. 133–150.

National Research Council of the National Academies (2006) *Status of Pollinators in North America*. National Academies Press, Washington, DC.

Naylor, R. and Ehrlich, P. (1997) Natural pest control services and agriculture. In: Daily, G. (ed.) *Nature's Services: Societal Dependence On Natural Ecosystems*. Island Press, Washington, DC, pp. 151–174.

Nee, S. (2004) More than meets the eye: earth's real biodiversity is invisible, whether we like it or not. *Nature* 429, 804–805.

Ostman, O. (2004) The relative effects of natural enemy abundance and alternative prey abundance on aphid predation rates. *Biological Control* 30, 281–287.

Ostman, O., Ekbom, B. and Bengtsson, J. (2001a) Landscape heterogeneity and farming practice influence biological control. *Basic and Applied Ecology* 2, 365–371.

Ostman, O., Ekbom, B., Bengtsson, J. and Weibull, A.C. (2001b) Landscape complexity and farming practice influence the condition of polyphagous carabid beetles. *Ecological Applications* 11, 480–488.

Ostman, O., Ekbom, B. and Bengtsson, J. (2003) Yield increase attributable to aphid predation by ground-living polyphagous natural enemies in spring barley in Sweden. *Ecological Economics* 45, 149–158.

Parker, F.D., Batra, S.W.T. and Tepedino, V.J. (1987) New pollinators for our crops. *Agricultural Zoology Reviews* 2, 279–304.

Pavuk, D.M. and Barrett, G.W. (1993) Influence of successional and grassy corridors on parasitism of *Plathypena scabra* (F) (Lepidoptera, Noctuidae) larvae in soybean agroecosystems. *Environmental Entomology* 22, 540–546.

Perfecto, I., Vandermeer, J.H., Bautista, G.L., Nunez, G.I., Greenberg, R., Bichier, P. and Langridge, S. (2004) Greater predation in shaded coffee farms: the role of resident neotropical birds. *Ecology* 85, 2677–2681.

Pickett, C.H. and Bugg, R.L. (1998) (eds) *Enhancing Biological Control: Habitat Management to Promote Natural Enemies of Agricultural Pests*. University of California Press, Berkeley, California.

Pimentel, D. (1998) Economic benefits of natural biota. *Ecological Economics* 25, 45–47.

Pimentel, D., Acquay, H., Biltonen, M., Rice, P., Silva, M., Nelson, J., Lipner, V., Giordano, S., Horowitz, A. and Damore, M. (1992) Environmental and economic costs of pesticide use. *Bioscience* 42, 750–760.

Polis, G.A. and Strong, D.R. (1996) Food web complexity and community dynamics. *American Naturalist* 147, 813–846.

Pontius, J., Dilts, R. and Bartlett, A. (2002) Rice IPM in Asia: ecological principles underlying the FFS. In: Pontius, J., Dilts, R. and Bartlett, A. (eds) *From Farmer Field Schools to Community IPM: Ten Years of IPM Training in Asia*. FAO, Bangkok.

Prasifka, J.R., Heinz, K.M. and Minzenmayer, R.R. (2004) Relationships of landscape, prey and agronomic variables to the abundance of generalist predators in cotton (*Gossypium hirsutum*) fields. *Landscape Ecology* 19, 709–717.

Prescott-Allen, R. and Prescott-Allen, C. (1990) How many plants feed the world? *Conservation Biology* 4, 365–374.

Priess, J., Mimler, M., Klein, A., Schwarze, S.T.T. and Steffan-Dewenter, I. (in press) Linking deforestation scenarios to pollination services and economic returns in coffee agroforestry systems. *Ecological Applications*.

Pritchard, K. (2005) *The Unseen Costs of Agricultural Expansion across a Rainforest Landscape: Depauperate Pollinator Communities and Reduced Yield in Isolated Crops*. Masters thesis, James Cook University.

Proctor, M., Yeo, P. and Lack, A. (1996) *The Natural History of Pollination*. Timber Press, Portland, Oregon.

Pywell, R.F., Warman, E.A., Hulmes, L., Hulmes, S., Nuttall, P., Sparks, T.H., Critchley, C.N.R. and Sherwood, A. (2006) Effectiveness of new agri-environment schemes in providing foraging resources for bumblebees in intensively farmed landscapes. *Biological Conservation* 129, 192–206.

Ricketts, T.H. (2004) Tropical forest fragments enhance pollinator activity in nearby coffee crops. *Conservation Biology* 18, 1262–1271.

Ricketts, T.H., Daily, G.C., Ehrlich, P.R. and Michener, C.D. (2004) Economic value of tropical forest to coffee production. *Proceedings of the National Academy of Sciences of the United States of America* 101, 12579–12582.

Riechert, S.E. and Bishop, L. (1990) Prey control by an assemblage of generalist predators: spiders in garden test systems. *Ecology* 71, 1441–1450.

Robertson, G.P. and Swinton, S.M. (2005) Reconciling agricultural productivity and environmental integrity: a grand challenge for agriculture. *Frontiers in Ecology and the Environment* 3, 38–46.

Rosenheim, J.A., Wilhoit, L.R. and Armer, C.A. (1993) Influence of intraguild predation among generalist insect predators on the suppression of an herbivore population. *Oecologia* (Heidelberg) 96, 439–449.

Rosenheim, J.A., Kaya, H.K., Ehler, L.E., Marois, J.J. and Jaffee, B.A. (1995) Intraguild predation among biological-control agents: theory and evidence. *Biological Control* 5, 303–335.

Roubik, D. (2001) Ups and downs in pollinator populations: when is there a decline? *Conservation Ecology* 5, 2.

Roubik, D.W. (1995) *Pollination of Cultivated Plants in the Tropics*. Food and Agriculture Organization, Rome, Italy.

Ruttan, V.W. (1999) The transition to agricultural sustainability. *Proceedings of the National Academy of Sciences of the United States of America* 96, 5960–5967.

Schmidt, M.H. and Tscharntke, T. (2005) Landscape context of sheetweb spider (Araneae: Linyphiidae) abundance in cereal fields. *Journal of Biogeography* 32, 467–473.

Schmidt, M.H., Lauer, A., Purtauf, T., Thies, C., Schaefer, M. and Tscharntke, T. (2003) Relative importance of predators and parasitoids for cereal aphid control. *Proceedings of the Royal Society of London Series B – Biological Sciences* 270, 1905–1909.

Settle, W.H., Ariawan, H., Astuti, E.T., Cahyana, W., Hakim, A.L., Hindayana, D., Lestari, A.S., Pajarningsih and Sartanto (1996) Managing tropical rice pests through conservation of generalist natural enemies and alternative prey. *Ecology* 77, 1975–1988.

Slaa, J. and Biesmeijer, K. (2005) Flower constancy. In: Dafni, A., Kevan, P.G. and Husband, B.C. (eds) *Practical Pollination Biology*. Enviroquestion, Cambridge, Canada.

Snyder, W.E. and Ives, A.R. (2003) Interactions between specialist and generalist natural enemies: parasitoids, predators, and pea aphid biocontrol. *Ecology* (Washington, DC) 84, 91–107.

Southwick, E.E. and Southwick, L. Jr (1992) Estimating the economic value of honey bees (*Hymenoptera: Apidae*) as agricultural pollinators in the United States. *Journal of Economic Entomology* 85, 621–633.

Steffan-Dewenter, I. and Tscharntke, T. (1999) Effects of habitat isolation on pollinator communities and seed set. *Oecologia* 121, 432–440.

Steffan-Dewenter, I., Münzenberg, U. and Tscharntke, T. (2001) Pollination, seed set and seed predation on a landscape scale. *Proceedings of the Royal Society of London Series B – Biological Sciences* 268, 1685–1690.

Steffan-Dewenter, I., Potts, S.G. and Packer, L. (2005) Pollinator diversity and crop pollination services are at risk. *Trends in Ecology and Evolution* 20, 1–2.

Steffan-Dewenter, I., Klein, A.-M., Gaebele, V., Alfert, T. and Tscharntke., T. (2006) Bee diversity and plant–pollinator interactions in fragmented landscapes. In: Waser, N.M. and Ollerton, J. (eds) *Specialization and Generalization in Plant–Pollinator Interactions*. University of Chicago Press, Chicago, Illinois, pp. 387–410.

Sunderland, K. and Samu, F. (2000) Effects of agricultural diversification on the abundance, distribution, and pest control potential of spiders: a review. *Entomologia Experimentalis et Applicata* 95, 1–13.

Symondson, W.O.C., Sunderland, K.D. and Greenstone, M.H. (2002) Can generalist predators be effective biocontrol agents? *Annual Review of Entomology* 47, 561–594.

Thies, C. and Tscharntke, T. (1999) Landscape structure and biological control in agroecosystems. *Science* 285, 893–895.

Thies, C., Steffan-Dewenter, I. and Tscharntke, T. (2003) Effects of landscape context on herbivory and parasitism at different spatial scales. *Oikos* 101, 18–25.

Thies, C., Roschewitz, I. and Tscharntke, T. (2005) The landscape context of cereal aphid–parasitoid interactions. *Proceedings of the Royal Society of London Series B – Biological Sciences* 272, 203–210.

Thomson, J.D. and Goodell, K. (2001) Pollen removal and deposition by honeybee and bumblebee visitors to apple and almond flowers. *Journal of Applied Ecology* 38, 1032–1044.

Thomson, J.D. and Thomson, B.A. (1992) Pollen presentation and viability schedules in animal-pollinated plants: consequences for reproductive success. In: Wyatt, R. (ed.) *Ecology and Evolution of Plant Reprocuction: New Approaches*. Chapman & Hall, New York, pp. 1–24.

Tilman, D., Lehman, C.L. and Bristow, C.E. (1998) Diversity–stability relationships: statistical inevitability or ecological consequence? *American Naturalist* 151, 277–282.

Tilman, D., Fargione, J., Wolff, B., D'Antonio, C., Dobson, A., Howarth, R., Schindler, D., Schlesinger, W.H., Simberloff, D. and Swackhamer, D. (2001) Forecasting agriculturally driven global environmental change. *Science* 292, 281–284.

Tixier, M.S., Kreiter, S., Auger, P. and Weber, M. (1998) Colonization of Languedoc vineyards by phytoseiid mites (Acari: Phytoseiidae): influence of wind and crop environment. *Experimental and Applied Acarology* 22, 523–542.

Townsend, G.F. (1974) Beekeeping and agricultural development. *World Animal Review* 12, 36–40.

Tscharntke, T., Gathmann, A. and Steffan-Dewenter, I. (1998) Bioindication using trap–nesting bees and wasps and their natural enemies: community structure and interactions. *Journal of Applied Ecology* 35, 708–719.

Tscharntke, T., Steffan-Dewenter, I., Kruess, A. and Thies, C. (2002) Contribution of small habitat fragments to conservation of insect communities of grassland–cropland landscapes. *Ecological Applications* 12, 354–363.

Tscharntke, T., Klein, A.M., Kruess, A., Steffan-Dewenter, I. and Thies, C. (2005) Landscape perspectives on agricultural intensification and biodiversity: ecosystem service management. *Ecology Letters* 8, 857–874.

USDA National Agriculture Statistics Service (1997) 1996 Honey production report. USDA, Washington, DC.

Varchola, J.M. and Dunn, J.P. (2001) Influence of hedgerow and grassy field borders on ground beetle (Coleoptera: Carabidae) activity in fields of corn. *Agriculture Ecosystems and Environment* 83, 153–163.

Vaughan, M., Shepard, M., Kremen, C. and Black, S.H. (2004) *Farming for Bees: Guidelines for Providing Native Bee Habitat on Farms*. The Xerces Society, Portland, Oregon.

Vazquez, D.P., Morris, W.F. and Jordano, P. (2005) Interaction frequency as a surrogate for the total effect of animal mutualists on plants. *Ecology Letters* 8, 1088–1094.

Vitousek, P.M., Mooney, H.A., Lubchenco, J. and Melillo, J.M. (1997) Human domination of Earth's ecosystems. *Science* 277, 494–499.

Weibull, A.C., Ostman, O. and Granqvist, A. (2003) Species richness in agroecosystems: the effect of landscape, habitat and farm management. *Biodiversity and Conservation* 12, 1335–1355.

Westphal, C., Steffan-Dewenter, I. and Tscharntke, T. (2003) Mass flowering crops enhance pollinator densities at a landscape scale. *Ecology Letters* 6, 961–965.

Williams, I. (1996) Aspect of bee diversity and crop pollination in the European Union. In: Matheson, A. (ed.) *The Conservation of Bees*. Academic Press for the Linnean

Society of London and the International Bee Research Association, London, pp. 63–80.

Williams, N.M., Minckley, R.L. and Silveira, F.A. (2001) Variation in native bee faunas and its implications for detecting community changes. *Conservation Ecology* 5, 7.

Wilson, E.O. (1987) The little things that run the world (the importance and conservation of invertebrates). *Conservation Biology* 1, 344–346.

Winfree, R., Griswold, T. and Kremen, C. (2007) Effect of human disturbance on bee communities in a forested ecosystem. *Conservation Biology* 21, 213–223.

With, K.A., Pavuk, D.M., Worchuck, J.L., Oates, R.K. and Fisher, J.L. (2002) Threshold effects of landscape structure on biological control in agroecosystems. *Ecological Applications* 12, 52–65.

Wolda, H. (1988) Insect seasonality: why? *Annual Review of Ecology and Systematics* 19, 1–18.

Wratten, S.D., Bowie, M.H., Hickman, J.M., Evans, A.M., Sedcole, J.R. and Tylianakis, J.M. (2003) Field boundaries as barriers to movement of hover flies (Diptera: Syrphidae) in cultivated land. *Oecologia* 134, 605–611.

Zabel, J. and Tscharntke, T. (1998) Does fragmentation of *Urtica* habitats affect phytophagous and predatory insects differentially? *Oecologia* 116, 419–425.

16 Insect Conservation in Agricultural Landscapes

TEJA TSCHARNTKE,[1] JASON M. TYLIANAKIS,[1,5] MARK R. WADE,[2] STEVE D. WRATTEN,[2] JANNE BENGTSSON[3] AND DAVID KLEIJN[4,6]

[1]*Agroecology, University of Göttingen, Waldweg 26, D-37073 Göttingen, Germany;* [2]*National Centre for Advanced Bio-Protection Technologies, PO Box 84, Lincoln University, Canterbury, New Zealand;* [3]*Department of Entomology (Landscape Ecology), Swedish University of Agricultural Sciences, PO Box 7044, SE-750-07 Uppsala, Sweden;* [4]*Former address: Nature Conservation and Plant Ecology Group, Wageningen University, Bornsesteeg 69, 6708 PD Wageningen, The Netherlands;* [5]*Current address: School of Biological Sciences, University of Canterbury, Private Bag 4800, Christchurch 8020, New Zealand;* [6]*Current address: Alterra, Centre for Ecosystem studies, PO Box 47, 6700 AA, Wageningen, The Netherlands*

1 Introduction

The demands of a growing human population have resulted in global-scale conversion of land to agriculture, and with the world population expected to increase by 50% to 9 billion by 2050, more than 1 billion hectares of natural systems will be converted to agriculture by this date (Tilman *et al.*, 2001). Consideration of the effects of agricultural land on biodiversity is therefore imperative, and although agricultural land use and biodiversity conservation were traditionally viewed as incompatible, recent recognition has increased that a restriction of conservation efforts to natural, undisturbed ecosystems is of limited value (Pimentel *et al.*, 1992; Bengtsson *et al.*, 2003). 'We are obviously past any point where strategies that focus on conservation of pristine habitat are sufficient for the job' (Novacek and Cleland, 2001). Agriculture can make important contributions to high diversity of landscapes, while it also benefits from sustainable ecosystem services provided by agricultural conservation management (Daily, 1997; Kremen, 2005; Tscharntke *et al.*, 2005). For example, enhanced biological pest control and improved crop pollination may directly increase the farmers' income (e.g. Östman *et al.*, 2003; Ricketts *et al.*, 2004; Olschewski *et al.*, 2006).

In this chapter, we review negative and positive effects of agriculture for biodiversity conservation and the role of biodiversity in multifunctional agriculture, including ecosystem services such as biological pest control. Biodiversity patterns change with the spatial and temporal scales considered, so integrating conservation in agriculture requires a multiscale landscape perspective. Further, we discuss political options for sustainable landscape planning. We argue that there is a need for a diversity of 'insurance' species, to support the landscape-wide capacity to reorganize after disturbance, and for tailoring agri-environment schemes at the landscape level.

2 Integrating Agriculture and Conservation

The greatest losses in biodiversity worldwide have been attributed to the expansion and intensification of agriculture (Robinson and Sutherland, 2002; Tilman *et al.*, 2002), and land use is often characterized by a transformation of complex ecosystems and landscapes into simple ones, resulting in reduced biodiversity (Fig. 16.1). The asymmetric losses of different taxa through habitat modification may also lead to changes in community structure or release of herbivores through reduced control by natural enemies (see Fig. 16.1).

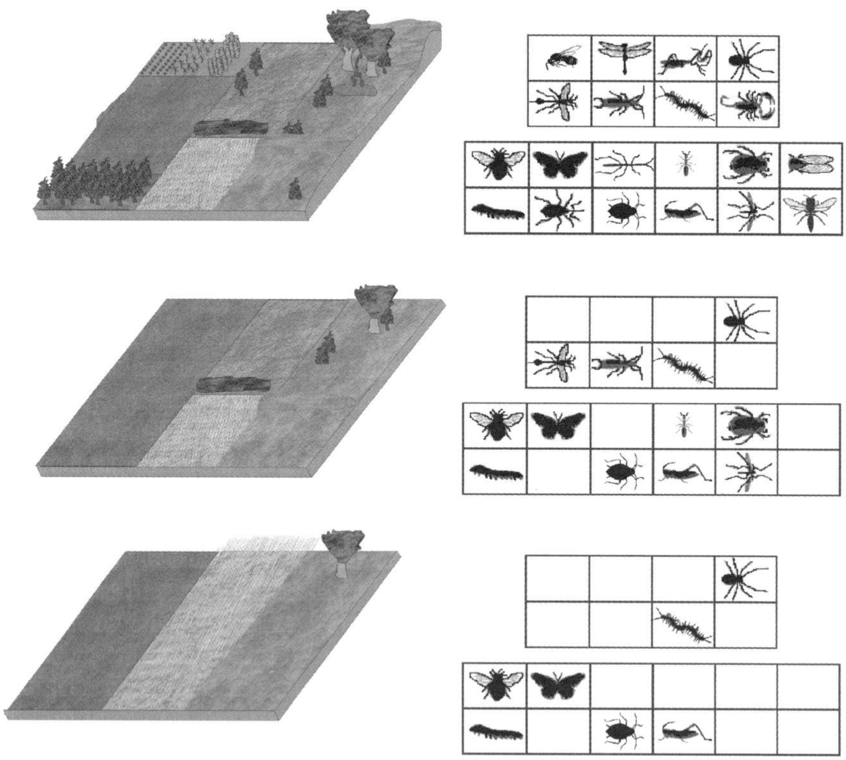

Fig. 16.1. Decreasing invertebrate biodiversity with decreasing landscape heterogeneity. From top to bottom, the agricultural landscape becomes simplified and the associated invertebrate communities become less diverse. For each landscape type, invertebrate symbols show herbivores (two bottom rows) and predators (two top rows). In the simplest landscape (bottom) only generalist species survive, pest outbreaks are common because higher trophic levels suffer more from simplification, and ecosystem services are limited.

However, agricultural land use does not just mean habitat destruction. Traditional and low-intensity land use practices may create species-rich habitats, an effect which is well known from Europe (e.g. high-diversity grasslands) but also known from the tropics (see Perfecto *et al.*, 1996). For example, coffee and cacao agroforestry, shaded by diverse natural or planted trees, represents the last remaining forested habitats in many tropical landscapes (Rice and Greenberg, 2000; Klein *et al.*, 2002; Perfecto and Armbrecht, 2003). Insect and bird species in these managed ecosystems may be as high as in the adjacent forest remnants, and the species composition can also be similar between forests and agroforests (Tylianakis *et al.*, 2005). However, tropical agroforests may be a special case, and several authors have argued that conserving high species richness does not often equate to conservation of endemics, if the diversity of agroecosystems comprises weedy, cosmopolitan species (Bos *et al.*, submitted; Lewis and Basset, Chapter 2, this volume). Nevertheless, many European species concerned with conservation depend on agricultural systems. For example, many arable weeds are actually classed as threatened species (Roschewitz *et al.*, 2005), and have associated rare invertebrate fauna. Ultimately, conversion of land to agriculture is inevitable for meeting global food demands (Tilman *et al.*, 2001, 2002), and this will of course affect biodiversity. Therefore, rather than downplay this harm, we will now discuss ways in which these impacts can be reduced at little cost to production.

Methods for increasing the conservation value of agricultural land at the habitat scale normally involve an attempt to decrease the intensity of management. Insect diversity and abundance can be enhanced by provision of woody borders or non-crop habitats adjacent to fields (e.g. Dyer and Landis, 1997; Holland and Fahrig, 2000), high-shade-tree diversity in agroforests (Stamps and Linit, 1997; Sperber *et al.*, 2004; Veddeler *et al.*, 2005) or through organic management (Hutton and Giller, 2003; Asteraki *et al.*, 2004; Wickramasinghe *et al.*, 2004; Bengtsson *et al.*, 2005; Morandin and Winston, 2005). As the invertebrate fauna depending on mature trees and associated coarse woody debris is usually depauparate on farmland, recent work in New Zealand is using untreated discs of pine wood to accelerate ecological succession. These discs provide many of the ecological functions of natural fallen logs and can harbour late-succession invertebrate communities which usually would not be present in highly modified farming landscapes (Bowie and Frampton, 2004). A similar example from New Zealand which again involves designing end-of-succession habitats is the use of 'Weta Hotels'. Wetas (Orthoptera: Anostostomatidae, Rhaphidophoridae) are large, iconic native insects that are usually associated with undisturbed forest landscapes. Weta hotels, which mimic coarse woody debris with cavities, harbour this specialist fauna on farmland in which this insect disappeared following forest clearance (Bowie *et al.*, unpublished data).

Despite the potential for integrating conservation with agriculture, the primary goal of land users is to maximize profit, and differences between those strategies that favour conservation and those that favour economic returns must be reconciled (Banks, 2004). In this sense, conservation efforts must be pragmatic, and search for strategies that are mutually beneficial for the landowners and for biodiversity. Several options are available for this type of mutual benefit. First of all, financial incentives can be offered to landowners in exchange for providing conservation services, such as unsprayed headlands or the agri-environment schemes discussed below. Second, reduced intensity through organic farming can be beneficial to a variety of taxa, and any extra costs associated with organic farming can be borne by the consumer, so that the financial burden of conservation is shared (e.g. Collins *et al.*, 1992).

3 Functional Biodiversity in Agroecosystems and Habitat Manipulation for Enhanced Biocontrol

Agricultural ecosystems are traditionally considered to be poor in biodiversity (see above). Consequently, the role of modified ecosystems in the form of agricultural and urban land in providing ecosystem services, such as biological control, climate regulation, food, fuel wood, gas regulation, pollination and soil formation, has received little attention. The pivotal paper by Costanza *et al.* (1997) which calculated the economic value of 17 ecosystem services in 16 biomes to be in the range of US$16–54 trillion per year, with an annual average of US$33 trillion, actually attributed no dollar value to managed ecosystems. This is patently not true, especially when 'ecological engineering' techniques are available to enhance ecosystem services, such as those performed by arthropods on farmland (reviewed by Gurr *et al.*, 2003, 2004). It was only a few decades ago that awareness of functional biodiversity (diversity that provides ecosystem services) in farmland was raised. A key paper by Potts and Vickerman (1974) linked the role of functional invertebrate biodiversity in cereal crops to pest population dynamics. In this work, cereal aphid populations were reduced when total arthropod diversity increased (see Fig. 11 in Potts and Vickerman, 1974). Parallel studies in cotton have shown that there is a large number of predacious fauna. There are estimated to be more than 250 species of predatory arthropods present in cotton in Australia, 41 of which are common (Room, 1979), and this number is even greater in the USA, where approximately 600 predatory species are present in that crop (Whitcomb and Bell, 1964).

Biodiversity of agroecosystems is largely sustained by the proximity of natural and managed ecosystems, and manipulation of these habitats may be used to enhance beneficial arthropods (Landis *et al.*, 2000; Gurr *et al.*, 2003, 2004) and their associated ecosystem services. We now examine the effect of such manipulations on agro-biodiversity, placing emphasis on species that are responsible for biological control.

Field boundaries adjacent to temperate crop fields may support colonization by insect natural enemies as well as the non-crop plants on which many parasitoids feed (e.g. Thies and Tscharntke, 1999). The value of field boundaries as refugia for beneficial arthropods such as Araneida and Coleoptera (Carabidae and Staphylinidae) has led to the development of 'beetle banks' – permanently vegetated strips across field centres that provide an alternative habitat for arthropods (Thomas *et al.*, 1991). This is useful in annual cropping systems or when pesticides are used, as it ensures the colonization, persistence and continuity of natural enemies to deliver the ecosystem service of pest control. A possible shortcoming of beetle banks and other types of refugia is the difficulty of ensuring that the individual arthropods they harbour actually colonize the adjacent fields in the spring as the crops begin to grow (see Cameron *et al.*, 1984).

Other recent work on functional biodiversity on farmland has explored the dynamics of arthropod dispersal, predation and parasitism rates, and population growth rates in more detail, and has emphasized the spatial scale at which these processes operate (see below). Some of this work has demonstrated the importance of uncultivated landscape elements in 'driving' these landscape effects (reviewed by Tscharntke *et al.*, 2005). These effects are clearly important, but policy changes are needed before this information can be used most effectively, as the spatial scales involved transcend the scales at which individual landowners operate. In highly modified 'colonial' landscapes, such as lowland New Zealand, Australia and perhaps in other relatively new farming landscapes, most native vegetation has been removed from the agroecosystem. These landscapes are much simpler than their equivalent in Europe, particularly in relation to the 'nodes' where field corners join. In Europe these nodes can be rich 'hot spots' of invertebrate, vertebrate and plant diversity (Fry, 1995; Fig. 16.2).

Adding functional biodiversity to agroecosystems can be a rapid process when particular predator–pest associations are targeted. For example, flowering plants such as alyssum (*Lobularia maritima* (L.) Desv, Brassicaceae), buckwheat (*Fagopyrum esculentum* Moench, Polygonaceae) and phacelia (*Phacelia tanacetifolia* Benth, Hydrophyllaceae) can be grown alongside horticultural crops which would otherwise be pure monocultures. These floral patches or strips provide nectar, pollen and shelter for beneficial arthropods, which can significantly improve the abundance, diversity and fitness of these arthropods, leading to marked reductions in crop pests (reviewed by Landis *et al.*, 2000; see also Berndt *et al.*, 2002; Tylianakis *et al.*, 2004; Berndt and Wratten, 2005; Lee and Heimpel, 2005). However, in this type of work the plant biodiversity (both crop and flowers) added is usually non-native, as are the pest and beneficial species involved. Habitat manipulation methods remain valid for enhancing the biodiversity of native species, as demonstrated by beetle banks (see above). However, the results are typically generated more slowly because many of these native taxa have specialist food requirements, reproduce slowly and/or have low mobility (including colonization abilities). Populations of pollinating bees and predatory wasps in agroecosystems can be enhanced with the introduction of suitable nesting sites (Gathmann *et al.*, 1994; Tscharntke *et al.*, 1998; Bosch and Kemp, 2001).

Much of the above work on restoring functional biodiversity has concentrated on increasing the contribution of natural enemies to pest control in order to minimize the negative effects associated with agricultural intensification, such as pesticide use. In summary, agroecosystems can support a rich

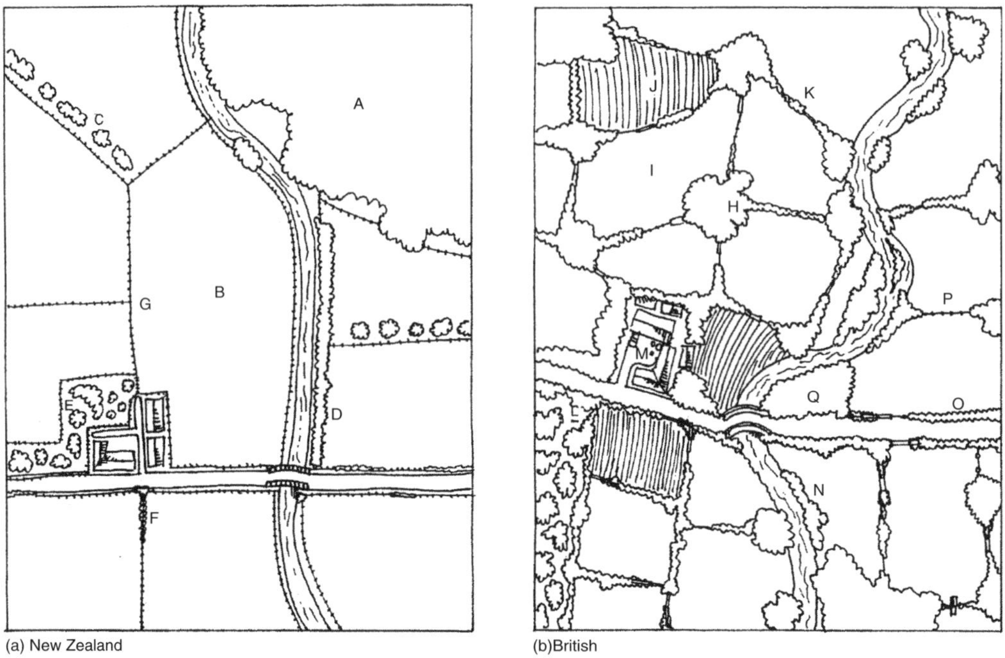

Fig. 16.2. Contrast between a homogeneous (or simple) and a heterogeneous (or complex) farm landscape, exemplified by some (a) New Zealand and (b) British landscapes. Note the heterogeneity of habitats and the connectedness features in the British landscape compared with the uniformity of the New Zealand farm landscape. A = indigenous plant reserve: these tend to be large tracts of land not integrated with farmland; B = pasture, exotic grasses; C = typical shelter belt (e.g. poplar, *Cupressus macrocarpa*, *Pinus radiata*); D = riparian vegetation (e.g. willow, grasses, some indigenous species); E = farmhouse garden; F = small areas of patchy gorse; G = wire fences: common field boundaries; H = small woodlot: a highly used but sustained feature; I = pasture; J = ploughed field; K = hedge fence; L = orchard; M = farmhouse garden; N = riparian vegetation; O = roadside vegetation, hedges, trees, etc.; P = wire fences or stone walls; and Q = woodland. (Illustration by Cor Vink, with permission from V. Keesing and S.D. Wratten, 1997.)

diversity of beneficial arthropods. Provided the land use areas are managed correctly, these arthropods should contribute to the ecosystem service of pest control. Habitat manipulation approaches, such as beetle banks and species-rich floral patches, provide the opportunity to enhance biological control, as well as to conserve important species of invertebrates, vertebrates and plants.

Habitat manipulation of farmland also offers great potential for conservation of non-arthropods. Many songbird populations have declined markedly in western Europe, as have those of the grey partridge, *Perdix perdix* (L.), and some mammals such as the harvest mouse, *Micromys minutes* (Pallas). Recent work by scientists at the Game Conservancy Trust in the UK has led to dramatic improvements in the populations of these species by the use of simple protocols. Again, beetle banks are prominent in this regard; not only do they harbour up to 1000 predatory invertebrate individuals per square metre in the winter, but the greatest nesting densities of grey partridge and the harvest mouse now occur in these 'island habitats' (Thomas *et al.*, 2001).

4 Biodiversity Patterns in Agricultural Landscapes

Despite this evidence that agricultural land can contribute greatly to biodiversity and ecosystem services, the extent of this contribution across different systems and scales is somewhat ambiguous (Tscharntke *et al.*, 2005). Previous studies have compared insect diversity at different management intensities within a particular land use type (e.g. DeVries *et al.*, 1997; DeVries and Walla, 2001; Klein *et al.*, 2002) or across a variety of land uses, comprising a gradient of intensity (e.g. Shahabuddin *et al.*, 2005; Tylianakis *et al.*, 2005). The results of these studies have not always been consistent. For example, insect diversity has been shown to increase (DeVries *et al.*, 1997; Lawton *et al.*, 1998; DeVries and Walla, 2001; Klein *et al.*, 2002; Tylianakis *et al.*, 2006), decrease (Di Giulio *et al.*, 2001; Maeto *et al.*, 2002; Sinclair *et al.*, 2002; Steffan-Dewenter *et al.*, 2002; Mas and Dietsch, 2003; Schulze *et al.*, 2004; Shahabuddin *et al.*, 2005; Tylianakis *et al.*, 2005) or not significantly differ (Steffan-Dewenter and Leschke, 2003) with increasing management intensity.

Agricultural intensification occurs at different spatial scales. Local intensification includes adverse effects such as shortened crop rotation cycles and increasing input of agrochemicals. On a landscape scale, fields have been amalgamated and enlarged, resulting in simplified landscapes without any non-crop habitats remaining (e.g. Swift and Anderson, 1993; Tscharntke *et al.*, 2005). While the effects of management intensity are frequently considered only at the farm scale, agricultural intensification has also modified entire landscapes, through the amalgamation of fields to improve efficiency, and the loss of natural habitats and mosaic quality of the landscape (Tscharntke *et al.*, 2005). Such homogenization of landscapes leads to fragmentation and isolation of natural habitats, ultimately causing species decline. For example, fragment size, shape and spatial configuration can all significantly affect insect diversity (Cane, 2001; Tscharntke *et al.*, 2002; Krauss *et al.*, 2003; Steffan-Dewenter, 2003; Stoner and Joern, 2004; Summerville and Crist, 2004; Tscharntke and Brandl, 2004; Ribas *et al.*, 2005). Isolation, the distance of a habitat from a natural insect source population (e.g. forest), can also be an important determinant of insect diversity, especially in functionally important groups such as bees and ants (Armbrecht and Perfecto, 2003; Klein *et al.*, 2003a,b; Ricketts *et al.*, 2004; but see Cunningham *et al.*, 2005). Proximity to natural habitats, a landscape-scale factor, may therefore partly mitigate harmful management practices at the habitat scale.

Given these different processes, it is not surprising that recent studies have shown scale- or landscape context-dependent variation in the responses of different taxa to habitat modification (Tscharntke and Brandl, 2004; Tylianakis *et al.*, 2006). As different forces will be structuring communities and populations at different scales, ranging from within the habitat, to the entire region or landscape, interacting species may experience these scales differently. Differences may arise through dispersal, feeding and life history strategies (Jonsen and Fahrig, 1997; Krauss *et al.*, 2003; Borges and Brown, 2004; Chust *et al.*, 2004; Stoner and Joern, 2004; Ribas *et al.*, 2005), and the influence of landscape-scale processes on habitat-scale interactions is frequently overlooked (Tscharntke *et al.*, 2005). It is therefore necessary to examine the effects of agricultural management on diversity at different scales and to consider approaches to integrating multiscale effects in order to gain a full understanding of the contribution of agricultural landscapes to overall biodiversity.

5 Factors Affecting Biodiversity at Different Spatial and Temporal Scales

The negative effects of agriculture at the habitat scale generally result from reduced plant diversity (e.g. Marshall *et al.*, 2003) or use of insecticides (e.g. Paoletti and Pimentel, 2000). However, the biodiversity of a habitat is ultimately limited by the total species pool within the landscape, and these two spatial scales are inherently linked. The diversity of species within a habitat is usually unsaturated with respect to the regional species pool (Holt and Gaston, 2002; Gaston and Spicer, 2004), and the degree of saturation can be assumed to be lower in agroecosystems due to their high disturbance frequency (Tscharntke *et al.*, 2005).

Although analyses at different spatial scales have received some attention, variation in diversity across different temporal scales is often neglected. Temporal heterogeneity in the biodiversity of different habitat types may result from temporal variation in resource availability (Wolda, 1978, 1988), e.g. during periods of mast flowering or anthropogenic sowing/harvesting. This can allow species to move between habitats, exploiting the availability of predictable, ephemeral resources (Wissinger, 1997; Bambaradeniya *et al.*, 2004). Studies that have explicitly examined temporal variation in insect diversity have concluded that small temporal sampling scales can lead to a serious underestimation of diversity (e.g. DeVries *et al.*, 1997; Summerville and Crist, 2005), or even a completely erroneous comparison of diversity between habitat types (Tylianakis *et al.*, 2005).

Therefore, rather than considering diversity in only one point in time and space, we need to also consider the turnover in species (beta diversity) between habitats within the landscape and across time. The challenge of assessing diversity at multiple scales can be met by partitioning diversity between different levels of a nested spatial and/or temporal hierarchy, thereby determining the scale across which the highest beta diversity occurs. This multiscale approach was used by Summerville *et al.* (2003) to evaluate lepidopteran diversity in temperate forests in Ohio, USA, and by Tylianakis *et al.* (2005, 2006) to compare the contribution of different land use types to Hymenoptera diversity in coastal Ecuador. Although beta diversity is often lower in agricultural systems than within or across natural habitats (Clausnitzer, 2003; Tylianakis *et al.*, 2005), species turnover between different patches of managed habitats can still make a significant contribution to regional biodiversity (Tscharntke *et al.*, 2002; Tylianakis *et al.*, 2005, 2006). Nevertheless, beta diversity at all scales is lost as homogeneity due to management intensity increases, such that homogeneous habitats have low turnover within fields, between fields across the landscape and through time (Tylianakis *et al.*, 2005).

On the basis of the varied responses of taxa to the environment at different scales, conservation tactics aimed at only one scale, or based on information from studies conducted at only one scale, may be misguided. Conservation policies should simultaneously target both the individual farm and landscape scales to maximize overall success (Östman *et al.*, 2001; Stoner and Joern, 2004; Tscharntke *et al.*, 2005).

At the landscape scale, insect biodiversity can be supported by moderating the effects of fragmentation and habitat loss, for example, by increasing connectivity between habitat types, to facilitate dispersal between local populations within metapopulations (e.g. Steffan-Dewenter, 2003) Additionally, although single large conservation areas are often advocated over several small areas, the spreading of several small fragments across a large geographic area may maximize beta diversity, as has been shown for butterflies and grasshoppers in calcareous grasslands (Tscharntke *et al.*, 2002; Peintinger *et al.*, 2003). Fragmentation effects may also be reduced by the availability of non-cultivated land within the landscape. For example, parasitoids of rape pollen beetles usually show higher densities near overwintering sites such as edge grassy strips, but these edge effects are overwhelmed by high overall densities of the parasitoids when a high percentage of non-crop area (>20%) remains in the landscape (Tscharntke *et al.*, 2002).

Despite the utility of maintaining non-crop habitat, landscape-scale conservation does not necessarily require large set-aside areas. The above example shows that insects may benefit from natural habitats; however, the resources provided by cultivated habitats may also be beneficial. Westphal *et al.* (2003) found that bumblebee densities did not respond to the proportion of natural habitat, but rather to the availability of rich floral resources (oilseed rape) within the landscape.

The detrimental effects of homogenization through landscape-scale intensification can be offset by maximizing the heterogeneity of the landscape by planting different crops, rather than monocropping over large areas. Landscape heterogeneity has been shown to be a good predictor of Collembola diversity (Chust *et al.*, 2003) and to enhance the diversity of a variety of other insect taxa (Steffan-Dewenter *et al.*, 2002; Dauber *et al.*, 2003; Krauss *et al.*, 2003; Kruess, 2003).

6 The Insurance Hypothesis and Sustainable Landscape Planning

In the recent debates on the relationship between biodiversity and ecosystem functioning (summarized in Hooper *et al.*, 2005) biodiversity can affect ecosystem services in two ways. First, the magnitude of ecosystem services, such as pollination or biological control, can be affected locally by diversity, mainly because species are complementary in their effects on ecosystem functioning. This idea has its roots in parts of classical niche theory and is based on the fact that the species differ in their impacts on the environment (Chase and Leibold, 2003) and on ecosystem properties. This effect has been examined in numerous studies of experimental grasslands (e.g. Hector *et al.*, 1999).

Second, and from the agricultural landscape perspective more importantly, diversity can affect the temporal variability and magnitude of ecosystem services because species differ in their responses to environmental conditions. The concept of the Hutchinsonian niche is applicable in this case (Chase and Leibold, 2003), as different species are expected to have different reaction norms or tolerance to resource levels or abiotic factors. Therefore, diverse ecosystems are expected to vary less, be more robust to external disturbances, and have a higher rate of ecosystem functioning when studied over time. The idea was termed the insurance hypothesis by Yachi and Loreau (1999). To separate the diversity in responses to the environment from diversity effects on ecosystem properties, Elmqvist *et al.* (2003) coined the term response diversity.

The effect of biodiversity on the magnitude of ecosystem services acts through mechanisms of local interactions, such as species sorting or positive interactions with mycorrhiza. However, most ecosystems, particularly agroecosystems, are subject to disturbances and environmental variation at different spatial and temporal scales, from local management of fields to climate change (Holling *et al.*, 1995; Bengtsson *et al.*, 2003). Recovery and reorganization of biodiversity and ecosystem functioning (termed resilience by, e.g. Gunderson, 2000) after such disturbances require a species pool at the landscape level from which species can recolonize, emphasizing the role of species dispersal and landscape structure for the stability, and ultimately also magnitude, of ecosystem services. The insurance hypothesis can have both a temporal aspect, as environmental conditions vary locally, and a spatial aspect when local patches are subjected to disturbances. It explicitly takes the interplay between local and regional landscape-level processes into account (Nyström and Folke, 2001; Loreau *et al.*, 2003). In a heterogenous mosaic landscape, species with different traits, tolerances and optimum requirements thrive in patches with different environmental conditions, ensuring that there will be at least some species available in source areas for recolonization after disturbances, and that species with optimum performance can more easily colonize patches of different quality (Bengtsson *et al.*, 2003; Rand *et al.*, 2006).

The theoretical background for the insurance hypothesis is, on one hand, island biogeography, metapopulation and metacommunity theory (see, e.g. Leibold *et al.*, 2004), and on the other hand theories for dynamic ecosystems (e.g. Holling *et al.*, 1995; Norberg *et al.*, 2001; Bengtsson *et al.*, 2003). There is considerable theoretical support for it (Yachi and Loreau, 1999; Norberg *et al.*, 2001; Loreau *et al.*, 2003), but it is intrinsically difficult to test explicitly, especially on the large spatial scale where it is likely to be most important for ecosystem services and society. The finding that mosaic agricultural landscapes, with a larger proportion of semi-natural areas, sustain higher levels of diversity in several organism groups (see, e.g. Tscharntke *et al.*, 2005) provides empirical support for one of its basic components. Additional support comes from experimental studies of the effects of spatial structure and landscape connectivity for diversity and biomass (Gonzales and Chaneton, 2002). On larger scales, Nyström and Folke (2001) highlighted the importance of source areas in the landscape for ecosystem or community recovery. Recent studies showing that semi-natural habitats enhance the ecosystem service of coffee pollination (Klein *et al.*, 2003b; Ricketts *et al.*, 2004; Olschewski *et al.*, 2006) imply that diversity on the landscape scale provides better ecosystem services to society. However, stringent tests of the insurance hypothesis need to vary local and regional diversity in landscapes of different structure in terms of, for example, connectivity and availability of natural habitats, and measure both diversity and the magnitude of ecosystem services over time, as environmental conditions vary and disturbances are allowed to occur. Although a daunting task, cleverly designed studies of agroecosystems can be good candidates for observational tests of the insurance hypothesis.

Irrespective of whether we can provide hard experimental evidence for the insurance hypothesis, the fact that diversity in agricultural habitats often depends on landscape structure (see above; Tscharntke *et al.*, 2005) as well as on local conditions has implications for the maintenance of diversity and ecosystem services in agroecosystems. There are, as stated above, good reasons to believe that maintaining landscape mosaics with different disturbance regimes and successional stages will contribute to biodiversity and a less variable delivery of ecosystem services, such as biological control or pollination. This is especially important in the face of climate change where the dispersal rates of species will determine how quickly ecosystems can respond. Low landscape diversity entails a small species pool and less genetic variation for adapting to new environmental conditions.

7 Agri-environment Schemes at a Landscape Scale and Political Options

Agri-environment schemes are one of the most important political and nature conservation instruments to safeguard or promote wildlife in agricultural landscapes (EEA, 2004). They stimulate farmers to adopt more environment-friendly farming practices and compensate them for any loss of income associated with scheme implementation. Conservation measures usually consist of a reduction or cessation of the use of agrochemicals and/or a reduction in stocking rates (Kleijn and Sutherland, 2003) but may also include the establishment and maintenance of landscape features such as ponds and hedges.

The conservation of insects or specific insect groups is rarely mentioned specifically as an objective of agri-environment schemes. However, many schemes aim to conserve biodiversity in general, and the class of insects is one of the most diverse taxa contributing to this area. A considerable number of studies have examined the response of insects to agri-environment schemes (Kleijn and Sutherland, 2003). Uncropped wildlife strips in the UK had positive effects on the species richness of carabid beetles and Heteroptera in cereal fields (Hassall *et al.*, 1992). Schemes reducing stocking rates resulted in higher species richness of Auchenorrhyncha, Heteroptera, Coleoptera and parasitic Hymenoptera in grasslands in northern Germany (Kruess and Tscharntke, 2002). Dutch management agreements reducing agrochemical applications and delaying the first seasonal agricultural activities in wet grasslands were found to have positive effects on species richness of bees and hoverflies (Kleijn *et al.*, 2001). Implementation of the English Environmentally Sensitive Area Scheme and the Countryside Stewardship Scheme led to positive population trends in habitat-specialist butterflies (Brereton *et al.*, 2002). The positive effects of a wide variety of schemes on a wide range of insect species groups suggest that insects in agricultural landscapes may be conserved easily and rapidly by means of agri-environment schemes.

Although not a scheme per se, organic farming may be encouraged by agri-environment schemes in Europe, and has mixed but mainly positive effects on diversity and density of insects (Bengtsson *et al.*, 2005). In a literature review, predatory arthropods generally increased in diversity and density in organic farming systems, while the responses of non-predatory insects were more heterogeneous and on average not different from neutral (Bengtsson *et al.*, 2005). The effects for arthropods were significantly positive on the smaller 'field' and 'plot' scales, but non-significant and highly heterogeneous on the farm scale (meta-analysis calculated from the arthropod subset of the data in the Appendices in Bengtsson *et al.*, 2005). Several recent studies show positive effects of organic farming on arthropod diversity or density (e.g. Kremen *et al.*, 2002; Hutton and Giller, 2003; Schmidt *et al.*, 2005), but there are also exceptions for one or both variables (e.g. Purtauf *et al.*, 2005). A common conclusion is that organic farming is less important for diversity than landscape factors. This highlights the problem of how to design agri-environmental schemes when the effects are likely to vary according to the heterogeneity of the landscape (Bengtsson *et al.*, 2005; Tscharntke *et al.*, 2005).

Most studies examining the response of insects to changes in farm management were conducted at small spatial scales, usually comparing fields with different types of management. Furthermore, virtually all studies examined responses of adult individuals. Since adults of many groups of insects are mobile, it is often impossible to determine with certainty whether the observed response is caused by an increase in population size or due to a foraging response resulting in a concentration of individuals in resource-rich patches in the landscape. An enhanced reproduction rate of insects on scheme fields relative to conventionally managed fields would be a clear indication that schemes have positive effects on population size. Only Gardener *et al.* (2001) examined effects of agri-environment schemes on juvenile insects. They found no significant differences in the densities of carabid beetle larvae on fields with and without agri-environment schemes.

Considering the importance of landscape context for insect species richness (Steffan-Dewenter *et al.*, 2002; Duelli and Obrist, 2003; Kleijn and van Langevelde, 2006), surprisingly little is known on whether the effects of agri-environment schemes depend on the structure of the surrounding landscape. Peter and Walter (2001) observed a positive effect of Swiss agri-environment schemes on the species richness of grasshoppers that was partially explained by distance from nature reserves. In the Netherlands, landscape structure did affect species richness of both bees and hoverflies but the effects of schemes did not depend on landscape context (Kleijn *et al.*, 2004). Summarizing the scant literature on the topic, Tscharntke *et al.* (2005) hypothesized that conservation measures on farmland are most effective in landscapes with intermediate complexity (Fig. 16.3). Very simple landscapes may be devoid of potential colonizers of scheme fields whereas very complex landscapes support overall high levels of species richness resulting in a continuous colonization from the surrounding landscape of even the most intensively farmed fields.

Despite the generally positive effects of agri-environment schemes on insects in general, they rarely promote endangered insect species. This is largely because contemporary farmland, with or without schemes, rarely

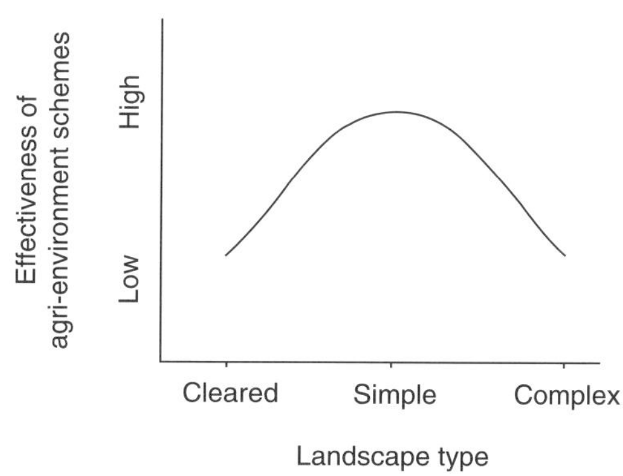

Fig. 16.3. Effectiveness of agri-environment schemes in relation to landscape type. Effectiveness is measured as biodiversity enhancement due to management, such as the conversion from conventional to organic farming (Roschewitz *et al.*, 2005) or the creation of crop-field boundaries (Thies and Tscharntke, 1999; Tscharntke *et al.*, 2002), compared to unmanaged control sites. Landscape type is classified as cleared (minimum diversity, <1% non-crop habitat), simple (low diversity, 1–20% non-crop habitat) and complex (high diversity, >20% non-crop habitat; see Andrén, 1994; Tscharntke *et al.*, 2002). The resulting hump-shaped relationship results from the different source pools in the surrounding landscape for recolonization of managed habitat. In cleared landscapes, the very few species are not a sufficient basis to result in a recognizable response to management changes. Similarly, in complex landscapes, management does not result in a significant effect, because biodiversity is high everywhere. In contrast, simple landscapes support intermediate species pools that allow a significant response to management. (Illustration with permission from Tscharntke *et al.*, 2005.)

hosts *Red Data Book* species (Kleijn *et al.*, 2006). Originally, a wide range of endangered insect species occurred in various types of agricultural habitats such as low productivity grasslands and arable fields or extensively managed vineyards and orchards. Due to agricultural intensification, this type of habitat has disappeared from north-western European farmland long before the introduction of the first agri-environment schemes, and these habitats are now almost exclusively restricted to nature reserves. Consequently, most endangered insect species have already disappeared from agricultural areas and are now restricted to nature reserves. *Red Data Book* species may therefore only be able to benefit from agri-environment schemes if schemes are implemented within colonizing distance of the refuges of these species (see, e.g. Peter and Walter, 2001). In any case, agri-environment schemes may prevent more common species from becoming endangered in the near future.

In conclusion, most contemporary agri-environment schemes and particularly those implemented in landscapes with intermediate complexity, promote richness of common insect species, thereby enhancing general biodiversity. This is a promising result since a more abundant and species-rich insect community may provide better ecosystem services such as pollination or pest control (Steffan-Dewenter and Tscharntke, 1999, but see Wilby and Thomas, 2002). On the other hand, the majority of agri-environment schemes fail to conserve the species that are most threatened by modern farming practices, usually because they are implemented in the wrong locations. It may therefore be prudent to differentiate the objectives of agri-environment schemes aimed at biodiversity conservation from schemes aimed to increase functional biodiversity and schemes aimed at the conservation of rare species. Schemes aimed at increasing functional biodiversity may then be implemented throughout the countryside and with little regard for the initial quality of the surrounding area. Schemes aimed at conserving endangered species should only be implemented in or near areas still hosting the target species.

8 Conclusions and Implications for Conservation in Agricultural Landscapes

Planning for future sustainable landscapes requires that several aspects of today's approach to conservation are modified, to incorporate the recent advances in spatial ecology and ecosystem dynamics. Although individual farmers always play a key role in conservation, incentive structures need to be targeted to ensure that diversity is maintained or improved at the larger landscape (regional) level. Some basic propositions for the new landscape approach to rural planning are:

1. Fields and small management units should not be used as a basic unit for conservation, but at least whole farms and preferably whole landscapes. A problem is that different organisms will respond to landscape structure at different spatial scales (Tscharntke *et al.*, 2005), but whole landscapes can be 'managed' taking this into account by imposing a variety of disturbances and management regimes, rather than single ones. Encouraging farmers to diversify and vary their land use is one way through which this may be accomplished.

2. Natural, semi-natural and semi-permanent managed areas allow many species to persist in the agricultural landscape, and should thus be the focus of most conservation efforts (Swift *et al.*, 2004). The conservation of semi-natural and natural areas will most likely enhance ecosystem services in intensely managed fields (Thies and Tscharntke, 1999; Östman *et al.*, 2001; Klein *et al.*, 2003; Ricketts *et al.*, 2004; Bianchi *et al.*, 2006).

3. Many current farming methods, especially organic or integrated farming, rely on species in less intensively managed ecosystems for biological control of pests, pollination of many crops, decomposition, etc. Recent discussions on the limits of oil reserves and increased oil prices imply that costs for agricultural inputs will increase. This means that in a future scenario where energy costs have increased dramatically, food production and security will rely more on natural ecosystem services. Hence, as an insurance against such a scenario, agricultural policies should strive to maintain and restore biodiversity in agricultural landscapes.

Agricultural systems may therefore harbour a significant diversity of insect species that can offer many services to landowners. In order for such ecosystem services to be maximized and sustained in the future, action must be taken by landowners and policy makers. This action needs to consider the effects of management at different spatial and temporal scales, and the identity (and potential utility) of the species concerned.

Acknowledgements

We thank Owen Lewis, Alan Stewart and an anonymous reviewer for insightful comments on the manuscript. Financial support for Teja Tscharntke and Jason M. Tylianakis came from the German Research Foundation (Deutsche Forschungsgemeinschaft, DFG), the German Ministry for Research and Education (Bundesministerium für Bildung und Forschung, BMBF; BIOTEAM and BIOLOG program) and the EU project EASY.

References

Andrén, H. (1994) Effects of habitat fragmentation on birds and mammals in landscapes with different proportions of suitable habitat: a review. *Oikos* 71, 355–366.

Armbrecht, I. and Perfecto, I. (2003) Litter ant's diversity and predation potential in two Mexican coffee matrices and forest fragments. *Agriculture, Ecosystems and Environment* 97, 107–115.

Asteraki, E.J., Hart, B.J., Ings, T.C. and Manley, W.J. (2004) Factors influencing the plant and invertebrate diversity of arable field margins. *Agriculture, Ecosystems and Environment* 102, 219–231.

Bambaradeniya, C.N.B., Edirisinghe, J.P., De Silva, D.N., Gunatilleke, C.V.S., Ranawana, K.B. and Wijekoon, S. (2004) Biodiversity associated with an irrigated rice agro-ecosystem in Sri Lanka. *Biodiversity and Conservation* 13, 1715–1753.

Banks, J.E. (2004) Divided culture: integrating conservation biology and agriculture. *Frontiers in Ecology and the Environment* 2, 537–545.

Bengtsson, J., Angelstam, P., Elmquvist, T., Emanuelsson, U., Forbes, C., Ihse, M., Moberg, F. and Nyström, M. (2003) Reserves, resilience and dynamic landscapes. *Ambio* 32, 389–396.

Bengtsson, J., Ahnström., J. and Weibull, A.C. (2005) The effects of organic agriculture on biodiversity and abundance: a meta-analysis. *Journal of Applied Ecology* 42, 261–269.

Berndt, L.A. and Wratten, S.D. (2005) Effects of alyssum flowers on the longevity, fecundity and sex ratio of the leafroller parasitoid *Dolichogenidea tasmanica*. *Biological Control* 32, 65–69.

Berndt, L.A., Wratten, S.D. and Hassan, P.G. (2002) Effects of buckwheat flowers on leafroller (Lepidoptera: Tortricidae) parasitoids in a New Zealand vineyard. *Agricultural and Forest Entomology* 4, 39–45.

Bianchi, F.J.J.A., Booij, C.J.H. and T. Tscharntke (2006) Sustainable pest regulation in agricultural landscapes: a review on landscape composition, biodiversity and natural pest control. *Proceedings of the Royal Society of London Series B – Biological Sciences* 273, 1715–1727.

Borges, P.A.V. and Brown, V.K. (2004) Arthropod community structure in pastures of an island archipelago (Azores): looking for local-regional species richness patterns at fine-scales. *Bulletin of Entomological Research* 94, 111–121.

Bos, M.M., Steffan-Dewenter, I., Tscharntke, T. (in press) The contribution of cacao agroforests to the conservation of lower canopy ant and beetle diversity in Indonesia. *Biodiversity and Conservation*.

Bosch, J. and Kemp, W. (2001) *How to Manage the Blue Orchard Bee as an Orchard Pollinator*. Sustainable Agriculture Network, Handbook series, Book 5. Beltsville, Maryland, pp. 20705–22351.

Bowie, M.H. and Frampton, C.M. (2004) A practical technique for non-destructive monitoring of soil surface invertebrates for ecological restoration programmes. *Ecological Management and Restoration* 5, 34–42.

Brereton,T., Stewart, K.S. and Warren, M.S. (2002) Developing a system for assessing the effect of agri-environment schemes on butterfly populations. Report to Department for Environment, Food and Rural Affairs, UK.

Cameron, P.J., Powell, W. and Loxdale, H.D. (1984) Reservoirs for *Aphidius ervi* Haliday (Hymenoptera: Aphidiidae): a polyphagous parasitoid of cereal aphids (Hemiptera: Aphididae). *Bulletin of Entomological Research* 74, 647–656.

Cane, J.H. (2001) Habitat fragmentation and native bees: a premature verdict? *Conservation Ecology* 5, 3. Available at: http://www.consecol.org/vol5/iss1/art3

Chase, J. and Leibold, M. (2003) Ecological niches: linking classical and contemporary approaches. University of Chicago Press, Chicago, Illinois.

Chust, G., Pretus, J.L., Ducrot, D. and Ventura, D. (2004) Scale dependency of insect assemblages in response to landscape pattern. *Landscape Ecology* 19, 41–57.

Clausnitzer, V. (2003) Dragonfly communities in coastal habitats of Kenya: indication of biotype quality and the need of conservation measures. *Biodiversity and Conservation* 12, 333–356.

Collins, J.K., Cuperus, G.W., Cartwright, B., Stark, J.A. and Ebro, L.L. (1992) Consumer attitudes on pesticide treatment histories of fresh produce. *Journal of Sustainable Agriculture* 3, 81–98.

Costanza, R., d'Arge, R., de Groot, R., Farber, S., Grasso, M., Hannon, B., Limburg, K., Naeem, S., O'Neill, R.V., Paruelo, J., Raskin, R.G., Sutton, P. and van den Belt, M. (1997) The value of the world's ecosystem services and natural capital. *Nature* 387, 253–260.

Cunningham, S.A., Floyd, R.B. and Weir, T.A. (2005) Do eucalyptus plantations host an insect community similar to remnant eucalyptus forest? *Austral Ecology* 30, 103–117.

Daily, G.C. (1997) *Nature's Services: Social Dependence on Natural Ecosystems*. Island Press, Washington, DC.

Dauber, J., Hirsch, M., Simmering, D., Waldhardt, R., Otte, A. and Wolters, V. (2003) Landscape structure as an indicator of biodiversity: matrix effects on species richness. *Agriculture Ecosystems and Environment* 98, 321–329.

DeVries, P.J. and Walla, T.R. (2001) Species diversity and community structure in neotropical fruit-feeding butterflies. *Biological Journal of the Linnaean Society* 74, 1–15.

DeVries, P.J., Murray, D. and Lande, R. (1997) Species diversity in vertical, horizontal, and temporal dimensions of a fruit-feeding butterfly community in an Ecuadorian rainforest. *Biological Journal of the Linnaean Society* 62, 343–364.

Di Giulio, M., Edwards, P.J. and Meister, E. (2001) Enhancing insect diversity in agricultural grasslands: the roles of management and landscape structure. *Journal of Applied Ecology* 38, 310–319.

Duelli, P. and Obrist, M.K. (2003) Regional biodiversity in an agricultural landscape: the contribution of seminatural habitat islands. *Basic and Applied Ecology* 4, 129–138.

Dyer, L.E. and Landis, D.A. (1997) Influence of noncrop habitats on the distribution of *Eriborus tenebrans* (Hymenoptera: Ichneumonidae) in cornfields. *Environmental Entomology* 26, 924–932.

EEA (2004) *High Nature Value Farmland: Characteristics, Trends and Policy Challenges*. European Environment Agency, Copenhagen.

Elmqvist, T., Folke, C., Nyström, M., Peterson, G., Bengtsson, J., Walker, B. and Norberg, J. (2003) Response diversity, ecosystem change, and resilience. *Frontiers in Ecology and the Environment* 1, 488–494.

Fry, G. (1995) Landscape ecology of insect movement in arable ecosystems. In: Glen, D.M., Greaves, M.P. and Anderson, H.M. (eds) *Ecology and Integrated Farming Systems* Wiley, Chichester, UK, pp. 177–202.

Gardener, S.M., Allen, D.S., Gundrey, A.L., Luff, M.L. and Mole, A.C. (2001) *Ground Beetles*. Technical appendix to ecological evaluation of the arable stewardship pilot scheme 1998–2000. ADAS, Wolverhampton, UK.

Gaston, K.J. and Spicer, J.I. (2004) *Biodiversity: An Introduction*. Blackwell Publishing, Oxford.

Gathmann, A., Greiler, H.-J. and Tscharntke, T. (1994) Trap-nesting bees and wasps colonizing set-aside fields: Succession and body size, management by cutting and sowing. *Oecologia* 98, 8–14.

Gonzales, A. and Chaneton, E.J. (2002) Heterotroph species extinction, abundance and biomass dynamics in an experimentally fragmented ecosystem. *Journal of Animal Ecology* 71, 594–602.

Gunderson, L. (2000) Ecological resilience: in theory and application. *Annual Review of Ecology and Systematics* 31, 425–439.

Gurr, G.M., Wratten, S.D. and Luna, J.M. (2003) Multi-function agricultural biodiversity: pest management and other benefits. *Basic and Applied Ecology* 4, 107–116.

Gurr, G.M., Wratten, S.D. and Altieri, M.A. (eds) (2004) *Ecological Engineering for Pest Management: Advances in Habitat Manipulation for Arthropods*. CSIRO Publishing, Melbourne, Australia.

Hassall, M., Hawthorne, A., Maudsley, M., White, P. and Cardwell, C. (1992) Effects of headland management on invertebrate communities in cereal fields. *Agriculture, Ecosystems and Environment* 40, 155–178.

Hector, A., Schmid, B., Beierkuhnlein, C., Caldeira, M.C., Diemer, M., Dimitrakopoulos, P.G., Finn, J.A., Freitas, H., Giller, P.S., Good, J., Harris, R., Hogberg, P., Huss-Danell, K., Joshi, J., Jumpponen, A., Korner, C., Leadley, P.W., Loreau, M., Minns, A., Mulder, C.P.H., O'Donovan, G., Otway, S.J., Pereira, J.S., Prinz, A., Read, D.J., Scherer-Lorenzen, M., Schulze, E.D., Siamantziouras, A.S.D., Spehn, E.M., Terry, A.C., Troumbis, A.Y., Woodward, F.I., Yachi, S. and Lawton, J.H. (1999) Plant diversity and productivity experiments in European grasslands. *Science* 286, 1123–1127.

Holland, J. and Fahrig, L. (2000) Effect of woody borders on insect density and diversity in crop fields: a landscape-scale analysis.

Agriculture, Ecosystems and Environment 78, 115–122.

Holling, C.S., Schindler, D.W., Walker, B.W. and Roughgarden, J. (1995) Biodiversity in the functioning of ecosystems: an ecological synthesis. In: Perrings, C.A., Mäler, K.G., Folke, C., Holling, C.S. and Jansson, B.O. (eds) *Biodiversity Loss: Economic and Ecological Issues*. Cambridge University Press, Cambridge, pp. 44–83.

Holt, A.R. and Gaston, K.J. (2002) Occupancy-abundance relationships and spatial distribution: a review. *Basic and Applied Ecology* 3, 1–13.

Hooper, D.U., Chapin, F.S., Ewel, J.J., Hector, A., Inchausti, P., Lavorel, S., Lawton, J.H., Lodge, D.M., Loreau, M., Naeem, S., Schmid, B., Setala, H., Symstad, A.J., Vandermeer, J. and Wardle, D.A. (2005) Effects of biodiversity on ecosystem functioning: a consensus of current knowledge. *Ecological Monographs* 75, 3–35.

Hutton, S.A. and Giller, P.S. (2003) The effects of the intensification of agriculture on northern temperate dung beetle communities. *Journal of Applied Ecology* 40, 994–1007.

Jonsen, I.D. and Fahrig, L. (1997) Response of generalist and specialist insect herbivores to landscape spatial structure. *Landscape Ecology* 12, 185–197.

Keesing, V. and Wratten, S.D. (1997) Integrating plant and insect conservation, In: Maxted, N., Ford-Lloyd, B.V. and Hawkins, J.G. (eds) *Plant Genetic Conservation*.Chapman & Hall, London, pp. 220–235.

Klein, A.-M., Steffan-Dewenter, I., Buchori, D. and Tscharntke, T. (2002) Effects of land use intensity in tropical agroforestry systems on coffee flower-visiting and trap-nesting bees and wasps. *Conservation Biology* 16, 1003–1014.

Klein, A.-M., Steffan-Dewenter, I. and Tscharntke, T. (2003a) Fruit set of highland coffee depends on the diversity of pollinating bees. *Proceedings of the Royal Society of London Series B – Biological Series* 270, 955–961.

Klein, A.-M., Steffan-Dewenter, I. and Tscharntke, T. (2003b) Flower visitation and fruit set of *Coffea canephora* in relation to local and regional agroforestry management. *Journal of Applied Ecology* 40, 837–845.

Kleijn, D., Berendse, F., Smit, R. and Gilissen. N. (2001) Agri-environment schemes do not effectively protect biodiversity in Dutch agricultural landscapes. *Nature* 413, 723–725.

Kleijn, D., Berendse, F., Smit, R., Gilissen, N., Smit, J., Brak, B. and Groeneveld, R. (2004) The ecological effectiveness of agri-environment schemes in different agricultural landscapes in the Netherlands. *Conservation Biology* 18, 775–786.

Kleijn, D., Baquero, R.A., Clough, Y., Díaz, M., De Esteban, J., Fernández, F., Gabriel, D., Herzog, F., Holzschuh, A., Jöhl, R., Knop, E., Kruess, A., Marshall, E.J.P., Steffan-Dewenter, I., Tscharntke, T., Verhulst, J., West, T.M. and Yela, J.L. (2006) Mixed biodiversity benefits of agri-environment schemes in five European countries. *Ecology Letters* 9, 243–254.

Kleijn, D. and van Langevelde, F. (2006) Interacting effects of landscape context and habitat quality on flower-visiting insects in agricultural landscapes. *Basic and Applied Ecology* 7, 201–214.

Kleijn, D. and Sutherland, W.J. (2003) How effective are European agri-environment schemes in conserving and promoting biodiversity? *Journal of Applied Ecology* 40, 947–969.

Krauss, J., Steffan-Dewenter, I. and Tscharntke, T. (2003) How does landscape context contribute to effects of habitat fragmentation on diversity and population density of butterflies? *Journal of Biogeography* 30, 889–900.

Kremen, C. (2005) Managing ecosystem services: what do we need to know about their ecology. *Ecology Letters* 8, 468–479.

Kremen, C., Williams, N.M. and Thorp, R.W. (2002) Crop pollination from native bees at risk from agricultural intensification. *Proceedings of the National Academy of Sciences of the United States of America* 99, 16812–16816.

Kruess, A. (2003) Effects of landscape structure and habitat type on a plant-herbivore-parasitoid community. *Ecography* 26, 283–290.

Kruess, A. and Tscharntke, T. (2002) Contrasting responses of plant and insect diversity to variation in grazing intensity. *Biological Conservation* 106, 293–302.

Landis, D.A., Wratten, S.D. and Gurr, G.M. (2000) Habitat management to conserve natural enemies of arthropod pests in agriculture. *Annual Review of Entomology* 45,175–202.

Lawton, J.H., Bignell D.E., Bolton, B., Bloemers, G.F., Eggleton, P., Hammond, P.M., Hodda, M., Holt, R.D., Larsen, T.B., Mawdsley, N.A., Stork, N.E., Srivastava, D.S. and Watt, A.D. (1998) Biodiversity inventories, indicator taxa and effects of habitat modification in tropical forest. *Nature* 391, 72–76.

Lee, J.C., Heimpel, G.E. (2005) Impact of flowering buckwheat on Lepidopteran cabbage pests and their parasitoids at two spatial scales. *Biological Control* 34, 290–301.

Leibold, M.S. Holyoak, M., Mouquet, N., Amarasekare, P., Chase, J.M., Hoopes, M.F., Holt, R.D., Shurin, J.B., Law, R., Tilman, D., Loreau, M. and Gonzalez, A. (2004) The metacommunity concept: a framework for multi-scale community ecology. *Ecology Letters* 7, 601–613.

Loreau, M., Mouquet, N. and Gonzales, A. (2003) Biodiversity as spatial insurance in heterogeneous landscapes. *Proceedings of the National Academy of Sciences of the Unites States of America* 100, 12765–12770.

Maeto, K., Sato, S. and Miyata, H. (2002) Species diversity of longicorn beetles in humid warm-temperate forests: the impact of forest management practices on old-growth forest species in southwestern Japan. *Biodiversity and Conservation* 11, 1919–1937.

Marshall, E.J.P., Brown, V.K., Boatman, N.D., Lutman, P.J.W., Squire, G.R. and Ward, L.K. (2003) The role of weeds in supporting biological diversity within crop fields. *Weed Research* 43, 77–89.

Mas, A.H. and Dietsch, T.V. (2003) An index of management intensity for coffee agroecosystems to evaluate butterfly species richness. *Ecological Applications* 13, 1491–1501.

Morandin, L.A. and Winston, M.L. (2005) Wild bee abundance and seed production in conventional, organic, and genetically modified canola. *Ecological Applications* 15, 871–881.

Norberg, J., Swaney, D.P., Dushoff, J., Lin, J., Casafgrandi, R. and Levin, S.A. (2001) Phenotypic diversity and ecosystem functioning in changing environments: a theoretical framework. *Proceedings of the National Academy of Sciences of the United States of America* 98, 11376–11381.

Novacek, M.J. and Cleland, E.E. (2001) The current biodiversity extinction event: scenarios for mitigation and recovery. *Proceedings of the National Academy of Sciences of the United States of America* 98, 5466–5470.

Nyström, M. and Folke, C. (2001) Disturbance and the spatial resilience of coral reefs. *Ecosystems* 4, 406–417.

Olschewski, R., Tscharntke, T., Benítez, P.C., Schwarze, S. and Klein, A.-M. (2006) Economic evaluation of pollination services comparing coffee landscapes in Ecuador and Indonesia. *Ecology and Society* 11(1), 7. Available at: http://www.ecologyandsociety.org/vol11/iss1/art7/

Östman, O., Ekbom, B., Bengtsson, J. and Weibull, A.C. (2001) Landscape complexity and farming practice influence the condition of polyphagous carabid beetles. *Ecological Applications* 11, 480–488.

Östman, Ö., Ekbom, B. and Bengtsson, J. (2003) Yield increase attributable to aphid predation by ground-living polyphagous natural enemies in spring barley in Sweden. *Ecological Economics* 45, 149–158.

Paoletti, M.G. and Pimentel, D. (2000) Environmental risks of pesticides versus genetic engineering for agricultural pest control. *Journal of Agricultural and Environmental Ethics* 12, 279–303.

Peintinger, M., Bergamini, A. and Schmid, B. (2003) Species-area relationships and nestedness of four taxonomic groups in fragmented wetlands. *Basic and Applied Ecology* 4, 385–394.

Perfecto, I. and Armbrecht, I. (2003) The coffee agroecosystems in the Neotropics: combining ecological and economic goals. In: Vandermeer, J.H. (ed.) *Tropical Agroecosystems*. CRC Press, Boca Raton, Florida, pp. 159–194.

Perfecto, I., Rice, R., Greenberg, R. and van der Volt, M. (1996) Shade coffee as refuge of biodiversity. *Bioscience* 46, 598–608.

Peter, B. and Walter, T. (2001) Heuschrecken brauchen ökologische Ausgleichsflächen. *Agrarforschung* 8, 452–457.
Pimentel, D., Stachow, U., Takacs, A., Brubaker, W., Dumas, A.R., Meaney, J.S., O'Neil, J.A.S., Onsi, D.E. and Corzilius, D.B. (1992) Conserving biological diversity in agricultural/forestry systems. *Bioscience* 432, 354–362.
Potts, R.W. and Vickerman, G.P. (1974) Studies on the cereal ecosystem. *Advances in Ecological Research* 8,107–197.
Purtauf, T., Roschewitz, I., Dauber, J., Thies, C., Tscharntke, T. and Wolters, V. (2005) Landscape context of organic and conventional farms: influences on carabid beetle diversity. *Agriculture, Ecosystems and Environment* 108, 165–174.
Rand, T.A., Tylianakis, J.M. and Tscharntke, T. (2006) Spillover edge effects: the dispersal of agriculturally subsidized insect natural enemies into adjacent natural habitats. *Ecology Letters* 9, 603–614.
Ribas, C.R., Sobrinho, T.G., Schroereder, J.H., Sperber, C.F., Lopes-Andrade, C. and Soares, S.M. (2005) How large is large enough for insects? Forest fragmentation effects at three spatial scales. *Acta Oecologica* 27, 31–41.
Rice, R.A. and Greenberg, R. (2000) Cacao cultivation and the conservation of biological diversity. *Ambio* 29, 167–173.
Ricketts, T.H., Daily, G.C., Ehrlich, P.R. and Michener, C.D. (2004) Economic value of tropical forest to coffee production. *Proceedings of the National Academy of Sciences of the United States of America* 101, 12579–12582.
Robinson, R.A. and Sutherland, W.J. (2002) Post-war changes in arable farming and biodiversity in Great Britain. *Journal of Applied Ecology* 39,157–176.
Room, P.M. (1979) Parasites and predators of *Heliothis* spp. (Lepidoptera: Noctuidae) in cotton in the Namoi Valley, New South Wales. *Journal of the Australian Entomological Society* 18, 223–228.
Roschewitz, I., Gabriel, D., Tscharntke, T. and Thies, C. (2005) Contrasting effects of landscape complexity on arable weed diversity in organic and conventional farming. *Journal of Applied Ecology* 42, 873–882.
Schmidt, M.H., Roschewitz, I., Thies, C. and Tscharntke, T. (2005) Differential effects of landscape and management on diversity and density of ground-dwelling farmland spiders, *Journal of Applied Ecology* 42, 281–287.
Schulze, C.H., Waltert, M., Kessler, P.J.A., Pitopang, R., Shahabuddin, Veddeler, D., Mühlenberg, M., Gradstein, S.R., Leuschner, C., Steffan-Dewenter, I. and Tscharntke, T. (2004) Biodiversity indicator groups of tropical land-use systems: comparing plants, birds, and insects. *Ecological Applications* 14, 1321–1333.
Shahabuddin, Schulze, C.H. and Tscharntke, T. (2005) Changes of dung beetle communities from rainforests towards agroforestry systems and annual cultures in Sulawesi (Indonesia) *Biodiversity and Conservation* 14, 863–877.
Sinclair, A.R.E., Mduma, S.A.R. and Arcese, P. (2002) Protected areas as biodiversity benchmarks for human impact: agriculture and the Serengeti avifauna. *Proceedings of the Royal Society of London Series B – Biological Sciences* 269, 2401–2405.
Sperber, C.F., Nakayama, K., Valverde, M.J. and Neves, F.D. (2004) Tree species richness and density affect parasitoid diversity in cacao agroforestry. *Basic and Applied Ecology* 5, 241–251.
Stamps, W.T. and Linit, M.J. (1997) Plant diversity and arthropod communities: implications for temperate agroforestry. *Agroforestry Systems* 39, 73–89.
Steffan-Dewenter, I. (2003) Importance of habitat area and landscape context for species richness of bees and wasps in fragmented orchard meadows. *Conservation Biology* 17, 1036–1044.
Steffan-Dewenter, I. and Leschke, K. (2003) Effects of habitat management on vegetation and above-ground nesting bees and wasps of orchard meadows in Central Europe. *Biodiversity and Conservation* 12, 1953–1968.
Steffan-Dewenter, I. and Tscharntke, T. (1999) Effects of habitat isolation on pollinator communities and seed set. *Oecologia* 121, 432–440.
Steffan-Dewenter, I., Münzenberg, U., Bürger, C., Thies, C. and Tscharntke, T. (2002)

Scale-dependent effects of landscape context on three pollinator guilds. *Ecology* 83, 1421–1432.

Stoner, K.J.L. and Joern, A. (2004) Landscape vs. local habitat scale influences to insect communities from tallgrass prairie remnants. *Ecological Applications* 14, 1306–1320.

Summerville, K.S. and Crist, T.O. (2004) Constrasting effects of habitat quantity and quality on moth communities in fragmented landscapes. *Ecography* 27, 3–12.

Summerville, K.S. and Crist, T.O. (2005) Temporal patterns of species accumulation in a survey of Lepidoptera in a beech-maple forest. *Biodiversity and Conservation* 14: 3393–3406.

Summerville, K.S., Boulware, M.J., Veech, J.A. and Crist, T.O. (2003) Spatial variation in species diversity and composition of forest Lepidoptera in eastern deciduous forests of North America. *Conservation Biology* 17, 1045–1057.

Swift, M.J. and Anderson, J.M. (1993) Biodiversity and ecosystem function in agricultural systems. In: Schulze, E.D. and Mooney, H.A. (eds) *Biodiversity and Ecosystem Function*. Springer, Berlin, Germany, pp. 15–41.

Swift, M.J., Izac, A.M.N. and van Noordwijk, M. (2004) Biodiversity and ecosystem services in agricultural landscapes: are we asking the right questions? *Agriculture, Ecosystems and Environment* 104, 113–134.

Thies, C. and Tscharntke, T. (1999) Landscape structure and biological control in agroecosystems. *Science* 285, 893–95.

Thomas, M.B., Wratten, S.D. and Sotherton, N.W. (1991) Creation of island habitats in farmland to manipulate populations of beneficial arthropods: predator densities and emigration. *Journal of Applied Ecology* 28, 906–917.

Thomas, S.R., Goulson, D. and Holland, J.M. (2001) Resource provision for farmland gamebirds: the value of beetle banks. *Annals of Applied Biology* 139, 111–118.

Tilman, D., Fargione, J., Wolff, B., D'Antonio, C., Dobson, A., Howarth, R., Schindler, D., Schlesinger, W.H., Simberloff, D. and Swackhamer, D. (2001) Forecasting agriculturally driven global environmental change. *Science* 292, 281–284.

Tilman, D., Cassman, K.G., Matson, P.A., Naylor, R. and Polasky, S. (2002) Agricultural sustainability and intensive production practices. *Nature* 418, 671–677.

Tscharntke, T. and Brandl, R. (2004) Plant-insect interactions in fragmented landscapes. *Annual Review of Entomology* 49, 405–430.

Tscharntke, T., Gathmann, A., Steffan-Dewenter, I. (1998) Bioindication using trap-nesting bees and wasps and their natural enemies: community structure and interactions. *Journal of Applied Ecology* 35, 708–719.

Tscharntke, T., Steffan-Dewenter, I., Kruess, A. and Thies, C. (2002) Contribution of small habitat fragments to conservation of insect communities of grassland-cropland landscapes. *Ecological Applications* 12, 354–363.

Tscharntke, T., Klein, A.-M., Kruess, A., Steffan-Dewenter, I. and Thies, C. (2005) Landscape perspectives on agricultural intensification and biodiversity: ecosystem service management. *Ecology Letters* 8, 857–874.

Tylianakis, J., Didham, R. and Wratten, S.D. (2004) Improved fitness of aphid parasitoids receiving resource subsidies. *Ecology* 85, 658–666.

Tylianakis, J.M., Klein, A.-M. and Tscharntke, T. (2005) Spatiotemporal variation in the effects of a tropical habitat gradient on Hymenoptera diversity. *Ecology* 86, 3296–3302.

Tylianakis, J.M., Klein, A.-M. and Tscharntke, T. (2006) Spatial scale of observation affects α, β and γ diversity of cavity-nesting bees and wasps across a tropical land use gradient. *Journal of Biogeography* 33, 1295–1304.

Veddeler, D., Klein, A.-M. and Tscharntke, T. (2005) Contrasting responses of bee communities to coffee flowering at different spatial scales. *Oikos* 112, 594–601.

Westphal, C., Steffan-Dewenter, I. and Tscharntke, T. (2003) Mass flowering crops enhance pollinator densities at a landscape scale. *Ecology Letters* 6, 961–965.

Whitcomb, W.H. and Bell, K. (1964) *Predaceous Insects, Spiders, and Mites of Arkansas Cotton Fields*. Agricultural Experiment Station, University of Arkansas. Division of

Agriculture, Fayetteville, Arkansas, Bulletin 690.

Wickramasinghe, L.P., Harris, S., Jones, G. and Jennings, N.V. (2004) Abundance and species richness of nocturnal insects on organic and conventional farms: effects of agricultural intensification on bat foraging. *Conservation Biology* 18, 1283–1292.

Wilby, A. and Thomas, M.B. (2002) Natural enemy diversity and pest control: patterns of pest emergence with agricultural intensification. *Ecology Letters* 5, 353–360.

Wissinger, S.A. (1997) Cyclic colonization in predictably ephemeral habitats: a template for biological control in annual crop systems. *Biological Control* 10, 4–15.

Wolda, H. (1978) Seasonal fluctuations in rainfall, food and abundance in tropical insects. *Journal of Animal Ecology* 47, 369–381.

Wolda, H. (1988) Insect seasonality: why? *Annual Review of Ecology and Systematics* 19, 1–18.

Yachi, S. and Loreau, M. (1999) Biodiversity and ecosystem productivity in a fluctuating environment: the insurance hypothesis. *Proceedings of the National Academy of Sciences of the United States of America* 96, 1463–1468.

17 Genetically Modified Crops and Insect Conservation

Ian P. Woiwod and Tanja H. Schuler

Plant and Invertebrate Ecology Division, Rothamsted Research, Harpenden, Hertfordshire AL5 2JQ, UK

1 Introduction

Although genetically modified (GM) or biotech crops are now widely grown throughout the world they have attracted a large amount of opposition, particularly in Europe. The reasons for this opposition are numerous and require a sociological analysis beyond the scope of this chapter, but a useful ecological perspective has been provided by Gray (2004). On the face of it, such crops could be welcomed and supported by conservationists and environmentalists alike, albeit with some caveats. The insect- and herbicide-resistant crops, of particular concern in this chapter, hold out the prospect of more focused pest control and reduced pesticide applications with a resulting benefit to many non-target species as well as being generally more environmentally benign in many respects. Although some of these benefits are already becoming apparent from crops already in commercial cultivation (e.g. Brookes and Barfoot, 2005), misgivings about GM crops still remain strong in Europe and elsewhere, and environmental risk assessment and management has become a big issue.

Initially, it may not be apparent that there is a direct relationship between biotech crops and insect conservation outside the generic and wider issue of agricultural intensification and insecticide use. Many of the rare and local species of most immediate concern to insect conservationists have very specific habitat requirements, often in natural or semi-natural areas that are unlikely to be more affected by the introduction of biotech crops than by agricultural intensification more generally, at least in the short term. However, if we are considering insect conservation in its wider context, where the object is to maximize non-pest insect abundance and diversity in the wider countryside, and optimize natural biological control in integrated pest management (IPM) systems, GM crops need careful consideration.

Over the past few years it has been quite surprising how often non-target insects have hit the headlines in relation to possible risks of GM crops. Perhaps the major example in this context is that of the monarch butterfly, *Danaus plexippus* L., which received massive publicity following the publication of a paper in *Nature* suggesting that the larvae were under threat from the cultivation of insect-resistant transgenic maize (Losey *et al.*, 1999). The resulting furore stimulated further research, which, when published, aroused much less publicity, perhaps because it failed to support the anti-GM cause. However, the monarch provides an interesting case study and will be discussed in more detail later.

In this chapter we propose to outline the current status of GM crops throughout the world. We will then take a critical and dispassionate look at some of the risk analyses and related studies that have been carried out and assess their relevance to the conservation of insect biodiversity. Finally we wish to consider some of the wider issues and look at possible future risks and benefits that might relate to GM crop development and cultivation.

2 Current Status of GM Crops Throughout the World

As 2005 was the tenth year of the commercial cultivation of GM crops it is perhaps a good time to take stock, and we are in fact beginning to get a very good idea of the worldwide rates of uptake of the technology. Fortunately widely accepted statistics are provided annually by the International Service for the Acquisition of Agri-biotech Applications (ISAAA) (e.g. James, 2005). These show that there continues to be a very high annual increase in the uptake of these crops, despite reluctance to accept their cultivation in certain regions. For example, there was a fairly typical 11% growth rate in the uptake of biotech crops worldwide in 2005, and it is estimated that there are now 8.5 million farmers in 21 countries growing 90 million hectares of such crops (James, 2005). Also it can no longer be claimed that crops are mainly being grown in the industrialized countries of North America, as the uptake is accelerating in developing countries such as India, China and Brazil – currently 23% increase per year compared to 5% increase per year in industrial countries such as USA, Canada and Australia (James, 2005). Even in Europe there are now modest areas of insect-resistant GM maize being grown in five European Union (EU) countries: Spain, Portugal, Germany, France and the Czech Republic in decreasing area of crop planted. Whatever else these figures show, they do suggest that many farmers are finding these crops a very attractive proposition, either because they increase gross margins (i.e. profit) or they just make life easier.

Despite many press reports of novel applications of the technology, there are still only two generic types of crop in widespread cultivation. These are the herbicide-tolerant (GMHT) crops, which resist broad-spectrum herbicides such as glyphosate and glufosinate-ammonium, and the insect-resistant (GMBt) crops, which express insecticidal toxins derived from the bacterium *Bacillus thuringiensis*. In some crops these two traits have been combined into a single cultivar. There are currently four main transgenic crops in cultivation, GMHT soybean, GMHT oilseed rape, GMHT and GMBt maize, and GMHT and GMBt cotton, with GMBt rice planted for the first time in 2005 (James, 2005).

The biotech versions of some of these crops now account for a very high proportion of planting; for example, 60% of all soybean cultivation in the world is now transgenic and so is 28% of the cotton crop (James, 2005). The days are therefore long gone when it would have been practical to consider returning to a GM-free world. Indeed a large number of people are now wearing GM-derived products if not actually eating them. Therefore, a more rational approach is needed rather than just condemnation if we are to ensure that appropriate regulatory processes are in place to minimize possible adverse effects and maximize environmental benefits. There is now a very large literature on all aspects of risk assessment and regulation of GM crops worldwide, so in this chapter we will be mainly concerned with environmental risk assessments of particular relevance to insect conservation in its wider context.

3 Herbicide-tolerant Crops

Crops with a genetically engineered trait for herbicide tolerance dominate the market and currently make up over 70% of biotech crops grown worldwide (James, 2005). Naturally occurring herbicide tolerance has long been used in modern agriculture; for example, chemicals which control only broadleaved weeds can often be used in cereal crops after the crop has emerged, and similarly chemicals which control grass species can sometimes be used in broadleaved crops after crop emergence. The problem arises when a farmer needs to control grasses in a growing cereal crop or broadleaved weeds in a growing broadleaved crop. This is where GMHT crops come in because they enable a broad-spectrum herbicide, such as glyphosate, to be used post emergence. This may make weed control simpler and often cheaper, particularly if insurance sprays can be eliminated and control applied only if and when a weed problem arises. There can also be environmental benefits if using a GMHT crop enables the farmer to use more benign herbicides with fewer applications, which can also mean reduced fossil fuel use and hence a smaller environmental footprint (e.g. Bennett *et al.*, 2004). A more benign herbicide, such as glyphosate, in this context means one that is less toxic, less mobile and less persistent than those commonly used before crop emergence (e.g. Dewar *et al.*, 2003). The wider implications of these issues will be discussed later in relation to farm-scale evaluations (FSEs).

So what is the relevance of GMHT crops to insect conservation concerns? There is very little chance of a direct insecticidal effect of the genetic modification itself and therefore any effects are likely to be indirect through changes in herbicide regimes and related cultivation practices. In the past this would have elicited very little concern within agriculture because weed control is something that all farmers strive for, including those engaged in organic farming. However, attitudes are changing, at least in Europe where agriculture is often the dominant land use, and loss of farmland biodiversity has become a big issue. This is notably the case in the UK because populations of a suite of formerly common farmland bird species such as the skylark, *Alauda arvensis*, have decreased greatly over the last 30 years (Fuller *et al.*, 1995). This has caused widespread public concern and as a result there are now Biodiversity Action Plans and government targets to halt and reverse these declines. The reasons for these farmland bird declines are species-specific but often seem to be related to reductions in their food supply of broadleaved weed seeds and insects at particular times of the year.

When the first GMHT crops were about to be released commercially into the UK, concerns began to be raised that widespread use of such crops might lead to even cleaner fields with even less bird food and hence hasten the decline of such farmland wildlife. Out of this concern came the FSEs, which were pioneering studies in environmental risk assessment and which are certainly relevant to insect conservation management, although perhaps not always in the ways commonly reported, as will be discussed later.

4 Insect-resistant Crops

One of the main aims of the development of insect-resistant GM plants was to provide a more targeted, and hence ecologically sustainable, means of pest control than is currently offered by insecticides. Although the use of insect-resistant crop plants obtained through conventional breeding has a long history, plant resistance obtained this way is often only partial and highly effective insect resistance genes have not been found for many crop species. Genetic engineering has therefore speeded up the breeding process and increased the pool of resistance genes available to breeders.

As with GMHT crops, the worldwide area planted with insect-resistant GM plants has grown dramatically since they were first introduced in the mid-1990s (Shelton *et al.*, 2002). In 2005 insect-resistant GM maize and cotton were grown on 26.3 million hectares worldwide (James, 2005). The only insect-resistant GM crop that has approval for cultivation in Europe is Bt maize and the main country that has taken up cultivation of GM maize to a significant extent in the EU is Spain, although four other countries in Europe are now growing small areas of the crop (James, 2005).

All insect-resistant GM plants commercialized to date express proteinaceous toxins derived from the common soil bacterium *B. thuringiensis*, widely referred to as Bt. Delta-endotoxins, also known as crystal or Cry proteins, are the most commonly used Bt toxins, although this bacterium also produces other toxins, such as vegetative insecticidal proteins (VIPs). These Bt toxins are not active on contact but need to be ingested by insects. After activation by serine gut proteases, Cry proteins bind to highly specific gut receptors in the insect's midgut and create pores in the gut wall leading to septicaemia and death in susceptible insects.

Each Bt toxin is active against a specific taxonomic group of insects. These toxins are therefore more specific in their activity than most synthetic, commercially available insecticides and many plant toxins (Czapla, 1997; Peferoen, 1997). Their efficacy against target pests, however, can rival or exceed that of synthetic pesticides and many conventionally bred resistant crop varieties.

The main Cry Bt toxins used so far for commercialized Bt plants are Cry1Ab, Cry1Ac, Cry3A, Cry3b, Cry1F and Cry9C. These toxins are only active against lepidopteran larvae with the exception of Cry3A and Cry3b, which are only active against some coleopteran species (particularly the Colorado potato beetle, *Leptinotarsa decemlineata* (Say), and corn root worms, *Diabrotica* spp.).

Some of these Bt toxins have been used in microbial spray formulations since the 1950s. Such Bt sprays (most of which contain a mixture of spores plus several delta proteins) are generally considered benign to beneficial insects, although exceptions have been reported (Glare and O'Callaghan, 2000). Bt sprays have therefore been favoured as a component of IPM programmes and, because they are regarded as a natural product, are also acceptable in organic farming.

In addition to Bt maize, Bt cotton, Bt potato and Bt rice, many other crops transformed to express Bt toxin genes are in an experimental stage. Potential insect resistance genes from sources other than Bt have also been transferred into crop plants, such as genes coding for inhibitors of digestive enzymes (including amylase and proteinase inhibitors) and lectins. The efficacy of these GM plants has so far not been very high and none have been commercialized (e.g. Gatehouse *et al.*, 1997, 1999; Altpeter *et al.*, 1999; Lee *et al.*, 1999; Stöger *et al.*, 1999, De Leo *et al.*, 2001; Delledonne *et al.*, 2001; Gatehouse, 2002). Although many of these Bt alternatives are plant-derived, their activity spectrum tends to be wider than that of Bt toxins. Research in both the public and private sectors, in developed as well as in developing nations, is currently directed towards identifying further genes active against pests and new plant–transgene combinations are continuously being developed and tested.

5 Regulatory Environmental Risk Assessment of GM Crops

It might be thought from popular reports that GM crops are being grown without responsible risk assessment and regulation. This is far from true and many countries throughout the world now have, or are developing, strict regulatory procedures for such crops. Here, as an example, we outline the situation in the EU and more locally within the UK, as it gives a good insight into how thorough the regulatory process has now become.

In the UK and all other EU member states, cultivation as well as importation of GM crops is controlled by strict legislation (EU Directives 2001/18 and 1829/2003). The regulation requires an environmental risk assessment for each GM crop, which takes into account not only immediate and direct effects but also delayed and indirect environmental impacts of the specific cultivation, management and harvesting techniques used for a GM plant. The regulatory regime is based on the Precautionary Principle as applied on a case-by-case basis, so the environmental impact of each GM crop is assessed individually. The EU legislation adopts a step-by-step approach to the assessment of GM crops. This means that their development is carried out initially in containment and they are only allowed to be cultivated outdoors if the risk assessment has shown that they are sufficiently safe. The initial releases are small and a gradual increase in scale occurs as more is learnt about each GM crop and its behaviour in the environment (further information is available at: http://www.defra.gov.uk/environment/gm/).

All applications to cultivate GM crops in the UK are scrutinized by the Advisory Committee on Releases to the Environment (ACRE), an independent scientific committee whose members include leading academic scientists. ACRE's role is to advise ministers on the risks to the environment associated with the deliberate release of genetically modified organisms (GMOs). The conditions of releases always remain open to reassessment in the light of new information. ACRE publishes its advice to ministers and guidance documents on the Internet (available at: http://www.defra.gov.uk/environment/acre/index.htm).

The potential of gene flow (through seed, pollen or crop volunteers) from the GM crop to wild relatives or habitats is one of the issues assessed as part of the environmental risk assessment for each GM crop. Applicants wishing to commercially cultivate a GM crop plant in the EU have to provide information relating to the likelihood of the GM plant becoming more persistent than the parental plants in agricultural habitats or more invasive in natural habitats. They also have to provide information showing whether genetic transfer from the GM plant would give any selective advantage or disadvantage to other sexually compatible plant species.

The case-by-case approach is particularly relevant in respect to gene flow as the risk of it occurring differs greatly between crop species and the introduced trait as well as regional flora and environmental conditions. Two contrasting examples particularly relevant for the UK are GM maize and oilseed rape. In the UK, maize has no wild relatives with which it can hybridize; nor does it produce volunteer plants because of the winter climate. In the UK the risk of gene flow from any GM maize to wild relatives or habitats is therefore minimal. Oilseed rape, on the other hand, is a crop that in many regions can hybridize with close wild relatives (in the UK with turnip rape, *Brassica rapa*; Wilkinson *et al.*, 2003) and oilseed rape seeds can survive for years in the soil, commonly resulting in volunteer plants (Pekrun *et al.*, 2005). The risk of transgene escape into wild relatives and habitats is thus greater for GM oilseed rape than with GM maize. Intense research efforts have therefore been directed at gene flow from oilseed rape over the past decades, not only aimed at quantifying the risk that gene flow occurs but particularly its consequences and at developing measures for avoiding or minimizing it (e.g. Wilkinson *et al.*, 2003; Damgaard and Kjellsson, 2005). So far, GM oilseed rape applications for cultivation in the EU have been limited to GMHT oilseed rape. As the herbicide tolerance trait only provides a fitness advantage to plants in situations where a herbicide is applied, the risk of negative consequences for wild habitats is very small (although herbicides are sometimes used in a conservation context) and there is no current evidence that the risk would be higher with herbicide-tolerant GM plants than for herbicide-tolerant conventionally bred plants (e.g. Hails *et al.*, 1997).

Wider biodiversity issues also have to be addressed by applicants wishing to commercialize a GM crop in the EU. They include an assessment of the direct and indirect, immediate and delayed effects of the management on all affected habitats, Biodiversity Action Plan (BAP) animal and plant species within these habitats, and key plant and invertebrate components of food webs (ACRE, 2001).

In addition, GM legislation requires post-market monitoring after a GM crop has been given consent for cultivation in the EU (ACRE, 2004). The objectives of the post-market monitoring are to: (i) confirm that any assumption in the environmental risk assessment regarding the occurrence and impact of potential adverse effects of the GMO or its use in the environmental risk assessment are correct; and (ii) identify the occurrence of adverse effects of the GMO or its use on human health and the environment which were not anticipated in the environmental risk assessment.

5.1 Exposure likelihood

The likelihood of exposure of non-target insects to transgene products (i.e. Bt toxins, herbicide-tolerant proteins, etc.) is an important issue in any risk assessment and will depend on where a non-target insect feeds and what it feeds on. Insects inhabiting crops are the most likely to be exposed to transgene products, although their exposure risk will naturally be higher if they feed on the GM crop itself rather than on weeds or arthropods within the crop. However, even herbivores feeding on a GM crop may not be exposed as not all plant parts contain the transgene product (see below).

The risk of exposure to transgene products is already dramatically lower in crop margins and is further reduced for insects in semi-natural farmland habitats. Insects in natural habitats have the lowest risk of exposure as they would only be exposed if a transgenic crop plant were to invade, a transgene were to escape into wild relatives or if pollen were transported to wild habitats in significant quantities. The regulatory process in the EU will only allow a GM plant to be released if all of the above risks are minimal or if the consequences of a gene escape are not considered to be hazardous to the environment, including non-target insects. Nearly all applications for commercial cultivation of GM plants in the EU so far have been limited to arable crops and the probability of them invading natural habitats is very small indeed. Oilseed rape is somewhat unusual as it has invasive characteristics and can hybridize with wild relatives. Again, this reinforces the point that the risks will vary from crop to crop.

A number of potential exposure routes have to be considered when assessing the risks of the transgene product, such as a Bt toxin, to non-target insects (Schuler *et al.*, 1999a). The main exposure route is by feeding on a GM plant. The transgene product is targeted at one or more herbivorous crop pests but other non-pest herbivores may also be exposed. However, not all plant tissues necessarily contain the transgene product. The risk to non-target herbivores is one of the risks considered by the biosafety committees in the EU for each application for commercial cultivation of a GM crop.

Pollinators can theoretically be exposed to transgene product when feeding on pollen or nectar. Like the other risks to herbivores described above, this risk is also considered during the application procedure and a GM plant that posed a significant risk to pollinators would not be permitted for cultivation in the EU.

Pollen from a GM plant can drop onto other plants potentially exposing non-target herbivores. This is mainly an issue with wind-pollinating plants which produce copious amounts of pollen, e.g. maize and willow. The case of Bt maize event 176, which was transformed with a pollen-specific promoter and the effects of its pollen on the monarch butterfly drew particular attention to this risk (see below). This particular maize event is being phased out. Other commercial Bt maize lines vary in their levels of pollen expression. The most widely grown lepidopteran active Bt maize events have zero or very low levels of Bt toxin in their pollen.

Apart from contamination with GM pollen, insects in natural or semi-natural habitats would only be exposed to a transgene product if a GM plant were to invade such a habitat or a transgene introgress and spread in a population of a wild relative. As discussed above, this risk is very low with most crop species and is one of the questions asked during the risk assessment for each and every GM plant application for cultivation.

A large number of studies have been conducted to investigate the exposure of natural enemies of herbivores to transgene products expressed by GM plants. Predatory insects can be exposed to transgene products when feeding on a range of herbivores on a GM crop. However, not all herbivores ingest the transgene product when feeding on GM plants. Lepidopteran larvae and spider mites have been shown to consume Bt toxin when feeding on Bt plants but not aphids (Dutton *et al.*, 2002, 2003; Schuler *et al.*, 2005). These studies show that the exposure of predators varies with the prey they consume. Predicting the exposure of parasitoids to transgene products is more complex, partly because their feeding behaviour as larvae is less well understood. In tolerant herbivore strains or species, Bt toxins are mostly limited to the alimentary system of the herbivore with the result that parasitoid larvae that feed on host haemolymph are unlikely to be exposed to Bt toxin in such hosts (Schuler *et al.*, 2005).

Post-harvest incorporation of Bt crop residues adds Bt toxins to soil. In some soils they are able to persist. However, no significant effects in soil of Bt toxin from plant biomass have been found on earthworms and nematodes or numbers of culturable protozoa, fungi or bacteria. Also, Bt toxins do not appear to be taken up from soil by other plant species (O'Callaghan *et al.*, 2005; ACRE, 2006).

6 GM Insect-resistant Crops and Non-target Insects

Insect-resistant plants, whether produced through conventional breeding or biotechnology, can potentially affect herbivores and their natural enemies in many different ways. Individual insects can be affected directly by insect-resistant GM plants if they are susceptible to the transgene product, or indirectly via changes in prey quality, prey behaviour or plant quality.

At a population level, natural enemies can also be affected by reductions in prey availability and through changes in crop management through the use of insect-resistant crops, such as reductions in pesticide use and planting of unsprayed refuges (Schuler *et al.*, 1999a, 2000).

A large number of studies have investigated the impact of insect-resistant GM crops on non-target arthropods. A proportion of these have been conducted for plants close to commercialization in support for regulatory risk assessments. In addition, other independent studies are available dealing with both GM plants close to commercialization and those in early stages of development. The latter are mainly restricted to the laboratory.

Several reviews of the effects of insect-resistant GM plants have been published, including: Schuler *et al.* (1999a, 2000); Hilbeck (2001); Obrycki *et al.* (2001); Malone and Pham Delegue (2001); Groot and Dicke (2002); Lovei and Arpaia (2005) and O'Callaghan *et al.* (2005). Most attention has focused on natural enemies of crop pests and pollinators as they are at most risk and their ecosystem functions are important in IPM systems and essential for sustainable crop production. In comparison, relatively few studies have considered soil insects although their important role as decomposers has been recognized (O'Callaghan *et al.*, 2005).

As already discussed, the risk of exposure for insects outside the crop environment is comparatively low and a large amount of information is already available for Bt toxins. Few studies for non-target insects that are not commonly thought of as associated with a crop environment are therefore available, with the exception of the monarch butterfly and the eastern black swallowtail, *Papilio polyxenes* Fabricius, which are discussed in more detail below.

Endangered species thought to be at risk would be unlikely to be used directly in any toxicity tests. However, knowledge of host plant distribution and the timing of larval emergence can be used to predict the likelihood of exposure and the potential for harmful effects. Such an approach has been used with the endangered Karner blue butterfly, *Lycaeides melissa samuelis* Nabokov, in the USA (O'Callaghan *et al.*, 2005). Field studies of the abundance of endangered insect species in and near large-scale GM field plots is another approach to predicting the impact of a GM plant on such rare species (O'Callaghan *et al.*, 2005) although the real problem with this approach is the low probability of finding truly rare species in the study area, unless there is a known local population under immediate threat.

The huge diversity of the insect fauna represents a dilemma for entomologists charged with assessing non-target effects of new crop protection approaches such as GM crops. As it is impossible to test impacts on every single non-target species, representative 'indicator' species or groups need to be chosen for each risk assessment.

Bt toxins have been used for decades in commercial Bt sprays and a large body of data is available for the effect of these products on a wide range of taxa of non-target insects. The review by Glare and O'Callaghan (2000) of the biology, ecology and safety of microbial Bt provides extensive lists of the studies conducted, the species tested and those shown to be susceptible to the various Bt strains and toxins.

Few studies with Bt plants have so far shown a direct effect on a predatory insect, reinforcing the previous knowledge about the specificity of the Bt toxins used. More effects have been reported for other transgene products, although some of the results are controversial and it is still not clear when or even whether any of these will be used commercially. Research projects on this topic have involved small-scale laboratory bioassays which investigate the interactions between GM plants and beneficial insects in detail under 'worst-case scenario' conditions (e.g. Dogan *et al.*, 1996; Pilcher *et al.*, 1997; Bell *et al.*, 1999; Birch *et al.*, 1999; Schuler *et al.*, 1999b; Zwahlen *et al.*, 2000; Bernal *et al.*, 2002; Dutton *et al.*, 2002; Sétamou *et al.*, 2002), behavioural choice studies (e.g. Schuler *et al.*, 1999b; Meier and Hilbeck, 2001), population-scale studies in controlled environments (e.g. Bell *et al.*, 2001; Schuler *et al.*, 2001, 2003) as well as field trials at various spatial scales (e.g. Hoffmann *et al.*, 1992; Johnson and Gould, 1992; Flint *et al.*, 1995; Orr and Landis, 1997; Pilcher *et al.*, 1997; Hoy *et al.*, 1998; Lozzia, 1999; Riddick *et al.*, 2000; Obrycki *et al.*, 2001; Wold *et al.*, 2001; Musser and Shelton, 2003; Naranjo *et al.*, 2005; Pons *et al.*, 2005). Large-scale field experiments represent the most realistic scenario but these are costly. Also, small arthropods such as parasitoids and predators are notoriously difficult to study in the field and the lack of control of interacting environmental variables often poses problems for data analysis and interpretation. Detailed laboratory studies represent an essential contribution to understanding and predicting developments in the field. In addition to investigating the interactions between GM plants and particular non-target organisms, these studies have contributed towards the development of risk assessment methodologies for incorporation into regulation procedures.

In the context of the ongoing GM debate, it is notable that the majority of studies have not found any unexpected or major negative effects of GM plants on arthropod natural enemies or non-target species. Effects that have been reported have generally been subtle and considerably less damaging than the effects of broad-spectrum synthetic insecticides (e.g. the 13 longer-term studies introduced by Naranjo *et al.*, 2005). This fosters optimism that GM crops can contribute to conserving biodiversity in farmland ecosystems and promote more sustainable crop protection strategies. However, it is important to guard against complacency since interactions between GM crops and arthropods depend on a range of factors (e.g. toxin specificity, expression levels, promoter, pleiotropic effects, level of plant resistance, pest pressure, associated non-target fauna, presence of refuges, crop size and crop management) that may differ considerably for different transgenic events, cropping systems and geographical regions. Some of the factors that must continue to be considered when assessing environmental risks are outlined below, but further information and comprehensive discussions are provided in more detailed reviews on the topic (e.g. Schuler *et al.*, 1999a, 2000; Hilbeck, 2001; Obrycki *et al.*, 2001; Groot and Dicke, 2002).

Knowledge about the specificity of the transgene product is fundamental in predicting the likelihood of direct toxic effects on non-target organisms, and has been one of the aspects investigated in small-scale laboratory experiments. These studies have not only assessed mortality but also investigated sublethal effects on developmental rates, size, fitness and fecundity. Bt toxins expressed in GM plants have so far not shown any direct negative effects on natural enemies with the exception of lacewing larvae of the species *Chrysoperla carnea* Stephens (Hilbeck *et al.*, 1998; Dutton *et al.*, 2002). When lacewing larvae were provided with Egyptian leafworm (*Spodoptera littoralis* Boisduval) larvae fed on Bt maize, lacewing mortality was increased by 38% compared to lacewings fed on *S. littoralis* larvae reared on conventional maize (Hilbeck *et al.*, 1998). In contrast, lacewing mortality was not increased when lacewings were provided with spider mites fed on Bt maize, despite mites having higher concentrations of toxins than *S. littoralis*, suggesting an interaction between prey and Bt toxin (Dutton *et al.*, 2002). Also, as aphids are the most abundant, suitable and preferred prey for *C. carnea* and have very low levels of Bt toxin, the worst-case scenario presented by Hilbeck *et al.* (1998) probably has little relevance to field situations (Dutton *et al.*, 2002).

Commercial Bt plants can result in mortality rates of target pests of close to 100%. Such high pest mortality can result in localized host depletion for parasitoids specific to the target pest, and possibly also in a reduction in prey for generalist natural enemies, when compared to populations in unsprayed non-GM crops. However, in addition to the target pests, crops support a range of other herbivores, which are not affected by the GM plant and which represent an alternative food source for many natural enemies. For example, aphids, whiteflies, heteropteran bugs and spider mites are not controlled by Bt crops. In contrast, broad-spectrum insecticides not only indiscriminately reduce prey availability for natural enemies, but are also often directly toxic on contact to both the larval and adult stages of arthropod natural enemies. Although partial plant resistance might be easier to incorporate into IPM programmes than the very high levels aimed for at present, the latter is a requirement for current management plans to delay pest adaptation to Bt toxins (Cohen *et al.*, 2000; Carrière *et al.*, 2001).

The behaviour of non-target insects can also play a part in determining how their populations will be affected by GM plants. Behavioural choice tests with Bt maize have shown that lacewing larvae prefer to prey on aphids rather than moth larvae, which will reduce their exposure to a Bt toxin (Meier and Hilbeck, 2001). Similarly, a parasitoid, *Cotesia vestalis* Haliday (until recently known as *C. plutellae*), of the diamondback moth (*Plutella xylostella* L.) was more attracted to wild-type oilseed rape damaged by Bt-susceptible moth larvae than to Bt oilseed rape damaged by such larvae (Schuler *et al.*, 1999b, 2003).

Risk assessment is an evolving process that changes in response to new scientific knowledge and new regulations. Early risk assessments of insect-resistant GM plants had to rely to a certain extent on guidelines developed for insecticides but the expanding knowledge base derived from studying the interactions of GM plants with non-target organisms increasingly allowed the fine-tuning of risk assessment methodologies. Based on several years of investigating the side effects of GM plants on non-target arthropods, a tiered testing scheme at different experimental scales has been recommended for the risk assessment of GM plants (Schuler *et al.*, 2000). International guidelines for the environmental risk assessment of GM plants are also being developed by a working group of the International Organization for Biological Control (IOBC). However, it is impossible to study all possible interactions in pre-approval studies, and practical protocols for further monitoring post-approval are being developed and implemented.

Any measures aimed at protecting crops from pests will reduce the food supply of organisms that use these pests as food and the effects of GM plants have to be interpreted in a realistic context. The level of pest pressure and extent of negative side effects of conventional control measures differ between crops and regions and these differences are in part responsible for disagreements between ecologists regarding the benefits of Bt plants. A comparison in the USA of the effects of Bt maize and Bt cotton on natural enemies illustrates this point. Both crops were developed for the control of major lepidopteran pests, and for both crops the only negative effects on natural enemies observed in field studies were lower populations of specialist parasitoids of the target pests (Orr and Landis, 1997; Wilson and Fitt, 2000; Obrycki *et al.*, 2001). However, the pest pressure in maize is usually not as heavy as in cotton and the intensity of conventional pest management differs dramatically between the two crops. The main pest targeted by Bt maize is the European cornborer, *Ostrinia nubilalis* Hübner, a species that causes major damage in some years. In the USA, damage by *O. nubilalis* was largely uncontrolled until the introduction of Bt maize as only ~5% of conventional field maize was treated with insecticides against this pest (Carpenter and Gianessi, 2001). In contrast, several sprays of broad-spectrum insecticides had to be applied to conventional cotton every year in most cotton-growing regions to control bollworms and budworms (*Heliothis virescens* Fabricius, *Pectinophora gossypiella* Saunders, *Helicoverpa armigera* Hübner or *H. zea* Boddie, depending on the region). The introduction of Bt cotton resulted in dramatic reductions in insecticide sprays. In Alabama, for example, insecticide sprays were reduced from 7/year in 1995 to 0.4 in 1999 (Carpenter and Gianessi, 2001). Similarly, in Australia sprays against bollworms were reduced from 10 to 4 in the 1996/97 season (Fitt, 2000). While maize entomologists criticize the prophylactic use of Bt maize in years when *O. nubilalis* populations are low (Obrycki *et al.*, 2001), cotton entomologists have generally welcomed Bt cotton since the reduction in broad-spectrum pesticide sprays now allows the use of natural enemies as part of IPM programmes in cotton (Fitt, 2000).

6.1 The monarch and Bt maize

The controversy surrounding the monarch butterfly and Bt crops is worth examining in some detail because it is the best example so far in which there was perceived risk of an insecticidal GM crop on a non-target species of wide conservation interest. The monarch, with its worldwide distribution, cannot be considered as endangered but it is certainly a flagship species in North America, and being so charismatic it has become an ideal educational resource and therefore important for instilling a greater interest in insects amongst children and the wider public. In addition, the spectacular Mexican overwintering site of the species, where millions of individuals cluster on groups of trees, has every right to be considered an endangered natural ecological phenomenon of international conservation concern (Brower, 1997). So it is hardly surprising that when Losey *et al.* (1999) published their paper suggesting that monarch larvae might be at risk from ingesting Bt maize pollen on their milkweed foodplants, there was massive international media coverage, ensuing outrage and calls, from both within and outside the USA, for an immediate ban on such crops. The paper soon started to receive criticism (Hodgson, 2000), perhaps with some justification because the research was from a single laboratory-based 'worst-case scenario', where larvae were given no choice but to eat unquantified amounts of Bt pollen dusted onto leaves of their foodplant or starve. The pollen was from event 176 Bt maize, an early cultivar now removed from the market, which contained particularly large amounts of the Bt toxin in its pollen. So it was not entirely surprising that the monarch larvae were also harmed, and indeed they grew more slowly, ate less and had higher mortality than those eating non-contaminated leaves. Such worst-case tests are important in the early stage of risk assessment but they are usually just a starting point to justify further more quantitative and realistic risk assessment. Without the GM tag line such studies do not normally get published in high-impact science journals such as *Nature*. The view that the study by Losey *et al.* (1999) might be unrealistic was endorsed and the conservation interest widened when a more complete field and laboratory study failed to detect any toxicity of Bt pollen on larvae of the eastern black swallowtail butterfly, *P. polyxenes* (Wraight *et al.*, 2000), although support for a real risk to the monarch was provided by the field study of Jesse and Obrycki (2000).

In hindsight, the media coverage served a useful purpose, as scientific experts got together with regulators and the biotech industry to assess all available information to identify what information was missing and to quantify and address the risk to the monarch population in North America. As a result, funding was made available to carry out a series of important studies that answered many of the outstanding scientific issues. These were peer-reviewed (important for such a contentious issue) and published together in a special section of *Proceedings of the National Academy of Sciences of the United States of America* (Hellmich *et al.*, 2001; Oberhauser *et al.*, 2001; Pleasants *et al.*, 2001; Sears *et al.*, 2001; Stanley-Horn *et al.*, 2001; Zangerl *et al.*, 2001). These papers completed a very thorough risk analysis and showed that in the most widely used commercial hybrids Bt expression in the pollen is low and would not produce acute toxic effects in the field. This fact, combined with the short period that pollen shed and monarch larval activity coincide, and the relatively small proportion of larvae that develop in the crop itself, suggests that Bt maize will have a negligible impact on monarch populations in North America (Sears *et al.*, 2001). Further toxicity tests combined with a modelling approach suggest that there might be an additional mortality to monarch populations over the entire Corn Belt equivalent to 0.6%. This intensive agricultural region is only 50% of the total breeding area of the species and the additional mortality is not considered to be a threat to the sustainability of monarch populations in North America (Dively *et al.*, 2004).

So is it all good news about the monarch and GM crops? Well perhaps not, according to knowledgeable monarch experts such as Lincoln Brower (2001) who point out that the GMHT crops now widely grown in North America may lead to farmers removing the monarch foodplants completely from maize and soya fields. This would be an indirect effect very similar to the one addressed in the UK FSE study (see below) and could be studied in a similar way by comparing paired GMHT and non-GMHT fields and quantifying milkweed densities and in-field breeding success in the two situations stratified across wide areas of the butterfly's breeding range. Whatever the results from such a study, the best long-term solution might still be to optimize monarch breeding habitat around field margins and in non-cultivated areas as discussed earlier for the UK butterfly situation.

Although the research paper that sparked the original concern about monarchs received massive public media coverage, the research papers that put the risk into proper perspective received much less attention. This scientific imbalance is always going to be a problem when worst-case results are publicized in the context of a controversial and polarized subject such as GM crops. However, if nothing else, the controversy has sparked new interest and research on the population dynamics of the monarch and emphasized the importance of agricultural practices throughout North America to its continued well-being.

It is worth noting that a recent study has shown that transgenic Bt maize pollen can have a negative effect on the larvae of another charismatic butterfly species, the European swallowtail, *P. machaon* L. (Lang and Vojtech, 2006). Again this is a worst-case scenario laboratory study but the results are interesting and suggest that further ecological studies may now be required to assess their importance for populations of this particular species at the field scale as the results from the monarch studies in North America may not be directly transferable to the European agricultural situation.

7 GM Herbicide-tolerant Crops and Non-target Insects

7.1 Farm-scale evaluations

By the end of 1998 four GMHT crops (maize, beet, spring and winter oilseed rape) had been through the rigorous regulatory risk assessment that existed in the UK at that time, and had been found to be safe for human consumption and likely to have no direct environmental impacts. However, when the question was raised about possible indirect environmental impacts of such crops, in terms of farmland biodiversity, it was realized that there was an important gap both in the risk assessment procedure and in our understanding of farmland ecology more generally. The FSE was funded to bridge that gap (Firbank *et al.*, 1999, 2003a; Woiwod *et al.*, 2000).

Details of the thinking behind the FSE, the sampling strategy and statistical methodology have been fully published elsewhere (Firbank *et al.*, 2003a; Perry *et al.*, 2003), so here we will just outline the main aspects of the experiments before discussing the entomological results of interest. The apparently simple purpose of the project was to test whether the four proposed GMHT crops would lead to a significant reduction in farmland biodiversity if they were to be commercialized and grown widely in the UK. This was done by sampling the weeds and invertebrates within fields, as these were considered to be the parts of the food chain most likely to be affected immediately by any change in herbicide regime associated with the GMHT crops. A split-field design was chosen where experimental fields were divided in two, with one half sown to the GMHT crop and the other half to its conventional equivalent. Cultivation of both halves was kept as near to usual practice as possible with the exception of the herbicide regime on the GM crop which followed industry guidelines. Between 60 and 70 fields were used for each crop type; in nearly all cases a field was only used for a single growing season as would be the case in a normal crop rotation of these particular break crops. Throughout the growing season a range of plant and invertebrate sampling protocols were done and some of the plant sampling was continued for up to 3 years after harvest. A range of invertebrates including Collembola, Carabidae, Heteroptera, Aranaea and Gastropoda were sampled using suction samplers, pitfall and bait traps. Bees and butterflies were sampled using transect counts (Firbank *et al.*, 2003a).

The large number of replicates in the experiment was required to cover the main arable areas of the UK with their wide range of climate, soil and farming practices and at the same time provide enough statistical power to detect significant ecological effects. As a result, this experiment became one of the largest agroecological risk assessments ever undertaken and is providing important information well beyond its original purpose.

The results have been published in detail (Brooks *et al.*, 2003; Haughton *et al.*, 2003; Hawes *et al.*, 2003; Heard *et al.*, 2003a,b; Roy *et al.*, 2003; Bohan *et al.*, 2005) and a commentary on the ecological implications of the results from the spring-sown crops is provided by Firbank *et al.* (2003b). The main conclusions were that for three of the four crops the broadleaved weed seed bank would indeed decline with the GM herbicide regime, with possible deleterious long-term effects on farmland biodiversity. However, for one crop, fodder maize, there was an actual increase in broadleaved weed populations.

Insects and other invertebrates were clearly affected by the weed populations but showed a mixture of responses. In all crops Collembola generally did better in the GM crops, probably because of the large pulse of dead vegetation as herbicides were applied later in the season than with conventional crops (Brooks *et al.*, 2003; Haughton *et al.*, 2003; Bohan *et al.*, 2005). There were some other significant effects on ground and plant surface-dwelling insects but they were rather idiosyncratic. For example, as might be expected, seed-feeding carabid beetles tended to decline, whereas Collembola-feeding species tended to do better in GM crops. When summed over taxa, e.g. total carabids, there were no significant differences between GM and non-GM crops, and species doing better were roughly balanced by those doing worse. The main exceptions to this were the adult bees and butterflies sampled visually along fixed transects. Generally, again with the exception of GM maize, there were significant declines in the numbers of bees and butterflies recorded in the GM half of the fields, at least during some of the sampling period. This applied both within the fields (Brooks *et al.*, 2003; Haughton *et al.*, 2003; Bohan *et al.*, 2005) and along the field margins (Roy *et al.*, 2003; Bohan *et al.*, 2005).

When the results were published it was perhaps not surprising that the positive Collembola result received much less media attention than the negative bee and butterfly results. Collembola are just not as charismatic for the general public. However, it is worth asking whether bee and butterfly populations would, as widely reported, really be threatened if such crops were to be widely planted in the UK or elsewhere in Europe. It has been suggested that, at least for butterflies, it would be very unlikely to cause a major effect (Woiwod, 2004). For a start, the between-treatment effects, although significant, were much smaller than any between-crop differences (Firbank *et al.*, 2003b) or the very large difference between crops and field margins. These results were also from transect counts of very mobile species and analysis suggests that they very much reflected a short-term response to nectar plants in the cultivated fields, mainly to Asteraceae such as *Cirsium* spp. and *Sonchus* spp. (Haughton *et al.*, 2003). If there were a real shortage of such nectar resources at the farm level, these results might have important implications for butterfly and bee populations, although whether farmers would ever welcome the requirement to cultivate or even tolerate thistles in their fields is another question. Data in respect of nectar deficits in farmland are inadequate but it is reasonable to propose that there is at present a real, and probably chronic, shortage of both nectar and many larval foodplants in most intensive arable areas. However, centres of arable fields are currently relatively inhospitable places for most invertebrates, and are likely to remain so, mainly because of herbicide use and annual cultivation. Therefore, rather than luring insects into fields with the promise of nectar and larval food resources, a better conservation option would be to provide such resources in permanent flower-rich field margins or other uncultivated areas around the farm, and indeed this is one of the approaches promoted to UK farmers under the current Defra Environmental Stewardship Schemes (available at: http://www.defra.gov.uk/erdp/schemes/es/default.htm).

Mainly as a result of the FSE findings and following advice from ACRE, the UK government decided to oppose cultivation of GMHT beet and GMHT oilseed rape if managed using current commercial guidelines. The UK government agreed to the cultivation of the particular GMHT maize under certain conditions (see: http://www.defra.gov.uk/corporate/ministers/statements/mb040309.htm). However, the application to the EU for the GMHT maize tested in the FSE has since been withdrawn for commercial reasons.

8 Wider Issues

8.1 Farm-scale evaluations

There are some wider implications of the FSE project that are worth outlining as they apply more generally to many environmental risk studies looking at indirect effects of GM crops. First of all it should be pointed out that the FSE was not, strictly speaking, a GM risk study but rather a herbicide trial. Herbicide-resistant crops can and have been produced by conventional breeding and exactly the same concerns should be expressed about such crops, but not surprisingly they generally have not been (although see ACRE (2006) for an exception).

In the case of sugarbeet a life cycle assessment (LCA) has been done using some of the agronomic results from the FSE (Bennett *et al.*, 2004). LCA is a methodology widely used and internationally accepted in industrial environmental impact studies which tries to estimate all environmental burdens of a particular process. In the case of sugarbeet production a comparison between GMHT and conventional cultivation showed clear environmental advantages of the GMHT crop in terms of energy requirement, global warming potential, ozone depletion, ecotoxicity, acidification and nutrification (Bennett *et al.*, 2004). In addition, for this crop there have been studies suggesting how the extra flexibility available through herbicide tolerance could, with further research, allow cultivation practices to be developed that would largely alleviate the negative aspects of GMHT beet found during the FSE study (Dewar *et al.*, 2003; May *et al.*, 2005). It has been argued that more long-term environmental damage may be done by not allowing cultivation of GM crops such as these, and that we need to take a much wider environmental perspective than has been evident so far by setting them against the well-known biodiversity damage caused by most conventional agriculture (Ammann, 2005).

There is now strong evidence that existing GM crops can have environmental benefits. GM crops have resulted in 172 million kg less pesticide being used worldwide since their introduction in 1996, and their overall environmental footprint is 14% lower than conventional crops (Brookes and Barfoot, 2005). The same authors calculated that GM crops have enabled an estimated 10 billion kg reduction in greenhouse gas emissions, equivalent to removing 5 million cars from the road for a year (Brookes and Barfoot, 2005). Despite this evidence, environmental impact assessment in the EU focuses entirely on the risks. ACRE (2006) suggests that a balancing of risks and benefits may be more appropriate in the future.

The instigation of the FSE marked an important turning point in the assessment of environmental impacts in an agricultural context in general. By accepting that the indirect, management-related, effect of GMHT crops on farmland biodiversity was to be included as an integral part of the risk assessment, the EU has in effect drawn a line in terms of the level of acceptable agricultural impact on farmland weeds, insects and other wildlife. As applied to weeds, this will require a complete change of attitude on the part of the farmer. Already research is underway to find out how to encourage 'good weeds' that provide environmental services whilst managing the 'bad weeds' that farmers must keep under control if they are to grow viable crops. This is perhaps a particularly European perspective in which agriculture is so integrated within the wider landscape that 'good weeds' are a really important component of the natural flora. By contrast, in Australia, weeds are largely noxious and exotic and the concerns addressed by the FSE are not regarded as so relevant (Londsdale *et al.*, 2003).

8.2 GM soya and the Amazon rainforest

One unexpected and potentially serious effect of GM crops on insect conservation has recently become apparent concerning a world insect biodiversity hotspot, the Amazon rainforest in Brazil. The rate of destruction of the rainforest for agricultural production suddenly jumped 40% in 2003 as a result of increased world demand for soya and beef cattle. Currently about 25,000 km^2 are being destroyed each year (often equated to an area the size of Wales or Belgium). It is reliably estimated that by 2050 current trends in agricultural expansion will eliminate 40% of the Amazon forests (Soares-Filho *et al.*, 2006). This will have serious effects on release of carbon into the atmosphere, with worldwide climate change implications. Spatial models also suggest very important conservation implications for the only animal group so far studied, terrestrial mammals, where a quarter of the 382 species analysed are predicted to lose critical areas of the forest within their Amazon ranges unless suitable environmental legislation is enacted and enforced across the Amazon basin (Soares-Filho *et al.*, 2006). Proportionally, the impact on insects is likely to be at least as severe as for mammals but affecting orders of magnitude of more species, most of which still remain to be described and studied.

Ironically, one of the drivers of the current destruction in Amazonia is the European demand for GM-free soya (McCarthy and Buncombe, 2005), which grows particularly well in the north of Brazil, as GM varieties are more suitable for cultivation further south. If nothing else, this underlines the difficulty in environmental risk analysis when large-scale sociopolitical effects come into play at a world scale.

8.3 Future crops

In this chapter we have concentrated on the two main types of biotech crops already under widespread cultivation worldwide. There are potentially many other crop developments which are possible using the technology such as pharmacological, biofuel, low N-input, disease-resistant, drought-tolerant and salt-tolerant crops. Providing good risk assessment and regulatory procedures are in place, which take into account both direct and indirect environmental impacts, most reasonable concerns should be addressed. However, it remains important that conservation biologists ensure that all their concerns about such crops are fed into any regulatory process in a constructive way.

The development of GM insect-resistant trees is one particular area of entomological interest and concern. Trees may pose additional risks because of their long lifespan and the ability of some tree families to hybridize freely with related wild and cultivated species. For ecological and economic reasons

China and Brazil are the most likely countries to make use of the technology for forestry plantations in the near future and indeed China has already developed a Bt poplar, *Populus nigra*, which has been planted in large experimental plots (Sedjo, 2005). The advantages of reduced requirement for conventional pesticides may counterbalance any non-target effects but comprehensive series of ecological studies and risk–benefit analyses, which include effects on non-target insects, should be done before such trees are grown more widely outside these experimental areas.

9 Conclusions

Scientific evidence strongly suggests that the GM crops grown so far are not in themselves more hazardous to the environment than those produced by conventional breeding (e.g. GM Science Review Panel, 2003; ACRE, 2006). Studies have repeatedly demonstrated that the significant factors are the trait and management of the crop, not the breeding method itself. This is why a case-by-case risk assessment approach is appropriate. At present, the focus in the EU is firmly on environmental risks and a more balanced approach that allows consideration of both environmental benefits and risks is recommended.

It is unavoidable that any measures aimed at reducing crop damage due to insects will have some knock-on effect on non-target insects, if only by reducing the number of prey available to predatory insects. The challenge for any new pest control measure or technique is to minimize negative effects on non-target insects as far as possible while allowing farmers to produce profitable crops.

The evidence so far in relation to non-pest insects is that the Bt crops are of overall conservation benefit (when the reductions in wide-spectrum insecticide use are set against the very small effects so far detected on non-target insect species). The insecticide reduction is particularly apparent in Bt cotton (Brookes and Barfoot, 2005; ACRE, 2006), as the pest pressure and damage potential in this crop, by a number of important Lepidoptera pests, is very high. As a result, large amounts of broad-spectrum insecticides are used routinely on most conventionally grown cotton, resulting in damage to non-target insects and widespread insect pesticide resistance problems. In contrast, the Bt insect-resistant cotton seems to provide environmental benefit in many areas where it is now grown. The widely predicted development of resistance to the Bt toxin has not so far occurred in this or any other Bt crop, suggesting that resistance management strategies have so far been very effective (Bates *et al.*, 2005).

There is an increasingly urgent need to balance food production and biodiversity conservation in our overpopulated world. We already sequestrate up to half of the entire planet's primary production for human food production (Vitousek *et al.*, 1986). GM crops, if developed carefully, tested rigorously and applied wisely, have the potential to provide at least some of the solutions to a more sustainable and environmentally benign world agriculture.

This does not mean complacency and the uncritical acceptance of all crops developed using the technology, but it does require a thoughtful case-by-case rational approach if we are to maximize the benefits of such crops to insect conservation while minimizing the risks.

Acknowledgements

Thanks to John Badmin and two anonymous referees for useful comments. Rothamsted Research receives grant-aided support from the Biotechnology and Biological Sciences Research Council.

References

ACRE (2001) *Guidance on the Assessment of the Impact on Wider Biodiversity from Proposed Cultivation of GM Crops*. Advisory Committee on Releases to the Environment, UK. Available at: http://www.defra.gov.uk/environment/acre/biodiversity/guidance/01.htm

ACRE (2004) *Guidance on Best Practice in the Design of Post-market monitoring plans in Submissions to the Advisory Committee on Releases to the Environment*. Advisory Committee on Releases to the Environment, UK. Available at: http://www.defra.gov.uk/environment/acre/postmarket/acre_postmarketmonitor-guidance.pdf

ACRE (2006) *Managing the Footprint of Agriculture: Towards a Comparative Assessment of Risks and Benefits for Novel Agricultural Systems*. Advisory Committee on Releases to the Environment, UK. Available at: http://www.defra.gov.uk/environment/acre/fsewiderissues/index.htm

Altpeter, F., Diaz, I., McAuslane, H., Gaddour, K., Carbonero, P. and Vasil, I.K. (1999) Increased insect resistance in transgenic wheat stably expressing trypsin inhibitor CMe. *Molecular Breeding* 5, 53–63.

Ammann, K. (2005) Effects of biotechnology on biodiversity: herbicide-tolerant and insect-resistant GM crops. *Trends in Biotechnology* 23, 388–394.

Bates, S.L., Zhao, J.Z., Roush, R.T. and Shelton, A.M. (2005) Insect resistance management in GM crops: past, present and future. *Nature Biotechnology* 23, 57–62.

Bell, H.A., Fitches, E.C., Down, R.E., Marris, G.C., Edwards, J.P., Gatehouse, J.A. and Gatehouse, A.M.R. (1999) The effect of snowdrop lectin (GNA) delivered via artificial diet and transgenic plants on *Eulophus pennicornis* (Hymenoptera: Eulophidae), a parasitoid of the tomato moth *Lacanobia oleracea* (Lepidoptera: Noctuidae). *Journal of Insect Physiology* 45, 983–991.

Bell, H.A., Fitches, E.C., Down, R.E., Ford, L., Marris, G.C., Edwards, J.P., Gatehouse, J.A. and Gatehouse, A.M.R. (2001) Effect of dietary cowpea trypsin inhibitor (CpTI) on the growth and development of the tomato moth *Lacanobia oleracea* (Lepidoptera: Noctuidae) and on the success of the gregarious ectoparasitoid *Eulophus pennicornis* (Hymenoptera: Eulophidae). *Pest Management Science* 57, 57–65.

Bennett, R., Phipps, R., Strange, A. and Grey, P. (2004) Environmental and human health impacts of growing genetically modified herbicide-tolerant sugar beet: a life-cycle assessment. *Plant Biotechnology Journal* 2, 273–278.

Bernal, C.C., Aguda, R.M. and Cohen, M.B. (2002) Effect of rice lines transformed with *Bacillus thuringiensis* toxin genes on the brown planthopper and its predator *Cyrtorhinus lividipennis*. *Entomologia Experimentalis et Applicata* 102, 21–28.

Birch, A.N.E., Geoghegan, I.E., Majerus, M., McNicol, J.W., Hackett, C., Gatehouse, A.M.R. and Gatehouse, J.A. (1999) Tri-trophic interactions involving pest aphids, predatory

2-spot ladybirds and transgenic potatoes expressing snowdrop lectin for aphid resistance. *Molecular Breeding* 5, 75–83.

Bohan, D.A., Boffey, C.W.H., Brooks, D.R., Clark, S.J., Dewar, A.M., Firbank, L.G., Haughton, A.J., Hawes, C., Heard, M.S., May, M.J., Osborne, J.L., Perry, J.N., Rothery, P., Roy, D.B., Scott, R.J., Squire, G.R., Woiwod, I.P. and Champion, G.T. (2005) Effects on weed and invertebrate abundance and diversity of herbicide management in genetically modified herbicide-tolerant winter-sown oilseed rape. *Proceedings of the Royal Society of London Series B – Biological Sciences* 272, 463–474.

Brookes, G. and Barfoot, P. (2005) GM Crops: the global economic and environmental impact: the first nine years 1996–2004. *AgBioForum* 8, 187–196.

Brooks, D.R., Bohan, D.A., Champion, G.T., Haughton, A.J., Hawes, C., Heard, M.S., Clark, S.J., Dewar, A.M., Firbank, L.G., Perry, J.N., Rothery, P., Scott, R.J., Woiwod, I.P., Birchall, C., Skellern, M.P., Walker, J.H., Baker, P., Bell, D., Browne, E.L., Dewar, A.J.G., Fairfax, C.M., Garner, B.H., Haylock, L.A., Horne, S.L., Hulmes, S.E., Mason, N.S., Norton, L.R., Nuttall, P., Randle, Z., Rossall, M.J., Sands, R.J.N., Singer, E.J. and Walker, M.J. (2003) Invertebrate responses to the management of genetically modified herbicide-tolerant and conventional spring crops. I. Soil-surface-active invertebrates. *Philosophical Transactions of the Royal Society of London Series B – Biological Sciences* 358, 1847–1862.

Brower, L.P. (1997) A new paradigm in conservation of biodiversity: endangered biological phenomena. In: Meffe, G.K. and Carroll, C.R. (eds) *Principles of Conservation Biology*, 2nd edn. Sinauer Associates, Sunderland, Massachusetts, pp. 115–118.

Brower, L.P. (2001) Canary in the cornfield: the monarch and the *Bt* corn controversy. *Orion* 20, 32–41.

Carpenter, J.E. and Gianessi, L.P. (2001) *Agricultural Biotechnology: Updated Benefit Estimates*. National Center for Food and Agricultural Policy, Washington, DC.

Carrière, Y., Dennehy, T.J., Pedersen, B., Haller, S., Ellers-Kirk, C., Antilla, L., Liu, Y.B., Willott, E. and Tabashnik, B.E. (2001) Large-scale management of insect resistance to transgenic cotton in Arizona: can transgenic insecticidal crops be sustained. *Journal of Economic Entomology* 94, 315–325.

Cohen, M.B., Gould, F. and Bentur, J.S. (2000) *Bt* rice: practical steps to sustainable use. *International Rice Research Notes* 25, 4–10.

Czapla, T.H. (1997) Plant lectins as insect control proteins in transgenic plants. In: Carozzi, N. and Koziel, M. (eds) *Advances in Insect Control: The Role of Transgenic Plants*. Taylor & Francis, London, pp. 123–138.

Damgaard, C. and Kjellsson, G. (2005) Gene flow of oilseed rape (*Brassica napus*) according to isolation distance and buffer zone. *Agriculture, Ecosystems and Environment* 108, 291–301.

De Leo, F., Bonade-Bottino, M., Ceci, L.R., Gallerani, R. and Jouanin, L. (2001) Effects of a mustard trypsin inhibitor expressed in different plants on three lepidopteran pests. *Insect Biochemistry and Molecular Biology* 31, 593–602.

Delledonne, M., Allegro, G., Belenghi, B., Balestrazzi, A., Picco, F., Levine, A., Zelasco, S., Calligari, P. and Confalonieri, M. (2001) Transformation of white poplar (*Populus alba* L.) with a novel *Arabidopsis thaliana* cysteine proteinase inhibitor and analysis of insect pest resistance. *Molecular Breeding* 7, 35–42.

Dewar, A.M., May, M.J., Woiwod, I.P., Haylock, L.A., Champion, G.T., Garner, B.H., Sands, R.J., Qi, A. and Pidgeon, J.D. (2003) A novel approach to the use of genetically modified herbicide-tolerant crops for environmental benefit. *Proceedings of the Royal Society of London Series B – Biological Sciences* 270, 335–340.

Dively, G.P., Rose, R., Sears, M.K., Hellmich, R.L., Stanley-Horn, D., Calvin D.D., Russo, J.M. and Anderson, P.L. (2004) Effects on monarch butterfly larvae (Lepidoptera: Danaidae) after continuous exposure to Cry1Ab-expressing corn during anthesis. *Environmental Entomology* 33, 1116–1125.

Dogan, E.B., Berry, R.E., Reed, G.L. and Rossignol, P.A. (1996) Biological parameters of convergent lady beetle (Coleoptera:

Coccinellidae) feeding on aphids (Homoptera: Aphididae) on transgenic potato. *Journal of Economic Entomology* 89, 1105–1108.

Dutton, A., Klein, H., Romeis, J. and Bigler, F. (2002) Uptake of *Bt* toxin by herbivores feeding on transgenic maize and consequences for the predator *Chrysoperla carnea*. *Ecological Entomology* 27, 441–447.

Dutton, A., Romeis, J. and Bigler, F. (2003) Assessing the risks of insect-resistant transgenic plants on entomophagous arthropods: *Bt*-maize expressing Cry1Ab as a case study. *BioControl* 48, 611–636.

Firbank, L.G., Dewar, A.M., Hill, M.O, May, M.J., Perry, J.N., Rothery, P., Squire, G.R. and Woiwod I.P. (1999) Farm-scale evaluation of GM crops explained. *Nature* 399, 727–728.

Firbank, L.G., Heard, M.S., Woiwod, I.P., Hawes, C., Haughton, A.J., Champion, G.T., Scott, R.J., Hill, M.O., Dewar, A.M., Squire, G.R., May, M.J., Brooks, D.R., Bohan, D.A., Daniels, R.E., Osborne, J.L., Roy, D.B., Black, H.I.J., Rothery, P. and Rothery, J.N. (2003a) An introduction to the farm-scale evaluations of genetically modified herbicide-tolerant crops. *Journal of Applied Ecology* 40, 2–16.

Firbank, L.G., Perry, J.N., Squire, G.R., Bohan, D.A., Brooks, D.R., Champion, G.T., Clark, S.J., Daniels, R.E., Dewar, A.M., Haughton, A.J., Hawes, C., Heard, M.S., Hill, M.O., May, M.J., Osborne, J.L., Rothery, P., Roy, D.B., Scott, R.J. and Woiwod, I.P. (2003b) The implications of spring-sown genetically modified herbicide-tolerant crops for farmland biodiversity: a commentary on the farm scale evaluations of spring-sown crops. Available at: http://www.defra.gov.uk/environment/gm/fse/results/fse-commentary.pdf

Fitt, G.P. (2000) An Australian approach to IPM in cotton: integrating new technologies to minimize insecticide dependence. *Crop Protection* 19, 793–800.

Flint, H.M., Henneberry, T.J., Wilson, F.D., Holguin, E., Parks, N. and Buehler, R.E. (1995) The effects of transgenic cotton, *Gossypium hirsutum* L., containing *Bacillus thuringiensis* toxin genes for the control of the pink bollworm, *Pectinophora gossypiella* (Saunders) Lepidoptera, Gelechiidae and other arthropods. *Southwestern Entomologist* 20, 281–292.

Fuller, R.J., Gregory, R.D., Gibbons, D.W., Marchant, J.H., Wilson, J.D., Baille, S.R. and Carter, N. (1995) Population declines and range contractions among lowland farmland birds in Britain. *Conservation Biology* 9, 1425–1441.

Gatehouse, A.M.R., Davidson, G.M., Newell, C.A., Merryweather, A., Hamilton, W.D.O., Burgess, E.P.J., Gilbert, R.J.C. and Gatehouse, J.A. (1997) Transgenic potato plants with enhanced resistance to the tomato moth, *Lacanobia oleracea*: growth room trials. *Molecular Breeding* 3, 49–63.

Gatehouse, A.M.R., Davison, G.M., Stewart, J.N., Gatehouse, L.N., Kumar, A., Geoghegan, I.E., Birch, A.N.E. and Gatehouse, J.A. (1999) Concanavalin A inhibits development of tomato moth (*Lacanobia oleracea*) and peach-potato aphid (*Myzus persicae*) when expressed in transgenic potato plants. *Molecular Breeding* 5, 153–165.

Gatehouse, J.A. (2002) Plant resistance towards insect herbivores: a dynamic interaction. *New Phytologist* 156, 145–169.

Glare, T.R. and O'Callaghan M. (2000) *Bacillus thuringiensis: Biology, Ecology and Safety*. Wiley, Chichester, UK.

GM Science Review Panel (2003) *GM Science Review First Report*. Available at: http://www.gmsciencedebate.org.uk/report/pdf/gmsci-report1-full.pdf DTI

Gray, A.J. (2004) Ecology and government policies: the GM crop debate. *Journal of Applied Ecology* 19, 363–366.

Groot, A.T. and Dicke, M. (2002) Insect-resistant transgenic plants in a multi-trophic context. *The Plant Journal* 31, 387–406.

Hails, R.S., Rees, M., Kohn, D.D. and Crawley, M.J. (1997) Burial and seed survival in *Brassica napus* subsp. *oleifera* and *Sinapis arvensis* including a comparison of transgenic and non-transgenic lines of the crop. *Proceedings of the Royal Society of London Series B – Biological Sciences* 264, 1–7.

Haughton, A.J., Champion, G.T., Hawes, C., Heard, M.S., Brooks, D.R., Bohan, D.A., Clark, S.J., Dewar, A.M., Firbank, L.G., Osborne, J.L., Perry, J.N., Rothery, P.,

Roy, D.B., Scott, R.J., Woiwod, I.P., Birchall, C., Skellern, M.P., Walker, J.H., Baker, P., Browne, E.L., Dewar, A.J.G., Garner, B.H., Haylock, L.A., Horne, S.L., Mason, N.S., Sands, R.J.N. and Walker, M.J. (2003) Invertebrate responses to the management of genetically modified herbicide-tolerant and conventional spring crops. II. Within-field epigeal and aerial arthropods. *Philosophical Transactions of the Royal Society of London Series B – Biological Sciences* 358, 1863–1877.

Hawes, C., Haughton, A.J., Osborne, J.L., Roy, D.B., Clark, S.J., Perry, J.N., Rothery, P., Bohan, D.A., Brooks, D.R., Champion, G.T., Dewar, A.M., Heard, M.S., Woiwod, I.P., Daniels, R.E., Young, M.W., Parish, A.M., Scott, R.J., Firbank, L.G. and Squire, G.R. (2003) Responses of plants and invertebrate trophic groups to contrasting herbicide regimes in the farm scale evaluations of genetically modified herbicide-tolerant crops. *Philosophical Transactions of the Royal Society of London Series B – Biological Sciences* 358, 1899–1913.

Heard, M.S., Hawes, C., Champion, G.T., Clark, S.J., Firbank, L.G., Haughton, A.J., Parish, A.M., Perry, J.N., Rothery, P., Scott, R.J., Skellern, M.P., Squire, G.R. and Hill, M.O. (2003a) Weeds in fields with contrasting conventional and genetically modified herbicide-tolerant crops. 1. Effects on abundance and diversity. *Philosophical Transactions of the Royal Society of London Series B – Biological Sciences* 358, 1819–1832.

Heard, M.S., Hawes, C., Champion, G.T., Clark, S.J., Firbank, L.G., Haughton, A.J., Parish, A.M., Perry, J.N., Rothery, P., Roy D.B., Scott, R.J., Skellern, M.P., Squire, G.R. and Hill, M.O. (2003b) Weeds in fields with contrasting conventional and genetically modified herbicide-tolerant crops. 2. Effects on individual species. *Philosophical Transactions of the Royal Society of London Series B – Biological Sciences* 358, 1833–1846.

Hellmich, R.L., Siegfried, B.D., Sears, M.K., Stanley-Horn, D.E., Daniels, M.J., Mattila, H.R., Spencer, T., Bidne, K.G. and Lewis, L.C. (2001) Monarch larvae sensitivity to *Bacillus thuringiensis* purified proteins and pollen. *Proceedings of the National Academy of Sciences of the United States of America* 98, 11925–11930.

Hilbeck, A. (2001) Implications of transgenic, insecticidal plants for insect and plant biodiversity. *Perspectives in Plant Ecology, Evolution and Systematics* 4, 43–61.

Hilbeck, A., Baumgartner, M., Fried, P.M. and Bigler, F. (1998) Effects of transgenic *Bacillus thuringiensis* corn-fed prey on mortality and development time of immature *Chrysoperla carnea* (Neuroptera: Chrysopidae). *Environmental Entomology* 27, 480–487.

Hodgson, J. (2000) Critics slam new Monarch *Bt*-corn data criticized. *Nature Biotechnology* 18, 1030.

Hoffmann, M.P., Zalom, F.G., Wilson, L.T., Smilanick, J.M., Malyj, L.D., Kiser, J., Hilder, V.A. and Barnes, W.M. (1992) Field evaluation of transgenic tobacco containing genes encoding *Bacillus thuringiensis* delta-endotoxin or cowpea trypsin inhibitor: efficacy against *Helicoverpa zea* (Lepidoptera: Noctuidae). *Journal of Economic Entomology* 85, 2516–2522.

Hoy, C.W., Feldman, J., Gould, F., Kennedy, G.G., Reed, G. and Wyman, J.A. (1998) Naturally occurring biological controls in genetically engineered crops. In: Barbosa, P. (ed.) *Conservation Biological Control.* Academic Press, London, pp. 185–205.

James, C. (2005) Global Status of Biotech/GM Crops in 2005. *ISAAA Briefs* No. 34. International Service for the Acquisition of Agri-biotech Applications. Ithaca, New York. Available at: www.isaaa.org/home.htm

Jesse, L.C.H. and Obrycki, J.J. (2000) Field deposition of *Bt* transgenic corn pollen: lethal effects on the monarch butterfly. *Oecologia* 125, 241–248.

Johnson, M.T. and Gould, F. (1992) Interaction of genetically engineered host plant resistance and natural enemies of *Heliothis virescens* (Lepidoptera: Noctuidae) in tobacco. *Environmental Entomology* 21, 586–597.

Lang, A. and Vojtech, E. (2006) The effects of pollen consumption of transgenic *Bt* maize on the common swallowtail, *Papilio machaon* L. (Lepidoptera, Papilionidae). *Basic and Applied Ecology* 7, 296–306.

Lee, S.I., Lee, S.H., Koo, J.C., Chun, H.J., Lim, C.O., Mun, J.H., Song, Y.H. and Cho, M.J. (1999) Soybean Kunitz trypsin inhibitor (SKTI) confers resistance to the brown planthopper (*Nilaparvata lugens* Stal) in transgenic rice. *Molecular Breeding* 5, 1–9.

Londsdale, M., Baker, G., Godfree, B., Hirsch, M., Williams, K. and Yates, D. (2003) *Findings from the UK Farmscale Evaluations of Genetically Modified Herbicide-tolerant Crops: An Appraisal of Their Implications for Australia*. CSIRO, Canberra, Australia. Available at: http://www.csiro.au/proprietary Documents/CSIRO_FSE_appraisal.pdf

Losey, J.E., Rayor, L.S. and Carter, M.E. (1999) Transgenic pollen harms monarch larvae. *Nature* 399, 214.

Lovei G.L. and Arpaia S. (2005) The impact of transgenic plants on natural enemies: a critical review of laboratory studies. *Entomologia Experimentalis et Applicata* 114, 1–14.

Lozzia, G.C. (1999) Biodiversity and structure of ground beetle assemblages (Coleoptera: Carabidae) in *Bt* corn and its effects on non-target insects. *Bollettino di Zoologia agraria e di Bachicoltura Serie II* 31, 37–58.

Malone, L.A. and Pham Delegue, M.H. (2001) Effects of transgene products on honey bees (*Apis mellifera*) and bumblebees (*Bombus* sp.). *Apidiologie* 32, 287–304.

May, M.J., Champion, G.T., Dewar, A.M., Qi, A.M. and Pidgeon, J.D. (2005) Management of genetically modified herbicide-tolerant sugar beet for spring and autumn environmental benefit. *Proceedings of the Royal Society of London Series B – Biological Sciences* 272, 111–119.

McCarthy, M. and Buncombe A. (2005) The rape of the rainforest and the man behind it. *The Independent* (London) 20 May 2005.

Meier, M.S. and Hilbeck, A. (2001) Influence of transgenic *Bacillus thuringiensis* corn-fed prey on prey preference of immature *Chrysoperla carnea* (Neuroptera: Chrysopidae). *Basic and Applied Ecology* 2, 35–44.

Musser, F.R. and Shelton, A.M. (2003) *Bt* sweet corn and selective insecticides: impacts on pests and predators. *Journal of Economic Entomology* 96, 71–80.

Naranjo, S.E., Head, G. and Dively, G.P. (2005) Field studies assessing arthropod nontarget effects in *Bt* transgenic crops: introduction. *Environmental Entomology* 34, 1178–1180.

Oberhauser, K.S., Prysby, M.D., Mattila, H.R., Stanley-Horn, D.E., Sears, M.K., Dively, G., Olson, E., Pleasants, J.M., Lam, W.F. and Hellmich, R.L. (2001) Temporal and spatial overlap between monarch larvae and corn pollen. *Proceedings of the National Academy of Sciences of the United States of America* 98, 11913–11918.

Obrycki, J.J., Losey, E.J., Taylor, O.R. and Jesse, L.C.H. (2001) Transgenic insecticidal corn: beyond insecticidal toxicity to ecological complexity. *Bioscience* 51, 353–361.

O'Callaghan M., Glare T.R., Burgess E.P. and Malone, L.A. (2005) Effects of plants genetically modified for insect resistance on nontarget organisms *Annual Review of Entomology* 50, 271–292.

Orr, D.B. and Landis, D.A. (1997) Oviposition of European corn borer (Lepidoptera: Pyralidae) and impact of natural enemy populations in transgenic versus isogenic corn. *Journal of Economic Entomology* 90, 905–909.

Peferoen, M. (1997) Insect control with transgenic plants expressing *Bacillus thuringiensis* crystal proteins. In: Carozzi, N. and Koziel, M. (eds) *Advances in Insect Control: The Role of Transgenic Plants*, Taylor & Francis, London, pp. 21–48.

Pekrun, C., Lane, P.W. and Lutman, P.J.W. (2005) Modelling seedbank dynamics of volunteer oilseed rape (*Brassica napus*). *Agricultural Systems* 84, 1–20.

Perry, J.N., Rothery, P., Clark, S.J., Heard, M.S. and Hawes, C. (2003) Design, analysis and statistical power of the farm-scale evaluations of genetically modified herbicide-tolerant crops. *Journal of Applied Ecology* 40, 17–31.

Pilcher, C.D., Obrycki, J.J., Rice, M.E. and Lewis, L.C. (1997) Pre-imaginal development, survival, and field abundance of insect predators on transgenic *Bacillus thuringiensis* corn. *Environmental Entomology* 26, 446–454.

Pleasants, J.M., Hellmich, R.L., Dively, G., Sears, M.K., Stanley-Horn, D.E., Mattila, H.R.,

Foster, J.E., Clark, P. and Jones, G.D. (2001) Corn pollen deposition on milkweeds in and near cornfields. *Proceedings of the National Academy of Sciences of the United States of America* 98, 11919–11924.

Pons, X., Lumbierres, B., López, C. and Albajes, R. (2005) Abundance of non-target pests in transgenic *Bt*-maize: a farm scale study. *European Journal of Entomology* 102, 73–79.

Riddick, E.W., Dively, G. and Barbosa, P. (2000) Season-long abundance of generalist predators in transgenic versus nontransgenic potato fields. *Journal of Entomological Science* 35, 349–359.

Roy, D.B., Bohan, D.A., Haughton, A.J., Hill, M.O., Osborne, J.L., Clark, S.J., Perry, J.N., Rothery, P., Scott, R.J., Brooks, D.R., Champion, G.T., Hawes, C., Heard, M.S. and Firbank L.G. (2003) Invertebrates and vegetation of field margins adjacent to crops subject to contrasting herbicide regimes in the farm scale evaluations of genetically modified herbicide-tolerant crops. *Philosophical Transactions of the Royal Society of London Series B – Biological Sciences* 358, 1879–1898.

Schuler, T.H., Poppy, G.M., Kerry, B.R. and Denholm, I. (1999a) Potential side effects of insect-resistant transgenic plants on arthropod natural enemies. *Trends in Biotechnology* 17, 210–216.

Schuler, T.H., Poppy, G.M., Potting, R.P.J., Denholm, I. and Kerry, B.R. (1999b) Interactions between insect-tolerant genetically modified plants and natural enemies. In: British Crop Protection Council (ed.), *Gene Flow and Agriculture: Relevance for Transgenic Crops*, Vol. 72. Keele University, UK, pp. 197–202.

Schuler, T.H., Poppy, G.M., Woiwod, I.P. and Denholm, I. (2000) Impact of GM plants on the non-target arthropod fauna. In: WHO (ed.), *WHO Seminar on Release of Genetically Modified Organisms in the Environment: Is It a Health Hazard?* World Health Organization, Rome, p. 7.

Schuler, T.H., Denholm, I., Jouanin, L., Clark, S.J., Clark, A.J. and Poppy, G.M. (2001) Population-scale laboratory studies of the effect of transgenic plants on non-target insects. *Molecular Ecology* 10, 1845–1853.

Schuler, T.H., Potting, R.P.J., Denholm, I., Clark, S.J., Clark, A.J., Stewart, C.N. and Poppy, G.M. (2003) Tritrophic choice experiments with *Bt* plants, the diamondback moth (*Plutella xylostella*) and the parasitoid *Cotesia plutellae*. *Transgenic Research* 12, 351–361.

Schuler, T.H., Clark, A.J., Clark, S.J., Poppy, G.M., Stewart, C.N. and Denholm, I. (2005) Laboratory studies of the effects of reduced prey choice caused by *Bt* plants on a predatory insect. *Bulletin of Entomological Research* 95, 243–247.

Sears, M.K., Hellmich, R.L., Stanley-Horn, D.E., Oberhauser, K.S., Pleasants, J.M., Mattila, H.R., Siegfried, B.D. and Dively, G. (2001) Impact of *Bt* corn pollen on monarch butterfly populations: a risk assessment. *Proceedings of the National Academy of Sciences of the United States of America* 98, 11937–11942.

Sedjo, R.A. (2005) Will developing countries be the early adopters of genetically engineered forests? *AgBioForum* 8, 205–212.

Sétamou, M., Bernal, J.S., Legaspi, J.C. and Mirkov, T.E. (2002) Effects of snowdrop lectin (*Galanthus nivalis* agglutinin) expressed in transgenic sugarcane on fitness of *Cotesia flavipes* (Hymenoptera: Braconidae), a parasitoid of the nontarget pest *Diatraea saccharalis* (Lepidoptera: Crambidae). *Annals of the Entomological Society of America* 95, 75–83.

Shelton, A.M., Zhao, J.-Z. and Roush, R.T. (2002) Economic, ecological, food safety, and social consequences of the deployment of *Bt* transgenic plants. *Annual Review of Entomology* 47, 845–881.

Soares-Filho, B.S., Nepstad, D.C., Curran, L.M., Cerqueira, G.C., Garcia, R.A., Ramos, C.A., Voll, E., McDonald, A., Lefebvre, P. and Schlesinger, P. (2006) Modelling conservation in the Amazon basin. *Nature* 440, 520–523

Stanley-Horn, D.E., Dively, G., Hellmich, R.L., Mattila, H.R., Sears, M.K., Rose, R., Jesse, L.C.H., Losey, J.E., Obrycki, J.J. and Lewis, L. (2001) Assessing the impact of Cry1Ab-expressing corn pollen on monarch butterfly larvae in field studies. *Proceedings of the National Academy of Sciences of the United States of America* 98, 11931–11936.

Stöger, E., Williams, S., Christou, P., Down, R.E. and Gatehouse, J.A. (1999) Expression of the insecticidal lectin from snowdrop (*Galanthus nivalis agglutinin*; GNA) in transgenic wheat plants: effects on predation by the grain aphid *Sitobion avenae*. *Molecular Breeding* 5, 65–73.

Vitousek, P.M., Ehrlich, P.R., Ehrlich, A.H. and Matson, P.A. (1986) Human appropriation of the products of photosynthesis. *Bioscience* 36, 368–373.

Wilkinson, M.J., Elliot, L.J., Allainguillaume, J., Shaw, M.W., Norris, C.E., Welters, R., Alexander, M., Sweet, J.B. and Mason, D.C. (2003) Hybridization between *Brassica napus* and *B. rapa* on a national scale in the United Kingdom. *Science* 302, 457–459.

Wilson, L.J. and Fitt, G.P. (2000) Non-target effects of *Bt*-cotton: a case study from Australia. In: *XXI International Congress of Entomology*, Vol. 1. Foz do Iguassu, Brazil, p. 314.

Woiwod, I., Dewar, A., Hawes, C. and Heard, M. (2000) Farm-scale trials for studying the effects of GM herbicide-tolerant crops on farmland biodiversity in the UK. *Antenna* 24, 66–68.

Woiwod, I.P. (2004) Butterflies and GM doom. *Butterfly* 85, 34.

Wold, S.J., Burkness, E.C., Hutchison, W.D. and Venette, R.C. (2001) In-field monitoring of beneficial insect populations in transgenic corn expressing a *Bacillus thuringiensis* toxin. *Journal of Entomological Science* 36, 177–187.

Wraight, C.L., Zangerl, A.R., Carroll, M.J. and Berenbaum, M.R. (2000) Absence of toxicity of *Bacillus thuringiensis* pollen to black swallowtails under field conditions. *Proceedings of the National Academy of Sciences of the United States of America* 97, 7700–7703.

Zangerl, A.R., McKenna, D., Wraight, C.L., Carroll, M., Ficarello, P., Warner, R. and Berenbaum, M.R. (2001) Effects of exposure to event 176 *Bacillus thuringiensis* corn pollen on monarch and black swallowtail caterpillars under field conditions. *Proceedings of the National Academy of Sciences of the United States of America* 98, 11908–11912.

Zwahlen, C., Nentwig, W., Bigler, F. and Hilbeck, A. (2000) Tritrophic interactions of transgenic *Bacillus thuringiensis* corn, *Anaphothrips obscurus* (Thysanoptera: Thripidae), and the predator *Orius majusculus* (Heteroptera: Anthocoridae). *Environmental Entomology* 29, 846–850.

18 Insect Conservation: Progress and Prospects

OWEN T. LEWIS,[1] TIMOTHY R. NEW[2] AND ALAN J.A. STEWART[3]

[1]*Department of Zoology, University of Oxford, South Parks Road, Oxford OX1 3PS, UK;* [2]*Department of Zoology, La Trobe University, Melbourne, Victoria 3086, Australia;* [3]*Department of Biology and Environmental Science, School of Life Sciences, University of Sussex, Falmer, Brighton BN1 9QG, UK*

1 Progress to Date

The chapters in this volume reflect continuing changes in perspectives and priorities in insect conservation biology. Some cover topics scarcely noticed at the time when their predecessor in this series of symposia was published, just 16 years ago (Collins and Thomas, 1991). They reflect the advent and establishment of insect conservation as a 'respectable' area of concern for our natural world, rather than the province of eccentrics. The wider trend, of which insects are now a firm component, is that many research programmes in ecology now address 'conservation' as a primary focus, rather than as a politically astute afterthought. Awareness of the vulnerability of the natural world to human cupidity is more widespread than ever before, although there remains a substantial gulf between the attitudes to insect conservation (and the capability to pursue it) in wealthy 'first-world' countries and much of the rest of the world.

There are three main drivers behind increased sympathy for insect conservation. First, insects are regarded increasingly as worthy of conservation in their own right. 'Insect conservation' originated mainly with concerns about population declines and local extinctions of individual species, particularly butterflies targeted by collectors, in the 19th century. Individual species, particularly those with popular appeal or 'charisma', remain a major tool for communication and progress in insect conservation. They are tangible entities to which the wider public can relate. Increasingly sophisticated and detailed knowledge of the intricacies of species biology has enabled conservation programmes to focus on the essentials of habitat and resource quality for individual species (Dennis *et al.*, Chapter 5, this volume), and to define threats as a basis for effective management of sites and populations (Warren *et al.*, Chapter 4, this volume). The second driver increasing the profile of insect conservation is a recognition of the important role played by many insect groups in key ecological processes, particularly

 Insect Conservation Biology (eds A.J.A. Stewart, T.R. New and O.T. Lewis)

pollination (Kremen and Chaplin, Chapter 15, this volume), and in parasitic and predaceous interactions associated with pest management (Memmott *et al.*, Chapter 10, this volume). Widespread application of biological control strategies to manage pest species, as substitutes for chemical control, has helped to emphasize the need to preserve and enhance insect natural enemy complexes and the semi-natural habitats that they require. Finally, insects are increasingly being used as signallers of environmental quality and as sentinels of environmental change (McGeoch, Chapter 7, this volume). Alarmingly rapid declines in certain well-studied insect taxa have drawn further attention to the possibility that the global biodiversity crisis may be even more severe than previously thought and may impact disproportionately on invertebrates compared with other better-studied and more popular taxa.

Partly for these reasons, the emphasis on single species is now complemented firmly by wider considerations. Recognition that species cannot be conserved outside the context of associated species and the landscapes in which they occur has increasingly encouraged insect conservation biologists to incorporate wider considerations, such as of the existence of metapopulations (Hanski and Pöyry, Chapter 8, this volume), in defining management strategies. A related development is the increasing realization that insect conservation cannot be achieved solely within relatively small reserved areas such as national parks, but must be undertaken at the wider landscape level (Samways, Chapter 6, this volume). In particular, the agricultural landscape, as the largest single component of terrestrial land use, is a vital arena to incorporate into wider endeavour (New, 2005; see also New, Chapter 13, this volume; Tscharntke *et al.*, Chapter 16, this volume).

2 New Attitudes and Expectations

Insect conservation was once considered a fringe subject of no real interest to wider audiences, including other conservationists. During the last decade in particular, attitudes have changed considerably, with insect conservation becoming the subject of increasing numbers of meetings, societies and publications. The specialist *Journal of Insect Conservation* was founded in 1997, with the British organization Butterfly Conservation as an associate. That organization now has more than 12,000 members in the UK, and, with partner organizations elsewhere in Europe, is leading the development of Butterfly Conservation Europe. Similarly, in North America the Xerces Society (named after an extinct lycaenid butterfly) has done much to publicize the need for insect conservation. Importantly, such organizations also promote the conservation of natural history, as an activity (Cheesman and Key, Chapter 14, this volume). Many national entomological societies have conservation committees or groups, and the richness of insect life forms and biology continues to be displayed through remarkable mass-media productions such as the recently acclaimed television series *Life in the Undergrowth* (Attenborough, 2005). Public perceptions of insects are a key facet of conservation – much of the value of butterflies as flagships stems simply from their intrinsic appeal for many people. Conversely, many people admit to not liking moths, cockroaches and other insects, and any moves to redress this perception imbalance through education (particularly of young people) are important. Living insect displays, sometimes in conjunction with important captive breeding programmes for conservation (Pearce-Kelly *et al.*, Chapter 3, this volume), are important components of this.

3 The Challenge of Insect Diversity

The progress in insect conservation interest and practice made in the northern temperate regions has been the foundation for work elsewhere, particularly in the southern temperate regions (Stewart and New, Chapter 1, this volume), and has led to recognition of the inadequacies of 'close focus' approaches to insect conservation in much of the tropics (Lewis and Basset, Chapter 2, this volume). There, greater levels of approximation are inevitable, because great insect diversity is matched by inadequate taxonomic knowledge and paucity of resident expertise. Even in the best documented parts of the world, considerable gaps in taxonomic documentation remain, with substantial numbers of insect species in poorly studied orders still undescribed or recognizable by only a handful of specialists. Continued reliance on 'focal groups' for conservation assessment (rather than attempts to document entire insect faunas) seems inevitable for the foreseeable future. There is an enormous need for handbooks and identification guides that can be used by non-specialists, exemplified by the AIDGAP series in the UK, that provide accessible yet authoritative introductions to taxa, the identification of which is regarded as 'difficult'. Increasingly popular 'short-cut' alternatives to conventional taxonomy and species richness estimation must be considered and interpreted with great caution.

The diversity of insects, both at taxonomic and ecological levels, is an asset as well as a hindrance in their conservation. The need to incorporate numerous undescribed taxa with largely inferred or unknown biological roles into inventory or conservation planning represents an enormous challenge, termed 'the taxonomic impediment' by Taylor (1983). Approaches such as using 'morphospecies' to facilitate inventory of diversity are popular, but are not always convincing substitutes for naming species; as Thompson (1998) put it, 'names are the key to information'. However, in this era of decreasing global support for descriptive taxonomy, such substitutes may provide the only possibility for documenting changes in insect assemblages over time and across sites; but they must not be used without solid quality control and the responsible deposition of voucher series for reference by future workers. It is ironic that at a time when calls to document the earth's biodiversity as a tool to help sustain it have never been more widespread, or seen as so significant, our capability to do this at the most basic level of diagnosing and describing species continues to diminish. We urge the training of more people capable of basic approaches to insect taxonomy, echoing sentiments expressed widely elsewhere (Lee, 2000; New, 2000).

We need to be clear that inventory (equated broadly to compiling lists of species) is not on its own a valid conservation goal, or a necessary prerequisite for conservation management. Perhaps the most urgent problem faced by insect conservation biologists is clarifying how to combine systematic and ecological assessments to: (i) select the insect groups of greatest value in assessing and monitoring the quality of natural environments, both terrestrial and freshwater; and (ii) focus taxonomic endeavour to ensure that these groups can indeed be employed reliably by field managers. In this context, alpha taxonomy and clear recognition of species are essential tools in practical conservation at a variety of ecological and geographical scales.

To address these challenges the discipline of insect conservation biology must continue to embrace new technologies and analytical techniques. Advances in digital imagery and barcoding systems increasingly facilitate handling the vast numbers of specimens that result from insect surveys (Oliver *et al.*, 2000; Janzen *et al.*, 2005). Increasingly rapid and user-friendly modern molecular analyses, allowing population structures and dynamics to be analysed in ways undreamed of a few decades ago (Thompson *et al.*, Chapter 12, this volume), and automated entity-recognition systems (even if those entities lack names!) are two developments likely to allow important future advances.

4 The Role of Research

The continuing development of insect conservation biology can, perhaps, only proceed through effective synergies and collaborations between researchers and practitioners. Research is indeed a vital component of, and often a necessary precursor to, conservation management. Basic aspects of insect autecology (food sources, natural enemies, dispersal strategies, etc.) are simply not known for many species, even amongst relatively well-studied taxa in countries with strong entomological traditions, yet are essential pieces of information for developing effective conservation strategies. Although preferable, manipulative field experiments may not be possible in many cases, because of low populations or vulnerability. Alternative observational approaches therefore need to be designed especially carefully to tease out the critical factors that limit the numbers and distributions of rare or endangered insects.

A fundamental distinction exists between research that is focused specifically on informing practical management and that which is more strategic in nature. For example, estimating the number of insect species that exist, or the number of extinctions that have occurred, may be important in conservation advocacy and may facilitate detection of diversity or endemicity 'hotspots', but such research has limited direct relevance to practical management. Furthermore, identification of hotspots cannot be a stand-alone strategy for insect conservation in the future. The criteria used by Myers *et al.* (2002) to designate global hotspots of biodiversity richness are simply not met for many insects, so that numerous significant insect taxa are not conserved by concentrating on the hotspots approach alone. More generally, there are many examples of habitats or particular sites that are species-poor, for insects or more widely, but which nevertheless support important rare or endemic insect species. Similarly, in both Britain and North America, the distribution of rare or threatened butterfly species showed poor correspondence with butterfly richness hotspots, so that protecting hotspots alone would be an inadequate strategy to conserve most of the species of high individual conservation concern (Prendergast *et al.*, 1993; Fagan and Kareiva, 1997).

5 Future Challenges

While changes in habitats, including their wholesale destruction, have been the most important factors affecting insects globally, particularly over the last century, the increasing diversity of human impacts continually presents us with new challenges. The impacts of genetically modified crops (Woiwod and Schuler, Chapter 17, this volume) and of climate change (Wilson *et al.*, Chapter 11, this volume) pose general questions for conservation need and management into the future. Insects will continue to serve as effective barometers of their impacts and of the success of any mitigating conservation actions. Population trends for single species, and of sets of species (Conrad *et al.*, Chapter 9, this volume), provide sensitive signals of the effects of such anthropogenic impacts, and long-term monitoring (although unfashionable and difficult to fund over the necessary long time periods) will remain a crucial task for insect conservation biologists.

Insect conservation biology must proceed on a variety of levels – some building on established approaches and practice, and others developing from novel technologies and analytical capabilities as these arise. The minutiae of understanding single-species biology and population dynamics (in both time and space), as the traditional 'fine filter' level of conservation, must proceed hand in hand with 'coarse filter' approaches appraising the needs of assemblages and communities. One or other level of approach may be more expedient or rewarding in any given situation.

Insect conservation must be visionary, recognizing the need to conserve by investing effort for the long-term future rather than simply for short-term political expediency. Insects are paramount amongst the organisms that can warn us of our effects in the wider environment. For humanity to benefit from the subtle information that insects give us about the natural world and our impacts on it, improved conservation protocols are indeed needed to assure their future, and that of the world we share. This volume is a collective contribution towards that assurance.

References

Attenborough, D. (2005) *Life in the Undergrowth*. BBC Publications, London.

Collins, N.M. and Thomas, J.A. (eds) (1991) *The Conservation of Insects and their Habitats*. Academic Press, London.

Fagan, W.F. and Kareiva, P.M. (1997) Using compiled species lists to make biodiversity comparisons among regions: a test case using Oregon butterflies. *Biological Conservation* 80, 249–259.

Janzen, D.H., Hajibabaei, M., Burns, J.M., Hallwachs,W., Remigio, E. and Hebert, P.N.D. (2005) Wedding biodiversity inventory of a large and complex Lepidoptera fauna with DNA barcoding. *Philosophical Transactions of the Royal Society of London Series B – Biological Sciences* 360, 1835–1845.

Lee, M.S.Y. (2000) A worrying systematic decline. *Trends in Ecology and Evolution* 15, 346.

Myers, N., Mittermaier, R.A., Mittermaier, C.G., da Fonseca, G.A.B. and Kent, J. (2002) Biodiversity hotspots for conservation priorities. *Nature* 403, 853–858.

New, T.R. (2000) The conservation of a discipline: traditional taxonomic skills in insect conservation. *Journal of Insect Conservation* 4, 151–152.

New, T.R. (2005) *Invertebrate Conservation and Agricultural Ecosystems*. Cambridge University Press, Cambridge.

Oliver, I., Pik, A., Britton, D., Dangerfield, J.M., Colwell, R.K. and Beattie, A.J. (2000) Virtual biodiversity assessment systems: the application of bioinformatics technologies to the accelerated accumulation of biodiversity information. *BioScience* 50, 441–450.

Prendergast, J.R., Quinn, R.M., Lawton, J.H., Eversham, B.C. and Gibbons, D.W. (1993) Rare species, the coincidence of diversity hotspots and conservation strategies. *Nature* 365, 335–337.

Taylor, R.W. (1983) Descriptive taxonomy: past, present and future. In: Highley, E. and Taylor, R.W. (eds) *Australian Systematic Entomology: A Bicentenary Perspective*. CSIRO, East Melbourne, pp. 93–134.

Thompson, F.C. (1998) Names: the keys to biodiversity. In: Reaka-Kudla, M.L., Wilson, D.E. and Wilson, E.O. (eds) *Biodiversity II. Understanding and Protecting Our Biological Resources*. Joseph Henry Press, Washington, DC, pp. 199–211.

Taxonomic Index

Page numbers in **bold** refer to illustrations and tables.

General Index

Page numbers in **bold** refer to illustrations and tables.